Biosimilars and Biologics in Cancer Treatment

Biosimilars and Biologics in Cancer Treatment

An Era in Therapeutic Strategies

Edited by

Kuttiappan Anitha
School of Pharmacy and Technology Management,
SVKM's Narsee Monjee Institute of Management Studies
(NMIMS) (Deemed to be University),
Shirpur, Maharashtra, India

Shvetank Bhatt
Department of Pharmaceutical Sciences,
School of Health Sciences and Technology,
Dr. Vishwanath Karad MIT World Peace University,
Pune, Maharashtra, India

Bandaru Nagaraju
Department of Pharmacology,
School of Pharmaceutical Sciences,
Sandip University, Nashik, Maharashtra, India

ELSEVIER

ACADEMIC PRESS
An imprint of Elsevier
elsevier.com/books-and-journals

Academic Press is an imprint of Elsevier

125 London Wall, London EC2Y 5AS, United Kingdom
50 Hampshire Street, 5th Floor, Cambridge, MA 02139, United States

ISBN: 978-0-443-43824-0

For Information on all Academic Press publications visit our website at https://www.elsevier.com/books-and-journals

Publisher: Megan Ball
Acquisitions Editor: Linda Buschman
Editorial Project Manager: Siddharth Khattri
Production Project Manager: Kiruthigadevi N
Cover Designer: Mark Rogers

Typeset by Aptara, New Delhi, India

Contents

Contributors

Chavan Aditya
Department of Pharmacology, School of Pharmacy and Technology Management, SVKM's Narsee Monjee Institute of Management Studies (NMIMS) (Deemed to be University), Shirpur, Maharashtra, India

Krishna Swaroop Akey
Department of Pharmaceutical Chemistry, Shri Vile Parle Kelavani Mandal's (SVKM's) College of Pharmacy, Shirpur, Maharashtra, India

Sravan Kumar Aligeti
Department of Pharmaceutics, Vaageswari College of Pharmacy, Karimnagar, Telangana, India

Deevan Paul Amarnath
Department of Pharmaceutics, Sri Ramachandra Faculty of Pharmacy, Sri Ramachandra Institute of Higher Education and Research (Deemed to be University), Porur, Chennai, Tamil Nadu, India

Kuttiappan Anitha
School of Pharmacy and Technology Management, SVKM's Narsee Monjee Institute of Management Studies (NMIMS) (Deemed to be University), Shirpur, Maharashtra, India

Prabakaran Annadurai
Department of Pharmaceutics, School of Pharmacy and Technology Management, SVKM's Narsee Monjee Institute of Management Studies (NMIMS) (Deemed to be University), Hyderabad, Telangana, India

Harika Balya
Department of Pharmacy, College of Dentistry & Health Sciences, Fujairah University, Fujairah, United Arab Emirates

Parag Behura
Department of Pharmaceutical Sciences, School of Health Sciences and Technology, Dr. Vishwanath Karad MIT World Peace University, Pune, Maharashtra, India

Shvetank Bhatt
Department of Pharmaceutical Sciences, School of Health Sciences and Technology, Dr. Vishwanath Karad MIT World Peace University, Pune, Maharashtra, India

Mohan Gandhi Bonthu
Department of Pharmaceutical Analysis, JSS College of Pharmacy, JSS Academy of Higher Education and Research, Ooty, Tamil Nadu, India

Sohini Chakraborty
Department of Biotechnology, School of Applied Sciences, REVA University, Bengaluru, Karnataka, India

Dishant Chaudhari
Department of Pharmacology, School of Pharmacy and Technology Management, SVKM's Narsee Monjee Institute of Management Studies (NMIMS) (Deemed to be University), Shirpur, Maharashtra, India

Anusha Lakshmi Cheetiyar
Department of Physiology, MVJ Medical College and Research Hospital, Bengaluru, Karnataka, India

Santenna Chenchula
Department of Pharmacology, All India Institute of Medical Sciences (AIIMS), Bhopal, Madhya Pradesh, India

Deepthi Cherukuri
SRM College of Pharmacy, SRM Institute of Science and Technology, Chengalpattu, Tamil Nadu, India

Sanidhya Chopra
Department of Pharmacology, School of Pharmacy and Technology Management, SVKM's Narsee Monjee Institute of Management Studies (NMIMS) (Deemed to be University), Shirpur, Maharashtra, India

Santoshi Roopa D
Department of Pharmacology, MVJ Medical College and Research Hospital, Bengaluru, Karnataka, India

Sudharshan Reddy Dachani
Department of Pharmacology and Pharmacy Practice, College of Pharmacy, Shaqra University, Al-Dawadmi Campus, Shaqra, Saudi Arabia

Deepanjana Dass
Department of Pharmacology, MGM Medical College, Navi Mumbai, Maharashtra, India

Ashwini Deshpande
Department of Pharmaceutics, School of Pharmacy and Technology Management, SVKM's Narsee Monjee Institute of Management Studies (NMIMS) (Deemed to be University), Hyderabad, Telangana, India

Yukta Dewangan
Department of Pharmaceutical Sciences, School of Health Sciences and Technology, Dr. Vishwanath Karad MIT World Peace University, Pune, Maharashtra, India

Jyoti Dwivedi
Department of Biochemistry, Era's Lucknow Medical College and Hospital, Era University, Lucknow, Uttar Pradesh, India

Mohan Krishna Ghanta
Department of Pharmacology, MVJ Medical College and Research Hospital, Bengaluru, Karnataka, India

Piyush Ghode
School of Pharmacy and Technology Management, SVKM's Narsee Monjee Institute of Management Studies (NMIMS) (Deemed to be University), Shirpur, Maharashtra, India

Hemanth Kumar Gorubala
Department of Pharmacology, Sheffield Hallam University, Sheffield, United Kingdom

Saravanan Govindaraj
Department of Pharmaceutical Chemistry, Vel Pharmacy College, VISTAS Off Campus Centre, Vels Institute of Science, Technology and Advanced Studies, Tiruvallur, Tamil Nadu, India

Raghavendra Kumar Gunda
Department of Pharmaceutics, Narasaraopeta Institute of Pharmaceutical Sciences (Autonomous), Narasaraopet, Andhra Pradesh, India

Simhadri Goud Gundrathi
School of Pharmacy and Technology Management, SVKM's Narsee Monjee Institute of Management Studies (NMIMS) (Deemed to be University), Hyderabad, Telangana, India

Prathap Madeswara Guptha
Department of Pharmaceutical Sciences, School of Biotechnology and Pharmaceutical Sciences, Vignan's Foundation for Science, Technology and Research (Deemed to be University), Guntur, Andhra Pradesh, India

Santosh C. Gursale
Department of Pharmacology, Maharishi Markandeshwar Institute of Medical Sciences & Research, Sadopur, Ambala, Haryana, India

Mamtha Helchel
Department of Pharmacology, RBVRR Women's College of Pharmacy, Hyderabad, Telangana, India

Ramesh Jadhav
Department of Pharmaceutics, Avanthi Institute of Pharmaceutical Sciences, Hyderabad, Telangana, India

Garvit Kanchan
Department of Pharmacology, School of Pharmacy and Technology Management, SVKM's Narsee Monjee Institute of Management Studies (NMIMS) (Deemed to be University), Shirpur, Maharashtra, India

Shriraam Karunakaran
Department of Biochemistry, MVJ Medical College and Research Hospital, Bengaluru, Karnataka, India

Saravanakumar Kasimedu
Department of Pharmaceutics, Seven Hills College of Pharmacy (Autonomous), Tirupati, Andhra Pradesh, India

Ruchi Keswani
Department of Pharmaceutical Sciences, School of Health Sciences and Technology, Dr. Vishwanath Karad MIT World Peace University, Pune, Maharashtra, India

Nawaz Mahammed
Department of Pharmaceutics, Raghavendra Institute of Pharmaceutical Education and Research (Autonomous), Anantapur, Andhra Pradesh, India

Tariq Mahmood
Department of Biochemistry, Era's Lucknow Medical College and Hospital, Era University, Lucknow, Uttar Pradesh, India

Pandhare Meeraraje
Department of Pharmacology, School of Pharmacy and Technology Management, SVKM's Narsee Monjee Institute of Management Studies (NMIMS) (Deemed to be University), Shirpur, Maharashtra, India

Aparna Misra
Department of Biochemistry, Era's Lucknow Medical College and Hospital, Era University, Lucknow, Uttar Pradesh, India

Rupesh More
Department of Pharmaceutics, School of Pharmacy and Technology Management, SVKM's Narsee Monjee Institute of Management Studies (NMIMS) (Deemed to be University), Shirpur, Maharashtra, India

Menaka Reddy Mukkala
Department of Pharmaceutical Chemistry, Vignan Pharmacy College, Jawaharlal Nehru Technological University, Guntur, Andhra Pradesh, India

Bandaru Nagaraju
Department of Pharmacology, Sree Dattha Institute of Pharmacy, Hyderabad, Telangana, India

Swapna R. Nayaka
Department of Pharmacology, MVJ Medical College and Research Hospital, Bengaluru, Karnataka, India

Patakamuri Niharika
Department of Pharmaceutical Sciences, School of Health Sciences and Technology, Dr. Vishwanath Karad MIT World Peace University, Pune, Maharashtra, India

Shanmugasundaram Palani
Department of Pharmaceutical Chemistry and Analysis, School of Pharmaceutical Sciences, Vels Institute of Science, Technology and Advanced Studies, Chennai, Tamil Nadu, India

Shrestha Palit
School of Pharmacy and Technology Management, SVKM's Narsee Monjee Institute of Management Studies (NMIMS) (Deemed to be University), Hyderabad, Telangana, India

Sagar Pamu
Department of Pharmacy Practice, Institute of Pharmacy, Nirma University, Ahmedabad, Gujarat, India

Balaji Pandiyan
Department of Pharmacology, School of Pharmaceutical Sciences, Vels Institute of Science, Technology and Advanced Studies, Chennai, Tamil Nadu, India

Prajakta Patil
Department of Pharmacology, School of Pharmacy and Technology Management, SVKM's Narsee Monjee Institute of Management Studies (NMIMS) (Deemed to be University), Shirpur, Maharashtra, India

Suryanarayana Polaka
Department of Pathology, Louisiana State University Health Shreveport, Shreveport, LA, United States

Nagaveni Pommala
Department of Pharmaceutics, Vemu College of Pharmacy & Health Sciences, Chittoor, Andhra Pradesh, India

Jalapa Pradhan
Department of Population Health and Leadership, School of Health Sciences, University of New Haven, West Haven, CT, United States

Sujith Kumar Pulukool
Department of Biosciences, Sri Sathya Sai Institute of Higher Learning, Prasanthi Nilayam, Anantapur, Andhra Pradesh, India

Venkata Sita Rama Raju
Department of Medical Biochemistry and Microbiology, Uppsala University, Uppsala, Sweden

Srinivasan Ramamurthy
Department of Pharmacy, College of Dentistry & Health Sciences, Fujairah University, Fujairah, United Arab Emirates

Kishore Rapolu
Department of Pharmaceutics, University College of Technology, Osmania University, Hyderabad, Telangana, India

Tadiparthi Reshma
Department of Pharmaceutics, Raghavendra Institute of Pharmaceutical Education and Research (Autonomous), Anantapur, Andhra Pradesh, India

Sujit Kumar Sah
Department of Pharmaceutical Sciences, School of Health Sciences and Technology, Dr. Vishwanath Karad MIT World Peace University, Pune, Maharashtra, India

Venkateswarlu Samunooru
Pharmacology and Drug Discovery MSc, Coventry University, West Midlands, United Kingdom

Bharat Kumar Reddy Sanapalli
Department of Pharmacology, School of Pharmacy and Technology Management, SVKM's Narsee Monjee Institute of Management Studies (NMIMS) (Deemed to be University), Hyderabad, Telangana, India

Vidyasrilekha Sanapalli
Department of Pharmaceutical Chemistry, School of Pharmacy and Technology Management, SVKM's Narsee Monjee Institute of Management Studies (NMIMS) (Deemed to be University), Hyderabad, Telangana, India

Krishna Sanka
Department of Pharmaceutics, School of Pharmacy and Technology Management, SVKM's Narsee Monjee Institute of Management Studies (NMIMS) (Deemed to be University), Shirpur, Maharashtra, India

Esha Sarkar
Department of Biochemistry, Era's Lucknow Medical College and Hospital, Era University, Lucknow, Uttar Pradesh, India

Bhagawati Saxena
Department of Pharmacology, Institute of Pharmacy, Nirma University, Ahmedabad, Gujarat, India

Neelufar Shama Shaik
Department of Pharmacognosy, Scient Institute of Pharmacy, Ibrahimpatnam, Telangana, India

Dilep Kumar Sigalapalli
Department of Pharmaceutical Chemistry, Vignan Pharmacy College, Jawaharlal Nehru Technological University, Guntur, Andhra Pradesh, India

Jay Prakash Soni
Department of Biochemistry, University of Washington, Seattle, WA, United States

Vijayaraj Surendran
Department of Pharmaceutical Analysis and Chemistry, Dr. Kalam College of Pharmacy, Thanjavur, Tamil Nadu, India

Shivani Suryavanshi
Department of Pharmaceutical Sciences, School of Health Sciences and Technology, Dr. Vishwanath Karad MIT World Peace University, Pune, Maharashtra, India

Vanrajsinh Thakor
Department of Pharmaceutical Chemistry, Institute of Pharmacy, Nirma University, Ahmedabad, Gujarat, India

Ramya Tokala
Athinoula A. Martinos Center for Biomedical Imaging, Department of Radiology, Massachusetts General Hospital, Harvard Medical School, Charlestown, MA, United States

Adhrit Chandra Pati Tripathi
Amity School of Communication, Amity University, Lucknow, Uttar Pradesh, India

Pratima Tripathi
National Institute of Pharmaceutical Education and Research Raebareli (NIPER-R), Lucknow, Uttar Pradesh, India; Kaustav's Careerica & Prakdrit Edtech Pvt. Ltd., Lucknow, Uttar Pradesh, India

Murugesan Vanangamudi
Institute of Pharmaceutical Research, GLA University, Mathura, Uttar Pradesh, India

Kasturi Vishwanathasetty Veerabhadrappa
Department of Pharmacognosy & Phytochemistry, School of Pharmacy, Kampala International University-Western Campus, Bushenyi, Uganda

Neethu Mangalasseril Vijayakumari
Department of Pharmacology, MET's College of Pharmaceutical Sciences and Research, Thrissur, Kerala, India

CHAPTER

Introduction to biosimilars and biologics

1

Ruchi Keswani, Parag Behura, Yukta Dewangan and Shvetank Bhatt
Department of Pharmaceutical Sciences, School of Health Sciences and Technology, Dr. Vishwanath Karad MIT World Peace University, Pune, Maharashtra, India

1.1 Introduction

Cancer is one of the most common causes of morbidity and mortality worldwide. Even in the 21st century, it is one of the major health challenges (Nwosu, 2024). Cancer is associated with cell proliferation, invasion of neighboring tissues, metastasis to other body parts, and resistance to programmed cell death (apoptosis). According to the World Health Organization, one in every six deaths around the world is because of cancer. The common types of cancer are liver, lung, prostate, and breast cancers. Due to the adverse effects, drug resistance, and limited effectiveness of benchmark chemotherapeutic agents, the treatment of cancer is still a challenge (Nwosu, 2024). Chemotherapy has been the main treatment for cancer for decades. These drugs decrease tumor size by targeting rapidly dividing cancerous cells. But they also damage healthy tissues, including hair follicles, the lining of the gastrointestinal tract, and bone marrow. The side effects of traditional chemotherapy, as well as the heterogeneity and complexity of malignant neoplasms, have made the need for more sophisticated and personalized treatment options more evident. The shift in paradigm has enhanced a shift away from nonselective cytotoxic agents to specific molecular markers of interest in oncogenesis and disease progression, which are now targeted by precision-oriented biologic modalities such as biological origin immunotherapies. This shift has changed the clinical focus to specific biological interventions that focus therapeutic pressure on individual tumor-associated antigens and signaling pathways that promote malignant proliferation and metastasis. Biological therapies, also known as biologics, are large and structurally complex molecules synthesized through the use of living systems, such as bacteria, yeast, and mammalian cells, through the recombinant DNA (rDNA) technology and any modalities of biotechnology. These agents bind with specific molecular targets, which include growth factor receptors, cytokines, and immune checkpoints, which are important in the survival and proliferation of the tumor cells. The biologics are contrasted to the traditional small-molecule drugs in that they have a highly elaborate three-dimensional structure, high molecular weight, and are highly sensitive to issues of manufacturing. Examples of biologics include

Biosimilars and Biologics in Cancer Treatment: An Era in Therapeutic Strategies.
DOI: https://doi.org/10.1016/B978-0-443-43824-0.00016-X

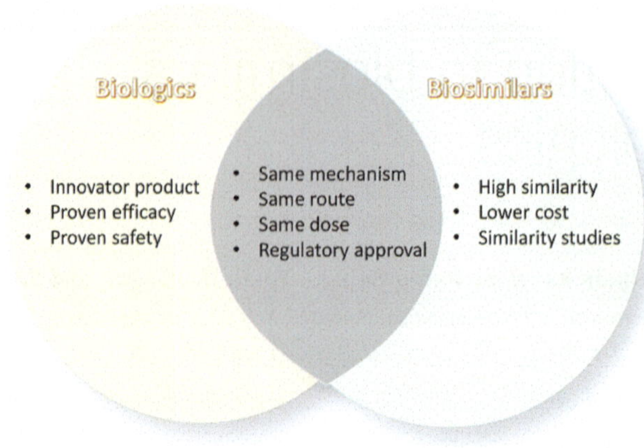

FIGURE 1.1

Comparison of key characteristics of biologics and biosimilars.

monoclonal antibodies (mAbs) (e.g., trastuzumab, rituximab), cytokines, recombinant proteins, and immune-checkpoint inhibitors like nivolumab, pembrolizumab, atezolizumab, and ipilimumab (Bowers et al., 2020; Revers & Furczon, 2010). On the other hand, the biosimilars are biologic medicinal products that are very similar, but not equivalent to an already licensed reference biologic. Since biologics are produced in natural environments and tend to be heterogeneous, biosimilars cannot be the same chemical products as traditional generics are expected to be. Based on this, they should prove a high degree of similarity in structural properties, effectiveness, safety, and immunogenicity by conducting exhaustive analytical, nonclinical, and clinical studies. Fig. 1.1 shows the comparison of key characteristics of biologics and biosimilars.

The regulatory agencies, such as the European Medicines Agency (EMA) and the United States Food and Drug Administration (USFDA), have made the approval systems to ensure that biosimilars meet the same therapeutic and safety standards as their reference counterparts (Allocati et al., 2022; Mascarenhas-Melo et al., 2024). Treatment of cancer has been revolutionized with the introduction of biologics, which present very specific therapeutic choices. Trastuzumab, rituximab, and other mAbs are direct adhesives to tumor-associated antigens and inhibit cell-signaling pathways and induce an immune response that targets malignant cells. Immunotherapies, like the checkpoint inhibitors nivolumab, pembrolizumab, atezolizumab, durvalumab, ipilimumab, and tremelimumab, have additionally revolutionized oncologic care by turning back to the activation of the immune system of the patient to help it recognize and destroy cancerous cells (Papież & Krzyściak, 2021). Biosimilars, equated to develop the clinical efficacy and safety profiles with those of existing biologic medicines, increase their affordability, and consequently, patients in low- and middle-income countries have increased access to life-saving treatments

(Allocati et al., 2022). Biologic cancer therapy emerged in the late 1900s, when interferon and interleukin began to be used as clinical therapy (Briukhovetska et al., 2021). The next milestone was reached after Kohler and Milstein made the breakthrough finding of monoclonal antibody technology in 1975 (Sagar et al., 2024). This was followed by the approval of rituximab in 1997, which propelled mAbs as a key ingredient in targeted therapy of cancer (Pierpont et al., 2018). Around the first decade of the 21st century, biosimilars, starting with the approval of the first product by the European Union in 2006, triggered both regulatory harmonization on a global scale and the accessibility of biosimilars in a broader marketplace. Biologics and biosimilars are among the booming markets of the pharmaceutical industry today. Their effects are not limited to oncology but also extend to autoimmune, metabolic, and infectious diseases. Biologics have significantly increased the survival rates and the quality of life of patients, and biosimilars have made healthcare more sustainable with the availability of biosimilar drugs at a comparatively lower price with similar clinical efficacy. Together, all these developments indicate an impressive leap forward in the pursuit of precision medicine, because treatment is becoming more personalized with reference to the genetic and molecular picture, providing a renewed hope in the worldwide battle against cancer (Revers & Furczon, 2010).

1.2 Biologics

1.2.1 Characteristics of biologics

Biologics are recognized for their global acceptance and are approved by the FDA worldwide. They are partly related to the field of oncology, where they are mostly used in the fields of chemotherapy, targeted therapy, and immunotherapy (Bowers et al., 2020). They are produced by reproducing, or growing, copies of an especially engineered living cell. There have been extensive studies going on in the field of biologics in cancer treatment. They are made up of complex molecules such as the mAbs, cytokines, vaccines, oncolytic viruses, or gene therapies (Zahavi & Weiner, 2020). They are generally produced from living systems like microorganisms, plant cells, or animal cells. Biologics mainly work by enhancing or redirecting the immune response (Henrickson et al., 2016; Jung & Kim, 2022). For instance, mAbs induce antibody-dependent cellular cytotoxicity (ADCC) or block signaling pathways; chimeric antigen receptors (CAR-T) cells directly lyse tumor cells via engineered receptors; vaccines present tumor antigens to stimulate T-cell responses; and oncolytic viruses replicate in cancer cells, releasing antigens to boost systemic immunity. They often target specific genetic aberrations or antigens, enabling personalized medicine through approaches like next-generation sequencing (Paccagnella et al., 2025). Biologic therapies are used in place of traditional chemotherapeutic methods in that, unlike the conventional methods, they are specific in targeting the antigens that are associated with the tumor, hence eliminating the risk of destruction of normal tissues and reducing adverse effects like systemic toxicity. These agents may be used alongside chemotherapy, radiotherapy, or other biologics such as mAbs, with an additional factor in biologic therapy, such as kinase inhibitors. Most recent changes in technology, such as next-generation sequencing, allow the design of personalized interventions, including the neoantigen-based vaccination, which are specific to the mutational setting of a specific patient (Letai & de The, 2025).

1.2.2 Sources and production methods

Biologics are sourced from various biological systems, as their starting materials come from living entities. Common sources include:

- *Human-derived:* Blood components, tissues, or cells from humans, such as plasma-derived proteins or human cell lines used in therapies.
- *Animal-derived:* Cells or materials from animals, including mammalian cell lines like Chinese hamster ovary (CHO) cells, which are widely used for producing mAbs and other proteins.
- *Microbial-derived:* Bacteria (e.g., *Escherichia coli*) or yeast, often genetically engineered to produce insulin or other recombinant proteins.
- *Plant or fungal-derived:* Less common but includes sources like plants or fungi for certain enzymes or vaccines.
- *Recombinant sources:* Genetically modified organisms where DNA is inserted to produce the desired biologic, often from any of the above categories.

1.3 Classification of biologics

The biologics are classified into the following types:

1.3.1 Monoclonal antibodies (e.g., trastuzumab, rituximab)

Monoclonal antibodies are engineered proteins designed to bind specific antigens with high affinity. Produced primarily in CHO cells, mAbs target disease-specific markers, such as Human Epidermal Growth Factor Receptor 2 (HER2) in breast cancer (trastuzumab) or CD20 in lymphomas (rituximab). Their production involves rDNA technology, cell culture, and rigorous purification to ensure homogeneity. Clinically, mAbs are pivotal in oncology and autoimmune disease management due to their precision (Lyu et al., 2022; Sathyan et al., 2025).

1.3.2 Cytokines and growth factors

Cytokines and growth factors form a group of signal proteins that control immune reactions and cell growth. Examples of representative agents are interferons, used to treat viral diseases, and erythropoietin (EPO), used to treat anemia. They are often produced by expressing heterologous vectors in microbial hosts, either in microorganisms like yeast or bacteria (usually E. *coli*), and fermentation factors must be carefully regulated to maintain functional bioactivity (Pouresmaeil & Azizi-Dargahlou, 2023). However, they have limited clinical efficiency due to a tendency to cause adverse events, in particular, systemic inflammation.

1.3.3 Immune checkpoint inhibitors

Immune checkpoint inhibitors, primarily mAbs, block inhibitory pathways (e.g., Programmed Cell Death Protein 1 [PD-1] or Cytotoxic T-Lymphocyte-associated Antigen 4 [CTLA-4]) to enhance T-cell activity against cancer cells (Buchbinder & Desai, 2016). Nivolumab and pembrolizumab (anti-PD-1), ipilimumab (anti-CTLA-4), and relatlimab (LAG-3 inhibitor). mAbs are used in melanoma and solid tumors. Their production is similar to that of mAbs but demands stringent quality control to mitigate immune-related adverse events (Khan et al., 2024). These biologics have transformed cancer immunotherapy.

1.3.4 Therapeutic cancer vaccines

Therapeutic cancer vaccines stimulate immune responses against tumor-specific antigens (Fan et al., 2023). Sipuleucel-T, used in prostate cancer, targets prostatic acid phosphatase (Handy & Antonarakis, 2018). The process of production is complex and includes patient-specific antigen-presenting cells. These vaccines are less common because of difficulties in scaling and effectiveness (Fan et al., 2023). In addition, potential side effects such as flu-like symptoms, injection site reactions, and rare allergic reactions provide significant challenges in the development of cancer vaccines.

1.3.5 Antibody-drug conjugates

Antibody-drug conjugates (ADCs) are the mAbs that have been conjugated with cytotoxic drugs to provide specific chemotherapy (Peters & Brown, 2015). Examples of this type include ado-trastuzumab emtansine, which is used to treat breast cancer that has the HER2-positive (*KADCYLA [Ado-trastuzumab emtansine] in HER2+ breast cancer*). The antibody and drug have to be conjugated through conjugation chemistry to manufacture, and they need to be highly purified to achieve stability and safety (Peters & Brown, 2015). Immunogenicity toward antibodies, linker instability, and controlled drug release at the target site are some of the main challenges in the development of ADCs.

1.3.5.1 Cellular therapies

Cellular therapies based on the use of CAR-T cells are another advanced treatment modality that uses autologous T-Lymphocytes that have been genetically engineered to produce antigen-specific receptors that bind malignant cells, e.g., CD19-specific CARs in the therapeutic product tisagenleucel (*I-Corps: Chimeric Antigen Receptor*

Mediated Immunosynapse Assay - National Science Foundation (Scordo et al., 2019). The manufacturing process includes leukapheresis, *ex vivo* genetic engineering, and final reinfusion, all of which have significant issues concerning the scalability, complexity of the process, and affordability (Almeida-Oliveira et al., 2025).

1.3.6 Development process of biologics with common challenges

Biologic therapeutics is a multidisciplinary, highly regulated process that combines the fields of research of advanced biotechnology, molecular biology, and bioprocess engineering. Unlike other traditional small-molecule drugs that are synthesized through the use of defined chemical reactions, biologics are synthesized in living biological systems through bacterial, yeast, or mammalian cell lines, which inherently lead to variability and heightened process complexity (Bowers et al., 2020; Revers & Furczon, 2010). As a result, all of the stages of target identification through large-scale production should be carefully monitored to achieve homogeneity, effectiveness, and safety of the resulting product. The process of development starts with the intensive determination and confirmation of a molecular target involved in the pathogenesis, most commonly receptors, enzymes, or signaling proteins (Sampathkumar & Kerwin, 2024). Examples of growth factor receptors (HER2), immune checkpoints (PD1 or PD-L1), and tumor-expanding angiogenic pathways are found in oncology (Schütz et al., 2017). After authentication of a target, the gene encoding the therapeutic protein is engineered into a useful expression vector and transduced into a host cell line, often CHO cells, *E. coli,* or yeast, which gives different advantages and disadvantages in terms of yield, fidelity of posttranslational modification, and regulation. Development of stable cell lines with sustained and high-level protein production is essential; the lines have to be carefully characterized in relation to productivity, genetic stability, and adherence to quality parameters to ensure long-term reproducibility and regulatory compliance (Sampathkumar & Kerwin, 2024). The upstream process in the manufacturing starts with the culture of genetically engineered microbial or mammalian cells in bioreactors under rigidly controlled environmental factors to increase the levels of recombinant proteins. The essential parameters such as temperature, pH, dissolved oxygen, and nutrient concentration are carefully attacked so as to achieve the highest possible expression without affecting the cell viability and the quality of the product. Before a clinical application is made, a thorough preclinical assessment is done. These include in vitro and in vivo animal studies to evaluate pharmacokinetic and pharmacodynamic characteristics, and also include toxicological profiling of biologics. The information obtained in these studies makes it possible to select dosage and risk avoidance programs in future human trials (Castan et al., 2018; Papież & Krzyściak, 2021; Sampathkumar & Kerwin, 2024). The process of human clinical investigation is carried out in consecutive, stage-based stages (Phases I, II, and III) to determine safety, efficacy, and immunogenicity to target populations. Once an acceptable clinical benefit has been demonstrated, the biologic product then comes under regulatory verification, where the manufacturing processes are expanded under Good Manufacturing Practice requirements to ensure analytical strength and reproducibility of processes.

Biologics face some daunting challenges to their developmental routes, although the therapeutic potential is considerable. Their macromolecular and conformational heterogeneity makes them sensitive to small changes in manufacturing inputs, which result in changes in folding patterns, glycosylation, and bioactivity. Achieving batch-in-batch consistency is therefore problematic in terms of manufacturing. In addition, special infrastructure, complex purification processes, and quality analysis requirements add to the cost of production. Cold-chain logistics are also required in the biologics, as they are thermodynamically unstable, and their proteinaceous nature tends to provoke unintended immune reactions, which require a comprehensive evaluation of immunogenicity. The regulatory environment that comes with biologics is very rigorous and extensive, so that filing of exhaustive analytical, preclinical, and clinical sets of data is required. To gain entry in the market, this cumulative body of evidence is required in order to prove safety, efficacy, and the overall quality of the product (Sathish et al., 2013).

1.3.7 Approved biologics

A wide range of biotherapeutic agents have been approved by the regulatory agencies to be used in clinics, hence highlighting their ability to disrupt precise molecular pathways with tremendous specificity. Bevacizumab is a mAb available commercially as Avastin, Mvasi, and Zirabev and is commonly used in the treatment of metastatic colorectal cancer, nonsmall-cell lung cancer (NSCLC), glioblastoma, renal cell carcinoma (RCC), cervical cancer, and ovarian cancer. Trastuzumab is a commercialized humanized mAb known as Herceptin, Ogivri, and Kanjinti, which forms the core of the treatment of breast cancer that is HER2-positive and gastric adenocarcinoma metastasis that is also HER2-positive (Findlay & Scholar, 2007). *Herceptin, Ogivri (Trastuzumab),* Rituximab, marketed as Rituxan, Riabni, Ruxience, and Truxima, is used as a selective-cytotoxic agent in the treatment of non-Hodgkin lymphoma (NHL) and chronic lymphocytic leukemia due to its selectivity to CD20-positive B-lymphocytes (Al-Katib & Mohamed, 2006). Immune-checkpoint inhibitors, comprised of pembrolizumab (Keytruda) and nivolumab (Opdivo), that counteract PD-1 have swept the therapeutic intervention of melanoma, NSCLC, Hodgkin lymphoma, urothelial carcinoma, and a line of different neoplasms characterized by unique biomarkers (Shiravand et al., 2022). Cetuximab (Erbitux) is approved for metastatic colorectal cancer of RAS -wild-type genotypes, and head-and-neck squamous-cell carcinoma (HNSCC) (Center, C). The anti-CTLA-4 human mAb, ipilimumab (Yervoy), is used to treat metastatic melanoma, RCC, and certain colorectal tumors (typically in combination with other immunotherapeutic agents) (Lipson & Drake, 2011). A novel type of T-cell-directed therapy, tebentafusp (Kimmtrak), is a bispecific T-cell-redirecting fusion protein approved in unresectable or metastatic uveal melanoma (Hassel et al., 2023). Denosumab, Xgeva brand, is used to prevent skeletal-associated occurrences in patients with bone metastases who have solid tumors and has proven effective in multiple myeloma as well as giant-cell tumor of bone (Cadieux et al., 2022). As an illustration of the growing applicability of bispecific biologics to hematologic malignancies, the bispecific T-cell engager Blincyto (blinatumomab) is approved to treat B-cell precursor acute lymphoblastic leukemia (ALL) (Burt et al., 2018). Taken together, these agents demonstrate the diversification of action, as well as the expansion of therapeutic activity of biologics in oncologic and immune-mediated diseases.

1.3.8 Some current preclinical and/or clinical studies using biologics

1.3.8.1 Preclinical studies using biologics

The preclinical studies on mecbotamab vedotin (BA3011) were aimed at the production of an AXL receptor tyrosine kinase (AXL)-specific, conditionally active ADC (AXL-targeted ADC) that would preferentially target the acidic tumor microenvironment. BA3011 was highly selective with a pH-dependent mode, where it has a high affinity to AXL-expressing tumor cells at a pH of 6.0, with a minimal level of affinity at the physiological pH of 7.4. Cytotoxic activity against AXL-positive human cancerous cell lines was confirmed in vitro. To determine AXL activity in vivo, the researchers used xenografts of human tumor cells being injected into murine hosts that expressed AXL; BA3011 was repeatedly found to exhibit severe and sustainable tumor-growth inhibition. Its safety profile was further supported using nonhuman primate models of cynomolgus monkeys, which could tolerate doses up to 5 mg/kg and even sustain stable linker-payload disposition in circulation. Even though no specific age and weight data of the primates were reported, the general results showed a potentially promising preclinical efficacy and safety, meaning that BA3011 might be more tumor selective with an easier off-afterwards toxicity attitude compared to other AXL-based biologics (Chang et al., 2025a). In another study by Chang et al. (2025b), the development of a biologic, ozuriftamab vedotin also known as BA3021, was based on a conditionally active biologic ADC that was designed to attack tumors expressing receptor tyrosine kinase-like orphan receptor 2 (ROR2) but sparingly interacts with normal tissue. BA3021 displayed pH repeat binding (high affinity with human and cynomolgus ROR2 at acidic tumor-microenvironment pH (6.0), but with high affinity greatly reduced at physiological pH (7.4). In vitro testing resulted in the strong cytotoxicity under low-pH conditions of the cell lines with ROR2 (e.g., SKMeL5, LCLC103H, HT1080) and weak cytotoxicity of the cell lines with ROR2 in low-pH conditions. BA3021 was used to treat female non-obese diabetic (NOD)/severe combined immunodeficiency (SCID) mice in the presence of SK MEL 5 or LCLC 103H xenografts intravenously (e.g., 6 mg/kg/day, six injections) when the tumor volumes had attained a relatively fixed point of about 100,150 mm^3, thereby generating an effect of CAM more than 100% and complete cures at high doses in vivo. Similar results with 6 mg/kg BA3021 were observed with patient-derived xenograft cultures of sarcoma, GIST, and melanoma, whereby significant tumor-growth inhibition was realized. Single and repeated IV doses of 10 mg/kg were well tolerated in cynomolgus monkeys, without lethality and reversible toxicological outcomes; pharmacokinetic analyses demonstrated that ADC and conjugated payload were appropriately stable. These findings showed that BA3021 showed great effectiveness on a wide range of tumor models with a good tolerability profile for further clinical development (Chang et al., 2025).

1.3.8.2 Clinical studies using biologics

Crenezumab, a humanized and IgG4 antiaggregated and soluble 1 peg-amyloid antibody, has been tested in phase II clinical trials in more than 450 schizophrenic patients with mild-moderate Alzheimer's disease (AD). The subjects were administered either intravenously or subcutaneously on a series of dosing schedules. Its pharmacokinetics were characterized by trials which showed dose-proportional increases in serum concentrations, a terminal half-life of about 20 to 23 returnees, and cerebrospinal fluid/plasma penetrations respectively of 0.3% and 0.3%. Pharmacodynamics showed a significant rise in total 8.0-amyloid concentrations in plasma and CSF, thus validating peripheral and central targets. The Phase II studies were not powered to identify clinical efficacy, but they furnished very important information regarding pharmacokinetics (PK)/pharmacodynamics (PD) and biomarkers and safety, facilitating dose choices and modelling procedures to follow in the further development of biologic treatments against amyloid pathology (Yoshida et al., 2020). N3pG-modified amyloid, which is an IgG1 mAb, was researched in the phase II TRAILBLAZER-ALZ-II randomized, placebo-controlled, double-blind study involving 1736 individuals with early symptomatic AD, mild cognitive impairment, and mild dementia confirmed using the amyloid PET scan. The subjects were given three monthly donanemab IV injections or a control every 18 months. Phase III trials revealed statistically significant reductions in the rate of cognitive and functional deterioration in the integrated AD Rating Scale and clinical dementia rating-sum of boxes, and significant declines in amyloid plaque burden and the translation of many patients into the amyloid-negative state. The most crucial risk associated with treatment highlighted by safety outcomes was amyloid-related imaging abnormalities. This methodologically intensive phase III trial generated strong clinical data on the existence of a therapeutic position of mAb biologics in the initial stages of AD (Sims et al., 2023).

1.4 Biosimilars

1.4.1 Biosimilars and differences between biosimilars and generics

Biosimilars are products that are very similar to the reference biologic product in terms of safety, quality, and efficacy, which is already approved. They are designed to offer similar therapeutic results without any clinically significant differences and are

Table 1.1 Key differences between biosimilars and generics.

Feature	Biosimilar	Generic
Molecule	Large and complex protein/peptide	Small and chemically defined compound
Manufacturing	Cell culture, fermentation, or recombinant DNA	Chemical synthesis
Structural variability	Minor, unavoidable differences (e.g., glycosylation)	None
Regulatory review	Stepwise: analytical, nonclinical, and clinical	Minimal clinical data (bioequivalence)
Pricing	Generally, 15%–30 % lower than the originator	80%–85% lower than brand (Revers & Furczon, 2010)

developed after the patent expiry of an original biologic (also known as the reference or innovator product). Similar to biologics, biosimilars are made in living systems like bacterial, yeast, or mammalian cells, unlike small-molecule medications, which are chemically synthesized and have a predetermined structure (Allocati et al., 2022; Mascarenhas-Melo et al., 2024). The development of biosimilars involves analytical characterization, nonclinical studies, and clinical trials to demonstrate similarity in pharmacokinetics, pharmacodynamics, safety, and immunogenicity. Regulatory bodies like the EMA and the USFDA have set up separate approval processes for biosimilars, focusing on a gradual approach to show similarity instead of carrying out extensive efficacy trials as they do for novel biologics (Car et al., 2025). The generic pharmaceuticals, on the other hand, are the same chemically congruent copies of the same reference drugs that are small molecules. They are characterized by their bioequivalence, essentially doing away with the requirement to engage in the preclinical or clinical tests. This leads to the occurrence of the fact that generics have the same active ingredient, strength, dosage form, and mode of administration as those reference products. Generics are easy to manufacture, unlike biologics, which are large and structurally complicated molecules that are sensitive to the manufacturing environment, because their chemical structures are defined and reproducible (Kaur et al., 2025). Table 1.1 represents the key differences between biologics and biosimilars.

Overall, the goal of both generics and biosimilars is to increase the affordability and accessibility of treatments after a patent expires, but because biologic molecules are complex, biosimilars need more rigorous development, manufacturing, and regulatory review (Revers & Furczon, 2010).

1.4.2 Sources and production methods

The development of biosimilars involves a highly intricate and tightly regulated biotechnological process that begins with selecting an appropriate biological source and extends through large-scale production under stringent quality control measures.

Because biosimilars are derived from living systems, factors such as the choice of source, cell line development, and process control are crucial in determining their stability, biological function, and overall safety profile (AL-Sabbagh et al., 2016). Most biosimilars are produced using rDNA technology, and their expression depends on the nature of the therapeutic protein being developed. Popular expression systems that are used to generate biologics are bacterial (e.g., *E. coli*), yeast (e.g., *Saccharomyces cerevisiae* and *Pichia pastoris*), mammalian cell lines (e.g., CHO and human embryonic kidney (HEK293) cells), and insect or plant cell systems. A high use of bacterial platforms has been to express nonglycosylated and relatively simple proteins, including insulin, interferons, and growth hormones. Even though bacteria do not have the tools to perform posttranslational modification, they are among the most beneficial in growth rate, scalability, and yield. In comparison, yeast systems offer a restricted glycosylation performance, and they form a sensible trade-off between an easy-to-manipulate system and the capability to carry out processing after translation. They have well-established fermentation processes; hence, they are viable in producing them economically and reproducibly. (Car et al., 2025). Mammalian cell lines are still the favored cell lines to be used in the production of complex therapeutic proteins, such as mAbs, cytokines, and fusion proteins. These have a stronger similarity to the human protein architecture, thus guaranteeing correct folding, assembly, and glycosylation, which are vital in therapeutic effectiveness and safety. Innovative platforms on plant and insect cells have also become promising. Insect cell systems include baculovirus expression vectors when expressed in insect cell systems (e.g., Sf9 and Sf21 based on *Spodoptera frugiperda*) (Damke et al., 2001). These systems facilitate precise folding and disulfide-bond formation; hence making them ideal for proteins that do not need complete human-like glycosylation (Huang et al., 2025). Plant-based systems, such as the use of Nicotiana benthamiana and *Arabidopsis thaliana*, have other advantages of low risks associated with contamination with human pathogens, lower production costs, and scalability that can be easily extended (Kulshreshtha et al., 2022). Glycoengineering has now made it possible to produce therapeutic proteins and vaccines with specific glycosylation patterns, but hurdles associated with variability in protein yield and uniform glycosylation in humans still need to be overcome (Dicker & Strasser, 2015). Finally, the choice of expression system is based on several variables, such as target protein molecular complexity, posttranslational modifications required, production efficiency, scalability, and regulatory requirements. Both the upstream (cell culture or fermentation) and downstream (purification and formulation) processes should be optimized carefully in order to generate biologically similar biosimilars (efficacy, purity, and safety) as compared to their reference biologics.

1.4.3 Classification of biosimilars

The following criteria can be used in classifying biosimilars based on the membership in a group, namely, structural complexity, therapeutic application, technological generation, regulatory status, and manufacturing source (Mascarenhas-Melo et al., 2024). All criteria present a unique view on the creation, assessment, and clinical use of biosimilars. Structurally, the biosimilars are represented by simple proteins, such as insulin, growth hormones, and insulin-like growth factors, and large, highly complex molecules, such as mAbs, cytokines, and fusion proteins. Simple proteins are usually easier to express and characterize since there are no complicated post-translational modifications, whereas more complicated proteins require advanced expression systems to ensure proper folding and assembly and avoid glycosylation to maintain biological activity and safety. The scope of therapeutic indications of biosimilars is wide. They are mainly used in endocrine and metabolic disorders in the form of hormonal biosimilars, such as recombinant human insulin, somatropin, and EPO (Krentz & Hompesch, 2015; Saenger, 2009). Infliximab and adalimumab mAbs are useful in the treatment of inflammatory and autoimmune diseases that are facilitated by immunomodulatory biosimilars. Filgrastim and pegfilgrastim, hematopoietic growth-factor biosimilars, assist bone-marrow recovery and help in the production of leukocytes in patients receiving chemotherapy (Rastogi et al., 2021). Another differentiation of biosimilars is the technological generation. The first-generation biosimilars typically include the earliest products of recombinant type, which are quite simply designed and made in microbial form. Second-generation biosimilars were developed that had better formulation, stability, and pharmacokinetics. Current third-generation biosimilars make use of modern and refined approaches of molecular engineering, including pegylation, fusion-protein formation, and glycoengineering to improve therapeutic effect, prolong systemic exposure, and mitigate immunogenicity (Mascarenhas-Melo et al., 2024). Regulatory bodies like the EMA and the USFDA categorize a biosimilar as a biological drug that is very similar to a previously approved reference drug, without clinically significant differences in terms of safety, purity, and potency. On the other hand, some biologic products in other markets that do not have extensive comparability information are categorized as intended copies or noncomparable biologics and are not considered bonafide biosimilars by leading regulators (Kaur et al., 2025).

Another level of classification of biosimilars is manufacturing source, since the system of expression significantly influences the nature of the product. The production of smaller, nonglycosylated proteins is normally done using the bacterial and the yeast systems, as they are cost and scale effective. Mammalian cell lines, on the other hand, are favored by intricate glycoproteins and mAbs since they are more capable of recapitulating human posttranslational alterations. Plant and insect cell systems are new, scalable, and economical systems, but there are still problems associated with yield consistency and human-compatible glycosylation (Konara et al., 2016). An in-depth application of biosimilar categorization using these multidimensional parameters can help researchers and clinicians to recognize the diversity and complexity of these therapeutic agents. It also highlights the importance of such analytical characterization and comparability studies to get the biosimilars to continuously fulfill the predetermined requirements of efficacy, safety, and quality similar to their reference biologics.

1.4.4 Approved biosimilars

In the last two decades, the world has witnessed the tremendous approval of biosimilars, indicating the speed and the increased embracement of biopharmaceutical development. The regulatory authorities like the EMA, the USFDA, and the Central Drugs Standard Control Organization (CDSCO) in India have developed detailed systems in order to approve biosimilars on comparability assessment based on rough comparisons. Europe has been leading in the field, and the first biosimilar approval was introduced in Europe in 2006 for somatropin, a recombinant human growth hormone sold as Omnitrope (Schiestl et al., 2017). Since then, a number of biosimilars have been granted across the therapeutic classes, such as hormones, mAbs, growth factors, and enzymes. The most popular types of biosimilars include those that treat chronic diseases, including diabetes, cancer, autoimmune diseases, and anemia. An example is the insulin glargine biosimilar and insulin lispro that have been approved to treat diabetes and which are sold by different trade names in different areas. Similarly, biosimilar epoetin alfa and filgrastim have been used regularly as a treatment of anemia and neutropenia (which is caused by chemotherapy), respectively (Barosi et al., 2011). The other aspect of the therapy that is changed by mAb biosimilars is that, being less costly than costly biologics, they present cost-effective treatment options. Remarkable ones are trastuzumab, adalimumab, bevacizumab, and rituximab biosimilars, having a number of approved indications in oncology and autoimmune (Mirjalili et al., 2021). Analytical, preclinical, and clinical analyses of these agents have been done to ensure that they are similar to the original reference products. India has reached substantial growth in the biosimilar market, increasing this country as being among the world leaders in biosimilar development and production (Meher et al., 2019). There are clear guidelines put into practice by the CDSCO in conjunction with the Department of Biotechnology in promoting biosimilar approval without compromising on international standards of quality (*CDSCO Guidelines for Biosimilars Explained: Ultimate Guide to India's Regulatory Pathway – Pharma Regulatory*). A number of biosimilars have also been launched by the Indian pharmaceutical firms, including Biocon, Dr. Reddy's Laboratories, and Intas Pharmaceuticals, in either the domestic or international markets (Panda et al., 2023). The most notable of them are the trastuzumab biosimilar (Canmab) of Biocon, pegfilgrastim biosimilar (Peg-Grafeel), and insulin glargine biosimilar (Basalog) (Priyarega & Natarajan, 2022). Such approvals have not only removed barriers to the access of biologic therapies for patients, but they have also given India a boost to its reputation as a cheap biotherapeutics hub. The first biosimilar regulatory approval in the United States took place in the year 2015 with filgrastim-sndz (Zarxio), which is a Neupogen biosimilar. FDA, since that landmark, the FDA has approved several biosimilars so far that represent therapeutic categories including oncology, immunology, and endocrinology. Such include adalimumab-atto (Amjevita), bevacizumab-awwb (Mvasi), and trastuzumab-dkst. The entry of these products has widened the choice of treatment available to the patient in addition to encouraging competitive pricing in the biologics market. The experience gained so far regarding approved biosimilars has strengthened their dependability and safety in the world. The postmarketing surveillance evidence continuously indicates biosimilars and their respective reference biologics have no significant difference in terms of clinical outcome and safety profile (Mascarenhas-Melo et al., 2024).

1.4.5 Development process of biosimilars with common challenges

The development of biosimilars is a very complex and multidisciplinary process that requires the combination of the fields of molecular biology, bioengineering, analyzing sciences, and clinical assessment (AL-Sabbagh et al., 2016). In contrast with small-molecule generics, which can be synthesized chemically to produce precise copies of the drug, biosimilars are given by biosystems, and thus they can only produce a high percentage of impact but not the exact replica of the reference biologic. This is started by the selection and the characterization of the reference product, and the reference product is the basis on which all the other development stages are based. Its structure, physicochemical properties, and biological and formulation properties are obtained by way of sophisticated analysis methods, including mass spectrometry, chromatography, and spectroscopy (AL-Sabbagh et al., 2016). After characterization, the development of cell lines is the basis of biosimilar production. rDNA technology involves the insertion of the gene that is, in the case of the desired therapeutic protein, into a suitable host cell, like bacterial, yeast, or even mammalian cells. High yield and stable expression are then optimized by selection of stable cell lines. Within this workflow, there are essential issues of consistency between production batches because even slight changes could affect the safety and efficacy of the biosimilar. Once the manufacturing process has been streamlined, analytical and preclinical studies in large numbers are conducted to compare the biosimilar to the reference product. Such studies evaluate molecular structure, bioactivity, pharmacokinetics, and pharmacodynamics to guarantee similar functionalities. It is only after obtaining enough evidence of similarity that the developers proceed to clinical trials, which are usually shorter than a novel biologic. The major aim of clinical studies is to support similar efficacy, safety, and immunogenicity as opposed to re-establishing therapeutic benefit with scratch. Although there has been improvement in the technological arena and regulatory directives, the development of biosimilars still faces a number of typical challenges. The main challenge is to replicate the complicated structural and functional properties of biologics that are generated in living cells; even slight alterations of the culture conditions, starting material, and purification strategies may introduce structural differences (Calvo & Zuniga, 2012). The issue of immunogenicity continues to be a major problem since even slight changes in glycosylation patterns or impurities can trigger immune reactions, which undermine safety or therapeutic effectiveness. Analytical comparability entails the availability of very sensitive and validated methods, which may be resource-intensive and technical (Scavone et al., 2017). Regulatory-wise, the development of biosimilars is further complicated by tough requirements to approve the same, which requires extensive data to show similarity on various fronts, such as analytical, preclinical, and clinical. The price and time of biosimilar development remain significantly higher than that of traditional generics, which may take a period of 8 to 10 years and may require significant financial resources (Kaur et al., 2025). The barrier to intellectual property, exclusivity rights on data, and the inability to harmonize global regulations may slow down the process of market entry. As a result, the biosimilars development process can be viewed as a fine balance between science and regulation (Kaur et al., 2025; Scavone et al., 2017).

1.4.6 Some current preclinical and clinical studies using biosimilars

1.4.6.1 Preclinical studies using biosimilars

The study by Wang et al. (2022) saw the use of a biosimilar monoclonal antibody (mAb), referred to as KA, and its biosimilarity with the originator omalizumab was just tested in the extensive preclinical characterization program. A detailed physicochemical analysis was used in the study, such as primary and secondary structure analysis and charge heterogeneity profiling, in vitro functions of IgE binding and cell-based inhibitor, and in vivo pharmacological and toxicological effects. The in vivo part was done in an allergic mouse model, whereby pharmacological efficacy was established, followed by a series of studies in cynomolgus monkeys to study pharmacokinetics, safety, and immunogenicity. Findings showed that there was little difference between KA and the reference, omalizumab: their structure profiles were very concordant, in vitro binding and functional inhibition were mostly overlapping, and in vivo animal experiments showed KA had a similar level of efficacy, pharmacokinetic behavior, and safety. These results were strong evidence of further development of KA for clinical testing as a biosimilar to omalizumab (Wang et al., 2022).

One of the studies conducted by Velasco-Velázquez et al. (2017) deserves mention as a commendable preclinical trial that was a thorough assessment to investigate an infliximab biosimilar, which is an mAb, and it is supposed to be the analogue of the reference product that is used to treat autoimmune diseases like rheumatoid arthritis and Crohn's disease, etc. The main goal of the trial was to prove the structural, functional, and pharmacological equivalency of the biosimilar and the parent molecule before human testing. The authors conducted a set of comparative in vitro and in vivo studies, including binding affinity to tumor necrosis factor-alpha, biological activity, and immunogenic potential. In the in vivo part, Sprague-Dawley rats were used as an animal model in order to compare the pharmacokinetics, tissue distribution, and safety of intravenous looked-at drug dosing of the reference infliximab and biosimilar candidate. The span of the treatment, excluding the follow-up, was 4 weeks; physiological, hematological, and biochemical parameters were closely monitored during this period. There were no significant differences between the two groups in terms of serum concentration time profile, systemic exposure, and adverse events. Histopathological analyses did not indicate any atypical presence or organ toxicity. In general, the research proved that the biosimilar recapitulated the reference infliximab with regard to its molecular structure, biological activity, pharmacokinetic characteristics, and safety, and, consequently, showed the substantiation of its advancement to clinical trials (Velasco-Velázquez et al., 2017).

Biosimilar candidate GP2013 was evaluated by comparison with reference rituximab through an intensive structural, functional, and in vivo pipeline in a focused preclinical development program to achieve pharmacological equivalence. A study was conducted to make sure that GP2013 possesses important features of rituximab, such as posttranslation modifications, binding to CD20, ADCC, and in vivo activity. Glycosylation profiling, quantifying the presence of particular glycan variants (in particular the α-fucosylated bG0 species), was used in the in vitro stage, and a distinct relationship between that particular glycan variant and ADCC potency was used to optimize cell line and process. In vivo efficacy was evaluated in SCID mouse xenograft models of NHL (SU-DHL-4 and Jeko-1 cell lines), in which GP2013 and rituximab were given at sub-therapeutic, although sensitized dosage regimens (i.e., every week, intraperitoneal injected), and showed overlapping antitumor activities. The cynomolgus monkeys were assessed in PK and PD studies: single-dose (5 mg/kg^{-1}) and multiple-dose (20 or 100 mg/kg/week for 17 weeks), along with follow-up in serum exposure to the drug and B-cell depletion (CD20+ subsets). Findings indicated similar area-under-curve values within prescribed ranges of equivalence, and both products exhibited similar kinetics in B-cell depletion/recovery. In total, this comparability exercise supported this conclusion that the pharmacological analogues of GP2013 and originator rituximab are structurally, functionally, in vivo-effector, and PK/PD-behaving analogues (da Silva et al., 2014).

1.4.6.2 Clinical studies using biosimilars

Kay et al. reported a clinical evaluation of a previously reported randomized, double-blind trial that has now been extended long-term to determine the continued activity, safety, and immunogenicity of PF-06438179/GP1111, a biosimilar of infliximab, in patients with rheumatoid arthritis. The extension involved patients who participated in the initial 54-week trial as they continued through weeks 54 to 78 and compared the outcomes of the two groups of patients who performed the reference infliximab and those who switched to the biosimilar. It was revealed that clinical response, disease activity scores, and safety outcomes showed no changes over the years and did not show a rise of adverse events and immunogenicity after the change of treatment. These results supported the idea of interchangeability of the biosimilar and the reference product. Zhou et al. tested HLX01, a rituximab biosimilar, in a phase III randomized, double, parallel-group clinical trial, which was done across various sites in China. The patients involved in the study were 407 patients between the ages of 18 and 80 years with the presence of CD20-positive diffuse large B-cell lymphoma (DLBCL) and who had not undergone any previous therapy. The participants were assigned randomly as cohorts to receive either HLX01 with CHOP chemo or reference rituximab with CHOP (combined chemotherapy drugs). Findings revealed similar efficacy outcomes between the two groups regarding overall response rate and progression-free survival, with safety and immunogenicity profiles also being consistent. Therefore, the results favored the application of the HLX01 as a cost-efficient replacement of originator rituximab that was clinically equivalent in the treatment of DLBCL (Shi et al., 2020). Marcus et al. used PF-05280586, which is a proposed rituximab biosimilar, in a mixed and randomized, comparative 52-week trial and evaluations of patients with untreated and previously untreated CD20-positive low-tumor-burden (LTB-FL) follicular lymphoma. The goal was to ensure biosimilarity to the European Union-approved rituximab. There were no significant differences noted between the biosimilar and the originator in the endpoints of efficacy, such as overall response rate and time to progression of the disease, during the duration of the study. The comparison of the safety and pharmacokinetic features and the immunogenicity was also quite similar, thus strong evidence that PF-05280586 can be used as an equivalent to the reference biologic in clinical practice. Lemos et al. published a pivotal randomized, double-blind, multicenter study that assessed the comparative efficacy and safety of PF-06438179/GP1111, another infliximab biosimilar, against reference infliximab (Remicade) in patients with moderate-to-severe rheumatoid arthritis who had an inadequate response to methotrexate. Over a 30-week treatment period, clinical equivalence was demonstrated based on the American College of Rheumatology (ACR20) response rates. Pharmacokinetic and immunogenicity assessments further confirmed high similarity between the two treatments, with no unexpected safety findings. The study provided critical clinical evidence supporting regulatory approval of this biosimilar for multiple inflammatory conditions (Cohen et al., 2018).

1.5 Mechanism of action of biosimilars and biologics in cancer treatment

Biologic agents have radically changed the face of oncology by offering very specific and targeted mechanisms of therapy, which is a marked contrast with the classical chemotherapeutic modalities (Bowers et al., 2020). Whereas cytotoxic chemotherapy exerts its action by nonselectively destroying proliferating cells rapidly, biologic therapeutics interact with targets in a specific manner, destabilizing tumor cell proliferation, existence, and avoidance of the immune action (Letai & de The, 2025). The assumption that biosimilars will assume an identical or almost identical mechanism of action is because their therapeutic effect will be a result of interaction with the same biological targets and epitopes of binding. The difference between the reference products and the biosimilars lies not in the different mechanisms but in their forms of molecular source and manufacturing, which all undergo a careful analytical and clinical comparability evaluation to prove similar biological activity (Shilpi et al., 2024). The most significant group of biologics used by oncologists is mAbs (Zahavi & Weiner, 2020). The engineered molecules are designed to uniquely identify and bind to antigens on malignant cells or within the tumor microenvironment, thereby inducing a repertoire of antitumor responses. As an example, early FDA-approved mAbs, including rituximab, whose therapeutic target, CD20, is found on B-cell lymphomas, have been shown to trigger tumor cell death by ADDC, complement-dependent cytotoxicity, and by apoptosis. In the same manner, the trastuzumab connects with the HER2 receptor in breast cancer cells, preventing receptor dimerization and receptor-mediated downstream means of signaling, such as PI3K/Akt and MAPK signals, which play key roles in the growth and survival of cells. The impediment of such signaling pathways eventually leads to growth arrest and programmed cell death (Valabrega et al., 2007). The other significant line of therapy is the angiogenesis inhibitors, which block the formation of the blood supply to the tumor that is necessary for its maintenance. The example is bevacizumab, which targets the vascular endothelial growth factor (VEGF) ligand so that it does not interact with the VEGF receptors of endothelial cells. This resembles inhibition of neovascularization (angiogenesis), which in turn leads to reduced oxygen and nutrient supply to tumors and a lack of ability to metastasize. Biosimilar agents with bevacizumab (e.g., bevacizumab-awwb) use the same VEGF-binding system and are of the same efficacy and safety as the original . Immune checkpoint inhibitors that mobilize the endogenous immune system to fight malignancy are also included in the category of biologics (Shiravand et al., 2022). Nivolumab and pembrolizumab are PD-1 receptor antagonists, and atezolizumab is a CAM programmed death-ligand 1 (PD-L1) antagonist (Khan et al., 2024). These interactions reinstate T-cell activity and promote immune-mediated tumor destruction. Despite the fact that the majority of checkpoint inhibitors do not have approved biosimilars to date, since most of them have intellectual property limitations, their mechanisms can be utilized to develop new biosimilars in the context of immuno-oncology (Scavone et al., 2017). The cancer therapeutic efficacy of both the biologics and the biosimilars can be elucidated by this particular point, though, highly specific molecular targeting and not indiscriminate cytotoxicity. These agents operate to stimulate long-lasting clinical responses and better patient outcomes through the modulation of cell signaling pathways, tumor angiogenesis alteration, and improved immune surveillance. They are also known as biosimilars that, via mechanically equivalent action, will increase the availability of these superior treatments, making sure that the advantages of targeted therapy and immune-based cancer treatment will be available to more patients at a lower price (Shilpi et al., 2024).

1.6 Applications of biologics and biosimilars

Biologics and biosimilars now form the primary approaches in the management of the various chronic and immune-mediated conditions and conditions in which conventional treatments have failed to be favorable (Revers & Furczon, 2010). TNF-2, IL-6, and B-cell-depleting mAb should be used in the management of rheumatoid arthritis to decrease the joint inflammation and structural damage; biosimilars of adalimumab, infliximab, and etanercept have substantially improved access and lowered the cost of treatment with no loss of efficacy. They also find application in inflammatory bowel disease, where anti-TNF, antiintegrin, and IL-12/23 blockers have been found useful in the healing of the mucosa and the maintenance of remission in moderate-to-severe Crohn's disease and ulcerative colitis; in this case, again, biosimilars have made medicine more affordable and enabled greater use of early biological-therapy approaches. Recombinant insulin analogues are still important in the management of type 1 and progressive type 2 diabetes as a glycemic control tool, and biosimilar insulins have become cost-effective alternatives that are therapeutically equivalent, but reduce the economic strain on patients and the medical system. The use of biologics is also groundbreaking in ophthalmology, where anti-VEGF antibodies, e.g., ranibizumab, aflibercept, and bevacizumab, prevent the abnormal neovascularization in age-related macular degeneration and diabetic macular edema to preserve visual function; the advent of anti-VEGF biosimilars has further increased access, especially where repeated intravitreal injections can be a significant economic burden. In oncology, mAbs that address pathways, including HER2, VEGF, and CD20, and immune checkpoint inhibitors have revolutionized the treatment of breast cancer and lymphomas and a number of other solid tumors by providing the opportunity to target the tumor and activate a stable immune reaction (Briukhovetska et al., 2021; Papież & Krzyściak, 2021). Biosimilar trastuzumab, bevacizumab, and rituximab have increased the reach of treatment and helped to provide fair access to life-saving therapies across the globe. Biologics, including PCSK9-inhibitory antibodies, have emerged in cardiovascular medicine and are noted to lower the levels of LDL-cholesterol in high-risk patients and significantly reduce the incidence of atherosclerotic cardiovascular events in combination with statins in patients with deep-seated dyslipidemia. All these applications help underscore the role of biologics to redefine disease management in any specialty and biosimilars to not only provide sustainability but also greater access to these novel therapeutics. Fig. 1.2 illustrates the clinical applications of biologics and biosimilars.

1.7 Regulatory guidelines/approval pathways

Since biosimilars and biologics are highly similar to a reference (original) product, with no to minimal differences in safety, purity, or potency, the regulators use a totality of evidence or comparability approach that is a stepwise demonstration via analytical, nonclinical, and clinical data. First, a licensed reference product must be chosen; in the EU, it is the product that is already approved in the EEA. The biosimilars are characterized on the basis of physicochemical properties (e.g., amino-acid sequence), purity, biological activity, aggregation, etc. The process of manufacturing must be consistent; the developer must produce a yield that is comparable to the reference product. Nonclinical and clinical studies require PK/PD testing, which shows pharmacokinetics and sometimes pharmacodynamics based on exposure and response. Immunogenicity testing is critical to measure antidrug antibodies and neutralizing potential, using validated assays. Earlier, the clinical efficacy trials were required, but

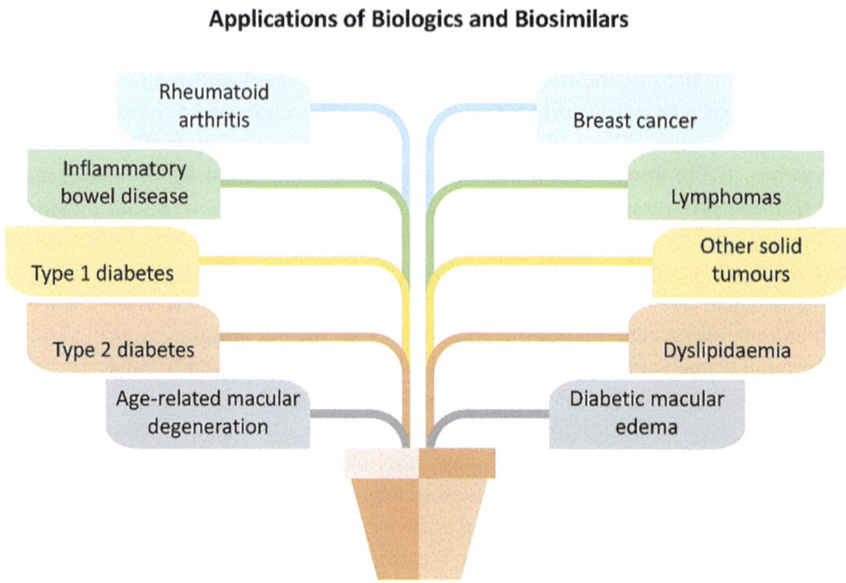

FIGURE 1.2

Clinical applications of biologics and biosimilars.

with time, regulators increasingly accept high-sensitivity analytics, robust PK/PD, and immunogenicity data. Glycosylation pattern, sialylation, fucosylation, etc., can affect half-life, immunogenicity, and effector functions. Regulators expect detailed glyco-profiling and functional assays showing that any differences do not alter clinical performance. Aggregates can increase immunogenicity risk; sensitive methods (SEC-MALS, orthogonal aggregation assays) and acceptance criteria are required. Potency must be biologically relevant and mechanism-of-action linked; multiple orthogonal potency assays strengthen the case for comparability. In the United States, the FDA has recommended switching studies to show evidence of interchangeability of a biosimilar, whereas the EU/EMA focuses on biosimilarity; physicians and national authorities manage switching or substitution practices (Car et al., 2025). In India, the CDSCO guidelines for similar biologics mandate reference-product selection, comparability, clinical/nonclinical studies, and quality control (Mascarenhas-Melo et al., 2024; Schiestl et al., 2017). In general, regulators around the world continue to work on improvements in the development of biosimilars and are attempting to reconcile the need to adhere to scientific standards and the need to allow broader accessibility. The fact that the analytical sensitivity has improved and that the comparability has been raised appears to have changed the basis on which approvals are evident and also reduced the value of extensive clinical trials. All these emerging regulatory systems are underlined in a manner that is conducive to the safe, effective, and wider use of biosimilars in oncologic management (Konara et al., 2016).

1.8 Conclusion

Biologics have significantly transformed oncologic treatments by enabling targeted, mechanism-based treatments that extend the lifespan and improve the quality of life. The example of the heterogeneity of these agents, including mAbs, cell therapies like CAR-T cells, and other biologically engineered modalities, reflects their ability to regulate complex oncogenic pathways. However, barriers of the manufacturing costs, complicated production processes, and limited global availability still exist, hence restricting the pervading nature of these therapies. Biosimilars, in this case, have emerged as an opportune solution, offering the same measure of safety, efficacy, and quality at a cheaper rate, hence extending access to advanced oncological care. These stringent analytical guidelines, clinical judgement, and regulation frameworks that support the development of biosimilars have enhanced confidence levels among the clinical and policy fraternity, hence making their integration into routine oncologic practices more feasible. Biosimilars and biologics are inseparable tools to achieve more balanced, sustainable, and effective cancer treatment, ensuring new therapeutic options can be available to a broader and more diverse group of patients.

References

Al-Katib, A. M., & Mohamed, A. N. (2006). Non-Hodgkin's lymphoma and chronic lymphocytic leukemia. *Principles of Molecular Medicine* (pp. 794–803). Humana Press. https://doi.org/10.1007/978-1-59259-963-9_82.

Allocati, E., Godman, B., Gobbi, M., Garattini, S., & Banzi, R. (2022). Switching among biosimilars: A review of clinical evidence. *Frontiers in Pharmacology, 13*. https://doi.org/10.3389/fphar.2022.917814.

Almeida-Oliveira, A., Bonamino, M. H., Simões, M., Villaça Martins, V. H., Borges Freire, D. P., Danowski, A., Catão, T. P., Varela, J. D. N. E. S., Mazzi, M. T., Gottgtroy, C. L., de Miranda Marques, A., Barbosa, M. M., & de Castro Fialho, B. (2025). Landscape of ex vivo gene therapies: Technological trends and future prospects. *Molecular Therapy Methods and Clinical Development, 33*(4). https://doi.org/10.1016/j.omtm.2025.101580.

AL-Sabbagh, A., Olech, E., McClellan, J. E., & Kirchhoff, C. F. (2016). Development of biosimilars. *Seminars in Arthritis and Rheumatism, 45*(5), S11. https://doi.org/10.1016/j.semarthrit.2016.01.002.

Barosi, G., Bosi, A., Abbracchio, M. P., Danesi, R., Genazzani, A., Corradini, P., Pane, F., & Tura, S. (2011). Key concepts and critical issues on epoetin and filgrastim biosimilars. A position paper from the Italian Society of Hematology, Italian Society of Experimental Hematology, and Italian Group for Bone Marrow Transplantation. *Haematologica, 96*(7), 937–942. https://doi.org/10.3324/haematol.2011.041210.

Bowers, R. L., Troyer, W. D., Mason, R. A., & Mautner, K. R. (2020). Biologics. *Techniques in Vascular and Interventional Radiology., 23*(4). https://doi.org/10.1016/j.tvir.2020.100704.

Briukhovetska, D., Dörr, J., Endres, S., Libby, P., Dinarello, C. A., & Kobold, S. (2021). Interleukins in cancer: From biology to therapy. *Nature Reviews Cancer, 21*(8), 481–499. https://doi.org/10.1038/s41568-021-00363-z.

Buchbinder, E. I., & Desai, A. (2016). CTLA-4 and PD-1 pathways similarities, differences, and implications of their inhibition. *American Journal of Clinical Oncology: Cancer Clinical Trials, 39*(1), 98–106. https://doi.org/10.1097/COC.0000000000000239.

Burt, R., Warcel, D., & Fielding, A. K. (2018). Blinatumomab, a bispecific B-cell and T-cell engaging antibody, in the treatment of B-cell malignancies. *Human Vaccines & Immunotherapeutics, 15*(3), 594–602. https://doi.org/10.1080/21645515.2018.1540828.

Cadieux, B., Coleman, R., Jafarinasabian, P., Lipton, A., Orlowski, R. Z., Saad, F., Scagliotti, G. V., Shimizu, K., & Stopeck, A. (2022). Experience with denosumab (XGEVA®) for prevention of skeletal-related events in the 10 years after approval. *Journal of Bone Oncology, 33*. https://doi.org/10.1016/j.jbo.2022.100416.

Calvo, B., & Zuniga, L. (2012). Therapeutic monoclonal antibodies: Strategies and challenges for biosimilars development. *Current Medicinal Chemistry, 19*(26), 4445–4450. https://doi.org/10.2174/092986712803251485.

Car, E., Barbier, L., Huys, I., Simoens, S., & Vulto, A. G. (2025). Evolving global regulatory landscape for approval of biosimilars: Current challenges and opportunities for convergence. *Expert Opinion on Biological Therapy, 25*(6), 649–668. https://doi.org/10.1080/14712598.2025.2507832.

Castan, A., Schulz, P., Wenger, T., & Fischer, S. (2018). Cell line development. *Biopharmaceutical processing: Development, design, and implementation of manufacturing processes* (pp. 131–146). Elsevier. https://doi.org/10.1016/B978-0-08-100623-8.00007-4.

Chang, H. W., Wang, J., Liu, H., Xing, C., Chen, J., Frey, G., Boyle, W. J., & Short, J. M. (2025a). Preclinical development of mecbotamab vedotin (BA3011), a novel, AXL-specific conditional active biologic antibody-drug conjugate. *Antibody Therapeutics, 8*(2), 145–156. https://doi.org/10.1093/abt/tbaf006.

Chang, H. W., Frey, G., Wang, J., Liu, H., Xing, C., Chen, J., Boyle, W. J., & Short, J. M. (2025b). Preclinical development of ozuriftamab vedotin (BA3021), a novel ROR2-specific conditionally active biologic antibody–drug conjugate. *mAbs, 17*(1). https://doi.org/10.1080/19420862.2025.2490078.

Cohen, S. B., Alten, R., Kameda, H., Hala, T., Radominski, S. C., Rehman, M. I., Palaparthy, R., Schumacher, K., Schmitt, S., Hua, S. Y., Ianos, C., & Sewell, K. L. (2018). A randomized controlled trial comparing PF-06438179/GP1111 (an infliximab biosimilar) and infliximab reference product for treatment of moderate to severe active rheumatoid arthritis despite methotrexate therapy. *Arthritis Research and Therapy, 20*(1). https://doi.org/10.1186/s13075-018-1646-4.

da Silva, A. D., Kronthaler, U., Koppenburg, V., Fink, M., Meyer, I., Papandrikopoulou, A., Hofmann, M., Stangler, T., & Visser, J. (2014). Target-directed development and preclinical characterization of the proposed biosimilar rituximab GP2013. *Leukemia and Lymphoma, 55*(7), 1609–1617. https://doi.org/10.3109/10428194.2013.843090.

Damke, H., Muhlberg, A. B., Sever, S., Sholly, S., Warnock, D. E., & Schmid, S. L. (2001). Expression, purification, and functional assays for self-association of dynamin-1. *Methods in Enzymology, 329*, 447–457. https://doi.org/10.1016/S0076-6879(01)29106-X.

Dicker, M., & Strasser, R. (2015). Using glyco-engineering to produce therapeutic proteins. *Expert Opinion on Biological Therapy, 15*(10), 1501–1516. https://doi.org/10.1517/14712598.2015.1069271.

Fan, T., Zhang, M., Yang, J., Zhu, Z., Cao, W., & Dong, C. (2023). Therapeutic cancer vaccines: Advancements, challenges and prospects. *Signal Transduction and Targeted Therapy, 8*(1). https://doi.org/10.1038/s41392-023-01674-3.

Findlay, V. J., & Scholar, E. (2007). Trastuzumab. *xPharm: The comprehensive pharmacology reference* (pp. 1–5). Elsevier Inc. https://doi.org/10.1016/B978-008055232-3.63738-9.

Handy, C. E., & Antonarakis, E. S. (2018). Sipuleucel-T for the treatment of prostate cancer: Novel insights and future directions. *Future Oncology, 14*(10), 907–917. https://doi.org/10.2217/fon-2017-0531.

Hassel, J. C., Berking, C., Forschner, A., Gebhardt, C., Heinzerling, L., Meier, F., Ochsenreither, S., Siveke, J., Hauschild, A., & Schadendorf, D. (2023). Practical guidelines for the management of adverse events of the T cell engager bispecific tebentafusp. *European Journal of Cancer, 191*, 112986. https://doi.org/10.1016/j.ejca.2023.112986.

Henrickson, S. E., Ruffner, M. A., & Kwan, M. (2016). Unintended immunological consequences of biologic therapy. *Current Allergy and Asthma Reports, 16*(6). https://doi.org/10.1007/s11882-016-0624-7.

Huang, N., Wei, Y., & Li, J. (2025). Insect cell expression system: Advances in applications, engineering strategies, and bioprocess development. *Journal of Biological Engineering, 19*(1). https://doi.org/10.1186/s13036-025-00562-4.

Jung, S. M., & Kim, W. U. (2022). Targeted immunotherapy for autoimmune disease. *Immune Network, 22*(1). https://doi.org/10.4110/in.2022.22.e9.

Kaur, S., Yadav, S., Sahu, V., Sharma, N., & Shukla, V. K. (2025). Biosimilar regulations: Current framework and future prospects. *Current Drug Safety*. https://doi.org/10.2174/0115748863360017250509063745.

Khan, B., Qahwaji, R. M., Alfaifi, M. S., & Mobashir, M. (2024). Nivolumab and ipilimumab acting as tormentors of advanced tumors by unleashing immune cells and associated collateral damage. *Pharmaceutics, 16*(6). https://doi.org/10.3390/pharmaceutics16060732.

Konara, C. S., Barnard, R. T., Hine, D., Siegel, E., & Ferro, V. (2016). The tortoise and the hare: Evolving regulatory landscapes for biosimilars. *Trends in Biotechnology, 34*(1), 70–83. https://doi.org/10.1016/j.tibtech.2015.10.009.

Krentz, A. J., & Hompesch, M. (2015). Biosimilar insulins: Current and future perspectives. *Diabetes Management, 5*(6), 405–409. https://doi.org/10.2217/dmt.15.33.

Kulshreshtha, A., Sharma, S., Padilla, C. S., & Mandadi, K. K. (2022). Plant-based expression platforms to produce high-value metabolites and proteins. *Frontiers in Plant Science, 13*. https://doi.org/10.3389/fpls.2022.1043478.

Letai, A., & de The, H. (2025). Conventional chemotherapy: Millions of cures, unresolved therapeutic index. *Nature Reviews Cancer, 25*(3), 209–218. https://doi.org/10.1038/s41568-024-00778-4.

Lipson, E. J., & Drake, C. G. (2011). Ipilimumab: An anti-CTLA-4 antibody for metastatic melanoma. *Clinical Cancer Research, 17*(22), 6958–6962. https://doi.org/10.1158/1078-0432.ccr-11-1595.

Lyu, X., Zhao, Q., Hui, J., Wang, T., Lin, M., Wang, K., Zhang, J., Shentu, J., Dalby, P. A., Zhang, H., & Liu, B. (2022). The global landscape of approved antibody therapies. *Antibody Therapeutics, 5*(4), 233–257. https://doi.org/10.1093/abt/tbac021.

Mascarenhas-Melo, F., Diaz, M., Gonçalves, M. B. S., Vieira, P., Bell, V., Viana, S., Nunes, S., Paiva-Santos, A. C., & Veiga, F. (2024). An overview of biosimilars—Development, quality, regulatory issues, and management in healthcare. *Pharmaceuticals, 17*(2). https://doi.org/10.3390/ph17020235.

Meher, B. R., Balan, S., Mohanty, R. R., Jena, M., & Das, S. (2019). Biosimilars in India; Current status and future perspectives. *Journal of Pharmacy and Bioallied Sciences, 11*(1), 12–15. https://doi.org/10.4103/JPBS.JPBS_167_18.

Mirjalili, S. Z., Sabourian, R., Sadeghalvad, M., & Rezaei, N. (2021). Therapeutic applications of biosimilar monoclonal antibodies: Systematic review of the efficacy, safety, and immunogenicity in autoimmune disorders. *International Immunopharmacology, 101*(Pt B), 108305. https://doi.org/10.1016/j.intimp.2021.108305.

Nwosu, N. (2024). Cancer: A disease of modern times? *Cureus, 16*(11). https://doi.org/10.7759/cureus.74666.

Paccagnella, M., Abbona, A., Denaro, N., Merlano, M. C., & Garrone, O. (2025). *Antibody-dependent cell-mediated cytotoxicity (ADCC) in cancer* (pp. 243–263). Springer Science and Business Media LLC. https://doi.org/10.1007/978-3-032-00749-0_11.

Panda, S., Singh, P. K., Mishra, S., Mitra, S., Pattnaik, P., Adhikary, S. D., & Mohapatra, R. K. (2023). Indian biosimilars and vaccines at crossroads—Replicating the success of pharmagenerics. *Vaccines, 11*(1). https://doi.org/10.3390/vaccines11010110.

Papież, M. A., & Krzyściak, W. (2021). Biological therapies in the treatment of cancer—Update and new directions. *International Journal of Molecular Sciences, 22*(21). https://doi.org/10.3390/ijms222111694.

Peters, C., & Brown, S. (2015). Antibody–drug conjugates as novel anti-cancer chemotherapeutics. *Bioscience Reports, 35*(4). https://doi.org/10.1042/bsr20150089.

Pierpont, T. M., Limper, C. B., & Richards, K. L. (2018). Past, present, and future of Rituximab-The world's first oncology monoclonal antibody therapy. *Frontiers in Oncology, 8*. https://doi.org/10.3389/fonc.2018.00163.

Pouresmaeil, M., & Azizi-Dargahlou, S. (2023). Factors involved in heterologous expression of proteins in *E. coli* host. *Archives of Microbiology, 205*(5). https://doi.org/10.1007/s00203-023-03541-9.

Priyarega, S., & Natarajan, R. (2022). An overview of biosimilars for cancer, diabetes mellitus, rheumatoid arthritis and other immune-mediated diseases approved between 2016 and 2021. *Results in Chemistry, 4*, 100356. https://doi.org/10.1016/j.rechem.2022.100356.

Rastogi, S., Kalaiselvan, V., Ali, S., Ahmad, A., Guru, S. A., & Sarwat, M. (2021). Efficacy and safety of filgrastim and its biosimilars to prevent febrile neutropenia in cancer patients: A prospective study and meta-analysis. *Biology, 10*(10). https://doi.org/10.3390/biology10101069.

Revers, L., & Furczon, E. (2010). An introduction to biologics and biosimilars. Part I: Biologics: What are they and where do they come from? *Canadian Pharmacists Journal, 143*(3), 134–139. https://doi.org/10.3821/1913-701X-143.3.134.

Saenger, P. (2009). Current status of biosimilar growth hormone. *International Journal of Pediatric Endocrinology, 2009*, 1–6. https://doi.org/10.1155/2009/370329.

Sagar, M. T., Rattan, A., & Prajapati, V. K. (2024). Unleashing the power of antibodies: Engineering for tomorrow's therapy. *Advances in Protein Chemistry and Structural Biology, 140*, 1–36. https://doi.org/10.1016/bs.apcsb.2023.12.009.

Sampathkumar, K., & Kerwin, B. A. (2024). Roadmap for drug product development and manufacturing of biologics. *Journal of Pharmaceutical Sciences, 113*(2), 314–331. https://doi.org/10.1016/j.xphs.2023.11.004.

Sathish, J. G., Sethu, S., Bielsky, M. C., De Haan, L., French, N. S., Govindappa, K., Green, J., Griffiths, C. E. M., Holgate, S., Jones, D., Kimber, I., Moggs, J., Naisbitt, D. J., Pirmohamed, M., Reichmann, G., Sims, J., Subramanyam, M., Todd, M. D., Van Der Laan, J. W., & Park, B. K. (2013). Challenges and approaches for the development of safer immunomodulatory biologics. *Nature Reviews Drug Discovery, 12*(4), 306–324. https://doi.org/10.1038/nrd3974.

Sathyan, A., Mohanapriya, M., Madhanraja, R., Indu Vadana, K. S., Indhuja, K., Sumathi, P., & Varshini, G. (2025). A comprehensive review of evidence and challenges in switching from originator drugs to biosimilars of monoclonal antibodies: Focus on rituximab and trastuzumab. *Journal of Drug Delivery and Therapeutics, 15*(2), 149–155. https://doi.org/10.22270/jddt.v15i2.7012.

Scavone, C., Rafaniello, C., Berrino, L., Rossi, F., & Capuano, A. (2017). Strengths, weaknesses and future challenges of biosimilars' development. An opinion on how to improve the knowledge and use of biosimilars in clinical practice. *Pharmacological Research, 126*, 138–142. https://doi.org/10.1016/j.phrs.2017.11.002.

Schiestl, M., Zabransky, M., & Sörgel, F. (2017). Ten years of biosimilars in Europe: Development and evolution of the regulatory pathways. *Drug Design, Development and Therapy, 11*, 1509–1515. https://doi.org/10.2147/DDDT.S130318.

Schütz, F., Stefanovic, S., Mayer, L., von Au, A., Domschke, C., & Sohn, C. (2017). PD-1/PD-L1 pathway in breast cancer. *Oncology Research and Treatment, 40*(5), 294–297. https://doi.org/10.1159/000464353.

Scordo, M., Lin, R. J., & Sauter, C. S. (2019). Hematopoietic-cell transplantation for lymphoma in the era of genetically engineered cellular therapy: It's not quite time to scrap the old vehicle for the new car. *Current Opinion in Hematology, 26*(4), 288–293. https://doi.org/10.1097/MOH.0000000000000515.

Shi, Y., Song, Y., Qin, Y., Zhang, Q., Han, X., Hong, X., Wang, D., Li, W., Zhang, Y., Feng, J., Yang, J., Zhang, H., Jin, C., Yang, Y. u, Hu, J., Wang, Z., Jin, Z., Su, H., Wang, H., … Zhang, X. (2020). A phase 3 study of rituximab biosimilar HLX01 in patients with diffuse large B-cell lymphoma. *Journal of Hematology & Oncology, 13*(1). https://doi.org/10.1186/s13045-020-00871-9.

Shilpi, S., Gurnany, E., Gupta, P. K., Sharma, K., Saini, K., Chimaniya, P., & Bhatt, S. (2024). Mechanism of biosimilars for the treatment of cancer. *Biosimilars for cancer treatment: A promising approach*, 35–54. https://doi.org/10.1007/978-981-97-3001-8_3.

Shiravand, Y., Khodadadi, F., Kashani, S. M. A., Hosseini-Fard, S. R., Hosseini, S., Sadeghirad, H., Ladwa, R., O'byrne, K., & Kulasinghe, A. (2022). Immune checkpoint inhibitors in cancer therapy. *Current Oncology, 29*(5), 3044–3060. https://doi.org/10.3390/curroncol29050247.

Sims, J. R., Zimmer, J. A., Evans, C. D., Lu, M., Ardayfio, P., Sparks, J. D., Wessels, A. M., Shcherbinin, S., Wang, H., Monkul Nery, E. S., Collins, E. C., Solomon, P., Salloway, S., Apostolova, L. G., Hansson, O., Ritchie, C., Brooks, D. A., Mintun, M., & Skovronsky, D. M. (2023). Donanemab in early symptomatic Alzheimer disease: The TRAILBLAZER-ALZ 2 randomized clinical trial. *JAMA., 330*(6), 512–527. https://doi.org/10.1001/jama.2023.13239.

Valabrega, G., Montemurro, F., & Aglietta, M. (2007). Trastuzumab: Mechanism of action, resistance and future perspectives in HER2-overexpressing breast cancer. *Annals of Oncology, 18*(6), 977–984. https://doi.org/10.1093/annonc/mdl475.

Velasco-Velázquez, M. A., Salinas-Jazmín, N., Hisaki-Itaya, E., Cobos-Puc, L., Xolalpa, W., González, G., Tenorio-Calvo, A., Piña-Lara, N., Juárez-Bayardo, L. C., Flores-Ortiz, L. F., Medina-Rivero, E., Pérez, N. O., & Pérez-Tapia, S. M. (2017). Extensive preclinical evaluation of an infliximab biosimilar candidate. *European Journal of Pharmaceutical Sciences, 102*, 35–45. https://doi.org/10.1016/j.ejps.2017.01.038.

Wang, Y., Zheng, C., Zhuang, C., Fu, Q., Zhang, B., Bian, Y., Qi, N., & Zhu, J. (2022). Characterization and pre-clinical assessment of a proposed biosimilar to its originator

Omalizumab. *European Journal of Pharmaceutical Sciences, 178*, 106292. https://doi.org/10.1016/j.ejps.2022.106292.

Yoshida, K., Moein, A., Bittner, T., Ostrowitzki, S., Lin, H., Honigberg, L., Jin, J. Y., & Quartino, A. (2020). Pharmacokinetics and pharmacodynamic effect of crenezumab on plasma and cerebrospinal fluid beta-amyloid in patients with mild-to-moderate Alzheimer's disease. *Alzheimer's Research and Therapy, 12*(1). https://doi.org/10.1186/s13195-020-0580-2.

Zahavi, D., & Weiner, L. (2020). Monoclonal antibodies in cancer therapy. *Antibodies., 9*(3), 34. https://doi.org/10.3390/antib9030034.

CHAPTER

Biosimilars and biologics in breast cancer

2

Simhadri Goud Gundrathi[a], Bharat Kumar Reddy Sanapalli[b], Ashwini Deshpande[c] and Vidyasrilekha Sanapalli[d]

[a] *School of Pharmacy and Technology Management, SVKM's Narsee Monjee Institute of Management Studies (NMIMS) (Deemed to be University), Hyderabad, Telangana, India,* [b] *Department of Pharmacology, School of Pharmacy and Technology Management, SVKM's Narsee Monjee Institute of Management Studies (NMIMS) (Deemed to be University), Hyderabad, Telangana, India,* [c] *Department of Pharmaceutics, School of Pharmacy and Technology Management, SVKM's Narsee Monjee Institute of Management Studies (NMIMS) (Deemed to be University), Hyderabad, Telangana, India,* [d] *Department of Pharmaceutical Chemistry, School of Pharmacy and Technology Management, SVKM's Narsee Monjee Institute of Management Studies (NMIMS) (Deemed to be University), Hyderabad, Telangana, India*

2.1 Introduction

Biologics have revolutionized the approach to many life-threatening illnesses, particularly cancer (Kroschinsky et al., 2017). These medications, derived from living cells or their products, act differently compared with conventional chemotherapy. They are able to target cancer cells more specifically, often resulting in improved outcomes with fewer side effects (Pérez-Herrero & Fernández-Medarde, 2015). Biologics not only increased people's life expectancy but also enhanced the quality of their lives. Biologics encompass a broad array of treatments such as monoclonal antibodies, cytokines, and growth factors. Each of these is configured to target a specific component of the immune system or environment of the cancer. Because they are very specific, they have fewer side effects than older treatments, which target both healthy and cancer cells (Naser-Tavakolian et al., 2025). In the treatment of cancer, biologics are also a crucial aspect now. In patients with curable, early-stage cancer, biologics are combined with other therapies to destroy remaining cancer cells, stop the cancer from recurring, and enhance long-term survival. In patients in whom the cancer has spread or is advanced, biologics control the disease, alleviate symptoms, and enable patients to live longer, frequently when combined with chemotherapy or other targeted therapies. Biologics also help a great deal with supportive care. They can lower the side effects of treatment, such as low white blood cell counts, anemia, or sores in the mouth (Georgakopoulou & Scully, 2015). By assisting the body in producing more blood cells or regulating the immune system, biologics keep patients strong enough to continue treatment. Monoclonal antibodies are perhaps the

Biosimilars and Biologics in Cancer Treatment: An Era in Therapeutic Strategies.
DOI: https://doi.org/10.1016/B978-0-443-43824-0.00010-9

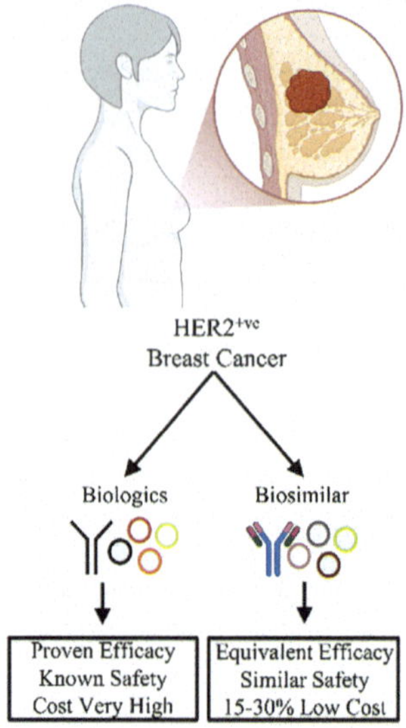

FIGURE 2.1

Comparative overview of biologics and biosimilars in HER2-positive breast cancer therapy.

most commonly used and effective form of biologics. They have become a central component of the treatment of some breast cancers, particularly those with high levels of HER2 protein (Barzaman et al., 2020). One such medication that has made a big impact for individuals with HER2-positive breast cancer is trastuzumab. It does this by binding to the HER2 receptor on cancer cells, hindering the signals that cause the cells to multiply (Plosker & Keam, 2006). Due to its success, other HER2-targeted drugs such as pertuzumab and ado-trastuzumab have also been created, with even more improved outcomes for patients with this cancer. Biosimilars have made cancer treatment more cost-effective, without losing the same advantages as the original biologic medications (Simoens, 2011). Though they are not identical, biosimilars are nearly identical regarding safety, quality, and effectiveness. They offer a more affordable option without compromising treatment outcomes (Gutka et al., 2018). This is particularly significant in cancer care, where biologic medications can be extremely costly and beyond the reach of many patients—particularly those in less resource-rich countries. Biosimilars enable more individuals to gain access to life-enhancing drugs without compromising the quality of care (Pathak et al., 2023). By reducing the cost to patients and healthcare systems, biosimilars have the potential to democratize access to life-saving therapies, enabling more patients to benefit from advances in the treatment of cancer Fig. 2.1.

The economic advantage of biosimilars is that the cost of production is much lower in comparison to the reference product, primarily because of lower research and development costs (Blackstone & Joseph, 2013). Unlike originator biologics that entail extensive clinical trials to establish safety and efficacy, biosimilars achieve a more streamlined approval process with the objective of establishing comparability to the reference product. Comparability with the reference product requires extensive analytical, nonclinical, and clinical studies to enable the determination that the biosimilar contains highly similar characteristics in the domain of structure, function, immunogenicity, and safety profile. Through the leverage of existing knowledge and information pertaining to the reference product, biosimilar manufacturers are able to save a lot of time and expense during the development process, leading to lower prices for patients and healthcare providers (Blandizzi et al., 2017). Recognizing the potential biosimilars hold for increasing access to affordable medication, the Biologics Price Competition and Innovation Act was signed into law with the expectation of sparking competition and innovation in the biologics market, ultimately cutting costs (Lu, 2014). The act provided an abbreviated pathway for biosimilars to gain regulatory approval, enabling the agents to come to market while maintaining high standards of safety and efficacy. Through providing an added level of competition in the market, the Act invites companies to make and sell biosimilars, cutting prices and opening opportunities for patient treatment. The additional level of availability of biosimilars benefits individual patients but also benefits healthcare systems overall by releasing resources to be utilized in other areas of need. The present work comprehensively reviews the current landscape of biologics and their biosimilars within breast cancer treatment. Table 2.1 offers a structured overview of key reference biologics, their targets, indications, representative approved biosimilars, the supporting clinical evidence, and regulatory status, including extrapolation.

2.1.1 Breast cancer as a primary area of application

Breast cancer is one of the diseases that has undergone a paradigm shift with the advent of biologic agents, especially the HER2-positive disease (Pondé et al., 2018). With the creation of trastuzumab as a monoclonal antibody specifically targeting the HER2 receptor, the survival of patients with this especially aggressive form of breast cancer has shot up (Kreutzfeldt et al., 2020). But, trastuzumab-based drugs and other HER2-based drugs are highly expensive, and the cost has emerged as one of the key limiting factors of access, not least to low-to middle-income countries (Uifălean et al., 2018). Such a solution to the problem would be the availability of trastuzumab biosimilars as an alternative that would be less expensive and would not lead to the loss of effectiveness and safety of medication (Uifălean et al., 2018). The fabrication of trastuzumab, the monoclonal antibody targeting the HER2 receptor, has revolutionized the treatment of HER2-positive breast cancer by delaying the survival of patients with metastatic breast cancer and early-stage breast cancer, the disease that is treated in combination with chemotherapy (Swain et al., 2023). Trastuzumab selectively inhibits the survival and proliferation pathways downstream of the HER2 receptor, and it causes cell destruction in the cancer cells through osmosis. The effectiveness of trastuzumab has already been proven, so it has turned into a standard

Table 2.1 Key biologics and biosimilars in breast cancer treatment.

Reference biologic	Target/class	Primary indication (s) in breast cancer	Representative approved biosimilars	Key evidence summary	Regulatory status and extrapolation
Trastuzumab	HER2 mAb	HER2+ early & metastatic BC	SB3 (Ontruzant), ABP 980 (Kanjinti), CT-P6 (Herzuma), PF-05280014 (Trazimera)	Phase III Trials (e.g., HERITAGE, LILAC)	Approved by the FDA and EMA. All HER2+ BC indications granted based on similarity.
Pertuzumab	HER2 mAb	HER2+ and HER2-	HLX11, QL1209	Phase III	Under review by FDA. Expected across indications.
Bevacizumab	VEGF mAb	Metastatic TNBC	ABP 215 (Mvasi), FKB238 (Equidacent) BCD-021 (Avegar)	Phase III trials	Approved by FDA (Mvasi), EMA (Equidacent). All approved indications, including mTNBC.

(continued on next page)

G-CSF (e.g., Filgrastim, Pegfilgrastim)	Supportive Care (Neutropenia prevention)	Prevention of chemotherapy-induced neutropenia	Filgrastim-sndz (Zarzio), Pegfilgrastim-jmdb (Fulphila), Pegfilgrastim-cbqv (Udenyca)	Equivalent efficacy in reducing neutropenia events/fever across cancers.	Approved by FDA and EMA. All approved indications.
CDK4/6 Inhibitors (e.g., Palbociclib, Ribociclib)	Small molecule Kinase inhibitor	HR+/HER2- metastatic BC	Ongoing	Emerges upon patent expiry, focus on bioequivalence.	It follows generic small molecule pathway (ANDA), not BLA biosimilar pathway.

Note: mAb, *Monoclonal antibody;* pCR, *pathological complete response;* PFS, *progression-free survival; OS, overall survival;* RWE, *real-world evidence;* TNBC, *triple-negative breast cancer;* HR+, *hormone receptor-positive;* G-CSF, *granulocyte colony-stimulating factor. Extrapolation: Regulatory approval for biosimilars in all indications of the reference product based on demonstrated similarity, typically tested in the most sensitive indication.CDK4/6 Inhibitors: While revolutionizing HR+ mBC treatment, they are chemically synthesized small molecules. Their postpatent market will involve generics ("similars" in function, but approved via ANDA, not the biosimilar BLA pathway). Included for context within the broader "biologics" discussion but marked distinctly.*

of care against HER2-positive breast cancer that has not been very readily available in most parts of the world because of its cost (Blackwell et al., 2018). Biosimilars of trastuzumab are appearing as one of the quicker methods of satisfying this need for lower-cost treatment, as they will substitute the originator biologic when treating breast cancer (Zucchetti et al., 2019). The biosimilars are developed in such a way that they are closely matched in terms of structure, performance, and activity with regard to trastuzumab and have perhaps the potential to provide the same clinical utility with reduced cost (Joshi & Rathore, 2020). To the extent that trastuzumab biosimilars have the potential of substantially expanding the degree of access to HER2 targeted therapy, particularly in situations with limited resources, to a level in which a greater number of patients can benefit by using the life-saving drug (Monga et al., 2025). In the given review, the paper is going to talk about clinical evidence of usage of the trastuzumab biosimilars, their health system-level impact, including the cost, and the challenges and opportunities of adopting biosimilars into clinical practice.

2.2 Understanding the biosimilars: Definition, development, and regulation

Another kind of biological medicine that has been shown to possess a significant degree of comparability with a reference medicine in terms of efficacy, safety, and purity, but not the same, includes biosimilars (Ahmad et al., 2016). This is a very high difference because the biologic drugs, because they are productions of living things in their own right, have a variance due to the fact that its production mechanism is complex. Compared with generic drugs, where they can be defined by the fact that they are produced chemically and hence they are molecularly equivalent to that of the reference drug, biosimilar is similar but not identical to that of the reference drug (Roger, 2006). This is because small variations in the manufacturing process that is, the environment (or the conditions under which a cell is cultured in) or the process by which one purifies the product may have an impact on the ultimate structure and activity of the biologic molecule. These differences indicate the peculiarities of regulation of biosimilars, where similarity and not identity is focused on. Regulatory bodies have put in place stringent approval processes where biosimilar manufacturers produce evidence that they are highly similar to the comparator product through large data base of analytical data, nonclinical and clinical trials, which poses a challenge to manufacturers of biosimilars (Mascarenhas-Melo et al., 2024). These clinical trials are meant to make sure that the biosimilar is comparable to the originator biologic in safety, efficacy and immunogenicity in order to ensure that patients can look forward to the same therapeutic effects without undue risk (Barbier et al., 2020).

The biosimilars are expected to be highly similar to their reference products, grounded in the awareness that the difference in molecular properties is not to be conflated with the difference in patient outcomes, which are of clinical significance (Leonard et al., 2019). The regulators are keen to demonstrate that differences between the biosimilar and reference product would not exercise any negative impact on efficacy or safety. This helps in the production of alternatives to expensive biologics in a cheaper fashion without compromising on quality and treatment effectiveness (Liu et al., 2025). The progression of the biosimilars is a careful and refined procedure, which is intended to demonstrate comparability to the reference biologic regarding the safety, quality, and efficacy. It entails a broader comparability exercise that encompasses the analytical, nonclinical, and clinical studies to determine the performance and behavior of the biosimilar against a series of parameters. The final solution is to demonstrate that the biosimilar can be conceived as being very similar to the reference product, and clinicians and patients would be guaranteed of its therapeutic similarity (Weise et al., 2012). The analysis that is significant in developing the biosimilar is a rigorous analytical characterization that involves a high level of comparison between the biosimilar and the reference product at a molecular level (Berkowitz et al., 2012). This involves assessment of the amino acid sequence, higher-order structure, posttranslational modifications, and other important quality features that can regulate the biological activity of the molecule. Sensitive methods of analysis, such as mass spectrometry, chromatography, and spectroscopy, are employed so as to detect even a minimal variation between the biosimilar and reference product (Falconer et al., 2011).

This is followed by the next stages of nonclinical tests, such as in vitro tests and in vivo tests, which are then done in order to establish the biological activity of the biosimilar, the pharmacological properties, and possible toxicity. The stated tests will be used to determine that the biosimilar will react to the same binding affinity of the target receptor, the same mechanism of action, and the same safety profile as the reference product (Kirchhoff et al., 2017). Further information on the pharmacokinetic and pharmacodynamic properties of a biosimilar, as well as its possible immunogenicity, may be provided by an in vivo test, which is usually performed in animal models. Lastly, clinical trials are conducted where the safety, efficacy, and immunogenicity of the biosimilar would be determined in human volunteers (Wang et al., 2020). The trials are usually a head-to-head comparison of the biosimilar and reference product in a population of patients that has a very high sensitivity to identify the presence of a clinically meaningful variation. The primary endpoint is normally an efficacy measure, safety is reflected in the adverse events, and immunogenicity is assessed by assessing the development of antidrug antibodies. EMA and FDA have established different regulatory paradigms that are aimed at ensuring that biosimilars will reach high levels of safety, efficacy, and quality. The two employ a science-driven procedure of authentication of analytical, nonclinical, and clinical data by subjecting the procedure to an intense review to establish biosimilarity with a reference product. Although the technical requirements and processes involved in the EMA and the FDA are slightly different, both processes share the identical goal of ensuring that the more widely utilized biological therapy treatment becomes more economical without forfeiting the safety or effectiveness of the drugs on a patient (Gherghescu & Delgado-Charro, 2020).

Starting with a thorough analytical characterization of biosimilarity and then proceeding to nonclinical and clinical studies, the EMA, having licensed biosimilars since the year 2006, adopts a step-by-step approach to the assessment of biosimilarity (Tsuruta et al., 2015). The agency requests that the biosimilar companies establish that their drug product is highly similar to the reference product in terms of quality, safety, and efficacy, and variations between them are not of clinical concern. Following, the EMA requires the potential of immunogenicity as well, and requires manufacturers to build effective pharmacovigilance planning to forecast the long-term safety of the biosimilar after its release into the market (Calvo et al., 2018). The FDA, which is now open to the approval of biosimilars, is following the same path as EMA in that biosimilar companies are asked to prove a high level of similarity of a reference product, utilizing the complete package of data. The FDA also has what is termed a biosimilar, which is considered a lot more interchangeable than its reference product, without necessarily involving a prescribing healthcare expert. In order to be interchangeable, a biosimilar company must provide further data demonstrating that the biosimilar has the same clinical effect as the reference product in any particular patient, and that interchangeability of the biosimilar with the reference product does not presume any increase in risks of safety. The concise ticketing procedure is utilized to promote innovation and competition in the biologics marketplace (Kabir et al., 2019).

2.3 Trastuzumab biosimilars: Healthcare-related evidence and effectiveness

2.3.1 Major clinical trials with equivalence

Another indication that trastuzumab biosimilars are similar to the reference product has been clinical trials (Cargnin et al., 2020). Clearly, clinical trials have shown that trastuzumab biosimilars have an analogous efficacy and safety to the reference (Cargnin et al., 2020). The trials are usually conducted in a randomized, double-blind manner, and head-to-head versus reference drug in HER2-positive breast cancer, with endpoints of interest being clinical endpoints such as overall response rate (ORR), progression-free survival (PFS), and overall survival (OS) (Pegram et al., 2019). The outcomes of these trials have been conclusive in establishing the effectiveness and safety of the trastuzumab biosimilars as equal to the reference drug, meaning a substituent is an acceptable alternative as an economically viable option when it comes to HER2-targeted therapy (Triantafyllidi & Triantafillidis, 2022). The Similar pathological complete response (pCR) rates in the neoadjuvant setting have also been reported in the clinical trials, which support the findings that trastuzumab biosimilars can eliminate the residual disease before surgery (Stebbing et al., 2018). The tests are commonly conducted by attaching the biosimilar or the reference trastuzumab with chemotherapy before the operation and analyzing the tumor tissue after the process to achieve the percentage of patients who would show complete pathological response

(Rugo et al., 2017). The fact that the levels of pCR are similar in the biosimilar and reference arms is strong evidence that the tumor size reduction and killing of cancer cells can be achieved by using trastuzumab biosimilars in the neoadjuvant setting (Lammers et al., 2018). The available evidence points to similarity throughout the metastatic disease context, providing studies of analogous objective response levels ORR of trastuzumab biosimilars to the reference product, thus implying that the drugs have potential to intervene on the disease process and enhance survival in patients (Barbier et al., 2019; Coory & Thornton, 2019). The studies involve treating patients with metastatic HER2-positive breast cancer with the biosimilar or reference trastuzumab, in combination with chemotherapy, and then an evaluation of the response to cancer with imaging techniques (Li et al., 2022). That comparable ORRs are documented in the biosimilar group and reference groups is reassuring that trastuzumab biosimilars have the capacity to slow down the development of tumors and enhance the survival of the metastatic patients. The clinical equivalence between the biosimilar trastuzumab and reference trastuzumab is also associated with similar rates of the ORR, which is proven by a phase III equivalence randomized trial of HLX02, a trastuzumab biosimilar. The fact that the trial where the patients with the repeat or metastatic HER2-positive breast cancer were recruited and randomly assigned HLX02 or reference trastuzumab and docetaxel confirms the fact that HLX02 and the reference trastuzumab are different drugs with the same therapeutic outcome. According to the results of the trial, there was almost no difference in ORR in either group by week 24.

2.3.2 Biosimilar efficacy in the real world

Real-life practice in the countries where trastuzumab biosimilars have been approved also provided the data on the safety and effectiveness of the biosimilars in environments closer to those of the routine clinical practice, and as such, the demonstration led to positions that they are a safe and cost-effective alternative to originator trastuzumab in HER2-positive breast cancer treatment (Triantafyllidi & Triantafillidis, 2022). Coupled with the help of the big patient historical databases, this type of research is able to reflect the experience of a heterogeneous cohort of patients treated in a high number of clinical locations, and could provide invaluable data about the actual world-of-use behavior of biosimilars. The Taiwanese clinical study revealed the effectiveness of trastuzumab biosimilars both in the metastatic and neoadjuvant setting, but this was just another confirmation of their use in everyday clinical practice (Leung et al., 2022). In the trial, when patients with HER2-positive breast cancer, who received treatment in a large hospital, were recruited, biosimilar trastuzumab was effective with pCR during neoadjuvant positions and disease maintenance during metastatic positions (Resende et al., 2021). Also, the study showed that even more effective disease control in a metastatic setting was seen when using the combination of trastuzumab with another agent against HER2, or, in other words, using the element of biosimilar second blockade. The safety and efficacy of biosimilar trastuzumab-dkst in adjuvant therapy in patients with early HER2-positive breast cancer is also proven by the real-world evidence of a Brazilian population (Gagliato et al., 2024). This observational and multicentric study was designed to enroll patients with trastuzumab-dkst in the real world and adhere to the major study outcomes, including the invasive disease-free survival (IDFS) and OS (Goutsouliak et al., 2020). The study results revealed high values of IDFS and OS rates, as well as the safety profile that was equivalent to comparator trastuzumab, not only further proving the efficiency of biosimilar trastuzumab-dkst use as a cost-effective oncology practice medication use but also indicating that further oncological research works should be done on the product (McBride et al., 2021).

2.3.3 Long-term effects and safety profiles

Although data on the safety and efficacy of trastuzumab biosimilars in the short term have been shown to be valid in clinical trials and observational studies, a long-term follow-up is needed to ensure likelihoods in the long-term outcomes of patients receiving the biologic molecule as adjuvant therapy because trastuzumab must continue to be administered to the patient over a prolonged period of time (Gion et al., 2021). A long-term study may give interesting data about long-term treatment response, development of resistance, and late-onset toxicity. Overall, biosimilar trastuzumab has met the criteria of a similar safety profile to that of the originator product, and similar rates of adverse events and serious adverse events were reported during clinical studies and real-world settings (Oliva et al., 2024). However, to monitor the long-term safety and efficacy of biosimilars, postmarketing surveillance and the acquisition of real-life data are required in this situation, which can identify rare or unforeseen adverse events that were not identified during the initial trials. These monitoring systems can also find possible risk factors specific to the occurrence of adverse events, and they can come up with effective measures to reduce the risks significantly (Nikitina et al., 2023). Assessment of safety information on the patients treated with the biosimilars, in other words, active pharmacovigilance, is a good aid in safety monitoring of these medicines, particularly after they have gained appropriate marketing. Such schemes may include a mix of any data source, including patient registries, electronic medical records, and spontaneous reporting systems, in order to complete the assessment of the safety profile of biosimilars in real-life clinical practice settings. By consistently monitoring the safety and efficacy of biosimilars, regulators and healthcare providers will be in a position to make certain that such medicine will continue to be a safe and effective therapeutic agent in patients with HER2-positive breast cancer (Casadevall et al., 2013).

2.4 Benefits of biosimilar implementation

The most optimistic application of biosimilars in the treatment of breast cancer is the minimization of the financial burden of the ailment on the patient and the healthcare system, since it will be relatively less expensive compared with the expensive reference biologics. Reduction of the cost of drugs, reduction in the cost of administration, and less consumption of healthcare through the existence of fewer side effects are some of the avenues through which the economic value of cost saving could be achieved. Biosimilars will be beneficial in that they will help improve access to drugs, as it will now become less expensive to offer care to more patients, and drugs that will help save their lives will also be accessible to more individuals (Giuliani & Bonetti, 2019). The savings of money that can be saved by using biosimilars involve massive savings of the cost in comparison to reference biologics and has been shown to be relevant, especially in treating HER2-negative breast cancer, where the HER2-targeted therapy will be made readily available. An example of a drug that is used in the treatment of this aggressive form of breast cancer is the drug trastuzumab which, like any other drug, poses a cost constraint (it can be very expensive) and hence in the context of limited resources that might be available in a particular environment, can in itself be a fancy toward the treatment process. The low-cost value of biosimilars is attractive, where the providers can achieve the same quality of service without the spending being as hard as on the initial treatment, on the profuse supply of the said treatment that saves lives (H) It had been also assumed that economic studies would always justify cost saving that would be occurred when using biosimilars and thus these drugs have been making such interventions economically affordable. These kinds of comparisons are most likely to be in opposition to the cost of biosimilar drugs and the drugs of comparators, healthcare usage, and management. The majority of the studies indicated the involvement of biosimilars in providing a significant decrease in expenses without having any impact on patient outcomes in any aspect and assisting in budgetary control of the healthcare market (Cesarec & Likić, 2017).

The advent of biosimilars is further likely to play a key role in helping change the shape of the healthcare cost structure and financial sustainability, and free up savings that can be put to other purposes where they are needed. The biosimilars will be able to take much financial pressure off the healthcare budgets of the cost of cancer treatment and thereby open the window to investing more in prevention and early detection, and other basic services. It has the potential to lead to better distribution of the healthcare resources, which is more equitable, also less expensive, to the benefit and advantage of individual patients and the population in general (Chai et al., 2022). The savings that will be achieved due to the utilization of the biosimilars will allow the release of such money that could be utilized in other fields of healthcare, such as research and development, and investment in available facilities and access to new treatment, hence enhancing the overall efficiency to become enhanced. It is possible to use these funds to conduct research and development of new approaches to treating cancer or improving the quality of care that is offered for cancer, as well as providing more access to healthcare provision that is confined to the vulnerable populations. The potential to enable fairer and make healthcare kinder and better for people and the environment by using healthcare resources more wisely can be supported by biosimilars (Trapani & Curigliano, 2021). It is vital that the managed care organizations and payers be enlightened in the procedure of managing the approval of biosimilars and stimulators of HER 2-targeted therapies, e.g., utilization of the biosimilar and the use of antibody conjugate to make an informative and data-driven decision regarding including formularies alongside their reimbursement policies. Payers can do the same by familiarizing themselves with the evidence base of biosimilar use and potential cost savings, and, additionally, by establishing the policy toward the use of biosimilars that will not only contribute to the limitation of costs but also guarantee that they will provide their patients with the best and most appropriate care. This needs to be addressed in the intersection between the practitioners of the healthcare sector, payors, and regulators to ensure biosimilar product usage in clinical practice is done in a format that has the optimal benefit to both patients and the healthcare system (Monga et al., 2025).

The possible contribution of biosimilars is also the increased availability of life-saving pharmaceuticals to patients, especially those in the low- and middle-income nations, where reference biologic price may be one of the cost-deterring factors (Godman et al., 2021). As are cost-effective solutions, biosimilars can provide more access to their patients, who did not obtain it in the first place, and consequently, reduce healthcare disparities and provide better healthcare outcomes. This is of particular interest in oncology, where the distinction between the timely treatment that is not accessible or at least accessible at a professional level, and the effective treatment can be the distinction between life and lack of life, life and quality of life. WHO acknowledges and advocates biosimilars and their importance to gain access to life-saving drugs by cancer patients with the request to add them to national formularies and treatment guidelines (Kang et al., 2024). There are few biosimilars listed on the WHO Model List of Essential Medicines, the proof of their worth in offering cost-saving and quality treatment of a variety of diseases, such as cancer. By stimulating the use of biosimilar solutions, the WHO will help to decrease the global burden of cancer and distribute equal health among all people (Nabhan et al., 2018). The availability of anti-HER2 therapy has increased since the advent of biosimilars, due to the lower costs, making it available to some more patients, who can obtain the life-saving treatment. Expanded access has also been particularly useful in low-resource settings where the price of trastuzumab previously restricted its usage. Biosimilars have mitigated disparity in the treatment of cancer and improved survival rates among patients with HER2-positive breast cancer by making anti-HER2 therapy more affordable.

2.5 Physician and patient notions on the use of biosimilars

Lack of education directed at the prescribers, when biosimilars of therapeutic oncology are not available in the clinical practice, is one of the major obstacles to applying biosimilars in the therapeutic oncology sector. Physicians can be misinformed about biosimilars, production, and evidence of their application, and therefore avoid prescribing biosimilars. It can be resolved through specific education that will enable doctors to make adequate decisions on the utilization of biosimilars by having the required information at their disposal (Sarnola et al., 2020). Another obstacle is that physicians are hesitant to switch patients on a drug that has been successful, thus there is a doubt when switching patients on the reference drug to the biosimilar. This dilemma can be avoided by clarifying to the physicians with concrete guidelines on substitution of procedures and by emphasizing the validity of demonstrating similarity in efficacy and safety of biosimilars. Also, the process of physician involvement in the decision-making process and the solution of their specific challenges can serve as a source of establishing trust and confidence in the use (Leonard et al., 2019). There should be official bodies of guidance and professional education that would improve the willingness and preparation of doctors to use biosimilars and equip them with the ability and knowledge of how to incorporate the agent into clinical activity. Such guidelines should be published by reputable organizations and they or their equivalent must rely on latest available scientific evidence. Professional education approaches have to offer physicians practical exposure to the usage of biosimilars and should attend to the frequently asked questions and the myths surrounding their use (Oskouei & Kusmierczyk, 2021).

One of the great obstacles that exist to the extensive use of biosimilars in the United States is the fact that patients are unaware of the existence of these drugs and they will hence fear switch and hesitate to move to biosimilars. Most of the patients will not be aware of the concept behind biosimilars and will be concerned with regard to their safety and efficacy. The stated ignorance may be overcome with help of education interventions focused on delivering specific, accurate, and simple information about biosimilars to the patients (Jacobs et al., 2016). A US researcher found a gap between patient self-report and the perceptions by oncologists of the patient experience with trastuzumab biosimilars that doctors underestimate the anxiousness of the patient and uncertainty. This finding highlight's the importance of an open dialogue and shared decision-making in the prescribing of biosimilars. Providers should attentively listen to the anxieties of their patients, answer the questions they have and involve them in the decision making (Ho, 2021). Competent and specific education is needed between oncologists and patients, to help the communication process in regards to trastuzumab biosimilars and to reassure the patient that they are safe and efficacious. Its contents must be unique with regards to the unique needs and challenges of the individual patient and consider the level of knowledge, the culture and his or her preference. It is also required that it be made available and captivating in many forms and shapes i.e. written information, video and interactive moments (Wu et al., 2023).

Information has also been a problem not only in the administration of trastuzumab biosimilars but also with the relationship between the doctor and the patient, and this may erode the feelings of trust and confidence in the treatment course. When patients are lost to confusion or even ignorance on treatment, then they can be worse to stick to this treatment regime, hence producing less satisfying results. Healthcare professionals are therefore compelled to place a lot of emphasis on educating the patient and in providing them with clear and evidence-based knowledge on biosimilars, their strengths and shortcomings (Class & Langis, 2012). Transparency and collective decision-making are the cornerstones of developing trust and a patient connection with biosimilar treatment; therefore, they also become agents of their healthcare. It also means establishing an amiable atmosphere where patients will not hesitate to raise their concerns and inquiries on matters of which they are not sure of. The medical workers are to be prepared to address these concerns using empathy and understanding, promising that they will be all right and make sure they have confidence in the treatment process. Patient oncology provides them with educational materials that are easy to read and understand to get to know more about biosimilars and their uses in the care of cancer. They have to be written in plain English and will not have jargon, and they will be easy to read and understand. They should also be cultural and sensitive to the needs of populations of particular groups of patients. Clinicians ensure that patients know all the information on which they can rely to make informed choices and, therefore which can support patients in staying on track, outcome, and satisfaction with care by educating them on the topic (Gasteiger et al., 2023).

2.6 Biosimilars safety and immunogenicity

An essential component of the development of biosimilars and their marketing approval is that they are established as comparable in immunogenicity to the reference drug. The immunogenicity helps to measure how the biological products have the potential of triggering the immune system of the patient to form antidrug antibodies (ADAs) that inhibit the pharmacology of the drug or lead to the adverse effects related to immunity. It is necessary to measure the immunogenic potential of the biosimilars in such a way that they will not be more prone to causing adverse immune responses than their reference products (Chamberlain & Kurki, 2018). Potential immunogenicity has many implications for individual patients, triggering a great number of controversies among the healthcare professionals, regulators, and patient groups, and making the situation require close monitoring and management of the risk. Although the majority of the patients are responsible toward biosimilars administration demonstrate no issues with the immunogenicity yet, there still is a certain number of individuals who can experience the emergence of ADAs, which will affect the effectiveness or even two safety of the treatment. As such, there should be the means of diagnosing and treating such patients (Schreitmüller et al., 2018). With monoclonal antibodies, such as trastuzumab, it might appear to have fewer risks of having side effects due to its immunogenicity potential compared with some others types of biologicals. This could be due to the fact that the monoclonal antibodies are usually of high purity and that would be devised in such a way that they would bind to specific receptors of the cancer cells, and would pose no danger of off-target effect that would cause a host immune response. However, in patients receiving trastuzumab biosimilars, they must, as well, be periodically monitored with respect to any indication of immunogenicity, i.e., incidence of infusion reactions or diminished response to treatment (Pineda et al., 2016).

Surveillance of adverse event during registration trials and also in the continuous process of the postmarketing surveillance is another important element of safety in all the biologics including the biosimilars. Observation of such activities would yield huge insights in the area of the risks that such drugs may be facing, and also help in determining any form of the out-of-the-ordinary and side effects that couldn't be seen before known about within the context of the first clinical development program. By continuously monitoring the safety of biosimilar drugs, the medical workers and the regulatory bodies are able to ascertain that the drugs continue to have an acceptable ratio of benefits and the risks to the patients (Sagi et al., 2023). The phase III trials were pooled together to establish that biosimilar filgrastim is effective and safe and its safety profile was similar to the earlier filgrastim trials in the prevention of febrile neutropenia in patients with breast cancer and assures the safety of this biosimilar in a sensitive clinical trial. A pooled study report of the different clinical studies showed that Biosimilar filgrastim was well tolerated and did not predetermine successfulness of incurring additional adverse events as opposed to the reference drug. This evidence advocates the use of biosimilar filgrastim as a rather safe and effective drug against febrile neutropenia prevention during myelosuppressive chemotherapy. It is also believed that, active pharmacovigilance is one of the most useful tools in monitoring the safety of biosimilars/biologics, more so in the postmarketing environment as this will ensure that any unusual or rare adverse events are identified that would not otherwise be identified during clinical trials. Such programs are very strict in collecting and analyzing the safety data of the patients using the biosimilars in their actual clinical practice via various sources like patient registry, electronic health records, spontaneous reporting program. monitoring the safety of the biosimilar, care givers and government agencies would be able to ensure that biosimilar drugs maintain a better benefit and risk profile to the patients (Singh et al., 2020).

Biosimilars have the advantage equal to biologics in managing toxicities associated with treatment and in supporting the overall well-being of the patients and in enabling them to remain on potentially life-saving/cure therapy. As an example, granulocyte colony-stimulating factors (G-CSFs) biosimilars, including filgrastim, can safely prevent and/or address neutropenia caused by chemotherapy treatment, among other things, and minimize febrile neutropenia development and its complications (Botteri et al., 2018). By regulating these treatment-related toxicities, biosimilars can assist in improving the quality of life if a patient has them and to enable them to continue with the entire cycle of cancer treatment (Mascarenhas-Melo et al., 2024). The use of supportive care remains relevant in the treatment of the possible side effects as well as the improvement of the quality of life of the patient, irrespective of whether they are using a reference biologic or a biosimilar. These are teaching the patient on possible side effects, close follow-up with regard to toxicity, and administration of suitable measures to suppress the side effects. Also supportive care can be the psychological and emotional counseling of the patient in such a way that they can handle the effects of cancer treatment. The healthcare professionals must have the capability to adopt and modify treatment models in response to negative events and QoL concerns, customizing all interventions with regard to the distinct needs of each patient. It could entail a change in dose of the intervention, inclusion or change of comfort medication, or initiation of lifestyle change to assist patients in coping with their symptoms. The proactive, patient-specific approach to treating treatment-related toxicities, which is explored in this case, allows healthcare professionals to maximize the overall experience of patients undergoing cancer treatment and allow them to retain the benefits of this treatment (Ngo & Chen, 2021).

2.7 Biosimilars in neoadjuvant, adjuvant, and metastatic breast cancer therapy

2.7.1 Neoadjuvant treatment: Pathological complete response

Biosimilar trastuzumab has performed well in the neoadjuvant setting when compared with innovator trastuzumab in p-CR, and therefore, biosimilar trastuzumab could be potentially safely used to reduce tumor burden prior to surgery. This treatment before surgery aims to make the tumor small enough, kill micrometastatic disease, and have a greater likelihood of attaining surgical resection: neoadjuvant therapy. The success of p-CR, the lack of residual invasive breast cancer and metastases to the axillary lymph nodes at the moment of operation, is a strong predictor of long-term survival without the disease (Pivot et al., 2018). The clinical efficacy of biosimilar trastuzumab in neoadjuvant treatment has been confirmed by real-world data. Each of the studies has been done in different patient groups and even in different clinical conditions, but have all come to agree that biosimilar trastuzumab has the ability to achieve the similar p-CR rates with those who received the innovator drug. This real-life evidence again proves the safety and efficacy of biosimilar trastuzumab as a therapeutic agent used in the context of neoadjuvant treatment of breast cancer that is positive of the HER2 factor. Trastuzumab effect, whether in the innovator form or the biosimilar, was observed in both axilla and breast pCR overall, which was highly significant in patients who underwent the treatment drug reporting about HER2-targeted therapy in this particular aspect. This is a finding that echoes the main benefit of trastuzumab as a sensitizer of the tumor to chemotherapy and in the elimination of residual disease prior to surgery. The similar p-CR levels obtained by using biosimilar trastuzumab further support its use as an overall cost-cutting option to the innovator medicine (Miller & Schwartzberg, 2019).

2.7.2 Adjuvant therapy: Maximizing disease-free survival

Biosimilar trastuzumab finds application in the adjuvant therapy of HER2-positive breast cancer to kill the remaining cancerous cells following the surgery and stop cancer recurrence. Adjuvant treatment, carried out following an operation, is meant to eliminate micrometastatic disease, which cannot able to be recognized through customary imaging tests. By eradicating these remaining cells of cancer, this adjuvant therapy is able to decrease the risk of recurrence to a great extent and develop long-term survival free of disease (Tesch & Gelmon, 2020). Experience in the real world has shown the similarity of safety and efficacy of biosimilar trastuzumab and Herceptin in adjuvant use with confidence in the clinical value of its application in that environment. Retrospective observational studies in large groups of patients prove in real life that biosimilar trastuzumab reduces the chance of recurrence, but does not increase the chance of adverse outcomes. The findings of the real-world evidence studies justify the application of biosimilar trastuzumab as an economical alternative instead of Herceptin during adjuvant HER2-positive breast cancer (Piezzo et al., 2021). A longer-term follow-up is needed to finish an evaluation of whether trastuzumab biosimilars will have an outcome in the adjuvant treatment setting since the implications of adjuvant medication will not be comprehended in absolute terms until several years. They can be helpful, as long-term studies can be informative as far as what the durability of the response to a treatment can be, what the resistance is, and what late adverse events might happen. By conducting continuous follow-up of patients receiving adjuvant trastuzumab biosimilars, clinicians and researchers will be in a better position to comprehend in detail how they affect the disease-free survival and survival. Healthcare providers and researchers will still be in a position to learn more about the long-term outcomes of trastuzumab biosimilars on disease-free survival and survival by following the patients for longer duration of time.

2.7.3 Metastatic disease: Maximizing control of disease and survival

Biosimilar trastuzumab also plays its role in the metastatic scenario, where ORR and CBRs are equal and similar to reference trastuzumab, an economically feasible form of therapy to treat advanced HER2-positive metastatic breast cancer. The objective of the treatment in the metastatic setting is to manage the disease, palliate, and exist longer. The treatment of metastatic HER2-positive breast cancer is based on the use of trastuzumab or a combination of trastuzumab and chemotherapy, and the biosimilar of the comparator product saves money (Yang et al., 2019). Biosimilars were also used in double blockade, which resulted in better disease control in the metastatic setting, and this means that a combination of trastuzumab biosimilars with other regimens against HER2 would improve the outcomes of patients with advanced illnesses. This finding is in agreement with the growing number of evidence showing support for the use of combination therapy in treating metastatic HER2-positive breast cancer. Combination treatments have a high probability of producing improved results than mono-agent treatment due to their ability to suppress multiple treatment pathways that drive tumor expansion and resistance (Triantafyllidi & Triantafillidis, 2022). Trastuzumab biosimilars are cost-effective treatment of breast cancer and an affordable medication to patients with advanced disease because they balance between the pharmacologic cost of drugs and the efficacy measure of TTF and pCR. The concept of cost-effectiveness is of particular importance to the metastatic setting since patients spend long periods under the treatment, meaning incurring numerous costs associated with healthcare. Biosimilars can not only increase access rates to life-changing therapies, but also lessen the economic burden of healthcare systems and patients since the cost of the HER2-targeted therapy is going to become lower (Simoens, 2011).

2.8 The extrapolation concept: Generalizing beyond biosimilar indications to other indications

Extrapolation refers to approving a biosimilar for clinical use without clinical experiments of such use, but based on the utilization of a comprehensive set of analytical and clinical data to establish biosimilarity of all tried and tested clinical uses of the reference product. It is an idea founded on the fact that, in case the biosimilar has the same structure, functionality, and clinical profile as the reference product, it will be safe and effective in all of the indications covered by the reference product. Extrapolation is cheaper and faster in the innovation and approval of biosimilars and introduction into the market and the fostering of competition (Agarwal & McBride, 2016; Calvo et al., 2018). It is a guideline that is science-based, applied by regulatory authorities like the EMA and FDA to make it possible to have the biosimilars approved without tampering with the safety and efficacy of the drug to patients. Governing agencies have laid down stringent guidelines based on which the suitability of extrapolation can be decided, such as the way the drug acts, its target receptors, and biosimilarity evidence across indicators are to be provided. The surveillance Organizations can take the necessary actions to guarantee that biosimilars can be provided to the patients early and as fast as possible through extrapolation. The FDA and EMA approve biosimilars to the indications of the reference product without clinical trials of all indications and lean on a relatively broad conclusion of biosimilarity to facilitate extension of approval to all indications included in the approval of the referenced product. The rationale behind this assumption is founded in the fact that in the event that a biosimilar possesses a high comparability with the reference product regarding their similarities in terms of structure, their functionality, and efficacy in the clinical arena, such a biosimilar is most likely safe and reliable across all of its indications approved in treating conditions similar to those posed by the reference product. The financial resources and time go to waste when it comes to the marketing authorization of biosimilars and induction of competition through extrapolation (Weise et al., 2014).

Each of such signs should be reasonably justified scientifically, and each of them should be considered independently by the regulator in such a way that the extrapolation could be established using an overall picture of the characteristics of the biosimilar as well as the clinical situation under which the biosimilar is going to be administered. Such a decision is the thought-out weighing of available data and possible differences between the biosimilar and the reference product, the consequences of which can be on the safety or efficacy of the biosimilar across indications. In the judgment on extrapolation, the regulatory authorities will also take into consideration the mechanism of action of the reference product as well as the risk of immunogenicity. Factors include history of use of the biologic, the mechanism of action, receptors involved, extent to which the functional units can be characterized analytically, which builds a broad template against which the rigor of extrapolation validity can be assessed. The regulatory authorities will also take into account the possibility of immunogenicity and information on postmarketing surveillance when coming up with the extrapolation determination. It is by striking a reasonable balance of all these elements that the regulatory bodies will ensure that they come up with safe and effective use of all the indications approved for the biosimilars. Health Canada has removed the term of extrapolation and replaced it with what is known under biosimilarity and attempts to be more explicit and transparent in regulation. This is all done in an effort to enhance communication and education between healthcare providers and patients in terms of biosimilars on a larger scale. Through simple and clear lingo, the regulators can ensure that all the parties concerned are aware to the latter extent of the science behind allowing biosimilars and the clinical implications of their use (Gherghescu & Delgado-Charro, 2020).

Filgrastim biosimilars got licensing priority in based on extrapolation into chemotherapy-induced neutropenia (CIN), but it demonstrated efficacy in other cancers, e.g., stem cell mobilization. Extrapolation was made possible by confirmatory trials of filgrastim in chemotherapeutic breast cancer patients where neutropenia occurred as a side effect. All these results were later confirmed by follow-up postmarketing real-world evidence, further reinforcing the concept of extrapolation. These are the grounds of successful extrapolation application in the field of oncology that allow approving biosimilars very effectively and operating them as well (Botteri et al., 2018).

2.9 Existing problems as well as hurdles in the uptake of biosimilars

In the United States, barriers to therapeutic oncology biosimilar adoption exist and include the absence of patient and prescriber education, unresolved efficacy issues and safety issues, and pay-for-delay, which have the potential to impact the introduction of biosimilars in the marketplace. Oncology biosimilar adoption problems arise with literally a plethora of concerns motivating a multifaceted environment, depending on majorly of cost of care, such as drug cost, patient access, formulary status, and treatment management algorithm, and further highlights the complexity of the biosimilar marketplace. Disagreement in the interchangeability and substitution policies amongst the regulatory agencies constitutes a concern with regard to the likely immunogenicity of individual patients alongside uncertainty amongst patients and prescribers (Gherghescu & Delgado-Charro, 2020). The utilization of oncology biosimilars is controlled through the price of the drug, availability of the patient, formulary status, and treatment-decision algorithms because it can potentially change the prescribing behavior and preference to the patients. Pay-for-delay may lead to less biosimilar uptake since they avoid entering the market and create limited competition. The biosimilar cost advantage over the originator, the amount of TTF saved a month, is huge, but is not always captured due to the forces of the market and the reimbursement strategies (Al Meslamani, 2024; Blackstone & Joseph, 2013). The lack of efficient specific training of both the medical providers and patients is also an obstacle to the implementation of therapeutic oncology biosimilars, since most other stakeholders might have insufficient data on such items and their benefits. Some of the most popular barriers are patient hesitancy to change to the biosimilar after using the reference product, poor knowledge about biosimilars, and patient-friendly information. The education that both patients and oncologists should receive must be focused and aimed at generating communication about trastuzumab biosimilars, dispelling fears with incorrect information, and reassuring patients of safety and effectiveness to ensure debate-free decision-making on an informed basis (Gibofsky et al., 2022).

2.10 Future research and conclusion

Biosimilar performance should also be monitored further via the postmarket surveillance programs to identify how far their effects end up having any long-term effect, as well as to confirm their efficacy safety even with a passage of time with the use of biosimilars. Further evidence is necessary to address such remaining questions as the use of biosimilars in relation to interchangeability and long-term safety of biosimilars in order to inform patients and clinicians with more arguments. Production of real-world evidence can be useful in validating biosimilar implementations and re-evaluations, which gives valuable data on their functionality in real embracement clinical practice. The switch rate of treatment regimens depending on biopsy results of a metastatic site needs to be more formally evaluated in future research because it can be used to inform decision to enhance patient outcomes. Comparative research into the long-term effects of biosimilars on patient outcomes, costs of care and patient access, and maximizing value also remains an area of concern and is on the wish list of the study. More studies should be conducted to determine the potential of biosimilars in conjunction with other therapies and in the other subtypes of breast cancer, to maximize their application and achieve better patient results. This should be implemented in the form of medical education programs to address the concerns and create awareness on biosimilars to enable the healthcare professional to make informed decisions on their usage. Indicating the specific guidelines and recommendations on how to use biosimilars could help in standardizing a clinical practice that would be equal in terms of patient care, decrease variability and lead to better outcomes. The paying systems and managed care organizations must be sensitized on a biosimilar approval process, stakeholders' issues (e.g., patients and providers), and the issues that determine the use of HER2-targeted therapies to be in a position to develop policies that foster the use of biosimilars appropriately. It may be possible to address these issues and optimize the use of biosimilars in practice to enable healthcare systems to achieve the full potential of these price-saving products and enhance access to necessary medications in breast cancer patients.

References

Agarwal, A. B., & McBride, A. (2016). Understanding the biosimilar approval and extrapolation process: A case study of an epoetin biosimilar. *Critical Reviews in Oncology/Hematology, 104*, 98–107.

Ahmad, A.-S., Olech, E., McClellan, J. E., & Kirchhoff, C. F. (2016). Development of biosimilars. *Seminars in Arthritis and Rheumatism*. doi:10.1016/j.semarthrit.2016.01.002.

Al Meslamani, A. Z. (2024). Short and long-term economic implications of biosimilars. *Expert Opinion on Biological Therapy, 24*(7), 567–570.

Barbier, L., Declerck, P., Simoens, S., Neven, P., Vulto, A. G., & Huys, I. (2019). The arrival of biosimilar monoclonal antibodies in oncology: clinical studies for trastuzumab biosimilars. *British journal of cancer, 121*(3), 199–210.

Barbier, L., Ebbers, H. C., Declerck, P., Simoens, S., Vulto, A. G., & Huys, I. (2020). The efficacy, safety, and immunogenicity of switching between reference biopharmaceuticals and biosimilars: A systematic review. *Clinical Pharmacology & Therapeutics, 108*(4), 734–755.

Barzaman, K., Karami, J., Zarei, Z., Hosseinzadeh, A., Kazemi, M. H., Moradi-Kalbolandi, S., Safari, E., & Farahmand, L. (2020). Breast cancer: Biology, biomarkers, and treatments. *International Immunopharmacology, 84*, 106535.

Berkowitz, S. A., Engen, J. R., Mazzeo, J. R., & Jones, G. B. (2012). Analytical tools for characterizing biopharmaceuticals and the implications for biosimilars. *Nature Reviews Drug Discovery, 11*(7), 527–540.

Blackstone, E. A., & Joseph, P. F. (2013). The economics of biosimilars. *American Health & Drug Benefits, 6*(8), 469.

Blackwell, K., Gligorov, J., Jacobs, I., & Twelves, C. (2018). The global need for a trastuzumab biosimilar for patients with HER2-positive breast cancer. *Clinical Breast Cancer, 18*(2), 95–113.

Blandizzi, C., Meroni, P. L., & Lapadula, G. (2017). Comparing originator biologics and biosimilars: A review of the relevant issues. *Clinical Therapeutics, 39*(5), 1026–1039.

Botteri, E., Krendyukov, A., & Curigliano, G. (2018). Comparing granulocyte colony-stimulating factor filgrastim and pegfilgrastim to its biosimilars in terms of efficacy and safety: A meta-analysis of randomised clinical trials in breast cancer patients. *European Journal of Cancer, 89*, 49–55.

Calvo, B., Martinez-Gorostiaga, J., & Echevarria, E. (2018). The surge in biosimilars: Considerations for effective pharmacovigilance and EU regulation. *Therapeutic Advances in Drug Safety, 9*(10), 601–608.

Cargnin, S., Shin, J. I., Genazzani, A. A., Nottegar, A., & Terrazzino, S. (2020). Comparative efficacy and safety of trastuzumab biosimilars to the reference drug: A systematic review and meta-analysis of randomized clinical trials. *Cancer Chemotherapy and Pharmacology, 86*, 577–588.

Casadevall, N., Edwards, I. R., Felix, T., Graze, P. R., Litten, J. B., Strober, B. E., & Warnock, D. G. (2013). Pharmacovigilance and biosimilars: Considerations, needs and challenges. *Expert Opinion on Biological Therapy, 13*(7), 1039–1047.

Cesarec, A., & Likić, R. (2017). Budget impact analysis of biosimilar trastuzumab for the treatment of breast cancer in Croatia. *Applied Health Economics and Health Policy, 15*, 277–286.

Chai, Q., Wen, H., Lang, Y., Zhang, L., Song, Y., & Liu, X. (2022). Budget impact analysis of the introduction of a trastuzumab biosimilar for HER2-positive breast cancer in China. *Clinical Drug Investigation, 42*(11), 937–947.

Chamberlain, P., & Kurki, P. (2018). Immunogenicity assessment of biosimilars: A multidisciplinary perspective. *Biosimilars: Regulatory, Clinical, and Biopharmaceutical Development, 34*, 489–542.

Class, J., & Langis, L. (2012). A patient centered paradigm for the biosimilars market. *GaBI Journal, 1*, 17–21.

Coory, M., & Thornton, K. (2019). Randomised clinical endpoint studies for trastuzumab biosimilars: a systematic review. *Breast Cancer Research and Treatment, 176*(1), 17–25.

Falconer, R. J., Jackson-Matthews, D., & Mahler, S. M. (2011). Analytical strategies for assessing comparability of biosimilars. *Journal of Chemical Technology & Biotechnology, 86*(7), 915–922.

Gagliato, D., Reinert, T., Rocha, C., Tavares, M., Pimentel, S., Fuzita, W., França, B., Magaton, É., Brito, N., Cardoso, A. C., & Castilho, V. (2024). Real-world study of adjuvant biosimilar trastuzumab-Dkst for HER2-positive breast cancer treatment in a Brazilian population. *Oncology and Therapy, 12*(3), 437–449.

Gasteiger, C., Perera, A., Yielder, R., Scholz, U., Dalbeth, N., & Petrie, K. J. (2023). Using family-centered communication to optimize patient-provider-companion encounters about changing to biosimilars: A randomized controlled trial. *Patient Education and Counseling, 106*, 142–150.

Georgakopoulou, E., & Scully, C. (2015). Biological agents: What they are, how they affect oral health and how they can modulate oral healthcare. *British Dental Journal, 218*(12), 671–677.

Gherghescu, I., & Delgado-Charro, M. B. (2020). The biosimilar landscape: An overview of regulatory approvals by the EMA and FDA. *Pharmaceutics, 13*(1), 48.

Gibofsky, A., Evans, C., & Strand, V. (2022). Provider and patient knowledge gaps on biosimilars: Insights from surveys. *American Journal of Managed Care, 28*(12), S227–S233.

Gion, M., Pérez-García, J. M., Llombart-Cussac, A., Sampayo-Cordero, M., Cortés, J., & Malfettone, A. (2021). Surrogate endpoints for early-stage breast cancer: A review of the state of the art, controversies, and future prospects. *Therapeutic Advances in Medical Oncology, 13* 17588359211059587. doi:10.1177/17588359211059587.

Giuliani, J., & Bonetti, A. (2019). The economic impact of biosimilars in oncology and hematology: The case of trastuzumab and rituximab. *Anticancer Research, 39*(7), 3971–3973.

Godman, B., Haque, M., Leong, T., Allocati, E., Kumar, S., Islam, S., & Jakovljevic, M. (2021). The current situation regarding long-acting insulin analogues, including biosimilars among African, Asian, European, and South American countries: Findings and implications for the future. *Frontiers in Public Health, 9*, 671961.

Goutsouliak, K., Veeraraghavan, J., Sethunath, V., De Angelis, C., Osborne, C. K., Rimawi, M. F., & Schiff, R. (2020). Towards personalized treatment for early-stage HER2-positive breast cancer. *Nature Reviews Clinical Oncology, 17*(4), 233–250.

Gutka, H. J., Yang, H., & Kakar, S. (2018). *Biosimilars: Regulatory, clinical, and biopharmaceutical development*: 34. Springer.

Ho, C. M. (2021). Biosimilar bias: A barrier to addressing American drug costs. *Denver Law Review, 99*, 517.

Jacobs, I., Singh, E., Sewell, K. L., Al-Sabbagh, A., & Shane, L. G. (2016). Patient attitudes and understanding about biosimilars: An international cross-sectional survey. *Patient Preference and Adherence*, 937–948. doi:10.2147/PPA.S104891.

Joshi, S., & Rathore, A. S. (2020). Assessment of structural and functional comparability of biosimilar products: Trastuzumab as a case study. *BioDrugs, 34*, 209–223.

Kabir, E. R., Moreino, S. S., & Sharif Siam, M. K. (2019). The breakthrough of biosimilars: A twist in the narrative of biological therapy. *Biomolecules, 9*(9), 410.

Kang, H. N., Wadhwa, M., Knezevic, I., Burns, C., & Griffiths, E. (2024). The importance of World Health Organization international reference standards in the product life cycle of biosimilars. *Annals of the New York Academy of Sciences, 1540*(1), 225–234.

Kirchhoff, C. F., Wang, X. Z. M., Conlon, H. D., Anderson, S., Ryan, A. M., & Bose, A. (2017). Biosimilars: Key regulatory considerations and similarity assessment tools. *Biotechnology and Bioengineering, 114*(12), 2696–2705.

Kreutzfeldt, J., Rozeboom, B., Dey, N., & De, P. (2020). The trastuzumab era: Current and upcoming targeted HER2+ breast cancer therapies. *American Journal of Cancer Research, 10*(4), 1045.

Kroschinsky, F., Stölzel, F., von Bonin, S., Beutel, G., Kochanek, M., Kiehl, M., & Group, O. P. C. (2017). New drugs, new toxicities: Severe side effects of modern targeted and immunotherapy of cancer and their management. *Critical Care, 21*, 1–11.

Lammers, P. E., Dank, M., Masetti, R., Abbas, R., Hilton, F., Coppola, J., & Jacobs, I. (2018). Neoadjuvant PF-05280014 (a potential trastuzumab biosimilar) versus trastuzumab for operable HER2+ breast cancer. *British Journal of Cancer, 119*(3), 266–273.

Leonard, E., Wascovich, M., Oskouei, S., Gurz, P., & Carpenter, D. (2019). Factors affecting health care provider knowledge and acceptance of biosimilar medicines: A systematic review. *Journal of Managed Care & Specialty Pharmacy, 25*(1), 102–112.

Leung, J. H., Tai, Y.-S., Wang, S.-Y., Fion, H.-T. Y., Tsung-Chin, H., & Chan, A. L. (2022). Cost-effectiveness of trastuzumab biosimilar combination therapy and drug wastage as first-line treatment for HER2-positive metastatic breast cancer. *The Breast, 65*, 91–97.

Li, R. K., Tokunaga, E., Adamchuk, H., Vladimirov, V., Yanez, E., Lee, K. S., & Ishikawa, T. (2022). Long-term safety and effectiveness of PF-05280014 (a trastuzumab biosimilar) treatment in patients with HER2-positive metastatic breast cancer: Updated results of a randomized, double-blind study. *BioDrugs, 36*(1), 55–69.

Liu, Y., Wang, Y., Wang, M., Zhai, S., Hou, C., Sun, F., & Jian, L. (2025). Evaluating biosimilars: Safety, efficacy, and regulatory considerations in clinical studies. *International Journal of Clinical Pharmacy, 47*(1), 232–236.

Lu, U. (2014). Biologics price competition and innovation act: Striking a delicate balance between innovation and accessibility. *Minnesota Journal of Law, Science & Technology, 15*, 613.

Mascarenhas-Melo, F., Diaz, M., Gonçalves, M. B. S., Vieira, P., Bell, V., Viana, S., & Veiga, F. (2024). An overview of biosimilars—Development, quality, regulatory issues, and management in healthcare. *Pharmaceuticals, 17*(2), 235.

McBride, A., MacDonald, K., Fuentes-Alburo, A., & Abraham, I. (2021). Cost-efficiency and expanded access modeling of conversion to biosimilar trastuzumab-dkst with or without pertuzumab in metastatic breast cancer. *Journal of Medical Economics, 24*(1), 743–756.

Miller, E. M., & Schwartzberg, L. S. (2019). Biosimilars for breast cancer: A review of HER2-targeted antibodies in the United States. *Therapeutic Advances in Medical Oncology, 11*, 1758835919887044.

Monga, A., Gagan, Jamwal, P., Sharma, S., & Kaur, A (2025). Biosimilars: A critical review of development, regulatory landscape, and clinical implications. *AAPS PharmSciTech, 26*(1), 46.

Nabhan, C., Parsad, S., Mato, A. R., & Feinberg, B. A. (2018). Biosimilars in oncology in the United States: A review. *JAMA Oncology, 4*(2), 241–247.

Naser-Tavakolian, A., Heard, J. R., & Ahdoot, M. (2025). *Biologics* translational urology (pp. 355–360). Elsevier.

Ngo, D., & Chen, J. (2021). A clinical review of biosimilars approved in oncology. *Annals of Pharmacotherapy, 55*(3), 362–377.

Nikitina, V., Santi Laurini, G., Montanaro, N., & Motola, D. (2023). Comparative safety profiles of oncology biosimilars vs. originators in Europe: An analysis of the EudraVigilance database. *Cancers, 15*(14), 3680.

Oliva, A., Scavone, C., Riccardi, C., Bernardi, F. F., Salvo, F., & Mascolo, A. (2024). Safety profile of trastuzumab originator vs biosimilars: A systematic review and meta-analysis of randomized clinical trials. *Clinical and Translational Oncology, 27*, 1826–1838.

Oskouei, S. T., & Kusmierczyk, A. R. (2021). Biosimilar uptake: The importance of healthcare provider education. *Pharmaceutical Medicine, 35*(4), 215–224.

Pathak, A., Singh, S. P., Tiwari, A., Tripathi, A. M., Jahan, T., & Singh, D. B. (2023). Biosimilar, biobetter, and biosuperior therapeutic proteins. In D. B. Singh, & T. Tripathi (Eds.), *Protein-based therapeutics* (pp. 325–353). Springer.

Pegram, M. D., Bondarenko, I., Zorzetto, M. M. C., Hingmire, S., Iwase, H., Krivorotko, P. V., & Aggarwal, R. (2019). PF-05280014 (a trastuzumab biosimilar) plus paclitaxel compared with reference trastuzumab plus paclitaxel for HER2-positive metastatic breast cancer: A randomised, double-blind study. *British Journal of Cancer, 120*(2), 172–182.

Pérez-Herrero, E., & Fernández-Medarde, A. (2015). Advanced targeted therapies in cancer: Drug nanocarriers, the future of chemotherapy. *European Journal of Pharmaceutics and Biopharmaceutics, 93*, 52–79.

Piezzo, M., D'Aniello, R., Avallone, I., Barba, B., Cianniello, D., Cocco, S., & Fusco, G. (2021). Uptake of trastuzumab biosimilars for the treatment of HER2-positive breast cancer: A real-world experience from a cancer center. *Pharmaceutics, 13*(5), 684.

Pineda, C., Castañeda Hernández, G., Jacobs, I. A., Alvarez, D. F., & Carini, C. (2016). Assessing the immunogenicity of biopharmaceuticals. *BioDrugs, 30*, 195–206.

Pivot, X., Bondarenko, I., Nowecki, Z., Dvorkin, M., Trishkina, E., Ahn, J.-H., & Wojtukiewicz, M. (2018). A phase III study comparing SB3 (a proposed trastuzumab biosimilar) and trastuzumab reference product in HER2-positive early breast cancer treated with neoadjuvant-adjuvant treatment: Final safety, immunogenicity and survival results. *European Journal of Cancer, 93*, 19–27.

Plosker, G. L., & Keam, S. J. (2006). Trastuzumab: A review of its use in the management of HER2-positive metastatic and early-stage breast cancer. *Drugs, 66*, 449–475.

Pondé, N., Brandão, M., El-Hachem, G., Werbrouck, E., & Piccart, M. (2018). Treatment of advanced HER2-positive breast cancer: 2018 and beyond. *Cancer treatment reviews, 67*, 10–20.

Resende, H. M., Ladislau, L., Cardoso, A. C. F., Brandão, J. D. P., Assis, B. R., Cardoso, P., & Castilho, V. (2021). Biosimilar use in breast cancer treatment: A national survey of Brazilian oncologists' opinions, practices, and concerns. *JCO Global Oncology, 7*, 1316–1324.

Roger, S. D. (2006). Biosimilars: How similar or dissimilar are they? *Nephrology, 11*(4), 341–346.

Rugo, H. S., Barve, A., Waller, C. F., Hernandez-Bronchud, M., Herson, J., Yuan, J., & Loganathan, S. (2017). Effect of a proposed trastuzumab biosimilar compared with trastuzumab on overall response rate in patients with ERBB2 (HER2)–positive metastatic breast cancer: A randomized clinical trial. *Jama, 317*(1), 37–47.

Sagi, S., Anjaneya, P., Kalsekar, S., Kottke, A., & Cohen, H. P. (2023). Long-term real-world post-approval safety data of multiple biosimilars from one marketing authorization holder after more than 18 years since their first biosimilar launch. *Drug Safety, 46*(12), 1391–1404.

Sarnola, K., Merikoski, M., Jyrkkä, J., & Hämeen-Anttila, K. (2020). Physicians' perceptions of the uptake of biosimilars: A systematic review. *BMJ Open, 10*(5), e034183.

Schreitmüller, T., Barton, B., Zharkov, A., & Bakalos, G. (2018). Comparative immunogenicity assessment of biosimilars. *Future Oncology, 15*(3), 319–329.

Simoens, S. (2011). Biosimilar medicines and cost-effectiveness. *ClinicoEconomics and Outcomes Research*, 29–36. doi:10.2147/CEOR.S12494.

Singh, A., Kalaivani, M., Srivastava, S., Goyal, R. K., & Gupta, S. K. (2020). Postmarketing safety of biosimilars: Current status, challenges, and opportunities in the spontaneous reporting system. *Therapeutic innovation & regulatory science, 54*(3), 667–680.

Stebbing, J., Baranau, Y., Manikhas, A., Lee, S. J., Thiruchelvam, P., Leff, D., & Esteva, F. J. (2018). Total pathological complete response versus breast pathological complete response in clinical trials of reference and biosimilar trastuzumab in the neoadjuvant treatment of breast cancer. *Expert Review of Anticancer Therapy, 18*(6), 531–541.

Swain, S. M., Shastry, M., & Hamilton, E. (2023). Targeting HER2-positive breast cancer: Advances and future directions. *Nature reviews Drug discovery, 22*(2), 101–126.

Tesch, M. E., & Gelmon, K. A. (2020). Targeting HER2 in breast cancer: Latest developments on treatment sequencing and the introduction of biosimilars. *Drugs, 80*(17), 1811–1830.

Trapani, D., & Curigliano, G. (2021). The global landscape of drug development of trastuzumab biosimilars. *Journal of Cancer Policy, 28*, 100273.

Triantafyllidi, E., & Triantafillidis, J. K. (2022). Systematic review on the use of biosimilars of trastuzumab in HER2+ breast cancer. *Biomedicines, 10*(8), 2045.

Tsuruta, L. R., Lopes dos Santos, M., & Moro, A. M. (2015). Biosimilars advancements: Moving on to the future. *Biotechnology progress, 31*(5), 1139–1149.

Uifălean, A., Ilieş, M., Nicoară, R., Rus, L. M., Hegheş, S. C., & Iuga, C.-A. (2018). Concepts and challenges of biosimilars in breast cancer: The emergence of trastuzumab biosimilars. *Pharmaceutics, 10*(4), 168.

Wang, J., Niu, S., Dong, W., Wei, L., Ou, L., Zhang, T., & Shen, T. (2020). A randomized phase I clinical trial comparing the pharmacokinetic, safety, and immunogenicity of potential biosimilar recombinant human HER2 monoclonal antibody for injection and trastuzumab in healthy Chinese adults. *Expert Opinion on Investigational Drugs, 29*(7), 755–762.

Weise, M., Bielsky, M.-C., De Smet, K., Ehmann, F., Ekman, N., Giezen, T. J., & Ho, K. (2012). Biosimilars: What clinicians should know. *Blood. The Journal of the American Society of Hematology, 120*(26), 5111–5117.

Weise, M., Kurki, P., Wolff-Holz, E., Bielsky, M.-C., & Schneider, C. K. (2014). Biosimilars: The science of extrapolation. *Blood. The Journal of the American Society of Hematology, 124*(22), 3191–3196.

Wu, Q., Wang, Z., Wang, X., Yu, H., & Sun, J. (2023). Patients' perceptions of biosimilars: A systematic review. *BioDrugs, 37*(6), 829–841.

Yang, J., Yu, S., Yang, Z., Yan, Y., Chen, Y., Zeng, H., & Zhang, Z. (2019). Efficacy and safety of anti-cancer biosimilars compared to reference biologics in oncology:A systematic review and meta-analysis of randomized controlled trials. *BioDrugs, 33*, 357–371.

Zucchetti, B. M., Nicolò, E., & Curigliano, G. (2019). Biosimilars for breast cancer. *Expert Opinion on Biological Therapy, 19*(10), 1015–1021.

CHAPTER

Biosimilars and biologics in gastric and colorectal carcinoma

3

Saravanakumar Kasimedu[a], Nawaz Mahammed[b], Tadiparthi Reshma[b], Nagaveni Pommala[c] and Kasturi Vishwanathasetty Veerabhadrappa[d]

[a]*Department of Pharmaceutics, Seven Hills College of Pharmacy (Autonomous), Tirupati, Andhra Pradesh, India,* [b]*Department of Pharmaceutics, Raghavendra Institute of Pharmaceutical Education and Research (Autonomous), Anantapur, Andhra Pradesh, India,* [c]*Department of Pharmaceutics, Vemu College of Pharmacy & Health Sciences, Chittoor, Andhra Pradesh, India,* [d]*Department of Pharmacognosy & Phytochemistry, School of Pharmacy, Kampala International University-Western Campus, Bushenyi, Uganda*

3.1 Introduction

3.1.1 Epidemiology and global burden of gastric and colorectal carcinoma

Gastric cancer (GC) and colorectal cancer (CRC) represent two of the most prevalent malignancies worldwide and are leading causes of cancer-related mortality (Wong et al., 2021). According to the GLOBOCAN 2020 estimates, gastric cancer ranks fifth in incidence and fourth in cancer-related deaths, with approximately 1.1 million new cases and 770,000 deaths annually (Iwu & Iwu-Jaja, 2023). Colorectal cancer, encompassing both colon and rectal cancers, is the third most commonly diagnosed malignancy and the second leading cause of cancer mortality, accounting for nearly 1.9 million new cases and 935,000 deaths globally each year (Pinheiro et al., 2024). Both cancers show significant geographical variation; gastric carcinoma is more common in East Asia, Eastern Europe, and parts of South America, whereas CRC demonstrates higher prevalence in developed regions such as North America, Europe, and Australia. Lifestyle factors such as dietary habits, obesity, alcohol consumption, smoking, and Helicobacter pylori infection contribute substantially to gastric cancer risk, while sedentary lifestyle, high-fat diets, and genetic predispositions such as Lynch syndrome are major determinants in CRC (Kendong et al., 2021). The late-stage diagnosis of both malignancies contributes to their high mortality, underscoring the need for early detection strategies and access to effective therapies. Together, gastric and CRC constitute a major global health burden requiring advanced, cost-effective therapeutic solutions.

Biosimilars and Biologics in Cancer Treatment: An Era in Therapeutic Strategies.
DOI: https://doi.org/10.1016/B978-0-443-43824-0.00007-9

3.1.2 Current therapeutic landscape: Chemotherapy versus targeted biologics

For decades, systemic chemotherapy has served as the backbone for the management of gastric and CRC. Agents such as fluoropyrimidines, platinum compounds, irinotecan, and taxanes remain standard first-line regimens. While chemotherapy provides modest survival benefits, its nonspecific cytotoxicity is often associated with significant toxicity and limited improvement in long-term outcomes (Jan et al., 2024). The advent of targeted biologics has revolutionized treatment paradigms, particularly in advanced or metastatic settings. In CRC, vascular endothelial growth factor (VEGF) inhibitors such as bevacizumab and aflibercept, and epidermal growth factor receptor (EGFR) inhibitors like cetuximab and panitumumab, have become integral to therapy, particularly in patients with RAS wild-type tumors (Gonzalez et al., 2024; Hansen et al., 2021). Similarly, in gastric cancer, ramucirumab (a VEGFR-2 antagonist) and trastuzumab (anti-HER2 antibody) have shown substantial benefits when combined with chemotherapy. More recently, immune checkpoint inhibitors (ICIs) targeting PD-1/PD-L1, including nivolumab and pembrolizumab, have demonstrated promising outcomes in microsatellite instability-high (MSI-H) or PD-L1-positive subsets of CRC and gastric cancer. Despite these advances, not all patients derive benefit, and resistance to biologics remains a clinical challenge, necessitating ongoing innovation and precision medicine strategies (Abdullah, 2023).

3.1.3 Rising healthcare costs due to biologics

The integration of monoclonal antibodies and other biologics into gastrointestinal oncology has considerably improved survival outcomes, but at the cost of escalating healthcare expenditures. Biologics are inherently expensive to produce due to their complex manufacturing processes involving living cell systems, stringent quality controls, and extended development timelines (Koustas et al., 2022). For instance, the annual cost of treatment with bevacizumab or cetuximab can exceed tens of thousands of dollars per patient, placing substantial strain on healthcare budgets, particularly in low and middle-income countries (Elshiekh et al., 2025). Moreover, as treatment paradigms increasingly incorporate combination regimens such as chemotherapy plus targeted therapy, or dual immunotherapy approaches, the cumulative costs rise further. In high-income countries, biologics account for a significant proportion of oncology drug spending, while in resource-limited settings, they remain inaccessible to the majority of patients. This disparity widens the treatment gap and contributes to inequities in cancer outcomes. Rising healthcare costs not only threaten the sustainability of national healthcare systems but also impose severe financial toxicity on individual patients and families. Therefore, addressing the affordability of biologics is essential to ensure equitable access to life-saving therapies (Salas-Benito et al., 2021).

3.1.4 Rationale for biosimilars: Access, affordability, sustainability

Biosimilars provide a scientifically robust and economically viable solution to the challenges posed by high-cost biologics. Defined as highly similar versions of reference biologic products with no clinically meaningful differences in safety, purity, or potency, biosimilars can significantly reduce costs through increased market competition while maintaining therapeutic equivalence. In oncology, the approval of bevacizumab, trastuzumab, and rituximab biosimilars has already demonstrated their potential to improve access to advanced therapies. For gastric and colorectal cancers, bevacizumab biosimilars are especially relevant, given their widespread use in metastatic CRC and gastric carcinoma. The availability of biosimilars can expand patient access in both developed and developing regions, thereby reducing disparities in cancer care. Importantly, healthcare systems stand to benefit from substantial cost savings, which can be reinvested into other areas of patient care or innovative therapies. Additionally, biosimilars promote long-term sustainability of oncology care by balancing innovation with affordability. While challenges remain regarding physician awareness, patient confidence, and regulatory harmonization, the increasing adoption of biosimilars represents a pivotal opportunity to democratize access to biologic therapies in gastric and CRC (Gunnam et al., 2025).

3.2 Biologics in gastric and colorectal carcinoma

The management of gastric and colorectal carcinoma has undergone a paradigm shift with the introduction of biologic therapies. Unlike conventional chemotherapy, which indiscriminately targets rapidly dividing cells, biologics specifically interfere with molecular pathways crucial to tumor growth and progression. This targeted approach not only enhances therapeutic efficacy but also reduces systemic toxicity. In gastrointestinal malignancies, biologics have primarily focused on inhibiting angiogenesis, blocking growth factor receptor signaling, and modulating immune checkpoints. Among the most established classes are VEGF inhibitors, EGFR inhibitors, and ICIs. Their integration into treatment algorithms has improved progression-free survival (PFS), overall survival (OS), and quality of life for patients with advanced disease. The following subsections explain the mechanism, approved indications, and clinical relevance of these biologic agents in gastric and CRC (Fig. 3.1).

3.2.1 Vascular endothelial growth factor inhibitors

Angiogenesis, the formation of new blood vessels, is essential for tumor growth beyond 1–2 mm in size, as neoplastic tissue requires a continuous nutrient and oxygen supply. VEGF is a central mediator of angiogenesis, and its overexpression in gastric and colorectal carcinomas correlates with aggressive disease and poor prognosis.

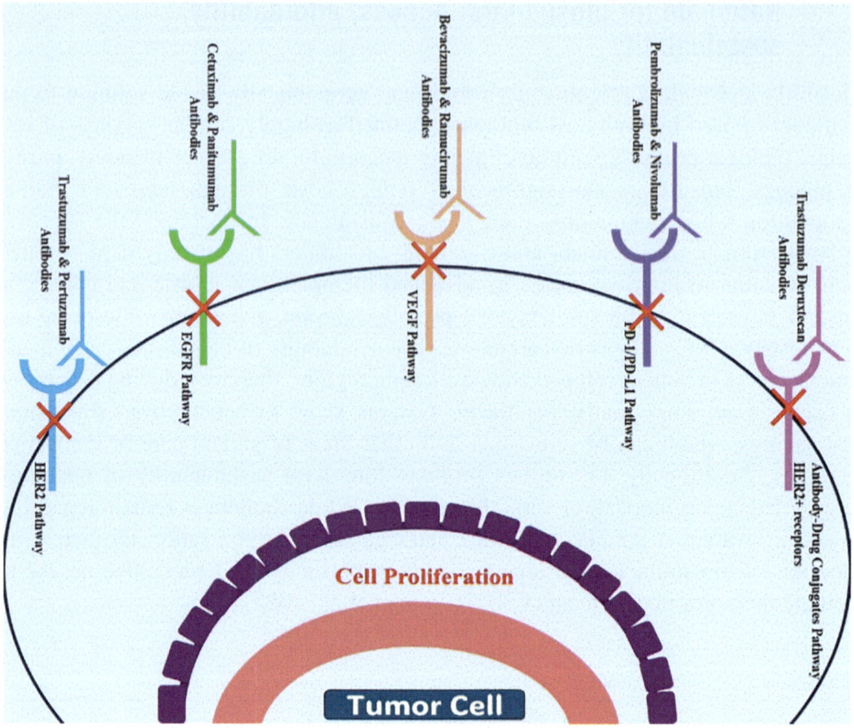

FIGURE 3.1

A schematic diagram illustrating mechanisms of action of biologics in gastric and colorectal cancer.

Targeting VEGF and its receptors has therefore become a rational therapeutic strategy (Brogowska et al., 2023; Królikowska et al., 2025).

Bevacizumab, a monoclonal antibody against VEGF-A, was the first anti-angiogenic agent approved for metastatic CRC. When combined with chemotherapy (e.g., FOLFOX, FOLFIRI), it significantly improved OS and PFS compared to chemotherapy alone. Its role in gastric carcinoma is more limited, as pivotal trials such as the AVAGAST study showed regional differences in efficacy, with greater benefit in non-Asian populations. Nonetheless, bevacizumab remains a cornerstone in metastatic CRC (Hansen et al., 2021).

Ramucirumab, an antibody targeting VEGFR-2, has demonstrated clinical efficacy in advanced gastric cancer. The RAINBOW and REGARD trials established its role as second-line therapy either in combination with paclitaxel or as monotherapy. In metastatic CRC, ramucirumab combined with FOLFIRI has shown survival benefits in patients progressing on first-line therapy (Zhang et al., 2024).

Aflibercept, a fusion protein acting as a decoy receptor for VEGF-A, VEGF-B, and placental growth factor, has been approved for use with FOLFIRI in metastatic CRC patients who previously failed oxaliplatin-based therapy.

VEGF inhibitors have improved outcomes in advanced gastric and colorectal cancers. However, their benefits are modest, often extending survival by only a few months, and are associated with adverse effects such as hypertension, bleeding, thromboembolic events, and gastrointestinal perforation. Despite these limitations, VEGF inhibition remains a critical therapeutic approach, particularly when integrated into multimodal treatment regimens (Lau et al., 2022).

3.2.2 Epidermal growth factor receptor inhibitors (cetuximab, panitumumab)

The EGFR is a transmembrane tyrosine kinase receptor that regulates cellular proliferation, differentiation, survival, and migration. Overexpression and aberrant signaling of EGFR contribute to tumorigenesis in colorectal carcinoma. Biologic agents targeting EGFR have significantly improved outcomes in selected patient populations (Cheng et al., 2021). Cetuximab, a chimeric monoclonal antibody, and panitumumab, a fully human monoclonal antibody, both act by binding to the extracellular domain of EGFR, thereby preventing ligand binding and subsequent receptor activation. This blockade inhibits downstream signaling through pathways such as RAS/RAF/MEK/ERK and PI3K/AKT, which are key drivers of cell proliferation and survival (Janani et al., 2022). In metastatic CRC, EGFR inhibitors have demonstrated efficacy primarily in patients with wild-type KRAS and NRAS genes. Mutations in these genes result in constitutive activation of downstream signaling independent of EGFR, rendering anti-EGFR therapy ineffective. Therefore, molecular testing for RAS mutations is a prerequisite before initiating cetuximab or panitumumab therapy. Clinical trials such as CRYSTAL and PRIME established the survival benefit of combining EGFR inhibitors with standard chemotherapy regimens in RAS wild-type tumors. Importantly, EGFR inhibitors are not effective in gastric cancer, as clinical trials have failed to demonstrate a significant survival benefit, possibly due to tumor heterogeneity and alternative signaling pathways. The most notable toxicities of EGFR inhibitors include skin rash, hypomagnesemia, and infusion reactions, which can impact patient quality of life but also serve as biomarkers of response in some cases. Thus, EGFR inhibition represents a paradigm of precision oncology in CRC, where treatment selection is guided by molecular profiling. Their successful incorporation highlights the importance of personalized medicine in gastrointestinal oncology (Nakamura et al., 2021).

3.2.3 Immune checkpoint inhibitors (nivolumab, pembrolizumab, dostarlimab, etc.)

Immune checkpoint pathways play a critical role in maintaining self-tolerance and preventing autoimmunity, but tumors exploit these mechanisms to evade immune surveillance. The PD-1/PD-L1 axis is particularly relevant in GIT cancers. Monoclonal antibodies targeting PD-1 (nivolumab, pembrolizumab, dostarlimab) or PD-L1 restore T-cell activity and enhance antitumor immune responses (Singh et al., 2023). In colorectal carcinoma, ICIs have demonstrated remarkable efficacy in tumors with microsatellite instability-high (MSI-H) or mismatch repair deficiency (dMMR). These tumors harbor high mutational burdens, generating numerous neoantigens that enhance immune recognition. The KEYNOTE-177 trial showed that pembrolizumab significantly prolonged PFS compared to chemotherapy in untreated MSI-H metastatic CRC, leading to its FDA approval as first-line therapy. Similarly, nivolumab, alone or in combination with ipilimumab (anti-CTLA-4 antibody), has shown durable responses in MSI-H CRC. In contrast, microsatellite-stable (MSS) CRC does not typically respond to ICIs, reflecting the need for novel combination strategies (Roth & Das, 2021). In gastric carcinoma, pembrolizumab has received approval for PD-L1-positive tumors (CPS ≥ 1) based on the KEYNOTE-059 trial, and for MSI-H/dMMR tumors regardless of PD-L1 status. Nivolumab has been approved in combination with chemotherapy as first-line treatment in advanced gastric or gastroesophageal junction cancers with PD-L1 expression. Dostarlimab, another PD-1 inhibitor, has shown promising activity in dMMR gastrointestinal cancers and may expand future treatment options (Narita & Muro, 2023). While ICIs have revolutionized the management of biomarker-selected subsets of gastric and CRC patients, challenges remain. Only a minority of patients exhibit durable benefit, resistance mechanisms are poorly understood, and immune-related adverse events (AEs) (e.g., colitis, pneumonitis, endocrinopathies) require careful management. Nevertheless, ICIs represent one of the most transformative advances in oncology, offering the possibility of long-term survival in cancers once considered incurable (Kelly et al., 2023).

3.3 Biosimilars: Concepts and regulatory framework

The advent of biosimilars has transformed the therapeutic landscape in oncology by providing cost-effective alternatives to expensive biologics, while maintaining comparable efficacy and safety. Unlike conventional small-molecule generics, biosimilars are complex, large-molecule biologic products that require stringent regulatory pathways for approval. Their development and integration into clinical practice rely on robust frameworks established by global regulatory authorities, emphasizing comparability, interchangeability, and pharmacovigilance. This section provides an overview of biosimilars in terms of definition, regulatory standards, and the key scientific principles governing their use in oncology, particularly for gastric and colorectal carcinoma (Malakar et al., 2024).

3.3.1 Definition and distinction from generics

Biosimilars are biologic medical products that are highly similar to already approved reference biologics, with no clinically meaningful differences in safety, purity, or potency. Unlike generics, which are chemically synthesized small molecules with identical active ingredients as their reference drugs, biosimilars are derived from living cells using complex biotechnological processes. This fundamental difference underscores their unique regulatory and clinical considerations. Generics can be exactly replicated due to their relatively simple chemical structure, and their approval typically requires only demonstration of bioequivalence. Biologics are large, heterogeneous molecules influenced by cell line variability, manufacturing conditions, and post-translational modifications. Hence, biosimilars cannot be exact copies but must instead demonstrate a high degree of similarity to reference products through comprehensive analytical, preclinical, and clinical evaluations. The distinction between biosimilars and generics extends to cost and development pathways. While generics are relatively inexpensive to develop and introduce, biosimilars involve significant investment in manufacturing infrastructure, rigorous characterization, and clinical trials, though still less costly than innovator biologics. Their introduction has profound implications in oncology, where biologics such as monoclonal antibodies and checkpoint inhibitors are critical yet financially burdensome. Thus, biosimilars occupy a unique space, offering affordable alternatives while demanding careful scientific and regulatory scrutiny to ensure therapeutic equivalence (Niazi, 2022).

3.3.2 Global regulatory pathways (EMA, FDA, WHO, DCGI)

The regulation of biosimilars is globally harmonized to ensure consistent standards of quality, safety, and efficacy. The European Medicines Agency (EMA) pioneered the first formal biosimilar pathway in 2005, establishing rigorous guidelines for comparability exercises involving analytical characterization, preclinical studies, and at least one confirmatory clinical trial. The EMA remains a leader in biosimilar approvals, with over 80 biosimilars authorized, including oncology agents like bevacizumab and trastuzumab biosimilars (Raposo et al., 2025). The U.S. FDA created the Biologics Price Competition and Innovation Act (BPCIA) of 2009, under which biosimilars are approved via the 351(k) pathway. This pathway emphasizes a stepwise demonstration of biosimilarity, with totality of the evidence being the cornerstone of evaluation. The FDA also provides guidance on interchangeability designations, which allow substitution at the pharmacy level without prescriber intervention, though this status is granted only after additional switching studies. The WHO has developed guidelines that serve as a reference for countries lacking robust biosimilar regulatory systems, ensuring global harmonization. Many national regulatory agencies, including those in Asia, Africa, and Latin America, adopt WHO standards. In India, the Drug Controller General of India (DCGI) and Central Drugs Standard Control Organization (CDSCO) have established biosimilar guidelines (2012, revised in 2016 and 2019). India has been a global leader in biosimilar development, being one of the first countries to approve trastuzumab and bevacizumab biosimilars. The country's relatively lower-cost production environment makes it pivotal for global access to oncology biosimilars (Rahalkar, 2021).

3.3.3 Importance of biosimilars

The scientific and regulatory foundation of biosimilars relies on four critical principles: comparability, extrapolation, interchangeability, and pharmacovigilance. Comparability involves rigorous head-to-head analytical studies between a biosimilar and its reference product to demonstrate similarity in structure, function, and clinical outcomes. Analytical similarity forms the cornerstone of biosimilar approval, often reducing the need for extensive phase III clinical trials compared to innovator biologics. Extrapolation refers to granting approval for all therapeutic indications of the reference biologic based on evidence from a limited set of indications, provided the mechanism of action, safety, and pharmacokinetics are similar across conditions. This principle is crucial in oncology, where a biosimilar tested in metastatic CRC may also be approved for use in gastric carcinoma if scientifically justified. Interchangeability is a higher regulatory designation, primarily in the United States, allowing pharmacists to substitute a biosimilar for the reference biologic without physician approval. Achieving interchangeability requires additional switching studies demonstrating that alternating between the reference and biosimilar does not affect safety, efficacy, or immunogenicity. Pharmacovigilance ensures ongoing monitoring of biosimilars once marketed. Given their complex nature and potential for immunogenicity, robust postmarketing surveillance, including registries and real-world studies, is essential. Pharmacovigilance builds confidence among healthcare providers and patients, reinforcing the long-term safety profile of biosimilars (Iskit, 2025; Mascarenhas-Melo et al., 2024). Collectively, these principles ensure that biosimilars not only reduce costs but also maintain the highest standards of patient safety and clinical efficacy. Their proper implementation fosters trust, enhances adoption, and strengthens the sustainability of healthcare systems globally.

3.4 Biosimilars in gastric and colorectal cancer: Current status

The biosimilars in gastric and CRC have significantly broadened treatment access and reduced the economic burden associated with high-cost biologics. While reference monoclonal antibodies such as bevacizumab, cetuximab, and panitumumab have established roles in the management of metastatic CRC and selected gastric cancers, their widespread use has been limited by cost constraints. Biosimilars provide a scientifically validated, clinically equivalent, and more affordable alternative. Regulatory agencies worldwide have approved several biosimilars targeting VEGF and EGFR pathways, with additional agents under development for immunotherapy. This section reviews the current status of biosimilars in gastric and CRC, focusing on available approvals, clinical equivalence data, and emerging research trends (Patel et al., 2024) (Table 3.1).

Table 3.1 Target site, mechanism of action and indications of biologics used in gastric and colorectal cancer.

Drug (type)	Target	Mechanism of action	Indications
Trastuzumab (mAb)	HER2/ERBB2	Binds HER2 receptor → inhibits downstream signaling (PI3K/AKT, MAPK) and mediates ADCC	HER2-positive metastatic gastric or gastroesophageal junction (GEJ) adenocarcinoma (with chemotherapy)
Ramucirumab (mAb)	VEGFR-2	Blocks VEGF ligand binding to VEGFR-2 → inhibits angiogenesis	Advanced/metastatic gastric or GEJ adenocarcinoma after progression; also with paclitaxel
Pembrolizumab (mAb)	PD-1	Immune checkpoint inhibitor restores T-cell activity against tumor cells	MSI-H/dMMR gastric or colorectal cancer; PD-L1-positive gastric cancer
Nivolumab (mAb)	PD-1	Blocks PD-1 receptor → enhances antitumor immune response	Advanced gastric/GEJ/esophageal adenocarcinoma (with chemo)
Bevacizumab (mAb)	VEGF-A	Binds VEGF-A → inhibits angiogenesis	Metastatic colorectal cancer (mCRC) in combination with chemotherapy
Cetuximab (mAb)	EGFR	Blocks EGFR activation → inhibits RAS/RAF/MAPK signaling	RAS wild-type metastatic colorectal cancer
Panitumumab (mAb)	EGFR	Fully human anti-EGFR antibody; inhibits downstream signaling	RAS wild-type metastatic colorectal cancer
Aflibercept (Fusion protein)	VEGF-A, VEGF-B, PlGF	Acts as a decoy receptor trapping VEGF ligands → antiangiogenic effect	mCRC resistant to oxaliplatin-based therapy (with FOLFIRI)

(*continued on next page*)

Table 3.1 Target site, mechanism of action and indications of biologics used in gastric and colorectal cancer—cont'd

Drug (type)	Target	Mechanism of action	Indications
Trastuzumab Deruxtecan (T-DXd) (ADC)	HER2	Antibody-drug conjugate delivering a topoisomerase I inhibitor to HER2-expressing cells	HER2-positive advanced gastric cancer after prior trastuzumab
Ziv-Aflibercept (VEGF Trap)	VEGF family	Neutralizes circulating VEGF ligands → suppresses tumor angiogenesis	Second-line metastatic colorectal cancer

→ *leads to.*

3.4.1 Bevacizumab biosimilars

Bevacizumab, a monoclonal antibody targeting VEGF-A, is a standard component of therapy for metastatic CRC, usually in combination with chemotherapy regimens such as FOLFOX or FOLFIRI. Although its role in gastric cancer is less robust, bevacizumab remains of clinical interest, particularly in Western populations, following evidence from the AVAGAST trial. Given its widespread use, bevacizumab was among the earliest oncology biologics to inspire biosimilar development.

Several FDA- and EMA-approved bevacizumab biosimilars are now available:

- Mvasi (bevacizumab-awwb; Amgen/Allergan, 2017) – the first bevacizumab biosimilar approved in the United States.
- Zirabev (bevacizumab-bvzr; Pfizer, 2019) – approved for multiple cancers, including CRC and nonsquamous NSCLC.
- Alymsys (bevacizumab-maly; Cinfa Biotech, 2021) – EMA-approved for colorectal and other cancers.
- Vegzelma (bevacizumab-adcd; Celltrion, 2022) – approved in both Europe and the United States.
- Avzivi (bevacizumab-tnjn; Biocon Biologics, 2022) – an additional U.S. FDA-approved biosimilar (Ngo & Chen, 2021).

Indications: These biosimilars have been approved for the same clinical indications as the reference Avastin^{TM}, including metastatic colorectal cancer in combination with chemotherapy, and for selected cases of advanced non-squamous NSCLC, glioblastoma, ovarian, and cervical cancers.

Clinical equivalence data: Multiple randomized phase III studies have confirmed the equivalence in efficacy, safety, and immunogenicity between bevacizumab biosimilars and the reference product. For example, the MAPLE study (evaluating ABP 215, marketed as Mvasi) demonstrated noninferiority to Avastin in patients with advanced non-squamous NSCLC, while similar efficacy was reported in metastatic CRC across multiple biosimilars. Toxicity profiles, including hypertension and proteinuria, were consistent with the reference product, and no new safety signals emerged. Real-world studies further reinforce their equivalence in treatment outcomes, supporting confidence among prescribers and regulatory authorities (Ngo & Chen, 2021). Collectively, bevacizumab biosimilars have become central to expanding access to antiangiogenic therapy in CRC, while their use in gastric cancer continues to be explored in ongoing trials and regional clinical practice.

3.4.2 Cetuximab and panitumumab biosimilars

Cetuximab and panitumumab are EGFR-targeting monoclonal antibodies approved for the treatment of RAS wild-type metastatic CRC. Both agents, when combined with chemotherapy or used as monotherapy in later lines, have demonstrated significant improvements in progression-free and OS. Their success has spurred the development of biosimilars, though these remain at an earlier stage compared to bevacizumab biosimilars.

Cetuximab biosimilars: Several biosimilar candidates are under development, with preclinical and early-phase clinical data showing promising results. Companies in Asia and Europe have advanced cetuximab biosimilars into phase I and phase III trials, aiming to demonstrate pharmacokinetic equivalence and therapeutic efficacy. However, as of now, no cetuximab biosimilar has received FDA or EMA approval. In India and China, some locally manufactured versions are available, reflecting regional regulatory differences.

Panitumumab biosimilars: Development is ongoing, with several biosimilars undergoing clinical trials. For example, early pharmacokinetic studies have confirmed equivalence in healthy volunteers, and phase III trials in metastatic CRC populations are underway. Panitumumab, being a fully human antibody, has a different immunogenicity profile compared to cetuximab, which may influence biosimilar development and uptake.

Current approvals and trials: While bevacizumab biosimilars are widely available, EGFR-targeted biosimilars are still in late-stage pipelines. Clinical equivalence data are expected to strengthen over the next decade, likely making cetuximab and panitumumab biosimilars accessible alternatives for RAS wild-type CRC patients (Doleschal et al., 2022; González et al., 2024). Thus, cetuximab and panitumumab biosimilars represent the next wave of biosimilar approvals in gastrointestinal oncology, with the potential to significantly reduce costs and improve access to precision-targeted therapy.

3.4.3 Emerging biosimilars in immunotherapy

ICIs, particularly PD-1/PD-L1 inhibitors such as nivolumab, pembrolizumab, and dostarlimab, have transformed the treatment landscape for subsets of CRC and gastric cancer. However, their high cost remains a major barrier to global accessibility. Consequently, the biosimilar development pipeline has now extended to immunotherapy agents, although most are still in preclinical or early clinical phases (Table 3.2).

Early-stage development: Several pharmaceutical companies are developing PD-1 biosimilars, with clinical trials already registered in China, Korea, and India. For instance, biosimilar versions of nivolumab and pembrolizumab are in phase I/III clinical studies assessing pharmacokinetics, safety, and efficacy. While none have yet achieved FDA or EMA approval, early clinical data suggest high similarity with reference drugs.

Research trends: The focus is on demonstrating that PD-1/PD-L1 biosimilars replicate the durable responses seen with reference products in MSI-H/dMMR colorectal cancer and PD-L1-positive gastric carcinoma. The complexity of immunotherapy responses, which involve long-term immune memory and heterogeneous biomarkers, makes biosimilar development more challenging than for VEGF/EGFR inhibitors.

Challenges: Regulatory frameworks are still evolving for immunotherapy biosimilars. Questions remain about the extrapolation of indications, given the diversity of tumor types treated with ICIs. Immunogenicity assessment is particularly critical, as even subtle structural differences could influence immune modulation. Moreover, designing equivalence trials is complex due to variable response rates and the long duration of benefit seen in responders (Monga, 2025).

3.5 Clinical evidence and trials

The clinical integration of biosimilars into the management of gastric and CRC requires rigorous demonstration of their similarity to reference biologics in terms of efficacy, safety, and immunogenicity. Regulatory approvals mandate comprehensive comparability exercises, but confirmatory clinical trials and post-marketing studies remain crucial to reinforce confidence among clinicians and patients. Evidence from randomized controlled trials (RCTs), meta-analyses, and real-world studies has consistently shown that biosimilars perform equivalently to innovator biologics. This section reviews the available comparative clinical studies, evaluates safety and immunogenicity outcomes, and highlights real-world data supporting the adoption of biosimilars in gastrointestinal (GI) oncology.

3.5.1 Comparative clinical studies in colorectal cancer and gastric cancer

Comparative clinical trials form the backbone of biosimilar evaluation, providing evidence that these agents achieve comparable therapeutic outcomes to their reference biologics. In CRC, bevacizumab biosimilars have been most extensively studied. For instance, the MAPLE trial compared ABP 215 (Mvasi) with reference bevacizumab

Table 3.2 Oncology biosimilars relevant to GC/CRC.

Reference biologic	Biosimilar name(s)	Region(s) approved	Indications covered (CRC/GC and Others)	Evidence basis (PK/PD, RCT, Extrapolation)	Notes (trial anchors, market year)	References
Bevacizumab	Mvasi (ABP 215); Zirabev (PF-06439535); Alymsys; Vegzelma; Avzivi	US, EU (varies by product)	mCRC; plus NSCLC, ovarian, cervical, glioblastoma per Avastin labels	Analytics + PK/PD; RCT equivalence (e.g., NSCLC) with indication extrapolation to CRC; RWE support	First US approval 2017 (Mvasi); subsequent launches 2019–2022; MAPLE/NSCLC anchors; broad uptake	Gao (2025)
Cetuximab	Multiple candidates (regional)	India, China (regional availability); no FDA/EMA approvals	mCRC (RAS WT) intent	Primarily analytics + PK; phase III ongoing; no major RCT approvals in US/EU	Late-stage development; immunogenicity considerations (chimeric mAb); regional versions in use	Candia et al. (2025)

(continued on next page)

Table 3.2 Oncology biosimilars relevant to GC/CRC—cont'd

Reference biologic	Biosimilar name(s)	Region(s) approved	Indications covered (CRC/GC and Others)	Evidence basis (PK/PD, RCT, Extrapolation)	Notes (trial anchors, market year)	References
Panitumumab	Multiple candidates	Trials ongoing; no FDA/EMA approvals	mCRC (RAS WT) intent	Analytics + PK equivalence shown; phase III trials in mCRC ongoing	Fully human mAb; potentially lower IRR risk vs cetuximab; market entries pending	Riedesser et al. (2022)
PD-1 inhibitors (nivolumab, pem-brolizumab)	Early-phase PD-1 biosimilars (various developers)	Asia (China, Korea, India trials); no FDA/EMA approvals	MSI-H/dMMR CRC; PD-L1+ GC contexts (future extrapolation)	Early analyt-ics + PK/PD; phase I/III in progress; extrapolation and immunogenicity under scrutiny	Complex equivalence design due to durable responses; regulatory expectations evolving	Chau (2023)

in nonsquamous non–small-cell lung cancer but included bridging evidence relevant for CRC. It demonstrated equivalent overall response rate (ORR), PFS, and safety profiles. Similarly, phase III trials of PF-06439535 (Zirabev) confirmed equivalence to bevacizumab in metastatic CRC, showing no significant differences in objective response or AE rates. In gastric cancer, fewer biosimilar trials exist, reflecting the limited use of biologics in this setting compared to CRC. However, biosimilars of ramucirumab are under development, and exploratory studies are evaluating bevacizumab biosimilars in advanced gastric carcinoma, particularly in non-Asian cohorts where the original bevacizumab showed benefit. Emerging checkpoint inhibitor biosimilars are being tested in MSI-H/dMMR subsets of gastric and CRC, though these remain early-phase investigations. Meta-analyses further support equivalence. For example, Xu et al. (2022) reported that bevacizumab biosimilars demonstrated comparable efficacy and safety in advanced CRC, with no statistically significant differences in PFS or OS. Trials for cetuximab and panitumumab biosimilars are ongoing, with preliminary pharmacokinetic data confirming similarity. Collectively, these comparative trials form the foundation for regulatory approvals and clinical confidence in biosimilars for GI cancers.

3.5.2 Safety, immunogenicity, and efficacy outcomes

The evaluation of biosimilars extends beyond efficacy to include safety and immunogenicity, both of which are critical in oncology, where patient populations are often immunocompromised. Across multiple trials, biosimilars of bevacizumab and other monoclonal antibodies have consistently demonstrated comparable safety profiles to their reference biologics. AEs such as hypertension, bleeding, proteinuria, thromboembolic events, and gastrointestinal perforations occurred at similar frequencies in biosimilar and reference arms, with no new toxicities identified. Immunogenicity remains a unique concern for biologics due to their protein-based structure, which can trigger antidrug antibody (ADA) formation. Clinical trials for bevacizumab biosimilars (e.g., Mvasi, Zirabev, Vegzelma) reported extremely low and comparable rates of ADA development versus reference bevacizumab. Importantly, the presence of ADAs did not correlate with reduced efficacy or increased toxicity, affirming biosimilar safety. For EGFR inhibitors, immunogenicity monitoring is ongoing, especially since cetuximab (a chimeric antibody) has higher immunogenic potential than panitumumab (a fully human antibody). Efficacy outcomes across RCTs and meta-analyses have consistently shown noninferiority of biosimilars. In CRC, biosimilars replicated response rates and OS benefits of innovator biologics when combined with chemotherapy. In gastric cancer, early-phase trials suggest equivalence, though data remain limited. Checkpoint inhibitor biosimilars are undergoing trials to establish efficacy in PD-L1-positive gastric carcinoma and MSI-H CRC, but long-term survival outcomes will be key in demonstrating similarity, given the unique durability of ICI responses.

3.5.3 Real-world data on biosimilars in gastrointestinal cancers

While RCTs provide robust efficacy and safety data, real-world evidence (RWE) is critical in confirming biosimilar performance across diverse populations and practice settings. Observational studies, registry data, and postmarketing surveillance provide insight into long-term outcomes, cost-effectiveness, and adoption trends. In CRC, real-world studies of bevacizumab biosimilars have demonstrated comparable effectiveness to reference Avastin in terms of OS and PFS, even in heterogeneous patient populations with multiple comorbidities. Data from European cancer registries indicate that the uptake of bevacizumab biosimilars has been associated with substantial cost savings, leading to broader patient access without compromising outcomes. For instance, a study in Germany reported that biosimilar use allowed treatment of more patients within existing budgets, effectively improving equity of care. Immunogenicity surveillance in real-world settings has reaffirmed the low incidence of ADAs, consistent with trial findings. Pharmacovigilance programs, mandated by regulatory bodies such as the EMA and FDA, continue to monitor rare or long-term AEs. Importantly, no new safety concerns have been identified since biosimilars entered clinical use. In gastric cancer, real-world data remain scarce due to the limited adoption of biologics. However, early reports suggest that bevacizumab biosimilars are being used in select cases, particularly in Western healthcare systems. As more checkpoint inhibitor biosimilars enter clinical practice, RWE will be vital to assess their true cost-effectiveness and patient quality-of-life outcomes, which are difficult to capture in controlled trials. From a health economics perspective, RWE strongly supports the value proposition of biosimilars. Cost savings from bevacizumab biosimilar adoption in the EU and U.S. are estimated to reach billions of dollars over the next decade. These savings can be reinvested in innovative therapies or healthcare infrastructure, underscoring the broader societal impact of biosimilar integration (Feng et al., 2022).

3.6 Adverse events and safety profile

The introduction of biologics and their biosimilar counterparts in gastric and CRC has significantly improved survival outcomes, yet their use is not devoid of risks. Unlike conventional chemotherapy, which causes predictable cytotoxicities, biologics exhibit distinct AE patterns linked to their mechanisms of action. Understanding these profiles is essential for safe clinical integration. While biosimilars undergo rigorous comparability assessments to ensure no clinically meaningful differences from their reference products, concerns about immunogenicity and long-term safety persist. Pharmacovigilance systems, including postmarketing surveillance and registries, therefore play a central role in monitoring their safety. This section reviews the common AEs associated with biologics, immunogenicity concerns specific to biosimilars, and pharmacovigilance strategies for ensuring safe clinical use.

3.6.1 Common adverse events of biologics

Biologics used in gastric and CRC primarily monoclonal antibodies targeting VEGF (e.g., bevacizumab, ramucirumab, aflibercept), EGFR (cetuximab, panitumumab), and immune checkpoints (nivolumab, pembrolizumab) exhibit distinct AE profiles based on their targets.

Infusion-related reactions (IRRs): These are among the most frequent AEs of biologics, occurring during or shortly after intravenous administration. Symptoms range from mild fever, chills, flushing, and dyspnea to severe anaphylaxis. Cetuximab is particularly associated with high rates of IRRs, especially in patients with pre-existing IgE antibodies against galactose-α-1,3-galactose, whereas panitumumab—being fully humanized—causes fewer reactions. Premedication with antihistamines and corticosteroids mitigates risk (Chow et al., 2022).

Hematologic toxicity: VEGF inhibitors are associated with hematologic side effects such as thrombocytopenia, neutropenia, and anemia, often exacerbated when combined with chemotherapy. While generally manageable, these toxicities necessitate regular blood count monitoring (Santorsola et al., 2024).

Gastrointestinal toxicities: Both VEGF and EGFR inhibitors may cause GI complications. Bevacizumab, for example, carries a rare but serious risk of gastrointestinal perforation, particularly in CRC patients with prior bowel surgeries. Diarrhea, nausea, and mucositis may occur, particularly with EGFR inhibitors. ICIs are associated with immune-mediated colitis, which can present with severe, life-threatening diarrhea if unrecognized (Adebayo et al., 2025).

Dermatological toxicities: EGFR inhibitors commonly cause acneiform rash, xerosis, pruritus, and paronychia. Interestingly, rash severity has been correlated with improved clinical response, making it both a therapeutic challenge and potential biomarker. ICIs may cause immune-mediated dermatitis. VEGF inhibitors occasionally induce delayed wound healing and skin ulcerations (Recuero et al., 2023).

3.6.2 Immunogenicity concerns with biosimilars

Immunogenicity refers to the ability of a biologic or biosimilar to elicit an unwanted immune response, typically through ADAs. While all biologics carry inherent immunogenicity risks due to their protein-based nature, biosimilars are scrutinized particularly closely to ensure their immunogenic potential does not exceed that of the reference product. The immunogenicity of biosimilars can be influenced by differences in manufacturing processes, posttranslational modifications, and impurities. Patient-related factors such as genetic background, immune status, and prior exposure to biologics also play important roles. ADAs can neutralize drug activity, reducing efficacy, or accelerate drug clearance, leading to subtherapeutic exposure. In some cases, ADAs may cross-react with endogenous proteins, posing additional risks. For instance, immunogenicity-related infusion reactions are a concern with cetuximab. Extensive comparative trials of bevacizumab, trastuzumab, and rituximab biosimilars demonstrate that ADA incidence is rare and comparable to reference biologics.

Importantly, ADA formation has not been associated with reduced efficacy or increased toxicity in these trials. Regulatory agencies require detailed immunogenicity assessments, including assays for ADA detection, neutralizing antibody activity, and clinical correlates of efficacy and safety. In gastric and CRC, where biologics are often combined with chemotherapy, immunogenicity risk is lower compared to supportive-care biologics (e.g., erythropoietin, filgrastim) used in immunocompetent settings. Nonetheless, continuous monitoring is warranted, particularly with EGFR inhibitors, which are known for higher immunogenic potential (Jarab et al., 2025).

3.6.3 Pharmacovigilance strategies

Given the complexity of biologics and biosimilars, pharmacovigilance is essential to ensure their safe integration into oncology practice. Regulatory agencies mandate robust risk management and post-marketing surveillance programs to detect rare, delayed, or population-specific adverse events. Biosimilars undergo continuous monitoring once approved. This includes mandatory periodic safety update reports (PSURs) submitted by manufacturers and spontaneous AE reporting systems maintained by agencies like the EMA, FDA, and DCGI. Real-world registries and observational studies complement RCTs by capturing data across diverse patient groups. Clear product labeling and pharmacovigilance coding ensure that any AE can be traced back to the specific product and batch. This is critical for biosimilars, given the existence of multiple versions of the same reference biologic in the market. Healthcare providers play a central role in pharmacovigilance by documenting AEs in clinical practice and reporting them through regulatory portals. Integration of electronic health records (EHRs) with pharmacovigilance systems further enhances the detection of AE patterns. WHO's pharmacovigilance program harmonizes global reporting, especially in low- and middle-income countries (LMICs). International cooperation is crucial for identifying rare safety signals, given the smaller individual patient populations exposed to specific biosimilars in certain regions. Successful pharmacovigilance depends on active participation from physicians, pharmacists, and patients. Education programs help dispel misconceptions about biosimilar safety and encourage reporting of potential AEs (Guerzoni et al., 2022).

3.7 Economic and accessibility impact

Biologics have revolutionized the management of gastric and CRC, but their high cost has posed substantial barriers to healthcare sustainability and equitable patient access. Biosimilars, as cost-effective alternatives, have emerged as a critical solution to mitigate financial toxicity while maintaining therapeutic efficacy and safety. Their adoption has potential not only to reduce drug expenditures but also to reallocate healthcare resources to other areas of need. The economic implications of biosimilars extend beyond direct cost savings, influencing treatment accessibility, patient adherence, and overall outcomes, especially in resource-limited settings. This section explores the cost-effectiveness of biosimilars in gastric and CRC, their potential contribution to healthcare savings, and their role in improving patient access globally, with emphasis on LMICs (Iqbal & Sadaf, 2022).

3.7.1 Cost-effectiveness of biosimilars in gastric and colorectal carcinoma

Cost-effectiveness analyses (CEA) evaluate whether the clinical benefits of a therapy justify its expense, typically expressed as cost per quality-adjusted life-year (QALY) gained. In the case of biologics for gastric and CRC, CEAs often highlight the high incremental cost-effectiveness ratios (ICERs), sometimes exceeding willingness-to-pay thresholds in many countries. Biosimilars have altered this equation by offering equivalent clinical efficacy at significantly reduced costs. For instance, bevacizumab biosimilars (Mvasi, Zirabev, Vegzelma, Avzivi) have consistently demonstrated therapeutic equivalence to Avastin in metastatic CRC. Health economic models across Europe and Asia show that the availability of bevacizumab biosimilars reduces the ICER substantially, making anti-VEGF therapy cost-effective in settings where the reference product was previously unaffordable. Similarly, ongoing evaluations of EGFR biosimilars (cetuximab and panitumumab) suggest that their introduction could shift cost-effectiveness in favor of widespread use for RAS wild-type CRC, especially when combined with chemotherapy. In gastric carcinoma, biologics like trastuzumab (HER2-positive cases) and ramucirumab have been associated with high costs and limited access. The anticipated availability of biosimilars for these agents could significantly improve cost-effectiveness ratios. For example, modeling studies predict that trastuzumab biosimilars could reduce costs by up to 50%, expanding treatment availability in both high- and middle-income countries. Ultimately, cost-effectiveness analyses consistently indicate that biosimilars allow healthcare systems to provide guideline-recommended biologic therapies to a larger proportion of patients at sustainable costs, improving both survival outcomes and health equity (Ball et al., 2023).

3.7.2 Potential for healthcare savings

One of the most compelling arguments for biosimilar adoption is their potential to generate significant healthcare savings. Biologics constitute a major proportion of oncology drug expenditures, often exceeding billions annually. The introduction of biosimilars creates market competition that drives down prices, benefiting both public and private healthcare systems. In the United States, projections estimate that widespread adoption of biosimilars could save up to $100–250 billion between 2020 and 2030 across oncology and immunology. In Europe, where biosimilar penetration is higher, bevacizumab and trastuzumab biosimilars have already generated substantial savings, with some reports indicating cost reductions of 20%–40% within the first few years of market entry. These savings not only free resources for reinvestment in innovative therapies but also enable broader patient coverage under existing healthcare budgets. For gastric and CRC, savings are particularly relevant. Bevacizumab biosimilars reduce treatment costs in metastatic CRC, allowing more patients to be treated without compromising outcomes. Trastuzumab biosimilars, used in HER2-positive gastric cancer, have demonstrated cost reductions of nearly 50% in some healthcare systems, dramatically lowering the financial barrier for first-line therapy. Furthermore, the ripple effect of biosimilar adoption extends beyond direct drug costs. Expanded treatment capacity leads to earlier and broader access to biologic therapies, potentially reducing downstream costs associated with disease progression, hospitalizations, and supportive care. Cost savings also facilitate clinical trial expansion and investment in newer treatment modalities such as immunotherapies and precision medicine approaches (Feng et al., 2024).

3.7.3 Patient access and affordability in low- and middle-income countries

Access to biologic therapies remains profoundly unequal, with patients in LMICs facing significant barriers due to high treatment costs, limited insurance coverage, and infrastructural constraints. In many LMICs, biologics are either unavailable or accessible to only a small fraction of patients who can afford out-of-pocket expenses. Consequently, a large proportion of patients with advanced gastric or CRC are deprived of therapies that could extend survival or improve quality of life. Biosimilars have the potential to bridge this access gap. Lower manufacturing costs and competitive pricing enable biosimilars to be offered at substantially reduced prices, making biologics affordable in regions where reference products are prohibitively expensive. India, for instance, has been a pioneer in biosimilar development, approving indigenous versions of trastuzumab and bevacizumab, thereby improving accessibility at a fraction of the cost. Similarly, in Latin America, local production and regional regulatory frameworks have supported the adoption of biosimilars for CRC and gastric cancer patients. However, challenges remain. Regulatory heterogeneity across LMICs may delay approvals. Limited pharmacovigilance infrastructure can hinder confidence in biosimilar use. Physician and patient awareness is often low, leading to hesitancy in prescribing biosimilars. Despite these barriers, successful examples of biosimilar adoption in oncology suggest that these obstacles can be overcome through targeted education, strong regulatory support, and international collaboration (Sariahmed et al., 2022). In addition, affordability at the patient level is critical. Biosimilars not only lower systemic healthcare costs but also reduce out-of-pocket expenditures, alleviating financial toxicity and improving adherence. Wider adoption in LMICs could reduce mortality disparities between high-income and low-resource countries by democratizing access to effective biologic therapies.

3.8 Ongoing research and future prospects

The development of biosimilars in gastric and CRC has already improved treatment affordability and accessibility, yet their full potential is far from realized. Current research efforts extend beyond simple replication of reference biologics, exploring novel paradigms such as precision oncology, rational drug combinations, global harmonization of approvals, and incorporation of patient-centered outcomes. These advances aim not only to maximize clinical efficacy but also to ensure long-term sustainability of cancer care worldwide. This section highlights the key areas of ongoing research and future prospects: personalized biosimilars, combination strategies, global trends in biosimilar approvals and uptake, and patient-reported outcomes (PROs) (Horgan et al., 2025).

3.8.1 Personalized biosimilars

The concept of personalized biosimilars is emerging from the intersection of precision oncology and biologic innovation. While current biosimilars are designed to replicate the reference biologic in all approved indications, future research may involve tailoring biosimilars to specific patient subgroups or molecular signatures that define gastric and CRC heterogeneity (Al-Doughaim et al., 2024). In CRC, KRAS/NRAS mutation status dictates responsiveness to EGFR inhibitors like cetuximab and panitumumab. While existing biosimilars of these drugs will be restricted to RAS wild-type tumors, research is exploring whether engineered biosimilars could provide differential benefits in subgroups with atypical mutations or rare polymorphisms. Similarly, in gastric cancer, HER2 status defines eligibility for trastuzumab therapy. The next generation of trastuzumab biosimilars may be optimized for specific HER2-positive subgroups or combined with companion diagnostics to enhance precision (Gonzalez et al., 2024). Advances in biomarker-driven treatment also support the notion of customized biosimilar deployment. MSI-H and dMMR subsets of CRC and gastric cancers have shown responsiveness to immune checkpoint inhibitors. In the future, PD-1/PD-L1 biosimilars may be developed with enhanced affinity or optimized formulations for patients harboring distinct immune phenotypes. Moreover, the integration of pharmacogenomics and real-world biomarker data could guide biosimilar usage, ensuring that patients most likely to benefit receive cost-effective biologics without unnecessary exposure. While the field is still conceptual, personalized biosimilars represent an exciting frontier that may redefine biosimilar research from replication to innovation.

3.8.2 Combination of biosimilars

Biosimilars are not restricted to monotherapy use; instead, they are increasingly being studied as part of combination regimens, replicating the strategies employed by their reference biologics. In metastatic CRC, bevacizumab biosimilars are being evaluated alongside standard chemotherapy regimens such as FOLFOX and FOLFIRI, demonstrating similar synergy and clinical outcomes. These findings mirror those of Avastin, confirming that biosimilars retain efficacy in multimodal regimens. Combination with other targeted agents is also gaining traction. EGFR inhibitors (cetuximab, panitumumab) biosimilars are under investigation in combination with VEGF inhibitors to explore potential synergistic antitumor effects. Trastuzumab biosimilars, already adopted in breast cancer, are being studied in HER2-positive gastric carcinoma with chemotherapy backbones, validating their role in combination regimens. A particularly promising area is the integration of biosimilars with immunotherapy. Preclinical and early-phase trials are investigating bevacizumab biosimilars in combination with checkpoint inhibitors like nivolumab or pembrolizumab, aiming to enhance immune infiltration into tumors by modulating angiogenesis. Similarly, HER2 biosimilars combined with PD-1 inhibitors may potentiate antitumor immunity in HER2-positive gastric carcinoma (Subramanian et al., 2024).

3.8.3 Global trends in approvals and uptake

Global uptake of biosimilars has been uneven, reflecting differences in regulatory frameworks, physician confidence, payer policies, and patient awareness. In Europe, where the EMA established the first biosimilar pathway (2005), adoption has been widespread, with bevacizumab and trastuzumab biosimilars achieving substantial market penetration. Competitive pricing has led to 20%–40% reductions in treatment costs, improving access across multiple oncology indications. In the United States, uptake has been slower due to market dynamics, rebate structures favoring reference biologics, and hesitancy around interchangeability. However, recent FDA approvals of bevacizumab biosimilars (Mvasi, Zirabev, Vegzelma, Avzivi) have accelerated adoption, with projected multi-billion-dollar savings over the next decade. Emerging markets show diverse trends. India has been a pioneer, with early approvals of trastuzumab and bevacizumab biosimilars under DCGI guidelines, providing affordable alternatives for gastric and CRC patients. Similarly, China and South Korea have rapidly growing biosimilar industries, contributing both to domestic access and global supply. WHO has played a pivotal role in guiding LMICs toward harmonized regulatory frameworks, though challenges in pharmacovigilance and traceability persist. Future global trends point to greater harmonization of regulatory standards, increased competition among biosimilar manufacturers, and accelerated adoption driven by payer incentives. As immunotherapy biosimilars enter the market, uptake patterns will likely mirror those of VEGF and EGFR biosimilars, albeit with added scrutiny due to immunogenicity concerns (Jarab et al., 2024).

3.8.4 Patient-reported outcomes

Beyond efficacy and safety, the future of biosimilars in gastric and CRC will be shaped by their impact on patient-reported outcomes (PROs). PROs include measures such as quality of life (QoL), treatment satisfaction, symptom burden, and financial well-being. Since biosimilars aim to deliver equivalent clinical outcomes at lower costs, their potential to improve patient-centered care is significant. Several studies have already demonstrated that patients treated with biosimilars report comparable QoL scores to those receiving reference biologics, particularly in domains like physical well-being, symptom management, and emotional support. Importantly, reduced out-of-pocket costs with biosimilars contribute to higher treatment satisfaction, as financial toxicity is a major source of distress in cancer care. In LMICs, where biologics are often unaffordable, biosimilar introduction may be the only pathway for patients to access advanced therapies. Thus, PROs are expected to show substantial improvement in these regions, not only in health-related QoL but also in psychosocial reassurance, knowing that effective therapies are within reach. Future research will increasingly integrate PRO endpoints into biosimilar clinical trials, aligning with the global movement toward patient-centered oncology. Additionally, real-world registries are expected to capture longitudinal PRO data, reinforcing the role of biosimilars in improving not just survival but also the lived experience of patients (Harris, 2023).

3.9 Critical appraisal: Opportunities versus challenges

The integration of biosimilars into the therapeutic landscape of gastric and CRC has ushered in a new era of affordability and accessibility in oncology care. While clinical evidence consistently supports their equivalence to reference biologics, biosimilars remain the subject of debate, owing to lingering concerns over safety, interchangeability, and long-term outcomes. This discussion critically examines the opportunities and challenges in biosimilar adoption, the barriers hindering their widespread use, and their transformative potential in reshaping therapeutic paradigms in GI cancers (Patel et al., 2024). The advent of biosimilars presents several opportunities for oncology. Chief among them is economic relief. Biologics constitute a large share of cancer treatment costs, and biosimilars priced 20%–50% lower enable broader access while alleviating healthcare budgets. For CRC, bevacizumab biosimilars already demonstrate cost savings without compromising clinical outcomes, while trastuzumab biosimilars extend HER2-targeted therapy to more gastric cancer patients. Another opportunity lies in global health equity, as biosimilars facilitate access in LMICs where reference biologics were often inaccessible. Biosimilars also offer an opportunity for innovation. Although initially designed as replicas, they can now be integrated into novel treatment combinations, investigated in personalized oncology, and serve as platforms for research in pharmacoeconomic and real-world outcomes. Their introduction stimulates market competition, driving down costs of reference biologics and fostering innovation across oncology. Despite these advantages, challenges remain. Scientific concerns include immunogenicity, pharmacovigilance gaps, and uncertainties about the extrapolation of indications. Regulatory differences between regions create inconsistencies in approvals, leading to unequal global adoption. Economic challenges include resistance from innovator companies and complex pricing negotiations that limit the cost advantage of biosimilars in some markets. Clinically, prescriber hesitancy persists due to limited long-term data, while patient acceptance is influenced by misconceptions that biosimilars are "inferior" or "cheap versions" of biologics.

3.9.1 Barriers to adoption: Physician awareness, patient acceptance, regulatory issues

Despite robust evidence of equivalence, the adoption of biosimilars in gastric and CRC therapy faces notable barriers. Physician awareness and confidence are among the most significant. Many oncologists remain cautious in prescribing biosimilars, particularly in curative or frontline settings, due to concerns over subtle structural variations and their impact on efficacy or safety. Surveys across Europe and the U.S. indicate that only 50%–70% of oncologists feel fully confident in initiating biosimilars for new patients, while even fewer are comfortable switching stable patients from reference biologics to biosimilars. Patient acceptance is another hurdle. Misconceptions that biosimilars are less effective or associated with greater side effects can lead to resistance. Informed consent discussions often require additional time and education to reassure patients of their equivalence. Furthermore, treatment satisfaction may be affected by financial perceptions, while lower costs may relieve financial stress, some patients perceive affordability as a marker of inferior quality. Regulatory issues exacerbate adoption gaps. While the EMA and FDA have developed rigorous approval frameworks, many LMICs lack robust regulatory infrastructure, leading to inconsistent quality standards. This raises concerns about "non-comparable biologics" or "intended copies" that do not meet stringent biosimilarity criteria, undermining trust. Additionally, the absence of universal policies on interchangeability, the ability to switch between biosimilars and reference biologics without additional risk, creates confusion among prescribers and payers. Economic factors also act as indirect barriers. In markets where rebate structures favor reference biologics, biosimilars struggle to gain traction despite lower list prices. Without clear payer incentives and hospital procurement policies, their adoption may remain limited. Overcoming these barriers requires multifaceted strategies: targeted education for physicians, awareness campaigns for patients, harmonization of regulatory frameworks, and policies that incentivize biosimilar use without compromising patient safety (Khelwatty et al., 2025; Subramanian et al., 2024).

3.9.2 Role in transforming therapeutic paradigms in gastrointestinal cancers

Biosimilars are poised to transform therapeutic paradigms in gastric and CRC, moving from cost-saving alternatives to integral components of cancer care strategies. Their availability ensures that biologics, once restricted to wealthier nations or select patient populations, can now be integrated into standard-of-care regimens globally. For example, bevacizumab biosimilars allow metastatic CRC patients in LMICs to receive anti-angiogenic therapy as part of first-line treatment, narrowing disparities in survival outcomes. Beyond access, biosimilars will influence treatment decision-making. As cost barriers diminish, oncologists may consider prolonged maintenance therapy or earlier introduction of biologics, strategies that were previously constrained by expense. Trastuzumab biosimilars, for instance, may expand HER2-targeted treatment for gastric carcinoma in healthcare systems where cost previously limited use. In the research domain, biosimilars are fostering novel combinations and trial designs, including integration with chemotherapy, other targeted agents, and immunotherapy. The availability of biosimilars also supports pragmatic real-world studies that generate evidence reflective of diverse populations, complementing traditional clinical trials. From a policy perspective, biosimilars drive a shift toward value-based oncology care, where therapeutic decisions consider not only clinical efficacy but also economic sustainability. Their widespread adoption may catalyze new reimbursement models, enhance competition, and accelerate innovation in biologic drug development. Ultimately, biosimilars represent more than cost-saving measures; they embody a paradigm shift in oncology care, ensuring that effective biologic therapies are democratized, sustainable, and adaptable to evolving treatment landscapes in gastric and colorectal cancer (Rama & Sousa, 2025).

3.10 Conclusion

Biosimilars represent a transformative force in the treatment of gastric and CRC, offering a strategic solution to the dual challenges of escalating cancer burden and the unsustainable costs associated with biologic therapies. By demonstrating clinical equivalence in terms of safety, efficacy, and immunogenicity, biosimilars have established themselves as reliable therapeutic alternatives to reference biologics. Their adoption has already begun to redefine treatment paradigms, enabling wider patient access to biologic-based therapies that were previously limited to select populations due to financial constraints. The central pillars underpinning biosimilar integration are safety, efficacy, affordability, and regulatory trust. Extensive preclinical comparability studies, confirmatory clinical trials, and post-marketing surveillance have consistently affirmed that biosimilars mirror their reference products in therapeutic performance while maintaining acceptable safety and immunogenicity profiles. This evidence, supported by robust regulatory frameworks from the EMA, FDA, WHO, and DCGI, has bolstered confidence in their clinical utility. Affordability is perhaps the most compelling driver, as biosimilars lower treatment costs by 20%–50%, yielding substantial healthcare savings and improving access in both high-income and LMICs. However, trust in regulatory rigor and sustained pharmacovigilance remains critical to ensuring clinician and patient acceptance. Looking forward, the success of biosimilars in gastric and CRC therapy will depend on continuous research, education, and stakeholder collaboration. Research must extend beyond replication to explore personalized biosimilars, novel combinations with chemotherapy and immunotherapy, and integration into precision oncology approaches. Educational initiatives targeting physicians, patients, and policymakers are essential to dispel misconceptions and enhance confidence in biosimilar prescribing. Collaborative efforts among regulators, healthcare providers, industry, and patient advocacy groups will be pivotal in harmonizing global standards, improving traceability, and expanding pharmacovigilance networks. In conclusion, biosimilars are more than cost-saving measures; they are catalysts for a new era of sustainable, equitable, and patient-centered cancer care. By ensuring affordability without compromising safety or efficacy, biosimilars empower healthcare systems to extend the benefits of biologic innovation to a broader population. With ongoing commitment to scientific rigor, regulatory transparency, and multidisciplinary collaboration, biosimilars will continue to transform the therapeutic landscape of gastric and colorectal carcinoma, ultimately narrowing global disparities and improving outcomes for patients worldwide.

References

Abdullah, Y. (2023). An overview of current biomarkers, the therapeutic implications, and the emerging role of hERG1 expression in gastric cancer: A literature review. *Cureus, 15*, e47501.

Adebayo, A. S., Adeleke, A., Adesina, S. K., Agbaje, K., & Olajubutu, O. (2025). Colorectal cancer: Therapeutic approaches and their complications. *Biomedicines, 13*(7), 1646.

Aldoughaim, M., Alsuhebany, N., Alzahrani, M., Alqahtani, T., Alghamdi, S., Badreldin, H., & Al Alshaykh, H. (2024). Cancer biomarkers and precision oncology: A review of recent trends and innovations. *Clinical Medicine Insights: Oncology, 18*, 11795549241298541.

Ball, G., Levine, M. A. H., Thabane, L., & Tarride, J.-E. (2023). Health technology reassessment: Addressing uncertainty in economic evaluations of oncology Drugs at time of reimbursement using long-term clinical trial data. *Current Oncology, 30*(7), 6596–6608. https://doi.org/10.3390/curroncol30070484.

Brogowska, K. K., Zajkowska, M., & Mroczko, B. (2023). Vascular endothelial growth factor ligands and receptors in breast cancer. *Journal of Clinical Medicine, 12*, 2412.

Candia, L., Cervantes, A., & Martinelli, E. (2025). Updated treatment recommendations for third and further lines of treatment in advanced colorectal cancer: From the ESMO Metastatic Colorectal Cancer Living Guideline. *Annals of Oncology, 36*, 342–344.

Chau, I. (2023). Pembrolizumab as a first-line treatment for advanced gastric cancer. *The Lancet Oncology, 24*(11), 1158–1159.

Cheng, W.-L., Feng, P.-H., Lee, K.-Y., Chen, K.-Y., Sun, W.-L., Van Hiep, N., Luo, C.-S., & Wu, S.-M. (2021). The role of EREG/EGFR pathway in tumor progression. *International Journal of Molecular Sciences, 22*, 12828.

Chow, T. G., Oppenheimer, J., & Joshi, S. R. (2022). A review of adverse reactions to biologics used in allergy-immunology practice. *The Journal of Allergy and Clinical Immunology: In Practice, 10*, 3075–3086.

Doleschal, B., Petzer, A., & Rumpold, H. (2022). Current concepts of anti-EGFR targeting in metastatic colorectal cancer. *Frontiers in Oncology, 12*, 1048166.

Elshiekh, C., Rudà, R., Cliff, E. R. S., Gany, F., & Budhu, J. A (2025). Financial challenges of being on long-term, high-cost medications. *Neuro-Oncology Practice, 12*, i49–i58.

Feng, K., Russo, M., Maini, L., Kesselheim, A. S., & Rome, B. N. (2024). Patient out-of-pocket costs for biologic drugs after biosimilar competition. *JAMA Health Forum, 5*, e235429.

Feng, X., Xie, H.-G., Malhotra, A., & Yang, C. F. (2022). *Biologics and biosimilars: Drug discovery and clinical applications* (p. 528). CRC Press.

Gao, Z., Wang, X., Song, T., Wu, S., & Jin, X. (2025). Real world effectiveness of chemotherapy plus bevacizumab with immunotherapy in colorectal cancer. *Scientific Reports, 15*(1), 29170.

Gonzalez, N. S., Marchese, P. V., Baraibar, I., Ros, J., Salva, F., Rodriguez, M., Salva, C., Vaghi, C., Alcaraz, A., & Garcia, A. (2024). Epidermal growth factor receptor antagonists in colorectal cancer: Emerging strategies for precision therapy. *Expert Opinion on Investigational Drugs, 33*, 613–625.

González, N. S., Ros, J., Baraibar, I., Salvà, F., Rodríguez-Castells, M., Alcaraz, A., García, A., Tabernero, J., & Élez, E. (2024). Cetuximab as a key partner in personalized targeted therapy for metastatic colorectal cancer. *Cancers, 16*, 412.

Guerzoni, S., Castro, F. L., Baraldi, C., Colella, G., & Pani, L. (2022). Pharmacovigilance of biological drugs. *Pharmacovigilance-Volume 2*. IntechOpen 10.5772/intechopen.105520.

Gunnam, S., Fathema, K., Nijhawan, M., Venna, R. S. A., & Aleti, R. (2025). A review on economic benefits and impact of biosimilars on healthcare system. *Journal of Pharma Insights and Research, 3*, 181–190.

Hansen, T. F., Qvortrup, C., & Pfeiffer, P. (2021). Angiogenesis inhibitors for colorectal cancer. A review of the clinical data. *Cancers, 13*, 1031.

Harris, C., Harris, R. J., Young, D., McDonnell, M., Clancy, B., Harvey, J., Araujo, C., Iria, I., Goncalves, J., Latter, S., & Cummings, J. F. (2023). Clinical outcomes and patient experience of biosimilar to biosimilar infliximab switching in patients with inflammatory bowel disease: A prospective, single-centre, phase IV interventional study with a nested qualitative study. *GastroHep, 2023*(1), 1248526.

Horgan, D., Tanner, M., Aggarwal, C., Thomas, D., Grover, S., Basel-Salmon, L., Dienstmann, R., Tan, T. J. Y., Park, W.-Y., & Abu Rasheed, H. M. (2025). Precision oncology: A global perspective on implementation and policy development. *JCO Global Oncology, 11*, e2400416.

Iqbal, Z., & Sadaf, S. (2022). Biosimilars: A comparative study of regulatory, safety and pharmacovigilance monograph in the developed and developing economics. *Journal of Pharmacy & Pharmaceutical Sciences, 25*, 149–182.

Iskit, A. B. (2025). Biosimilars and interchangeability: Regulatory, scientific, and global perspectives. *European Journal of Pharmaceutical Sciences, 213*, 107224.

Iwu, C. D., & Iwu-Jaja, C. J. (2023). Gastric cancer epidemiology: Current trend and future direction. *Hygiene, 3*, 256–268.

Jan, N., Sofi, S., Qayoom, H., Shabir, A., Haq, B. U., Macha, M. A., Almilaibary, A., & Mir, M. A. (2024). Metronomic chemotherapy and drug repurposing: A paradigm shift in oncology. *Heliyon, 10*, e24670.

Janani, B., Vijayakumar, M., Priya, K., Kim, J. H., Prabakaran, D., Shahid, M., Al-Ghamdi, S., Alsaidan, M., Othman Bahakim, N., & Hassan Abdelzaher, M. (2022). EGFR-based targeted therapy for colorectal cancer: Promises and challenges. *Vaccines, 10*, 499.

Jarab, A. S., Abu Heshmeh, S. R., & Al Meslamani, A. Z. (2024). Bridging the gap: The future of biosimilars regulations. *Human Vaccines & Immunotherapeutics, 20*, 2362450.

Jarab, A. S., Abu Heshmeh, S. R., & AL Meslamani, A. Z. (2025). Biosimilars and immunogenicity: A matter of concern? *Expert Opinion on Drug Safety, 24*, 519–527.

Kelly, R. J., Bever, K., Chao, J., Ciombor, K. K., Eng, C., Fakih, M., Goyal, L., Hubbard, J., Iyer, R., & Kemberling, H. T. (2023). Society for Immunotherapy of Cancer (SITC) clinical practice guideline on immunotherapy for the treatment of gastrointestinal cancer. *Journal for Immunotherapy of Cancer, 11*, e006658.

Kendong, S. M. A., Ali, R. A. R., Nawawi, K. N. M., Ahmad, H. F., & Mokhtar, N. M. (2021). Gut dysbiosis and intestinal barrier dysfunction: Potential explanation for early-onset colorectal cancer. *Frontiers in Cellular and Infection Microbiology, 11*, 744606.

Khelwatty, S. A., Puvanenthiran, S., Seddon, A. M., Bagwan, I., Essapen, S., & Modjtahedi, H. (2025). Advancements in targeted therapies for colorectal cancer: Overcoming challenges and exploring future directions. *Cancers, 17*, 2810.

Koustas, E., Trifylli, E.-M., Sarantis, P., Papadopoulos, N., Karapedi, E., Aloizos, G., Damaskos, C., Garmpis, N., Garmpi, A., & Papavassiliou, K. A. (2022). Immunotherapy as a therapeutic strategy for gastrointestinal cancer: Current treatment options and future perspectives. *International Journal of Molecular Sciences, 23*, 6664.

Królikowska, K., Kurman, N., Błaszczak, K., Ławicki, S., Gudowska-Sawczuk, M., & Zajkowska, M. (2025). The significance of neuropilins in gastrointestinal cancers. *International Journal of Molecular Sciences, 26*, 4937.

Lau, D. K., Mencel, J., & Chau, I. (2022). Safety and efficacy review of aflibercept for the treatment of metastatic colorectal cancer. *Expert Opinion on Drug Safety, 21*, 589–597.

Malakar, S., Gontor, E. N., Dugbaye, M. Y., Shah, K., Sinha, S., Sutaoney, P., & Chauhan, N. S. (2024). Cancer treatment with biosimilar drugs: A review. *Cancer Innovation, 3*, e115.

Mascarenhas-Melo, F., Diaz, M., Gonçalves, M. B. S., Vieira, P., Bell, V., Viana, S., Nunes, S., Paiva-Santos, A. C., & Veiga, F. (2024). An overview of biosimilars—development, quality, regulatory issues, and management in healthcare. *Pharmaceuticals, 17*, 235.

Monga, A., Gagan, Jamwal, P., Sharma, S., & Kaur, A. (2025). Biosimilars: A critical review of development, regulatory landscape, and clinical implications. *AAPS PharmSciTech, 26*(1), 46.

Nakamura, Y., Kawazoe, A., Lordick, F., Janjigian, Y. Y., & Shitara, K. (2021). Biomarker-targeted therapies for advanced-stage gastric and gastro-oesophageal junction cancers: An emerging paradigm. *Nature Reviews Clinical Oncology, 18*, 473–487.

Narita, Y., & Muro, K. (2023). Updated immunotherapy for gastric cancer. *Journal of Clinical Medicine, 12*, 2636.

Ngo, D., & Chen, J. (2021). A clinical review of biosimilars approved in oncology. *Annals of Pharmacotherapy, 55*, 362–377.

Niazi, S. K. (2022). The coming of age of biosimilars: A personal perspective. *Biologics, 2*, 107–127.

Patel, M., Nandpal, M., Parekh, N., Gunvantbhai Patel, S., & Patel, A. (2024). Biosimilars in colorectal cancer. In *Biosimilars for cancer treatment: A promising approach* (p. 300). Springer.

Pinheiro, M., Moreira, D. N., & Ghidini, M. (2024). Colon and rectal cancer: An emergent public health problem. *World Journal of Gastroenterology, 30*, 644.

Rahalkar, H. (2021). *[Doctoral Thesis.* University of Hertfordshire].

Rama, N. J., & Sousa, I. (2025). Bridging the gap: The role of technological advances in shaping gastrointestinal oncological outcomes. *World Journal of Gastrointestinal Oncology, 17*, 101752.

Raposo, M. C., Feiteira, C., & Ribeiro, M. H. (2025). Regulatory and clinical aspects in biosimilar medicines: Comparability, extrapolation, interchangeability, and safety. *Drugs & Therapy Perspectives, 41*, 1–15.

Recuero, J. K., Fitz, J. R., Pereira, A. A., & Bonamigo, R. R. (2023). EGFR inhibitors: Clinical aspects, risk factors and biomarkers for acneiform eruptions and other mucosal and cutaneous adverse effects. *Anais Brasileiros de Dermatologia, 98*, 429–439.

Riedesser, J. E., Ebert, M. P., & Betge, J. (2022). Precision medicine for metastatic colorectal cancer in clinical practice. *Therapeutic Advances in Medical Oncology, 14*, 17588359211072703.

Roth, M. T., & Das, S. (2021). Pembrolizumab in unresectable or metastatic MSI-high colorectal cancer: Safety and efficacy. *Expert Review of Anticancer Therapy, 21*, 229–238.

Salas-Benito, D., Pérez-Gracia, J. L., Ponz-Sarvisé, M., Rodriguez-Ruiz, M. E., Martínez-Forero, I., Castañón, E., LóPEZ-Picazo, J. M., Sanmamed, M. F., & Melero, I. (2021). Paradigms on immunotherapy combinations with chemotherapy. *Cancer Discovery, 11*, 1353–1367.

Santorsola, M., Capuozzo, M., Nasti, G., Sabbatino, F., Di Mauro, A., Di Mauro, G., Vanni, G., Maiolino, P., Correra, M., & Granata, V. (2024). Exploring the spectrum of VEGF inhibitors' toxicities from systemic to intra-vitreal usage in medical practice. *Cancers, 16*, 350.

Sariahmed, K., Kurian, J., Singh, A. K., Leyton, C., Minuti, A., Jerschow, E., Arora, S., & Jariwala, S. P. (2022). Social, political, and economic determinants of access to biologics: A scoping review of structural determinants in the clinical disparities literature. *Research in Social and Administrative Pharmacy, 18*, 4038–4047.

Singh, S., Singh, N., Baranwal, M., Sharma, S., Devi, S. K., & Kumar, S. (2023). Understanding immune checkpoints and PD-1/PD-L1-mediated immune resistance towards tumour immunotherapy. *3 Biotech, 13*, 411.

Subramanian, A., Muthukumarasamy, G. D., Venkataraman, S., Rangasamy, K., Kanade, A. N., Arulrajan, P. V., Saravanan, K., Shanmuganathan, S., Gopal, G., & Ambi, S. V. (2024). Future scope of biosimilars for the treatment of various cancers. In *Biosimilars for cancer treatment: A promising approach* (pp. 287–309). Springer.

Wong, M. C., Huang, J., Chan, P. S., Choi, P., Lao, X. Q., Chan, S. M., Teoh, A., & Liang, P. (2021). Global incidence and mortality of gastric cancer, 1980–2018. *JAMA Network Open, 4*, e2118457.

Xu, X., Zhang, S., Xu, T., Zhan, M., Chen, C., & Zhang, C. (2022). Efficacy and safety of bevacizumab biosimilars compared with reference biologics in advanced non-small cell lung cancer or metastatic colorectal cancer patients: A network meta-analysis. *Frontiers in Pharmacology, 13*, 880090. https://doi.org/10.3389/fphar.2022.880090.

Zhang, X., Zhou, L., Zhou, C., & Shen, L. (2024). Real-world effectiveness and safety of ramucirumab as a second-line treatment for patients with unresectable advanced or metastatic gastric/gastroesophageal junction adenocarcinoma in Japan and South Korea: A systematic literature review. *Advances in Therapy, 41*, 2112–2132.

CHAPTER

Biosimilars and biologics in cervical cancer

4

Vidyasrilekha Sanapalli[a], Shrestha Palit[b], Simhadri Goud Gundrathi[b], Ashwini Deshpande[c], Krishna Swaroop Akey[d], Prajakta Patil[e] and Bharat Kumar Reddy Sanapalli[f]

[a]*Department of Pharmaceutical Chemistry, School of Pharmacy and Technology Management, SVKM's Narsee Monjee Institute of Management Studies (NMIMS) (Deemed to be University), Hyderabad, Telangana, India,* [b]*School of Pharmacy and Technology Management, SVKM's Narsee Monjee Institute of Management Studies (NMIMS) (Deemed to be University), Hyderabad, Telangana, India,* [c]*Department of Pharmaceutics, School of Pharmacy and Technology Management, SVKM's Narsee Monjee Institute of Management Studies (NMIMS) (Deemed to be University), Hyderabad, Telangana, India,* [d]*Department of Pharmaceutical Chemistry, Shri Vile Parle Kelavani Mandal's (SVKM's) College of Pharmacy, Shirpur, Maharashtra, India,* [e]*Department of Pharmacology, School of Pharmacy and Technology Management, SVKM's Narsee Monjee Institute of Management Studies (NMIMS) (Deemed to be University), Shirpur, Maharashtra, India,* [f]*Department of Pharmacology, School of Pharmacy and Technology Management, SVKM's Narsee Monjee Institute of Management Studies (NMIMS) (Deemed to be University), Hyderabad, Telangana, India*

4.1 Introduction

One of the most widespread cancerous diseases that occurs in women globally is cervical cancer, which has been a major cause of cancer morbidity and mortality, especially in low and middle-income countries (Arbyn et al., 2020). Even with the improved screening, vaccination, and treatment, cervical cancer remains a serious burden on the population. Human papillomavirus (HPV) is the primary cause of the disease, and the disease occurs mainly when the person has been infected with the high-risk types of HPV, specifically HPV-16 and HPV-18, which cause about 70% of all cases (Okunade, 2020). These oncolytic viral strains cause malignant transformation through integration in the host genome and upheaval of key regulatory pathways in cell cycle control, apoptotic, and genomic stability (Chianese et al., 2021). Cervical cancer is a multi-stage disease that occurs as a result of HPV infection, which leads to cervical intraepithelial neoplasia that eventually advances to invasive carcinoma (Kabir & Ribeiro, 2025). Through this development, viral

Biosimilars and Biologics in Cancer Treatment: An Era in Therapeutic Strategies.
DOI: https://doi.org/10.1016/B978-0-443-43824-0.00015-8

oncoproteins E6 and E7 induce degradation in tumor suppressor proteins like p53 and retinoblastoma (pRb), leading to uncontrolled cellular division and antiapoptotic effects (Manocha et al., 2021). Angiogenesis, immune evasion, and tumor microenvironmental changes also contribute to tumor progression, along with viral oncogenesis (Dzobo, 2021). Modern biologic therapies have therapeutic targets of interest provided by these biological processes. The conventional treatment of cervical cancer consists of surgery, radiotherapy, and cytotoxic chemotherapy, which are still effective in early-stage cancer (Guimarães et al., 2022). Yet, recurring, persistent, or metastatic cervical cancer patients do not have many treatment options of treatment and tend to die. Traditional chemotherapy is not very specific and has been noted to be very toxic, thus making it important to develop more effective and targeted methods of therapy (Kaur et al., 2023). The progress of molecular biology and immunology has led to the creation of biologics that inhibit tumor-specific pathways selectively and enhance clinical response with a decreased level of systemic toxicity (Croft et al., 2024). The therapeutics of cervical cancer have been changed by the introduction of biologics such as monoclonal antibodies, immune checkpoint inhibitors, and antibody-drug conjugates (ADCs) (Wang et al., 2022). These agents have the effect of inhibiting angiogenesis, reinstating antitumor immune responses, or delivering cytotoxic agents to tumor cells (Huinen et al., 2021). The first biologic to be demonstrated to increase the survival of patients with cervical cancer at advanced stages was bevacizumab, a monoclonal antibody against vascular endothelial growth factor (VEGF) (Bizzarri et al., 2016). In more recent times, the use of immune checkpoint inhibitors, including pembrolizumab and cemiplimab, has proven to be very effective by increasing the efficacy of immune-mediated tumor destruction, especially in tumors that express programmed death-ligand 1 (PD-L1) (Sezer et al., 2021). Also, antibody drug conjugates, like tisotumab vedotin, are a new type of targeted therapy, a combination of the specificity of antibodies and the cytotoxic effect of chemotherapy (De Bono et al., 2019). Although these therapies are now available, their high cost is still a significant barrier to large-scale usage, particularly in those areas where the occurrence of cervical cancer is most prevalent (Broutet et al., 2022). Biosimilars are biologic products that are very closely related to an approved reference biologic; they have come into the picture as a strategy to enhance affordability and accessibility (Farhat et al., 2018). These agents are subjected to strict comparability tests with the aim of achieving the same efficacy, safety, and immunogenicity. The cost reduction of healthcare and the increased availability of treatment as a result of the introduction of biosimilars, especially bevacizumab and supportive biologics, can be realized (Yang et al., 2021).

The chapter will give an in-depth description of the use of biologics and biosimilars in the treatment of cervical cancer, their mechanisms of action, use, regulation, and future use as therapeutic agents. These advances need to be understood to enhance the treatment outcomes and provide fair access to the use of modern cancer therapies.

4.2 Molecular pathogenesis of cervical cancer and therapeutic targets

The pathogenesis of cervical cancer develops in a multistage process, under the influence of which a primary role is played by persistent infection with the oncogenic types of HPV, leading to the development of genetic, epigenetic, and microenvironmental changes (Pavelescu et al., 2025). These molecular processes interfere with cell cycle regulation, facilitate immune evasion, and facilitate tumor growth, angiogenesis, and metastasis. The elucidation of these processes has facilitated the discovery of certain molecular targets of biologic drugs, specifically, immune checkpoint inhibitors and antiangiogenic agents (Li & Li, 2025).

4.2.1 Role of high-risk human papillomavirus in carcinogenesis

HPVs are associated with carcinogenesis in humans, but no specific virus causes this disease in humans. The key etiologic role in cervical carcinogenesis is the persistent HPV infection with high-risk HPV types, particularly HPV-16 and HPV-18. HPV enters the basal cells of the epithelium of the cervical transformation zone, which is especially susceptible to the virus because of the active cell turnover (Bonello & Blundell, 2016). The HPV infection is short-lived and disappears in the majority of people. Integration of the viral DNA into the host genome is, however, a vital process in malignant transformation, but is only possible through persistent infection. HPV has oncogenic capabilities whose major mediating role is played by viral oncoproteins E6 and E7 (Tomaić, 2016). E6 protein enhances the degradation of tumor suppressor protein, p53, via ubiquitin-mediated proteolysis. Apoptosis, DNA repair ability, and genetic mutation accumulation are compromised in scenarios where the p53 function is lost. At the same time, the E7 protein attaches to the retinoblastoma protein (pRb), which is a major influence on the G1S cell cycle checkpoint. When pRb is inactivated, E2F transcription factors are liberated, leading to unregulated cell cycle and cell proliferation (Giacinti & Giordano, 2006). The result of these molecular upheavals is genomic instability, chromosomal abnormalities, and progressive remodeling of normal cervical epithelium into cervical intraepithelial neoplasia (CIN) that could ultimately develop into invasive carcinoma (Volkova et al., 2021). The continuous existence of HPV oncoproteins also aids in resisting apoptosis and increasing the survival of tumor cells, and so, HPV-related pathways are significant therapeutic topics of interest.

4.2.2 Cell cycle and tumor suppressor pathways dysregulation

Besides a direct viral impact, host cell regulatory pathways change during cervical cancer progression. Deactivation of tumor suppressors, activation of oncogenes, and cell cycle checkpoints are some of the factors that facilitate the promotion of malignant transformation (Molinari, 2000). The p53 pathway is extremely important in ensuring genomic stability, which results in repairing or causing apoptosis of the cellular damage (Zhivotovsky & Kroemer, 2004). P53 is degraded through HPV, thus inhibiting genetically damaged cells from undergoing apoptosis, leading to their further growth. Also, pRb inhibition ablates essential regulation of cell cycle advancement. All these are further stimulated by overexpression of cyclins, cyclin D, and cyclin E, which further stimulate the cell cycle, leading to unregulated proliferation. There are other signaling pathways that are usually dysregulated in cervical cancer, and they include the phosphatidylinositol-3-kinase (PI3K)/AKT/mTOR pathway that controls cell survival, growth, and metabolism (Zhang et al., 2015). The stimulation of this pathway increases the proliferation and apoptotic resistance of tumor cells. These changes in the molecular structure offer possible targets for new biologic and targeted therapy.

4.2.3 Angiogenesis and VEGF signaling pathway

Angiogenesis, or the process of developing new blood vessels out of the existing ones, is a prerequisite for the further development of the tumor beyond a certain size or the further dissemination of the tumor in metastases (Bielenberg & Zetter, 2015). Angiogenic factors are more expressed in cervical tumors, especially VEGF. VEGF promotes the proliferation, migration, and development of endothelial cells in order to maintain sufficient oxygen and nutrient supply to the tumor (Carmeliet, 2005). The hypoxia in the tumor microenvironment also enhances the expression of VEGF by the activation of hypoxia-inducible factor-1 alpha (HIF-1α). The high level of VEGF is linked to high levels of metastatic potential, high disease stage, and low prognosis in patients with cervical cancer (Shi & Fang, 2004). The VEGF signaling pathway has emerged as a therapeutic target. VEGF-neutralizing monoclonal antibodies can prevent tumor angiogenesis, tumor vascularization, and disease progression. Angiogenesis is an alternative approach that has also shown clinical usefulness, especially in recurrent and advanced cervical cancer (Chan et al., 2025).

4.2.4 Immune evasion and PD-1/PD-L1 immune checkpoint axis

The immune system is very important in the recognition and destruction of HPV-infected and transformed cells (Stanley, 2015). But cervical cancer cells devise means to avoid immunological surveillance, enabling further tumor growth. The programmed death-1 (PD-1) receptor and PD-L1 ligand are among the mechanisms of the first most significant mechanisms (Guan et al., 2017). PD-L1 is commonly overexpressed on both cervical tumor cells and tumor-associated immune cells. PdL1 binding to PD-1 receptors on activated T lymphocytes suppresses the growth of T cells, production of cytokines, and cytotoxic effects (Saunders et al., 2005). This contact inhibits immune functions to tumor cells and maintains immune tolerance. Sustained PD-L1 expression and immune suppression in the tumor microenvironment are associated with chronic HPV infection. The PD-1/PD-L1 blockade improves

T-cell activities, boosts the destruction of tumors by the immune system, and lies at the center of immune checkpoint inhibitor therapy in cervical cancer (Ge et al., 2022).

4.2.5 Cancer microenvironment and immune modulation

The microenvironment of cervical tumors comprises tumor cells, immune cells, stromal cells, cytokines, and extracellular matrix factors (Peltanova et al., 2019). This microclimate is very important in the development and therapeutic responses of tumors. They involve tumor-associated macrophages, regulatory T cells, and immunosuppressive cytokines like transforming growth factor- beta (TGF-β) and interleukin 10, which play a role in suppressing the immune system (Mirlekar, 2022).

Persistent HPV infection causes chronic inflammation that enhances cell proliferation, angiogenesis, and DNA damage (Georgescu et al., 2018). Inflammatory mediators increase the growth of tumors and promote metastasis. These inflammatory and immune elements are significant biologic therapy targets in an attempt to repurify immune functions (Fernandes et al., 2015).

4.2.6 Epidermal growth factor receptor and related signaling pathways

Epidermal growth factor receptor (EGFR) has been overexpressed in a large percentage of cervical cancer and has been linked to tumor aggressiveness and adverse prognosis. The stimulation of EGFR activates downstream signaling pathways, such as PI3K/AKT, mitogen-activated protein kinase (MAPK), which stimulate cell proliferation, survival, and apoptotic resistance. Other receptor-mediated pathways promote tumor progression, such as HER family receptors, angiogenic signaling pathways, among others. Even though the targeted therapies of such receptors are still under investigation, they can be considered as possible future treatment modes (Tian et al., 2016).

4.2.7 Therapeutic implications of molecular targets

The discovery of the molecular pathways that take part in the progression of cervical cancer has made it possible to generate targeted biologic therapies. Some of the important therapeutic targets are VEGF, which is an angiogenic target, and immune checkpoints, like PD-1/PD-L1 which are immune targets used to restore antitumor immunity. Another target of ADCs has been tissue factor, which is overexpressed in cervical cancer cells. With these molecular tools, it is possible to specifically suppress tumor growth without causing as much harm to normal tissues. This molecularly directed method has achieved much in terms of treatment in both progressive and recurring cervical cancer and has propagated the creation of new biologic and biosimilar medications (Manzo-Merino et al., 2014).

4.3 Biologics: Definition, characteristics, and classification

Biologics are drugs that are made in the body, using improved biotechnology, and are based on a living organism or a biological system. They are monoclonal antibodies, recombinant proteins, and immune-modulating molecules, which are directed to highly specific molecular targets of disease progression. However, as opposed to traditional small-molecule drugs, which generally affect the general processes in cells, biologics are tailored to cause specific exposure of tumor-related pathways, enhancing the prediction of treatment and minimizing systemic toxicity. Angiogenesis, immune checkpoint pathways, and tumor-associated antigens are the main targets of biologics in cervical cancer. The discovery of these molecular targets and specifically oncogenesis and immunological evasion of HPV mediated by these molecular targets has facilitated the development of biologic therapies that enhance recurrent and metastatic disease outcomes (Willmott et al., 2010).

4.3.1 Structural and functional characteristics of biologics

Biologics are unlike the conventional chemotherapeutic agents with respect to their structure, production, and mode of action. They are generally huge, multifaceted proteins with hundreds or thousands of amino acids organized in accurate 3-dimensional folds. The binding specificity of their highly specific structure to a target molecule, e.g., growth factors, receptors, or immune checkpoint proteins, is possible. Biologics are made through recombinant DNA technology in living host cells; that is, in mammalian, bacterial, or yeast. Gene insertion, expression of the protein, purification, and much quality control are carried out in the manufacturing process to achieve biological activity and safety. Since biologics are vulnerable to environmental factors, the stability of the products requires storage and handling under controlled conditions. The difference between pharmacokinetics between biologics and small molecules lies in the fact that they are typically delivered by parenteral route, most frequently by infusion through the IV or by injection into the subcutaneous space. Their high molecular weight does not allow them to be absorbed by the mouth, and penetrates some biological barriers. The long half-lives of biologics are usually because they are not easily cleared by the kidneys and degraded (Greenman et al., 2024).

4.3.2 Mechanism of action of biologics in cervical cancer

Biologics have a high specificity in their therapeutic mechanisms because they are dependent on molecular targets that mediate tumor growth and immune regulation. These have inhibition of angiogenesis, immune responses, and specific delivery of cytotoxic. Antiangiogenic biologic agents are specific for the VEGF or its receptors; these inhibit the development of new blood vessels needed in tumor growth and metastasis. This decreases the tumor vascularization, resulting in a decrease in the oxygen and nutrient supply and finally tumor regression or inhibition of tumor growth. Cytokine blockers that involve the immune checkpoints are used against immunosuppressive pathways. These biologic agents rejuvenate T-cell activation and increase tumor cell destruction mediated by the immune system by blocking inhibitory receptors, including programmed death-1 (PD-1) and PD-L1. The mechanism becomes quite topical in cervical cancer because of the viral nature of the illness and its immunogenic properties. Another significant type of biologics is the ADCs. These agents consist of a cytotoxic drug and a monoclonal antibody, with selective delivery of chemotherapy to tumor cells that express particular surface antigens. This specific method enhances the treatment efficacy with reduced system toxicity (Vanajothi et al., 2022).

4.3.3 Benefits of biologics over traditional chemotherapy

Biologics offer a number of benefits over traditional cytotoxic chemotherapy. Their specificity of action decreases the harm to normal tissues, which leads to better tolerability and decreases nonspecific side effects like severe myelosuppression and gastrointestinal toxicity. Biologics are also therapeutic agents that offer patients with recurrent or metastatic cervical cancer a limited response to chemotherapy. The other significant benefit is that they can control the immune system (Cohen et al., 2020). The immune checkpoint inhibitors have the potential to offer lasting effects by providing permanent immune monitoring of the tumor cells. This leads to long-term disease containment in some patients despite the fact that the therapy is withdrawn. Biologics are also capable of being used to complement chemotherapy or radiotherapy in order to make treatment more effective. Such combination therapies enhance response rates and survival of patients with advanced cervical cancer (Jung et al., 2023).

4.3.4 Limitations of biologics

Biologics also have a number of limitations, even though they have therapeutic advantages. Their manufacturing is complicated and expensive, and treatment costs are very high, thus making them inaccessible, especially in low-resource environments where cervical cancers are greatest. Their use is also complicated because of the necessity of parenteral administration and special storage conditions (Colonna, 2025). Immune activation or inhibition of normal physiological pathways is also a unique adverse effect that can be produced by biologics. As an example, antiangiogenic medication can lead to hypertension and thrombo-embolic events, and immune checkpoint medication can lead to immune-related adverse events affecting other body parts, the thyroid, liver, lungs, and skin. Moreover, biologic therapy does not affect all patients. The variability in treatment effects is attributed to tumor heterogeneity, changes in the expression of biomarkers, and the variation in immune response. The discovery of predictive biomarkers is also a field of research (Rider et al., 2016).

4.3.5 The classification of biologics utilized in cervical cancer

The Biologics applied in cervical cancer may be broadly divided according to their mechanism of action:

- Antiangiogenic monoclonal antibodies: These are drugs that inhibit VEGF signaling and inhibit tumor angiogenesis.
- Immune checkpoint inhibitors: These biologics promote the antitumor immunity by inhibiting PD-1 and PD-L1, among other inhibitory immune pathways.
- Antibody drug conjugates: These are agents in which the cytotoxic drug is directly delivered to the tumor cells that express certain antigens.
- Investigational therapeutic vaccines and immunotherapies: These therapies are targeting stimulating immune responses to HPV-associated tumor antigens.

Each of the classes has a specific biological process, which takes place in the development of cervical cancer, which provides several therapeutic options to manage the disease (Sharma et al., 2020).

4.3.6 Biologics in the management of cervical cancer in contemporary times

Biologics have dramatically changed the treatment level of advanced cervical cancer. There is a significant benefit in using antiangiogenic agents in combination with chemotherapy, and immune checkpoint inhibitors in selected patients have the potential to offer lasting effects. ADCs provide targeted cytotoxic therapy for treatment-resistant disease patients. Biologics have become part of treatment recommendations when dealing with recurrent, persistent, and metastatic cervical cancer, especially among patients who do not qualify to undergo curative surgery or radiotherapy. Current studies are being done in an effort to identify new biologic targets, combination strategies, and biomarkers in an attempt to enhance therapeutic outcomes (Nandi et al., 2025) (Table 4.1).

4.4 Antiangiogenic biologics in cervical cancer

Angiogenesis plays a very important role in the growth and development of cervical cancer. Tumors have a need of sufficient supply of blood to supply oxygen and nutrients and to eliminate wastes. Hypoxic conditions arise in the tumor microenvironment as cervical tumors grow in response to the production of pro-angiogenic factors. Primary among them, there is the VEGF, the most significant of the angiogenic regulators of tumors. High levels of VEGF expression have been closely linked with high stages of the disease, lymph node metastasis, and adverse prognosis in cervical cancer. It is now a viable therapeutic approach to target the VEGF signaling pathway,

Table 4.1 Antiangiogenic biologics and their biosimilars relevant to cervical cancer.

Reference biologic	Target	Mechanism of action	Clinical Use in cervical cancer	Examples of biosimilars	Key adverse effects	References
Bevacizumab	VEGF-A	Binds VEGF-A and prevents activation of VEGFR-1 and VEGFR-2, inhibiting angiogenesis and tumor vascularization	Approved in combination with chemotherapy for recurrent, persistent, or metastatic cervical cancer	Bevacizumab-awwb, Bevacizumab-bvzr, Bevacizumab-bvzl, Bevacizumab-adcd	Hypertension, proteinuria, bleeding, thromboembolism, impaired wound healing	Quintanilha et al. (2022), Kazazi-Hyseni et al. (2010)
Investigational anti-VEGF antibodies	VEGF or VEGFR	Block angiogenic signaling pathways to prevent tumor blood vessel formation	Under clinical investigation in advanced cervical cancer	Biosimilars under development	Similar to bevacizumab, including vascular and renal effects	Hsu and Wakelee (2009), Li et al. (2020)

especially in recurrent and metastatic disease (Tomao et al., 2014). Monoclonal antibodies or recombinant proteins, which inhibit angiogenesis by neutralization of VEGF or by blocking interaction with endothelial cell receptors, are called antiangiogenic biologics. These agents inhibit the formation of new blood vessels, limiting the growth of tumors, lowering their likelihood of metastasis, and increasing the activity of chemotherapy (Luvero et al., 2017).

4.4.1 VEGF signaling pathway in cervical cancer

VEGF is a glycoprotein secreted to induce the proliferation, migration, and survival of endothelial cells by binding to its receptors, namely VEGFR-1 and VEGFR-2, present on vascular endothelial cells. Stimulation of these receptors initiates intracellular cascades of signaling, such as PI3K/AKT and MAPK pathways, which induce angiogenesis and vascular permeability (Chen et al., 2014). Hypoxia in cervical tumors leads to a rise in the hypoxia-inducible factor-1 alpha (HIF-1α) expression that upregulates the production of VEGF. Increased VEGF is also caused by persistent HPV infection and inflammatory cytokines. The resultant abnormal tumor vasculature aids in tumor growth and the spread of cancer cells to other body organs. Due to the key role of VEGF in the vascularization of tumors, VEGF signaling inhibition is an effective way of disrupting tumor growth and progression (Chen et al., 2023).

4.4.2 Bevacizumab: Mechanism of action, clinical use, pharmacological characteristics, and limitations

Bevacizumab is a humanized monoclonal antibody of recombinant nature that specifically attaches to VEGF-A and inhibits the binding of VEGF-A to the VEGF receptor on the endothelial cells. Bevacizumab prevents endothelial cell activation and inhibits the new blood vessel formation by neutralizing the circulating VEGF (Fig. 4.1). This suppression of the vascular density of tumors, the decrease in the oxygen and nutrient supply, and the constrained tumor growth. Besides reducing angiogenesis, bevacizumab is able to normalize dysfunctional tumor vasculature, enhancing the delivery and efficacy of similarly administered chemotherapeutic drugs (Garcia et al., 2020). The overall effect of this is the enhancement of treatment outcomes in advanced cervical cancer. Bevacizumab is indicated to be used with the platinum-based chemotherapy in patients with recurrent, persistent, or metastatic cervical cancer. Bevacizumab, when combined with chemotherapy, is significantly beneficial in terms of overall survival and progression-free survival as compared to chemotherapy (Bizzarri et al., 2016). Antiangiogenic therapy is also of great use in progressive disease with minimal curative treatment. It is usually given intravenously at specific dosages at fixed time intervals with chemotherapeutic drugs like cisplatin and paclitaxel or carboplatin-based regimens. Bevacizumab is not mostly applied to the treatment of early cervical cancer, but it has a significant role in the treatment of advanced and resistant cases. The route of administration of Bevacizumab is

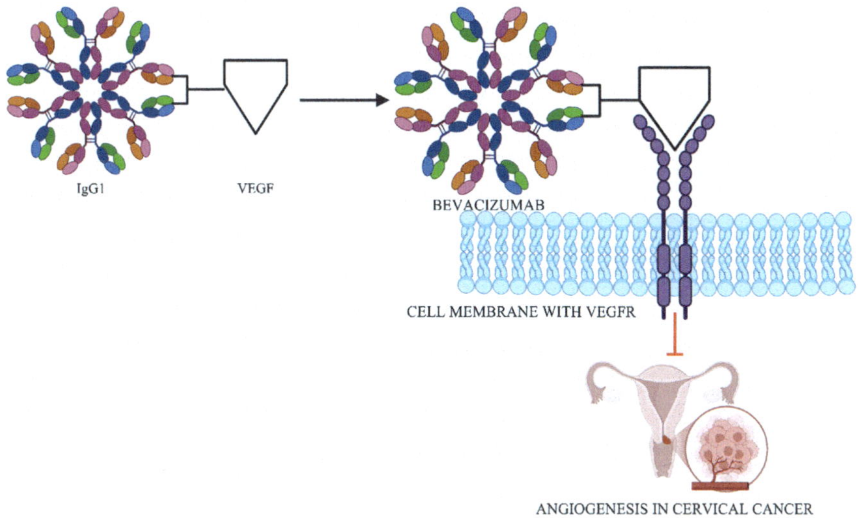

FIGURE 4.1

Mechanism of action of bevacizumab in arresting angiogenesis of cervical cancer.

intravenous, as it has a high molecular weight that does not allow it to be absorbed through the mouth. Its half-life is long and thus can be administered after every 2–3 weeks. Proteolytic degradation of the drug to small peptides and amino acids is the main method of eliminating the drug. Because of its mechanism of action, bevacizumab influences the physiological processes associated with the formation of blood vessels, which is the subject of its side effects as well as its medicinal activity (Keating, 2014). Even though bevacizumab does not have severe side effects, its antiangiogenic effect may have certain side effects related to vascular function. Hypertension is the most frequent side effect that is caused by the decreased production of nitric oxide and the heightened resistance of the vessels. Proteinuria can be caused by injury to the endothelial cells of glomeruli. Bevacizumab patients should have their blood pressure and renal function monitored. Bleeding complications, delayed wound healing, and thromboembolic events may be the other possible negative events. Serious complications are rare, such as gastrointestinal perforation and fistula, which occur especially in patients who have undergone radiotherapy. To reduce these risks, it is necessary to carefully select patients and monitor them. The use of bevacizumab is a significant breakthrough in the biologic therapy of cervical cancer. Bevacizumab also targets a major tumor progression mechanism of angiogenesis (Shord et al., 2009). It has been clinically proven that bevacizumab is superior to chemotherapy as an adjunct to the treatment of recurrent and metastatic disease. Another avenue through which biosimilars can be developed is based on antiangiogenic therapy, as this approach assists in increasing accessibility to treatment, since it is less expensive and equally effective and safe.

4.4.3 New antiangiogenic therapies

Other antiangiogenic agents are also being tested to treat cervical cancer, besides bevacizumab. These are monoclonal antibodies against VEGF receptors and multi-target inhibitors against angiogenic signaling pathways. There are also efforts underway to use combination approaches with antiangiogenic biologics and immune checkpoint inhibitors, as angiogenesis inhibition can increase immune response in the tumor microenvironment. The further investigation of angiogenic processes and targeted therapy is anticipated to enhance the treatment outcomes and increase the possible treatment options of patients with advanced cervical cancer (Lopes-Coelho et al., 2021).

4.5 Immune checkpoint inhibitors

Immune checkpoint inhibitors have turned out to be one of the biggest breakthroughs in biologic therapy of cervical cancer, and that is augmenting the immune system of the host in opposition to tumor cells (Xie et al., 2022) (Table 4.2). Cervical cancer is one cancer for which immunotherapy is very applicable in the sense that it is a persistent infection of oncogenic HPV that expresses viral antigens, which the immune system can recognize. Tumor cells, however, devise ways in which they can reduce immune activity, the main ones being immune checkpoint pathways, which block the actions of T cells (Han et al., 2022) (Fig. 4.2).

The PD-1 receptor on activated T lymphocytes and the PD-L1 receptor on tumor cells and immune cells are one of the most vital immune regulatory approaches. The binding of PD-L1 to PD-1 inhibits T-cell stimulation, lowers cytokine synthesis, and stops cytotoxic functions. This enables the tumor cells to avoid being destroyed by the immune and they keep growing. PD-L1 has been noted to be overexpressed in a significant percentage of cervical cancer and is also linked to the disease evolution and unfavorable clinical results (Minion & Tewari, 2018). The immune checkpoint inhibitors are monoclonal antibodies that inhibit the antagonistic receptors like PD-1, thus reinstating the activity of T cells and facilitating the destruction of tumor cells by the immune system. These biologics induce self-directed elimination of malignant cells by harnessing the innate immune response, unlike cytotoxic chemotherapy, which destroys malignant cells by directly attacking rapidly dividing cells. This strategy is capable of generating lasting treatment effects since activated immune cells have the capacity to maintain the capacity to identify and destroy tumor cells with time (De Felice et al., 2018). The first immune checkpoint inhibitor that was broadly used in treating cervical cancer was pembrolizumab. It is applied in the case of recurrent, persistent, or metastatic disease in patients, especially when the tumor expresses PD-L1 or has a high mutational burden. Clinical trials have demonstrated that pembrolizumab has the potential to induce long-lasting tumor effects and enhance disease control in patients who have advanced through chemotherapy. It is also being applied with chemotherapy to help improve therapeutic effects (Colombo et al., 2021).

Table 4.2 Immune checkpoint inhibitors approved or investigated for cervical cancer.

Drug	Target	Year of key approval/clinical adoption	Major clinical setting	Typical use strategy	References
Pembrolizumab	PD-1	2018 (accelerated approval; later expanded indications)	Recurrent or metastatic cervical cancer, especially PD-L1-positive tumors	Monotherapy or combined with chemotherapy	Borcoman and Tourneau (2020)
Nivolumab	PD-1	Investigational/clinical trial use	Advanced or treatment-resistant cervical cancer	Primarily evaluated in clinical trials and combination regimens	Santin et al. (2020)
Cemiplimab	PD-1	2021 approval for cervical cancer	Recurrent cervical cancer after platinum-based chemotherapy	Monotherapy as second-line immunotherapy	Tewari et al. (2022)

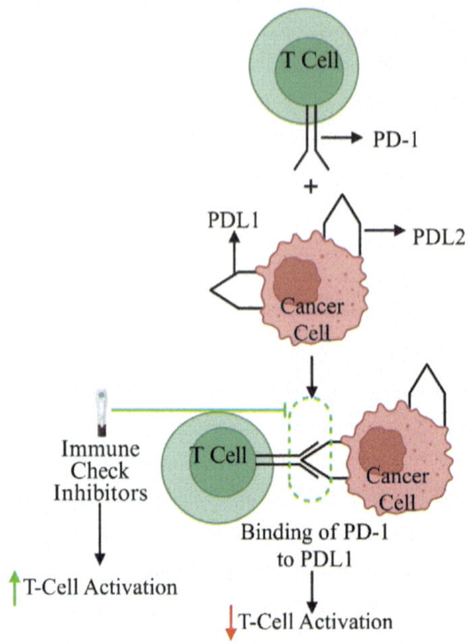

FIGURE 4.2

Mechanism of action of cancer cells in inactivation of T-cell and role of immune check inhibitors in activation of T-cell.

Another PD-1-inhibitor that has shown clinical activity in advanced cervical cancer is nivolumab. Even though its application is still lower in comparison to pembrolizumab, clinical trials have demonstrated meaningful and long-term responses in selected patients. Continued studies are measuring its efficacy when used together with other biologics and chemotherapy regimens (Huang et al., 2022). Cemiplimab has become a significant intervention to be used in patients with recurrent cervical cancer who develop disease progression after using platinum-based chemotherapy. Cemiplimab enhances overall patient survival by reinstating immune functions as opposed to conventional chemotherapy. Its approval has increased the facilities of immunotherapy available to patients with advanced disease (Wu & He, 2024). Compared to the usual chemotherapy, immune checkpoint inhibitors are tolerated better, although they can have immune-related side effects owing to augmented immune response. These reactions can potentially have an impact on endocrine organs, the liver, lungs, the gastrointestinal tract, and skin. It is necessary to detect and manage it early to avoid severe complications. Immune checkpoint inhibitors have integrated into the management of cervical cancer and have largely enhanced the quality of treatment, especially in advanced and treatment-resistant cervical cancer. Future research is in the context of predictive biomarker identification, combination therapy optimization, and the role of immunotherapy in the preliminary cervical cancer.

4.6 Antibody-drug conjugates in cervical cancer

ADCs represent a new generation of biologic agents that target the delivery of cytotoxic agents to tumor cells with minimal systemic toxicity. The latter agents are a combination of the selectivity of monoclonal antibodies and the lethal cell-destroying power of chemotherapeutic agents. The antibody portion is attached to particular antigens on the surface of the tumor cells, and the attached cytotoxic entity is discharged into the interior of the target cell, leading to specific cell death of the tumor cells. This selective method enhances the therapeutic effect and prevents excessive harm of normal tissues in comparison with traditional chemotherapy (Zhang et al., 2025). It is common to find tumor-associated antigens in cervical cancer cells, which can be used as a targeted therapy. Tissue factor is one of the most significant targets of a transmembrane glycoprotein that plays a role in blood clotting and in cancer development. The tissue factor is the protein that is highly expressed in cervical cancer cells and relates to tumor growth, angiogenesis, invasion, and metastasis, and is an ideal target of the ADC-based therapy. Tisotumab vedotin is the first antibody-drug conjugate that was approved for managing recurring or metastatic cervical cancer. It is a monoclonal antibody that targets the tissue factor conjugated to the cytotoxic agent, monomethyl auristatin E (MMAE), a microtubule-disrupting drug. Once ADC binds to tissue factor on the tumor cell surface, it is then internalized by receptor-mediated endocytosis. The linker is then cleaved to release MMAE, which disrupts cell cycle assembly and causes cell death by arresting the cell cycle and triggering apoptosis within the cell. This is a targeted intracellular delivery method that increases antitumor efficacy and reduces systemic exposure to the cytotoxic agent (Markham, 2021). Tisotumab vedotin has proven to be of great clinical value in patients with advanced cervical cancer who have gone through the development of the disease after chemotherapy. Clinical trials have demonstrated significant levels of tumor response and long-term disease control in highly pretreated patients. This therapy offers a significant alternative to patients with few treatment options and is a significant step forward in targeted biologic therapy. Besides its direct cytotoxic action, the targeting of tissue factor can also disrupt tumor-related angiogenesis and can result in tumor metastasis. This twofold process will add to better treatment results. ADC therapy is, in general, a therapy that is given intravenously at regular intervals, enabling continuous therapeutic exposure but with manageable toxicity. The safety profile of tisotumab vedotin is different compared to the use of conventional chemotherapy because of its targeted mechanism. The most unpleasant side effects are fatigue, nausea, peripheral neuropathy, and eye-related complications, such as conjunctivitis and dry eye. Due to the low expression of tissue factor in some normal tissues, careful monitoring is needed to ensure that the off-target effects are minimized. During treatment, prophylactic eye care and periodic medical examination are advised. The emergence of ADCs is a significant milestone towards the precision of cervical cancer oncology. These agents offer selective cytotoxic therapy that is well targeted and has better selectivity and index of therapy. Current studies aim at the discovery of more tumor-specific antigens, linker stability, and the use of next-generation ADCs with better efficacy and safety. The role of ADCs in the treatment of advanced cervical cancer may become even more significant in those cases when the patient fails to respond to chemotherapy or immunotherapy (Luu et al., 2023).

4.7 Therapeutic vaccines and emerging biologics in cervical cancer

Therapeutic vaccines are a new biologic therapy that aims at triggering the host immune system to identify and destroy already existing HPV-related cervical cancer cells. In contrast to prophylactic HPV vaccines, which prevent infection, therapeutic vaccines attack tumor cells that express viral oncoproteins, especially E6 and E7, which are expressed all the time in malignant cervical epithelial cells and are vital to cell survival (Table 4.3). These viral antigens are perfect immunologic targets since they are tumor-specific and not found in normal tissues.

Vaccines used as a treatment option work by increasing antigen presentation and the ability of cytotoxic cells of the CD8+ T cell to identify and kill tumor cells infected with HPV. Different vaccine platforms have been established, and they include: DNA vaccines, peptide-based vaccines, viral vector vaccines, and dendritic cell-based vaccines. HPV oncogenes are encoded in DNA vaccines and, upon introduction into host cells, result in antigenic production of proteins, which result in the production of cellular immunity. Vaccines that deliver antigenic fragments directly, which activate T cells, include peptide vaccines; whereas vaccines that deliver HPV antigen genes efficiently into the host cells are viral vectors. Dendritic cell vaccines entail the loading of antigen-presenting cells that have been obtained from the patient with HPV antigens ex vivo and the subsequent reinfusion of the patient with the loaded cells to induce a specific immune response (Fontenot et al., 2024). A number of therapeutic vaccines have shown promising levels of immunogenicity and clinical activity in pre-clinical trials, and have been shown to produce positive results in patients with earlier-stage lesions and advanced cervical cancer. These vaccines have demonstrated the capacity to induce long-term immunity and, in others, remission of the tumor or stabilization of the disease. Their activity seems to be increased in the case of the use of immune checkpoint inhibitors because the checkpoint inhibition prevents tumor-induced inhibition of the vaccine-activated T cells. Besides therapeutic vaccines, new biologics inhibiting new molecular and immune pathways are being explored. These are next-generation immune checkpoint blockers of other regulatory molecules, bispecific antibodies with the capability to bind tumor antigens (and immune cells) simultaneously, and engineered monoclonal antibodies with improved immune-activating capabilities. Antibody engineering has been enhanced to enhance specificity, binding affinity, and immune effector functions, and has been found to add to their potential as therapeutic agents. Emerging biologic approaches, such as allowing engineered T-cell therapies against HPV-associated antigens, are

Table 4.3 Therapeutic vaccines and emerging biologic strategies in cervical cancer.

Therapy type	Target antigen/pathway	Platform	Mechanism	References
DNA vaccines (e.g., HPV E6/E7 DNA vaccines)	HPV E6 and E7 oncoproteins	Plasmid DNA	Host cells express viral antigens, activating cytotoxic T-cell immune response	Wlazlo et al. (2004)
Peptide vaccines	HPV E6/E7 antigen peptides	Synthetic peptide fragments	Direct stimulation of antigen-specific T lymphocytes	Khan et al. (2024)
Viral vector vaccines	HPV oncogenes	Modified viral vectors (e.g., adenovirus, vaccinia virus)	Efficient antigen delivery and strong immune activation	Ji et al. (2024)
Dendritic cell vaccines	HPV tumor antigens	Patient-derived immune cells	Activated dendritic cells present tumor antigens to T cells, enhancing immune response	Silva et al. (2021)
Bispecific antibodies	Tumor antigen and immune cell receptor	Engineered monoclonal antibodies	Redirect immune cells to tumor cells for targeted killing	Parola and Colombo (2025)

also under investigation as adoptive cell therapies. These are methods that include altering the cells of the patient's immune system and making them have greater tumor-targeting capacity, and then they can be reintroduced back into the patient, leading to increased tumor cell recognition and destruction (Zheng et al., 2025). In general, therapeutic vaccines and emerging biologics provide effective approaches to the better treatment of cervical cancer through advancements in the immune response to specific tumors and disease-specific molecular mechanisms. Further investigation of the clinical research and technological developments is likely to increase their contribution to the treatment of cervical cancer.

4.8 Biosimilars: Definition and regulatory considerations

Biologics have revolutionized the treatment of cervical cancer and are rather costly, making them unaffordable in numerous healthcare systems. The solution to this dilemma came in the form of biosimilars that offered cheaper alternatives to the same without compromising on clinical performance. The biosimilars are biological products that are very close to an already approved reference biologic and do not have a clinically significant difference in safety, purity, and efficacy (Kirchhoff et al., 2017). Biologics are synthesized using living cells, which, unlike small-molecule drugs, lead to complex molecular structures that are impossible to synthesize in the exact form. Thus, biosimilars are not perfect copies but need to prove the high level of similarity to the reference product by large analytical and clinical assessments. Small variations in production procedures are also allowed as long as they do not impact clinical performance (Rathore & Bhargava, 2020). Biosimilars development is achieved in a step-wise process of comparability. Analytical research considers structural features, biological activity, and physicochemical properties. Functional tests verify target association and mechanism of action. Pharmacodynamics and pharmacokinetics studies make the absorption of drugs, their distribution, and their biological activity comparable. Equivalent efficacy, safety, and immunogenicity have also been proven by clinical trials. The regulatory bodies like the United States Food and Drug Administration, European Medicines Agency, and other global authorities have set up certain approval schemes for biosimilars. These routes give less focus on independent clinical benefit demonstration, as the reference biologic has already proven its effectiveness and safety (Gottlieb, 2008). One such regulatory notion is extrapolation of indications, where a biosimilar that is approved for one indication of the reference biologic can be approved for other indications, provided there is scientific justification. This opens the possibility of using the biosimilars of the antiangiogenic biologics like bevacizumab in the management of cervical cancer after the comparability is proven (Declerck et al., 2017). To guarantee safety and effective continued use, biosimilars are closely monitored in terms of quality and postmarketing. The immunogenicity test is vital, especially since the biologics can provoke immune reactions that influence the therapeutic effectiveness or safety. The emergence of biosimilars has brought about a great deal of accessibility to biologic treatments by lowering the cost of treatment and enhancing availability. This is particular to cervical cancer, whereby the disease burden is greatest in low- and middle-income countries. Biosimilars are essential in increasing the availability of increased access to biologic treatment with high clinical care (Halimi et al., 2020).

4.9 Biosimilars in cervical cancer therapy

The emergence of biosimilars has greatly increased access to biologic treatments in the field of oncology by lowering direct cost of treatment without affecting the efficacy and safety. The first biologic target, on which biosimilars are offered in cervical cancer, is the VEGF, which is blocked by the reference bevacizumab monoclonal antibody. Since bevacizumab is already acknowledged as an essential part of the treatment of recurrent, persistent, or metastatic cervical cancer, its biosimilars represent a valuable possibility to increase access to treatment, especially in healthcare systems with a limited number of resources (Morparia et al., 2024). The biosimilars of Bevacizumab are designed to be similar in terms of their structural, functional, and clinical features to the reference product. Analytical research has demonstrated that protein structure, receptor binding affinity and biological activity are similar. Similar distribution and elimination are shown by pharmacokinetic studies and similar efficacy and safety are shown by clinical trials in approved oncology indications. On this basis, regulatory bodies allow extrapolating indications on the basis of which these biosimilars will be used in cervical cancer once such similarity has been established in other malignancies (Sharma & An, 2024) (Table 4.4).

Bevacizumab biosimilars have the same mechanism of action as their reference biologic. These agents attach themselves to the circulating VEGF-A, which prevents the binding of VEGF receptors on endothelial cells and angiogenesis. This leads to decreased vascularization of tumors, retarded tumor growth, and increased efficiency of chemotherapy. VEGF inhibition continues to be a viable treatment option because angiogenesis is a key factor in cervical cancer development. Bevacizumab biosimilars have similar clinical advantages, such as, similar tumor response, progression-free survival, and overall survival, in combination with chemotherapy. They also have a similar safety profile to the reference product, and adverse effects are mostly due to the vascular inhibitor effect, such as hypertension and proteinuria. Close clinical observation is needed to ensure proper usage (Casak et al., 2018). The cost of biosimilars is one of the greatest benefits over reference biologics. Lower costs of treatment enhance access and enable wider use of biologic treatment in areas with less access to healthcare. This is more so in cervical cancer, which is disproportionately burdened in low- and middle-income countries disproportionately. The rising use of biosimilars has reinforced the sustainability of healthcare by minimizing the financial obligation, and treatment quality does not diminish. Biosimilars will continue to gain momentum as more of them are introduced, increasing their role in treatment regimens and enhancing access to effective biologic therapy across the globe.

Table 4.4 Approved bevacizumab biosimilars relevant to oncology and cervical cancer developed based on the reference biologic avastin.

Biosimilar name	Brand name	Manufacturer	Clinical relevance	References
Bevacizumab-awwb	Mvasi	Amgen	Used as a cost-effective alternative for VEGF inhibition in oncology, including cervical cancer	Jin et al. (2023)
Bevacizumab-bvzr	Zirabev	Pfizer	Comparable efficacy and safety; used in multiple solid tumors	Yang et al. (2022)
Bevacizumab-aybi	Aybintio	Samsung Bioepis	Approved based on biosimilarity studies demonstrating equivalent clinical performance	Park et al. (2020)
Bevacizumab-adcd	Vegzelma	Celltrion, Inc	Expands access to anti-angiogenic therapy, especially in developing countries	Park (2022)

4.10 Advantages of biologics and biosimilars in cervical cancer

Biologic therapies have also provided a more specific mode of treatment in cervical cancer to target a particular molecular and immunological pathway of tumor progression. Biologics, in contrast to conventional cytotoxic chemotherapy, which targets both malignant and normal rapidly dividing cells, bind specific molecular targets, including vascular growth factors, immune checkpoints, and tumor-associated antigens. This selectivity enhances the efficacy of therapy with minimal non-selective cell destruction. Consequently, biologics have proven helpful in achieving clinically significant gains in survival, especially in patients with recurrence, persistent, or metastatic disease where therapy choices are limited (Ahmad, 2025). Regulation of the tumor microenvironment is one of the greatest benefits of biologics. Antiangiogenic drugs interfere with the tumor vascular supply, depriving the tumor of nutrient and oxygen supply, and slow down tumor growth and metastasis. Immune checkpoint inhibitors also reverse tumor-related immunosuppression to reinstate antitumor immunity, where cytotoxic T cells attack and kill malignant cells. ADCs also improve therapeutic precision in that cytotoxic agents are only delivered to tumor-specific antigens, and systemic exposure is minimized enhancing the therapeutic index. Biologics also give a possibility of biomarker-informed therapy, which will support individualized treatment options. Biomarker expression of PD-L1, tissue factor, and angiogenic factors can be used in identifying the patients who are most likely to respond to targeted biologic interventions. This method enhances the efficiency of treatment, dissipates unnecessary toxicity, and leads to the optimization of clinical decisions. Although these benefits may be seen, high manufacturing costs and low supply have historically made the use of biologics widespread, especially in low- and middle-income areas where there is the greatest burden of cervical cancer. Biosimilars respond to this issue by providing many more similar therapeutic options with similar structural, functional, and clinical performance. The biosimilar introduction leads to lowering the cost of treatment, enhancing the sustainability of healthcare, and increasing the availability of life-saving biologic therapies without any negative effects on their efficacy and safety. Biosimilars also enhance the level of competition in the pharmaceutical market, in addition to enhancing affordability, which facilitates increased use of targeted therapies in day-to-day clinical practices. This increased access is highly significant to resource-constrained healthcare systems, where cost may often dictate the availability of treatments. Biosimilars also promote long-term treatment plans, which allow healthcare providers to implement biologic therapy earlier in the disease course when it is clinically suitable. On the whole, biologic drugs and biosimilars are a significant development in cervical cancer management as they enhance the target of treatment, increase the capacity of immunotherapy to combat tumors, and have given more people access to innovative treatment. They have been incorporated into mainstream treatment regimens and have led to better results, and are now part of the current treatment of cervical cancer (Shilpi et al., 2024).

4.11 Limitations and challenges of biologics and biosimilars in cervical cancer

Even though biologics and biosimilars provide substantial therapeutic benefits, many clinical, economic, and regulatory issues restrict their optimal application in the treatment of cervical cancer. The variability of patient response is one of the major limitations. Biologic therapies are not equally effective on all patients due to the fact that tumor heterogeneity, variations in immune status, and variations in the expression of molecular targets may affect the efficacy of treatment. In other cases, the tumors have intrinsic resistance and acquired resistance in the course of treatment with activation of other signaling pathways, immune evasion mechanisms, or tumor microenvironment adaptive changes. The other significant issue is the possibility of the negative effects related to immune and vascular modulation. Excessive activation of the immune response can result in immune-related adverse events in the thyroid, liver, lungs, skin, and gastrointestinal tract of the body caused by immune checkpoint inhibitors. The normal vascular functioning can be disrupted by antiangiogenic agents, resulting in hypertension, delayed wound healing, thromboembolic events, and proteinuria. Though these toxicities can be managed, they need close surveillance and possibly even a treatment break or termination in extreme situations. Economic reasons are also a significant obstacle, especially in low-resource environments where the occurrence rate of cervical cancer is the greatest. Biologics are produced using complicated cell-based production, highly rigorous purification, and in-depth quality control, leading to high treatment costs. Although biosimilars have led to a reduction in the total costs, the cost-effectiveness and reimbursement restrictions keep limiting their universal implementation in most healthcare systems. Facilities such as cold chain storage, specialized administration facilities, and trained healthcare personnel are also infrastructure requirements and contribute to accessibility issues. There are also regulatory and clinical issues in the acceptance of biosimilars. Even though biosimilars pass through stringent comparability tests, the issues raised by clinicians and patients about interchangeability, safety in the long run, and immunogenicity are likely to delay their uptake. It is important to use postmarketing surveillance to monitor actual safety and create confidence in the usage of biosimilars. Moreover, varying regulatory adequacies between nations can slow down approvals and supply. Inadequate predictive biomarkers that are universally reliable to select biologic therapy are also another constraint. Although the presence of certain biomarkers like PD-L1 gives some indication, their predictive value is not absolute, and there are patients who do not have biomarker expression but still can respond to treatment. More studies are required to come up with strong biomarkers that would effectively predict response to treatment and ideal patient selection (Subramanian et al., 2024). On the whole, the use of biologics and biosimilars has enhanced the survival rates in cervical cancer patients; however, issues regarding resistance, toxicity, cost, regulatory acceptance, and limitations of biomarkers have to be overcome to make the modalities the most useful in clinical practice and provide universal access to people.

4.12 Biomarkers and patient selection

Detection of predictive biomarkers can be a key to fully exploiting biologics and biosimilars in treating cervical cancer, as these therapeutic agents are best applied to specific patient groups. Biomarkers are useful in estimating the success of treatment, informing the clinician of the treatment decisions, and minimizing the exposure of patients to ineffective treatment of the disease. New developments in molecular diagnostics and immunohistochemistry have facilitated the assessment of tumor-specific and immune-related markers that have an effect on biologic therapy. Programmed death ligand-1 (PD-L1) expression is considered to be one of the most clinically important biomarkers. PD-L1 is expressed on tumor cells and on immune cells that invade the tumor, where it prevents T cell-mediated immunity. The elevation of PD-L1 is linked to immune evasion and has been used as a response predictor to immune checkpoint inhibitors. PD-L1-positive tumors are more likely to respond to PD-1 or PD-L1 inhibitors, and therefore, PD-L1 testing is a valuable part of treatment planning. Another emerging biomarker is tumor mutational burden (TMB), which is an indicator of the count of somatic mutations in tumor DNA fragments. High mutational burden tumors generate more neoantigens, which augment immune defense and raise immunotherapy responsiveness. Likewise, the microsatellite instability (MSI) is a defect in DNA mismatch repair machinery and has been linked to greater immunogenicity and to enhance the reaction to immune checkpoint blockade, though the MSI-high phenotype is less typical in cervical cancer than in other malignancies. Angiogenesis-associated biomarkers, including VEGF and microvessel density, can also be used to predict response to antiangiogenic therapies. High levels of VEGF correlate with tumor hyperactivity and enhanced responsiveness to VEGF-based biologic therapy. Also, tissue factor expression has been identified as an applicable biomarker in cervical cancer, especially in relation to ADCs against tumor-associated antigens. Molecular markers associated with HPV, DNA integration, and expression of viral oncoproteins E6 and E7 are also linked to tumor immunogenicity and could affect immunotherapy and therapeutic vaccines' responsiveness. Further investigations are underway on gene expression patterns, immune cell infiltration patterns, and circulating tumor DNA as possible biomarkers to further and more precisely narrow down biologic therapy choice. Personalized treatment approach has been facilitated by the introduction of biomarker testing into clinical practice, which helps clinicians to choose the most suitable biologic therapy according to the tumor biology. It can be anticipated that future studies in the area of molecular profiling and precision oncology will potentially enhance the prediction and treatment outcomes in cervical cancer (Song, 2022).

4.13 Combination strategies involving biologics and biosimilars in cervical cancer

One of the new strategies that has been found to achieve a greater clinical effect is combination therapy to add to the clinical benefit of biologics and biosimilars in cervical cancer. When specific biologic agents are combined with chemotherapy, radiotherapy, or other immunotherapies, there can be synergistic antitumor effects when the combination of the two agents targets multiple pathways that contribute to tumor growth, immune evasion, and metastasis. This combined method has enhanced control of tumors and could slow the occurrence of therapeutic resistance. Antiangiogenic biologics in combination with platinum-based chemotherapy have been shown to have a strong clinical benefit in advanced cervical cancer. Antiangiogenic factors regulate the abnormal tumor vascularity to normalize the blood vessels and enhance the delivery of drugs and oxygen to the tumor microenvironment. This increases the cytotoxic efficacy of chemotherapy and adds to better treatment response and survival. These combinations are currently part of the routine treatment of recurrent and metastatic disease. Immune checkpoint inhibitors have been demonstrated as well to be more effective in conjunction with chemotherapy. Cytotoxic drugs have the ability to elevate the release of tumor antigens and raise immune surveillance, which strengthens immunotherapy. The immune checkpoint inhibitor may also be more effective in activating the cytotoxic T cells when chemotherapy decreases the number of immunosuppressive cells in the tumor microenvironment. Another promising method is radiotherapy with the use of biologic agents. Radiation can cause the death of tumor cells and enhance the presentation of antigens, which may trigger immunogenicity. This effect can be enhanced by using immune checkpoint inhibitors, a mechanism called the abscopal effect. Antiangiogenic drugs can also enhance tumor oxygenation, making them radiosensitive and radiosensitive to radiotherapy. Moreover, various biologic combinations that address various pathways are also under study in order to beat resistance mechanisms and enhance the outcomes of therapy. Indicatively, the use of antiangiogenic drugs in combination with immune checkpoint inhibitors can increase immune cells that enter tumors by promoting the functional state of the vasculature and suppressing immunosuppressive cues. In such combinations, biosimilars may be added as a substitute to reference biologics and offer equal clinical advantage, enhancing the affordability of treatment. Combination therapy helps the use of biosimilars to enhance the effective use of advanced treatment in healthcare systems with limited financial means. Altogether, mixed approaches in using biologics and biosimilars have a better therapeutic effect through complementary mechanisms, immune responses, and a better drug delivery system, which leads to improved clinical outcomes in the management of cervical cancer (Morparia et al., 2024).

4.14 Future perspectives

It is projected that with the rapid new developments in tumor immunology, molecular biology, and biotechnology, the role of biologics and biosimilars in treating cervical cancer will continue to grow. Among other developments, there is the identification of new immune checkpoint targets other than the PD-1 and PD-L1 molecules, which include the CTLA-4, LAG-3, and TIGIT molecules. These other immune regulatory pathways can be targeted to enhance response rates and overcome resistance mechanisms raised by the current immunotherapies. Another promising approach is a therapeutic vaccine against antigens of HPV. In contrast to preventive vaccines, therapeutic vaccines are developed so that they provoke the formation of cytotoxic T-cell responses against HPV-infected and malignant cells that express viral oncoproteins. New developments in DNA vaccines, in peptide vaccine platforms, and viral vector platforms have demonstrated promising results in preclinical trials, especially in combination with immune checkpoint inhibitors. It is also being actively researched through ADCs and next-generation targeted biologics. These agents allow highly selective targeting of cytotoxic payloads or immune-modulating molecules to tumor-specific antigens and enhance therapeutic specificity and reduce systemic toxicity. The advances in linker technology and antibody engineering should lead to an increase in drug stability, tumor penetration, and clinical efficacy. Individualized medicine solutions utilizing genomic and proteomic studies will tend to become of growing significance in informing the choice of biologic therapy. Molecular characterization of tumors is able to identify targets to act on, is predictive of treatment responses, and aids in the planning of individual treatment. Noninvasive modalities, such as liquid biopsy methods, such as circulating tumor DNA, might be of use in assessing treatment response and resistance. The growth of biosimilar development will go a long way in enhancing global access to biologic therapies. With the expiry of patents of the major biologics, there will be more biosimilars and thus more competition that will lead to a lower cost of treatment. This is especially so in low and middle-income nations because affordability is a significant impediment to advanced cancer treatment in these nations (Kate, 2024). The ongoing progress in combination therapy, the better identification of biomarkers, and further innovative development of biologic drugs are supposed to contribute to the increased efficiency of the treatments and increase the range of available therapeutic options. This will lead to increased ease of management, customization, and efficiency in cervical cancer in the future.

4.15 Conclusion

The landscape of cervical cancer has seen a huge advancement in the form of biologics and biosimilars, offering both targeted and immune-mediated therapy system, which covers major pathways and mechanisms of tumor development. These agents, unlike traditional cytotoxic treatments, selectively disrupt the signaling pathways that include angiogenesis and immune checkpoint control, leading to better disease control and patient survival in patients with recurrent or metastatic cervical cancer. Upcoming treatment options have been further enhanced by the development of ADCs and immunotherapeutic agents, which allow very specific targeting of tumor-associated antigens in addition to increasing the antitumor response. With the introduction of biosimilars, the availability of biologic therapy has been critical. Biosimilars can lower economic barriers that have restricted the use of biologics in the past by providing similar quality, safety, and clinical efficacy at a lower cost. It is especially relevant in areas where cervical cancer happens to be extremely prevalent and where the prevalence of advanced treatment methods is limited due to financial and healthcare infrastructure factors. The identification of biomarkers, molecular diagnostics, and personalized medicine is advancing to enhance patient selection and maximize patient therapeutic outcomes. Biologic, chemotherapy, and radiotherapy combinations have proved to be more effective as they affect various mechanisms at the same time. Current efforts in the development of new biologic agents, therapeutic vaccines, and next-generation targeted therapies are likely to enhance the treatment outcomes. Altogether, the biologics and biosimilars are a necessary part of the contemporary cancer treatment of the neck. Further development in the field of drugs, biomarkers, and access to biosimilars will not only lead to a more precise treatment of the disease but also improve patient outcomes and publicize effective cervical cancer treatments globally.

References

Ahmad, S. (2025). Biosimilars in gynecologic cancers: Basic principles and new horizons. *Frontiers in Bioscience-Elite, 17*(4), 33415.

Arbyn, M., Weiderpass, E., Bruni, L., De Sanjosé, S., Saraiya, M., Ferlay, J., & Bray, F. (2020). Estimates of incidence and mortality of cervical cancer in 2018: A worldwide analysis. *The Lancet Global Health, 8*, e191–e203.

Bielenberg, D. R., & Zetter, B. R. (2015). The contribution of angiogenesis to the process of metastasis. *The Cancer Journal, 21*, 267–273.

Bizzarri, N., Ghirardi, V., Alessandri, F., Venturini, P. L., Valenzano Menada, M., Rundle, S., Leone Roberti Maggiore, U., & Ferrero, S. (2016). Bevacizumab for the treatment of cervical cancer. *Expert Opinion on Biological Therapy, 16*, 407–419.

Bonello, K., & Blundell, R. (2016). The role of the human papillomavirus (HPV) in cervical cancer: A review about HPV-induced carcinogenesis and its epidemiology, diagnosis, management and prevention. *International Journal of Medical Students, 4*, 26–32.

Borcoman, E., & Le Tourneau, C. (2020). Keynote-158 study, FDA granted accelerated approval of pembrolizumab for the treatment of patients with advanced PD-L1-positive cervical cancer. *Annals of Translational Medicine, 8*, 1611.

Broutet, N., Jeronimo, J., Kumar, S., Almonte, M., Murillo, R., Huy, N. V. Q., Denny, L., Kapambwe, S., Bhatla, N., & Sebitloane, M. (2022). Implementation research to accelerate scale-up of national screen and treat strategies towards the elimination of cervical cancer. *Preventive Medicine, 155*, 106906.

Carmeliet, P. (2005). VEGF as a key mediator of angiogenesis in cancer. *Oncology, 69*, 4–10.

Casak, S. J., Lemery, S. J., Chung, J., Fuchs, C., Schrieber, S. J., Chow, E. C., Yuan, W., Rodriguez, L., Gwise, T., & Rowzee, A. (2018). FDA's approval of the first biosimilar to bevacizumab. *Clinical Cancer Research, 24*, 4365–4370.

Chan, E., Fan Gaskin, J., & Chan, E. C. (2025). Corneal neovascularisation and anti-VEGF therapy. *Targets, 3*, 9.

Chen, B., Zhang, C., Dong, P., Guo, Y., & Mu, N. (2014). Molecular regulation of cervical cancer growth and invasion by VEGFa. *Tumor Biology, 35*, 11587–11593.

Chen, X., Lin, L., Wu, Q., Li, S., Wang, H., & Sun, Y. (2023). Tumor necrosis factor-α promotes the tumorigenesis, lymphangiogenesis, and lymphatic metastasis in cervical cancer via activating VEGFC-mediated AKT and ERK pathways. *Mediators of Inflammation, 2023*, 5679966.

Chianese, A., Santella, B., Ambrosino, A., Stelitano, D., Rinaldi, L., Galdiero, M., Zannella, C., & Franci, G. (2021). Oncolytic viruses in combination therapeutic approaches with epigenetic modulators: Past, present, and future perspectives. *Cancers, 13*, 2761.

Cohen, A. C., Roane, B. M., & Leath, C. A., III (2020). Novel therapeutics for recurrent cervical cancer: Moving towards personalized therapy. *Drugs, 80*, 217–227.

Colombo, N., Dubot, C., Lorusso, D., Caceres, M. V., Hasegawa, K., Shapira-Frommer, R., Tewari, K. S., Salman, P., Hoyos Usta, E., & Yañez, E. (2021). Pembrolizumab for persistent, recurrent, or metastatic cervical cancer. *New England Journal of Medicine, 385*, 1856–1867.

Colonna, G. (2025). Overcoming barriers in cancer biology research: Current limitations and solutions. *Cancers, 17*, 2102.

Croft, M., Salek-Ardakani, S., & Ware, C. F. (2024). Targeting the TNF and TNFR superfamilies in autoimmune disease and cancer. *Nature Reviews Drug Discovery, 23*, 939–961.

De Bono, J. S., Concin, N., Hong, D. S., Thistlethwaite, F. C., Machiels, J.-P., Arkenau, H.-T., Plummer, R., Jones, R. H., Nielsen, D., & Windfeld, K. (2019). Tisotumab vedotin in patients with advanced or metastatic solid tumors (InnovaTV 201): A first-in-human, multicentre, phase 1–2 trial. *The Lancet Oncology, 20*, 383–393.

DE Felice, F., Marchetti, C., Palaia, I., Ostuni, R., Muzii, L., Tombolini, V., & Panici, P. B. (2018). Immune check-point in cervical cancer. *Critical Reviews in Oncology/Hematology, 129*, 40–43.

Declerck, P., Danesi, R., Petersel, D., & Jacobs, I. (2017). The language of biosimilars: Clarification, definitions, and regulatory aspects. *Drugs, 77*, 671–677.

Dzobo, K. (2021). The role of viruses in carcinogenesis and molecular targeting: From infection to being a component of the tumor microenvironment. *Omics: A Journal of Integrative Biology, 25*, 358–371.

Farhat, F., Torres, A., Park, W., De Lima Lopes, G., Mudad, R., Ikpeazu, C., & Abi Aad, S. (2018). The concept of biosimilars: From characterization to evolution—A narrative review. *The Oncologist, 23*, 346–352.

Fernandes, J. V., Cobucci, R. N. O., Jatobá, C. A. N., De Medeiros Fernandes, T. A. A., De Azevedo, J. W. V., ARAÚJO, DE, & J. , M. G (2015). The role of the mediators of inflammation in cancer development. *Pathology & Oncology Research, 21*, 527–534.

Fontenot, V. E., Francoeur, A., & Tewari, K. S. (2024). Review of emerging biological therapies for recurrent and advanced metastatic cervical cancer. *Expert Opinion on Biological Therapy, 24*, 709–713.

Garcia, J., Hurwitz, H. I., Sandler, A. B., Miles, D., Coleman, R. L., Deurloo, R., & Chinot, O. L. (2020). Bevacizumab (Avastin®) in cancer treatment: A review of 15 years of clinical experience and future outlook. *Cancer Treatment Reviews, 86*, 102017.

Ge, Y., Zhang, Y., Zhao, K.-N., & Zhu, H. (2022). Emerging therapeutic strategies of different immunotherapy approaches combined with PD-1/PD-L1 blockade in cervical cancer. *Drug Design, Development and Therapy, 16*, 3055–3070.

Georgescu, S. R., Mitran, C. I., Mitran, M. I., Caruntu, C., Sarbu, M. I., Matei, C., Nicolae, I., Tocut, S. M., Popa, M. I., & Tampa, M. (2018). New insights in the pathogenesis of HPV infection and the associated carcinogenic processes: The role of chronic inflammation and oxidative stress. *Journal of Immunology Research, 2018*, 5315816.

Giacinti, C., & Giordano, A. (2006). RB and cell cycle progression. *Oncogene, 25*, 5220–5227.

Gottlieb, S. (2008). Biosimilars: Policy, clinical, and regulatory considerations. *American Journal of Health-System Pharmacy, 65*, S2–S8.

Greenman, M., Chang, Y. E., Mcnamara, B., Mutlu, L., & Santin, A. D. (2024). Unmet needs in cervical cancer–can biological therapies plug the gap? *Expert Opinion on Biological Therapy, 24*, 995–1003.

Guan, J., Lim, K. S., Mekhail, T., & Chang, C.-C. (2017). Programmed death ligand-1 (PD-L1) expression in the programmed death receptor-1 (PD-1)/PD-L1 blockade: A key player against various cancers. *Archives of Pathology & Laboratory Medicine, 141*, 851–861.

Guimarães, Y. M., Godoy, L. R., Longatto-Filho, A., & Reis, R. D (2022). Management of early-stage cervical cancer: A literature review. *Cancers, 14*, 575.

Halimi, V., Daci, A., Ancevska Netkovska, K., Suturkova, L., Babar, Z.-U.-D., & Grozdanova, A (2020). Clinical and regulatory concerns of biosimilars: A review of literature. *International Journal of Environmental Research and Public Health, 17*, 5800.

Han, X., Chang, W.-W., & Xia, X. (2022). Immune checkpoint inhibitors in advanced and recurrent/metastatic cervical cancer. *Frontiers in Oncology, 12*, 996495.

Hsu, J. Y., & Wakelee, H. A. (2009). Monoclonal antibodies targeting vascular endothelial growth factor: Current status and future challenges in cancer therapy. *BioDrugs, 23*, 289–304.

Huang, W., Liu, J., Xu, K., Chen, H., & Bian, C. (2022). PD-1/PD-L1 inhibitors for advanced or metastatic cervical cancer: From bench to bed. *Frontiers in Oncology, 12*, 849352.

Huinen, Z. R., Huijbers, E. J., Van Beijnum, J. R., Nowak-Sliwinska, P., & Griffioen, A. W. (2021). Anti-angiogenic agents—Overcoming tumor endothelial cell anergy and improving immunotherapy outcomes. *Nature Reviews Clinical Oncology, 18*, 527–540.

Ji, T., Liu, Y., Li, Y., Li, C., & Han, Y. (2024). Viral vector-based therapeutic HPV vaccines. *Clinical and Experimental Medicine, 24*, 199.

Jin, R., Ogbomo, A. S., Accortt, N. A., Lal, L. S., Bishi, G., Sandschafer, D., & Goldschmidt, J. H. (2023). Real-world outcomes among patients with metastatic colorectal cancer treated first line with a bevacizumab biosimilar (bevacizumab-awwb). *Therapeutic Advances in Medical Oncology, 15*, 17588359231182386.

Jung, J. M., Kim, Y.-J., Chang, S. E., Lee, M. W., Won, C. H., & Lee, W. J. (2023). Cancer risks in patients with psoriasis administered biologics therapy: A nationwide population-based study. *Journal of Cancer Research and Clinical Oncology, 149*, 17093–17102.

Kabir, I. M., & Ribeiro, R. I. M. D. A. (2025). HPV and cervical cancer: From molecular diagnostics to emerging treatment approaches. *World Academy of Sciences Journal, 7*, 58.

Kate, M. O. A. (2024). Biosimilars: Revolutionizing treatment for serious illnesses. *International Journal of All Research Writings, 7*, 7–20.

Kaur, R., Bhardwaj, A., & Gupta, S. (2023). Cancer treatment therapies: Traditional to modern approaches to combat cancers. *Molecular Biology Reports, 50*, 9663–9676.

Kazazi-Hyseni, F., Beijnen, J. H., & Schellens, J. H. (2010). Bevacizumab. *The Oncologist, 15*, 819.

Keating, G. M. (2014). Bevacizumab: A review of its use in advanced cancer. *Drugs, 74*, 1891–1925.

Khan, M. A., Ali, K., & Rizvi, M. (2024). Human papillomavirus type 16 based L1, L2, E6, and E7 peptide microspheres induce encapsulated peptide mixture specific cytotoxic T lymphocytes and tumor regression in a murine model of cervical cancer. *Cancer Screening and Prevention, 3*, 85–90.

Kirchhoff, C. F., Wang, X. Z. M., Conlon, H. D., Anderson, S., Ryan, A. M., & Bose, A. (2017). Biosimilars: Key regulatory considerations and similarity assessment tools. *Biotechnology and Bioengineering, 114*, 2696–2705.

Li, J., & Li, S. (2025). From viral infection to genome reshaping: The triggering role of HPV integration in cervical cancer. *International Journal of Molecular Sciences, 26*, 9214.

Li, N., Wang, Z., Yuan, G., Sun, Y., Zhang, R., Li, X., Li, N., Wang, J., & Wu, L. (2020). An oral small molecule VEGFR2 inhibitor, apatinib, in patients with recurrent or refractory cervical cancer. A real world study. *Journal of Oncology, 2020*, 3852373.

Lopes-Coelho, F., Martins, F., Pereira, S. A., & Serpa, J. (2021). Anti-angiogenic therapy: Current challenges and future perspectives. *International Journal of Molecular Sciences, 22*, 3765.

Luu, K., Chu, A., & Chang, B. (2023). A review of the novel tissue factor antibody–drug conjugate: Tisotumab vedotin. *Journal of Oncology Pharmacy Practice, 29*, 441–449.

Luvero, D., Plotti, F., Lopez, S., Scaletta, G., Capriglione, S., Montera, R., Antonelli, G., Ciuffreda, S., Carassiti, R., & Oliveti, A. (2017). Antiangiogenics and immunotherapies in cervical cancer: An update and future's view. *Medical Oncology, 34*, 115.

Manocha, E., Caruso, A., & Caccuri, F. (2021). Viral proteins as emerging cancer therapeutics. *Cancers, 13*, 2199.

Manzo-Merino, J., Contreras-Paredes, A., Vázquez-Ulloa, E., Rocha-Zavaleta, L., Fuentes-Gonzalez, A. M., & Lizano, M (2014). The role of signaling pathways in cervical cancer and molecular therapeutic targets. *Archives of Medical Research, 45*, 525–539.

Markham, A. (2021). Tisotumab vedotin: First approval. *Drugs, 81*, 2141–2147.

Minion, L. E., & Tewari, K. S. (2018). Cervical cancer–State of the science: From angiogenesis blockade to checkpoint inhibition. *Gynecologic Oncology, 148*, 609–621.

Mirlekar, B. (2022). Tumor promoting roles of IL-10, TGF-β, IL-4, and IL-35: Its implications in cancer immunotherapy. *SAGE Open Medicine, 10*, 20503121211069012.

Molinari, M. (2000). Cell cycle checkpoints and their inactivation in human cancer. *Cell Proliferation, 33*, 261–274.

Morparia, S., Suvarna, V., & Murahari, M. (2024). Biosimilar in cervical cancer. In S. Bhatt, et al. (Eds.), *Biosimilars for cancer treatment: A promising approach* (pp. 75–95). Springer. https://doi.org/10.1007/978-981-97-3001-8_5.

Nandi, A., Ghosh, A., & Sil, S. K. (2025). The interplay between angiogenesis and cervical cancer: A comprehensive review. *Expert Review of Anticancer Therapy, 25*, 633–642.

Okunade, K. S. (2020). Human papillomavirus and cervical cancer. *Journal of Obstetrics and Gynaecology, 40*, 602–608.

Park, B. 2022. Bevacizumab biosimilar vegzelma gets FDA approval. MPR Monthly Prescribing Reference.

Park, D., Kim, J., Yun, J., & Park, S. J. (2020). Evaluation of the physico-chemical and biological stability of SB8 (Aybintio), a proposed biosimilar to bevacizumab, under ambient and in-use conditions. *Advances in Therapy, 37*, 4308–4324.

Parola, S., & Colombo, I. (2025). Bispecific antibodies in the treatment of epithelial ovarian, endometrial, and cervical cancer: An overview of current applications, challenges, and emerging opportunities. *Exploration of Targeted Anti-tumor Therapy, 6*, 1002345.

Pavelescu, L. A., Mititelu-Zafiu, N. L., Mindru, D. E., Vladareanu, R., & Curici, A. (2025). Molecular insights into HPV-driven cervical cancer: Oncoproteins, immune evasion, and epigenetic modifications. *Microorganisms, 13*, 1000.

Peltanova, B., Raudenska, M., & Masarik, M. (2019). Effect of tumor microenvironment on pathogenesis of the head and neck squamous cell carcinoma: A systematic review. *Molecular Cancer, 18*, 63.

Quintanilha, J. C., Wang, J., Sibley, A. B., Jiang, C., Etheridge, A. S., Shen, F., Jiang, G., Mulkey, F., Patel, J. N., & Hertz, D. L. (2022). Bevacizumab-induced hypertension and proteinuria: A genome-wide study of more than 1000 patients. *British Journal of Cancer, 126*, 265–274.

Rathore, A. S., & Bhargava, A. (2020). Biosimilars in developed economies: Overview, status, and regulatory considerations. *Regulatory Toxicology and Pharmacology, 110*, 104525.

Rider, P., Carmi, Y., & Cohen, I. (2016). Biologics for targeting inflammatory cytokines, clinical uses, and limitations. *International Journal of Cell Biology, 2016*, 9259646.

Santin, A. D., Deng, W., Frumovitz, M., Buza, N., Bellone, S., Huh, W., Khleif, S., Lankes, H. A., Ratner, E. S., & O'cearbhaill, R. E (2020). Phase II evaluation of nivolumab in the treatment of persistent or recurrent cervical cancer (NCT02257528/NRG-GY002). *Gynecologic Oncology, 157*, 161–166.

Saunders, P. A., Hendrycks, V. R., Lidinsky, W. A., & Woods, M. L. (2005). PD-L2: PD-1 involvement in T cell proliferation, cytokine production, and integrin-mediated adhesion. *European Journal of Immunology, 35*, 3561–3569.

Sezer, A., Kilickap, S., Gümüş, M., Bondarenko, I., Özgüroğlu, M., Gogishvili, M., Turk, H. M., Cicin, I., Bentsion, D., & Gladkov, O (2021). Cemiplimab monotherapy for first-line treatment of advanced non-small-cell lung cancer with PD-L1 of at least 50%: A multi-centre, open-label, global, phase 3, randomised, controlled trial. *The Lancet, 397*, 592–604.

Sharma, H., & An, S. S. A. (2024). VEGF/VEGFR inhibition for cancer treatment using bevacizumab and its biosimilars: A review. *Biomedicine & Pharmacotherapy, 70*, 115956.

Sharma, S., Deep, A., & Sharma, A. K. (2020). Current treatment for cervical cancer: An update. *Anti-cancer Agents in Medicinal Chemistry, 20*, 1768–1779.

Shi, Y.-H., & Fang, W.-G. (2004). Hypoxia-inducible factor-1 in tumor angiogenesis. *World Journal of Gastroenterology, 10*, 1082.

Shilpi, S., Gurnany, E., Gupta, P. K., Sharma, K., Saini, K., Chimaniya, P., & Bhatt, S. (2024). Mechanism of biosimilars for the treatment of cancer. In *Biosimilars for cancer treatment: A promising approach* (pp. 35–54). Springer. https://doi.org/10.1007/978-981-97-3001-8_3.

Shord, S. S., Bressler, L. R., Tierney, L. A., Cuellar, S., & George, A. (2009). Understanding and managing the possible adverse effects associated with bevacizumab. *American Journal of Health-System Pharmacy, 66*, 999–1013.

Silva, M. O., Almeida, B. S., Sales, N. S., Diniz, M. O., Aps, L. R., Rodrigues, K. B., Silva, J. R., Moreno, A. C., Porchia, B. F., & Sulczewski, F. B. (2021). Antigen delivery to DEC205+ dendritic cells induces immunological memory and protective therapeutic effects against HPV-associated tumors at different anatomical sites. *International Journal of Biological Sciences, 17*, 2944.

Song, J. (2022). Biologics and biosimilars: Clinical applications and biomarker testing. In X. Feng, H.-G. Xie, A. Malhotra, & C. F. Yang (Eds.), *Biologics and biosimilars* (pp. 405–466). CRC Press.

Stanley, M. (2015). Immunology of HPV Infection. *Current Obstetrics and Gynecology Reports, 4*, 195–200.

Subramanian, A., Muthukumarasamy, G. D., Venkataraman, S., Rangasamy, K., Kanade, A. N., Arulrajan, P. V., Saravanan, K., Shanmuganathan, S., Gopal, G., & Ambi, S. V. (2024). Future scope of biosimilars for the treatment of various cancers. In *Biosimilars for cancer treatment: A promising approach.* (pp. 287–309). Springer. https://doi.org/10.1007/978-981-97-3001-8_16.

Tewari, K. S., Monk, B. J., Vergote, I., Miller, A., De Melo, A. C., Kim, H.-S., Kim, Y. M., Lisyanskaya, A., Samouëlian, V., & Lorusso, D (2022). Survival with cemiplimab in recurrent cervical cancer. *New England Journal of Medicine, 386*, 544–555.

Tian, W.-J., Huang, M.-L., Qin, Q.-F., Chen, Q., Fang, K., & Wang, P.-L. (2016). Prognostic impact of epidermal growth factor receptor overexpression in patients with cervical cancer: A meta-analysis. *PLoS One, 11*, e0158787.

Tomaić, V (2016). Functional roles of E6 and E7 oncoproteins in HPV-induced malignancies at diverse anatomical sites. *Cancers, 8*, 95.

Tomao, F., Papa, A., Rossi, L., Zaccarelli, E., Caruso, D., Zoratto, F., Benedetti Panici, P., & Tomao, S. (2014). Angiogenesis and antiangiogenic agents in cervical cancer. *OncoTargets and Therapy, 7*, 2237–2248.

Vanajothi, R., Srikanth, N., Vijayakumar, R., Palanisamy, M., Bhavaniramya, S., & Premkumar, K. (2022). HPV-mediated cervical cancer: A systematic review on immunological basis, molecular biology, and immune evasion mechanisms. *Current Drug Targets, 23*, 782–801.

Volkova, L. V., Pashov, A. I., & Omelchuk, N. N. (2021). Cervical carcinoma: Oncobiology and biomarkers. *International Journal of Molecular Sciences, 22*, 12571.

Wang, A.-J., Gao, Y., Shi, Y.-Y., Dai, M.-Y., & Cai, H.-B. (2022). A review of recent advances on single use of antibody-drug conjugates or combination with tumor immunology therapy for gynecologic cancer. *Frontiers in Pharmacology, 13*, 1093666.

Willmott, L. J., Sumner, D. A., & Monk, B. J. (2010). Biologics in cervical cancer therapy. *Journal of the National Comprehensive Cancer Network, 8*, 1417–1423.

Wlazlo, A. P., Deng, H., Giles-Davis, W., & Ertl, H. C. (2004). DNA vaccines against the human papillomavirus type 16 E6 or E7 oncoproteins. *Cancer Gene Therapy, 11*, 457–464 .

Wu, Y., & He, X. (2024). Immunotherapy for recurrent and metastatic cervical cancer: A review. *Clinical and Experimental Obstetrics & Gynecology, 51*, 155.

Xie, Y., Kong, W., Zhao, X., Zhang, H., Luo, D., & Chen, S. (2022). Immune checkpoint inhibitors in cervical cancer: Current status and research progress. *Frontiers in Oncology, 12*, 984896.

Yang, J., Kelton, J. M., Thompson, J., Jimenez Alvir, J. M., Maculaitis, M. C., & Shelbaya, A. (2022). Real-world usage of bevacizumab-bvzr biosimilar in US oncology practice. *American Journal of Managed Care, 28*, S164–S171.

Yang, J., Liu, R., Ektare, V., Stephens, J., & Shelbaya, A. (2021). Does biosimilar bevacizumab offer affordable treatment options for cancer patients in the USA? A budget impact analysis from US commercial and medicare payer perspectives. *Applied Health Economics and Health Policy, 19*, 605.

Zhang, L., Wu, J., Ling, M. T., Zhao, L., & Zhao, K.-N. (2015). The role of the PI3K/Akt/mTOR signalling pathway in human cancers induced by infection with human papillomaviruses. *Molecular Cancer, 14*, 87.

Zhang, Y., Ding, Y., Liao, Y., Shu, J., & Gong, Y. (2025). Unlocking the potential of antibody-drug conjugates in cervical cancer: Emerging targets and clinical trials. *Frontiers in Pharmacology, 16*, 1636120.

Zheng, Q., He, M., Mao, Z., Huang, Y., Li, X., Long, L., Guo, M., & Zou, D. (2025). Advancing the fight against cervical cancer: The promise of therapeutic HPV vaccines. *Vaccines, 13*, 92.

Zhivotovsky, B., & Kroemer, G. (2004). Apoptosis and genomic instability. *Nature Reviews Molecular Cell Biology, 5*, 752–762.

CHAPTER

5 Biosimilars and biologics in lung cancer

Chavan Aditya [a], Hemanth Kumar Gorubala [b], Pandhare Meeraraje [a] and Suryanarayana Polaka [c]

[a] *Department of Pharmacology, School of Pharmacy and Technology Management, SVKM's Narsee Monjee Institute of Management Studies (NMIMS) (Deemed to be University), Shirpur, Maharashtra, India,* [b] *Department of Pharmacology, Sheffield Hallam University, Sheffield, United Kingdom,* [c] *Department of Pathology, Louisiana State University Health Shreveport, Shreveport, LA, United States*

5.1 Lung cancer: Epidemiology and risk factors

5.1.1 Introduction

With an estimated 2.2 million new cases and 1.8 million deaths from lung cancer in 2020, it is the leading cause of cancer-related deaths worldwide. It accounts for around 18% of all cancer-related deaths and is the second most common type of cancer diagnosed globally. The two primary kinds of lung cancer are small-cell lung cancer (SCLC) and nonsmall-cell lung cancer (NSCLC), with NSCLC being further differentiated into giant cell carcinoma, squamous cell carcinoma (SCC), and adenocarcinoma (Araghi et al., 2023). With over 80% of cases occurring in high-income areas, tobacco use is the primary risk factor for lung cancer. But other environmental elements, including radon, asbestos, air pollution, and biomass fuels, also have a significant impact, especially in low-income nations (Rodak et al., 2021). Certain treatments can target genetic alterations like epidermal growth factor receptor (EGFR) and lymphoma kinase (ALK) seen in adenocarcinoma, which is highly common in nonsmokers, especially women. Some patient groups have shown improved results from recent therapeutic developments, such as immunotherapies and tyrosine kinase inhibitors (TKIs). The 5-year survival rate for lung cancer worldwide, however, is still quite poor, ranging from 7% to 25% (Chapman et al., 2016). Differences in smoking habits, environmental exposures, and genetic predispositions all have an impact on regional variations in lung cancer incidence, mortality, and histological subtypes. Understanding these epidemiological patterns and risk factors is essential to developing successful preventative and therapeutic strategies that can lessen the burden of lung cancer worldwide (Thandra et al., 2021). The discovery of EGFR and ALK mutations is one of the molecular advances that has significantly advanced the treatment of NSCLC. These findings have made it easier to implement individualized treatment plans and targeted medicines. Furthermore, genome sequencing has improved

Biosimilars and Biologics in Cancer Treatment: An Era in Therapeutic Strategies.
DOI: https://doi.org/10.1016/B978-0-443-43824-0.00006-7

diagnosis and produced more accurate histopathological classification (Sung et al., 2021). Generational risk factors, environmental exposures, and differences in smoking habits all have an impact on the global epidemiological trends of NSCLC. Analysis of birth cohorts provides important information about these shifts. Furthermore, survivors of primary lung cancer are more likely to acquire second primary lung cancer (SPLC), but there aren't many thorough, scientifically supported suggestions for keeping an eye on SPLC. These elements highlight how important it is to continue conducting molecular and epidemiological studies in order to improve NSCLC outcomes (Huang et al., 2022).

5.1.2 Epidemiology

With almost 2.2 million new cases and 1.8 million deaths from lung cancer in 2020, it remains the leading cause of cancer-related deaths worldwide. Regional variations in smoking rates, exposures to the environment, and access to healthcare all have an impact on the incidence of lung cancer. Developing nations are seeing an increase in incidence rates, in contrast to industrialized places such as Western Europe and North America, where death rates are lower because of more developed healthcare systems. High smoking rates and greater exposure to air pollution in these areas are the main causes of this increase (Zhang et al., 2021). The most common form of lung cancer worldwide, particularly in women and nonsmokers, is adenocarcinoma. Early detection may be possible with recent developments in diagnostic methods, such as miRNA profiling (Tao & Imaging, 2019). Age plays a crucial role in lung cancer incidence, with rates rising sharply after the age of 45. Although smoking cessation has led to a decrease in lung cancer rates among men in developed countries, the increasing incidence among women and in regions such as China highlights the necessity for targeted prevention strategies and equitable access to advanced diagnostic and therapeutic technologies Fig. 5.1 (Luo et al., 2023).

Lung cancer is the primary cause of cancer-related mortality for women in 28 nations, including China and the United States, and for men in 93 countries. The incidence and mortality rates of lung cancer among women are rising in the United States and Europe, in contrast to the trend for males. Lung cancer has lately surpassed breast cancer as the leading cause of cancer-related mortality among women in these areas (Bray et al., 2024). In the U.S., the 5-year survival rate for lung cancer has shown significant improvement, increasing from 11.5% in 1975 to 20.5% between 2010 and 2016. This progress is largely attributed to advancements in early detection methods, such as CT screening, and the introduction of targeted and immune therapies (Lu et al., 2019). Survival rates for lung cancer differ significantly based on the stage at diagnosis, with localized disease having a 59% survival rate, regional disease at 31.7%, and metastatic disease at a much lower 5.8%. A major challenge persists due to late-stage diagnoses, as approximately 57% of lung cancer cases in the U.S. are identified after the disease has spread. This underscores the critical need for early detection and ensuring equitable access to advanced treatment options (Leiter et al., 2023).

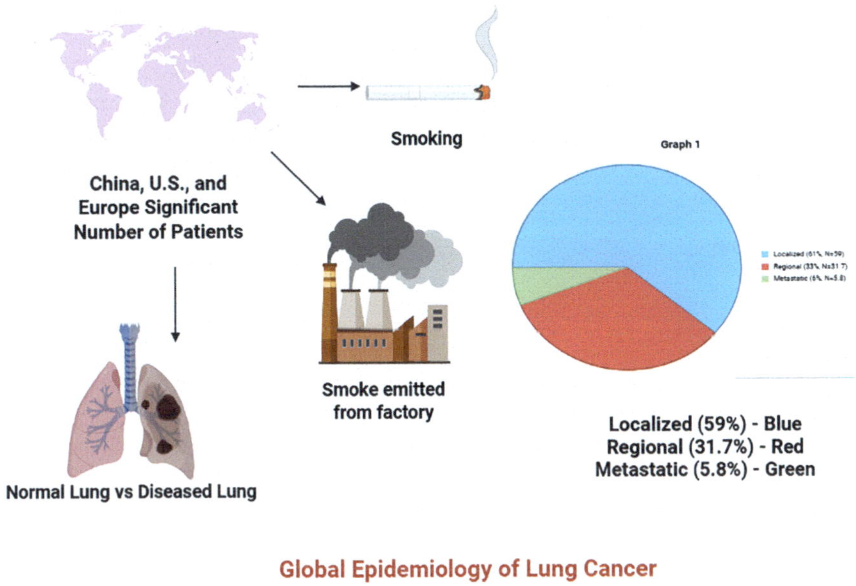

FIGURE 5.1

Epidemiology of lung cancer.

5.1.3 Risk factors

5.1.3.1 Nonmodifiable risk factors

5.1.3.1.1 Age

The average age at which lung cancer is diagnosed is 70 years old, and most cases—53%—occur in people aged 55 to 74% and 37% in people aged 75 and beyond. The most common cause of death for women over 59 and males over 40 is lung cancer. The process of tumor formation often begins decades after exposure to harmful substances, and it is influenced by biological aging factors such as telomere shortening, decreased NAD+ levels, and diminished cellular DNA repair and surveillance capabilities (Kanwal et al., 2017). Although uncommon, individuals under 55 years old make up about 10% of lung cancer cases, frequently presenting with advanced adenocarcinoma. This younger demographic tends to be predominantly female, nonsmokers, and their disease is often driven by inherited genetic mutations rather than environmental factors Fig. 5.2 (Aredo et al., 2021).

5.1.3.1.2 Gender

The higher incidence of smoking among males is the main reason why men are more than twice as likely as women to be diagnosed with and die from lung cancer worldwide. In affluent nations, lung cancer rates are rising among women despite declining among males, which is indicative of historical smoking habits

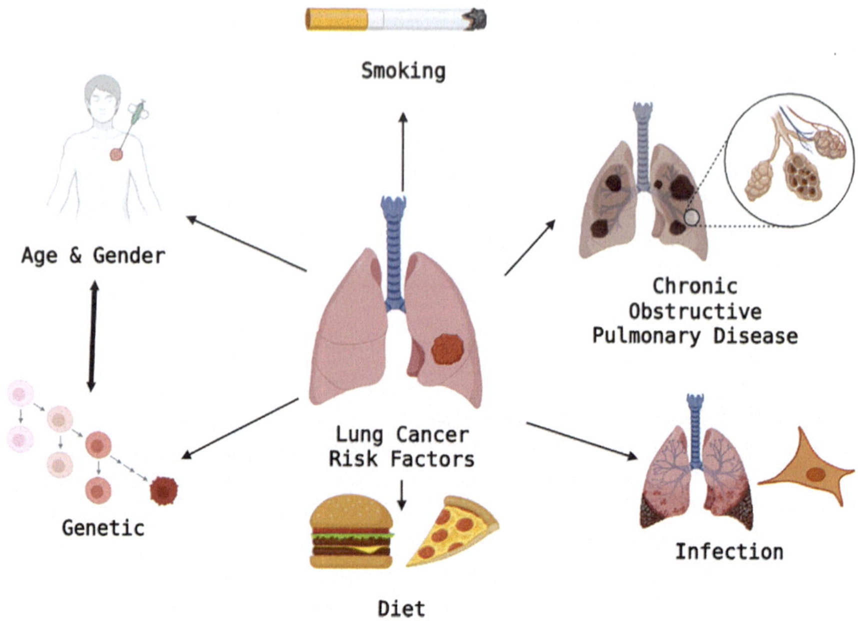

FIGURE 5.2

Risk factor of lung cancer.

(Yamamoto et al., 2018). Women who do not smoke demonstrate higher rates of EGFR-positive adenocarcinomas and mutations in genes like CY1A1, TP53, ALK, and BRAF compared to men. Familial risk also plays a more prominent role in women, even when accounting for smoking history. Hormonal factors, such as the overexpression of α-estrogen receptors in adenocarcinomas and differences in estrogen exposure, may contribute to lung cancer susceptibility. However, hormone replacement therapy has not been shown to significantly affect risk (Atchley et al., 2021).

5.1.3.1.3 Genetic

Somatic mutations that drive lung cancer development vary significantly based on geographical location, patient demographics, and environmental factors. For instance, EGFR mutations are more frequently found in East Asia, among women, nonsmokers, and in adenocarcinoma cases. KRAS mutations are more common in smokers and show regional differences. Nonsmokers often exhibit mutations in genes such as ALK, ROS1, and ERBB2. These variations highlight the importance of considering regional and demographic factors in lung cancer diagnosis and treatment (de Alencar et al., 2020). Genetic predisposition is a significant factor in lung cancer, with a family history increasing the risk by 1.7 to 4 times, especially among first-degree relatives. Genome-wide association studies have identified several risk-related genetic loci, such as 5p15 (TERT), 6p21, and 15q25–26. However, unlike breast cancer, which has specific germline mutations like BRCA1/2, lung cancer does not have equivalent mutations. These findings suggest the potential for personalized detection and treatment strategies that incorporate both genetic and environmental factors (Gee & Yendamuri, 2024).

5.1.3.2 Modifiable risk factors

5.1.3.2.1 Smoking

In Western nations, cigarette smoking is responsible for over 80% of lung cancer occurrences, making it the leading preventable cause of death worldwide. While tobacco combustion creates more than 60 carcinogens, such as polycyclic aromatic hydrocarbons (PAHs) and N-nitrosamines, nicotine encourages addiction by altering brain receptor function. These chemicals damage DNA and cause cancer to form. The World Wars caused a sharp rise in smoking rates, which peaked in the middle of the 20th century. However, epidemiological research in the 1950s was necessary to demonstrate the link between smoking and lung cancer (Fig. 5.2) (Malhotra et al., 2016). In affluent nations, smoking rates have significantly decreased as a result of public health campaigns since the 1960s, which has decreased the incidence of lung cancer. But in low and middle-income nations, smoking is still common, which adds to the rising incidence of lung cancer worldwide. Because secondhand smoke exposure raises the risk of lung cancer by 20%–30%, laws restricting smoking in public places have been implemented. Concerns regarding carcinogenic compounds in aerosols and their possible influence on starting smoking are added to the complexity of the increased use of e-cigarettes and vaping among young people. More than a billion people continued to smoke in 2019, despite international initiatives like the WHO Framework Convention on Tobacco Control (FCTC), underscoring the difficult task of tackling tobacco-related health problems (Young et al., 2011).

5.1.3.2.2 Environmental exposure

Lung cancer risk is significantly increased by occupational and environmental exposures. Carcinogens, including benzene and PAHs, are released when half of the world's population uses unprocessed biomass fuels for cooking and heating. This increases the risk of lung cancer, particularly for coal users in Asia. Because they are exposed to higher amounts of indoor air pollution from biomass fuels, women are disproportionately harmed. Furthermore, known carcinogens that cause lung cancer worldwide include radon in homes and workplaces, particulate matter from air pollution, and exposure to arsenic through tainted food and water (Laguna et al., 2024). Lung cancer is significantly increased by occupational exposures to chemicals such as asbestos, diesel exhaust, and heavy metals like nickel and cadmium, especially in industrial and developing nations with laxer safety standards. Even with the advancement of regulatory measures like improved ventilation systems, lower diesel emissions, and asbestos prohibitions, there are still significant concerns in low and middle-income nations. In order to reduce these hazards, initiatives like the Diesel Emissions Reduction Act and the U.S. National Radon Action Plan emphasize how crucial it is to impose stronger international air quality and occupational safety regulations (Mustafa et al., 2016).

5.1.3.2.3 Infection

Immunosuppression and respiratory infections are major factors in the development of lung cancer. The risk of lung cancer is 1.76 times higher in people who have tuberculosis (TB), the most common infectious cause of mortality worldwide. Lung cancer is the leading cause of non-AIDS-related cancer deaths among individuals living with HIV in the United States. Similarly, HIV infection raises the risk of lung cancer by up to 2.5 times, independent of smoking status. Long-term immunosuppression, lung inflammation, and virus-induced cancer development are some of the factors that are linked to the increased risk (Akhtar & Bansal, 2017). Smoking significantly elevates lung cancer risk in people with HIV, with the disease frequently emerging at younger ages even among those with less cumulative tobacco exposure. Limited healthcare access and comorbidities like cardiovascular disease further reduce survival rates in this population. Immunosuppressive conditions, including therapies following organ transplantation, are also linked to higher lung cancer incidence. Emerging evidence indicates severe respiratory infections such as COVID-19 might elevate long-term lung cancer risk through mechanisms like chronic inflammation, cytokine dysregulation, and lung tissue damage from acute respiratory distress. While not directly addressed in the provided studies, this aligns with broader research on inflammation-driven carcinogenesis (Turner et al., 2020).

5.1.3.2.4 Chronic obstructive pulmonary disease

Regardless of smoking history, chronic obstructive pulmonary disease (COPD) is a significant independent risk factor for lung cancer. According to research, smokers with COPD are between 6 and 13 times more likely to get lung cancer than smokers without the disease. Higher incidence and mortality rates of lung cancer are linked to more severe COPD, which is characterized by increasing emphysema, decreased lung function, and more severe airway obstruction (Taucher et al., 2022). According to research, there are several molecular pathways that link lung cancer and COPD. Carcinogenesis is aided by the protumor microenvironment created by persistent inflammation in COPD, which is fueled by cytokines such as NF-κB and immune cell activation (macrophages, T cells). Reactive oxygen species (ROS) are produced by contaminants and tobacco smoke, which overwhelm antioxidant defenses. Especially in people with COPD, this results in oxidative DNA damage and compromised repair systems. Both disorders exhibit common genetic loci (e.g., CDKN2A, PTEN) and epigenetic alterations (e.g., hypermethylation of tumor suppressor genes). These alterations encourage carcinogenesis and interfere with cell cycle regulation (Parris et al., 2019).

5.1.3.2.5 Diet and metabolic factors

Obesity, diabetes, and dietary practices all significantly affect the risk of lung cancer. According to epidemiological studies, eating a lot of red and processed meat can raise your risk of developing lung cancer, but eating a lot of fruits and vegetables, which are rich in antioxidants, can help prevent the disease. Different vitamins and supplements have different effects: Methionine and vitamin B6 may lower the risk of lung cancer, whereas vitamins E and β-carotene may raise it, particularly in smokers (Grosso et al., 2017). It is advised against using supplements containing β-carotene and vitamin A as a preventative measure because randomized trials have shown that they raise the risk of lung cancer. An "obesity paradox" exists in relation to obesity, whereby a higher body mass index is paradoxically associated with a reduced incidence of lung cancer and better outcomes. This could be because of confounding variables like smoking status. However, a higher risk of lung cancer is linked to insulin resistance and central obesity, which is defined by an increased waist circumference. This implies that the development of cancer may be influenced by endogenous insulin rather than high blood sugar levels (Bossé et al., 2018).

5.2 Symptoms and causes of lung cancer

The leading cause of cancer-related death worldwide, lung cancer is typified by the unchecked proliferation of aberrant cells in the lungs. With over 85% of instances under nonsmall cell lung cancer (NSCLC) and almost 15% under small cell lung cancer (SCLC), it is essentially divided into two main forms. With different histological and clinical characteristics, NSCLC consists of subtypes including adenocarcinoma, SCC, and giant cell carcinoma. Lung cancer has a complex etiology; tobacco smoking is the most important risk factor. Because tobacco smoke contains carcinogens such as PAHs and nitrosamines, smoking accounts for 80–85% of lung cancer cases reported worldwide. Other risk factors include radon gas, asbestos, air pollution, and occupational dangers, including diesel exhaust and heavy metals. Increased lung cancer risk has also been linked to genetic inclination and diseases, including HIV and TB. Many times, in its early stages, lung cancer remains asymptomatic, which results in late diagnosis. When symptoms do strike, they can be a chronic cough, hemoptysis, coughing up blood, chest pain, dyspnea, and unexplained weight loss. Advanced stages could show systemic symptoms related to metastases, such as neurological impairments or bone discomfort. Another manifestation is paraneoplastic disorders brought on by hormone-like molecules generated by tumors. Lung cancer's pathophysiology consists of genetic and molecular changes producing malignant transformation. Common mutations in NSCLC are those in the EGFR, KRAS, and ALK genes; SCLC usually results from changes in the MYC, BCL2, and p53 genes. These mutations upset the balance of normal cell cycle control, therefore encouraging unchecked cell growth and tumor development. Lung cancer pathophysiology consists of genetic and molecular changes causing malignant transformation. Common mutations in NSCLC are those in the EGFR, KRAS, and ALK genes; SCLC usually results from changes in the MYC, BCL2, and p53 genes. These mutations upset the balance of normal cell cycle control, therefore encouraging unchecked cell growth and tumor development (Athey et al., 2018).

5.3 Current benchmark drugs for the treatment of lung cancer

Current clinical guidelines for metastatic NSCLC differentiate treatment approaches based on the presence or absence of targetable driver mutations. Patients without these mutations often receive immunotherapy either alone or combined with chemotherapy. Those with specific mutations, such as EGFR or ALK, are typically treated with molecularly targeted therapies. This distinction is due to the lower tumor mutational burden in these cases and their generally limited response to immune checkpoint inhibitors (ICIs), making targeted therapies more effective for patients with these mutations (Hirsch et al., 2016). Emerging evidence suggests immunochemotherapy combinations may provide clinical benefits even for NSCLC patients with targetable mutations (e.g., EGFR/ALK), who typically receive molecularly targeted therapies. HER2-mutated NSCLC (HER2mu NSCLC), a rare subtype accounting for approximately 3% of NSCLC cases, is more common in nonsmokers, women, and adenocarcinoma subtypes. Although no targeted therapies are FDA-approved for HER2mu NSCLC, early-phase studies report TKI (e.g., poziotinib) and antibody-drug conjugates (e.g., trastuzumab deruxtecan) have demonstrated potential efficacy in heavily pretreated patients (Board, 2024). The effectiveness of ICIs in this subgroup is variable, with response rates varying widely from under 10% to nearly 30%. Moreover, there is a lack of data on their use as a first-line treatment, emphasizing the necessity for additional research to inform treatment choices (Saalfeld et al., 2021).

New methods to immunotherapy for NSCLC emphasize the value of biomarker-driven tactics to enhance patient outcomes and treatment personalization. Although tumor-infiltrating lymphocytes, tumor mutational burden (TMB), and PD-L1 expression continue to be fundamental predictors of ICI efficacy, tumor heterogeneity and the need for repeated tissue samples limit their therapeutic relevance. Without the need for invasive biopsies, circulating tumor DNA (ctDNA) analysis has become a viable minimally invasive option for dynamic TMB monitoring and the identification of actionable mutations (such as EGFR and ALK). By addressing the issues of temporal and spatial variability, this method could improve treatment decisions made in real time. For instance, during treatment, ctDNA can detect clonal evolution or resistance mutations (such as KRAS or MET amplifications) (Coleman et al., 2023). Combination immunotherapy strategies for NSCLC, such as dual checkpoint blockade targeting CTLA-4 or TIGIT in conjunction with PD-1/PD-L1 inhibitors, have demonstrated mixed clinical success. While these approaches can enhance antitumor immune responses, they are often limited by immune-related adverse events (e.g., colitis, pneumonitis) and a lack of reliable biomarkers to predict patient response. For example, the Check Mate 227 trial showed long-term survival benefits with nivolumab (anti-PD-1) + ipilimumab (anti-CTLA-4) in advanced NSCLC, but treatment-related toxicities necessitated careful management. Engineered cytokine therapies, including modified IL-2 and IL-15 variants, aim to amplify T-cell activation while minimizing systemic toxicity. These therapies focus on improving tumor-specific immunity through enhanced cytokine receptor binding or reduced off-target effects (Howlader et al., 2020). Prospective biomarker-based selection is crucial for personalizing these therapies to specific subgroups of NSCLC patients. This approach helps avoid unnecessary treatment by identifying those most likely to benefit, thereby optimizing clinical outcomes and reducing overtreatment (Wang et al., 2021).

Patients with extensive stage SCLC who have already received treatment have shown encouraging antitumor activity in response to tarlatamab, a bispecific T-cell engager immunotherapy created by Amgen. The disease's limited second-line therapy choices and often dismal survival results are its defining characteristics. In order to promote T-cell-mediated lysis of tumor cells, tarlatamab targets CD3 on T cells and delta-like ligand 3 (DLL3) on cancer cells. It was shown to be effective in the DeLLphi-301 trial (ClinicalTrials.gov ID NCT05060016), providing SCLC patients with a new therapy option (Lahiri et al., 2023). Tarlatamab, used at a dose of 10 mg every 2 weeks, produced a noteworthy objective response rate of 40% in the phase 2 DeLLphi-301 study. Of those who replied, 55% had continuous replies at the time of data cutoff, and 59% had durable responses that lasted for at least 6 months. At 9 months, the overall survival (OS) rate was 68%, with a median progression-free survival of roughly 4.9 months. These findings demonstrate the effectiveness and possible longevity of tarlatamab in the treatment of SCLC that has already received treatment (PDQ Adult Treatment Editorial Board, 2002). Cytokine-release syndrome was the most commonly reported adverse event linked to tarlatamab, and it was primarily mild to moderate (grade 1 or 2). It is noteworthy that just 3% of patients discontinued their medication as a result of adverse effects. These results highlight the potential of tarlatamab as a new and efficient therapeutic option for individuals with SCLC that expresses DLL3 (Ahn et al., 2023).

Advances in lung cancer screening and treatment have revolutionized the management of NSCLC, allowing for early interventions such as surgical resection, stereotactic body radiation therapy, and ablation for early-stage disease. Chemoradiotherapy remains a cornerstone for unresectable cases. However, emerging therapies, including targeted mutation treatments and checkpoint blockade immunotherapy, are hindered by high failure rates and high costs. This underscores the importance of improving patient selection to optimize outcomes and ensure that these innovative treatments are used effectively (Nguyen et al., 2014). Radiomics, a sophisticated computational approach, extracts detailed quantitative features from medical images such as CT and PET scans, uncovering tumor characteristics that surpass human visual capabilities. By analyzing features like shape, intensity, texture, and wavelet-based metrics, radiomics facilitates the development of predictive models for tumor histology, genetic mutations, treatment outcomes, and prognosis. This technology enhances precision medicine in lung cancer by addressing limitations in traditional diagnostic methods and optimizing personalized patient care (Kneuertz et al., 2022).

With a median progression-free survival of more than 5 years, Lorlatinib, a third-generation ALK TKI, has demonstrated exceptional effectiveness in treating advanced ALK-positive NSCLC. For this patient group, it is therefore a very successful targeted therapy. For patients with ALK-positive NSCLC who are candidates for surgical resection, the FDA has approved the use of alexinib (Alecensa), another ALK inhibitor.

For a particular group of patients, its approval offers a new therapy alternative that may help them manage their illness (Lovly, 2024). The FDA has approved the use of amivantamab (Rybrevant) in combination with lazertinib to treat NSCLC that has certain EGFR mutations. For some individuals, this combination provides an alternative to chemotherapy, offering a new way to treat this illness. Checkpoint inhibitors, such as atezolizumab (Tecentriq) and pembrolizumab (Keytruda), are now essential parts of the treatment of lung cancer. These immunotherapy medications greatly increase patient survival rates and quality of life when administered either alone or in conjunction with chemotherapy (Chen et al., 2023).

5.4 Biosimilars and biologics agent in lung cancer treatment

Biosimilars are biologic products that are highly similar to FDA-approved reference biologics, with no significant differences in safety, purity, or potency. These products are produced from living cells using advanced biotechnology and undergo rigorous testing, including analytical, pharmacokinetic, and clinical evaluations, to ensure their similarity to the original products (Table 5.1). While biologics have transformed cancer treatment, their high cost poses a substantial financial strain on healthcare systems and patients, making biosimilars a promising alternative to increase accessibility and affordability (Malakar et al., 2024). The emergence of biosimilars has alleviated the financial burden associated with biologic therapies by fostering competition, improving accessibility, and offering cost-effective options without sacrificing safety or efficacy. As patents for pivotal oncology biologics like rituximab and trastuzumab have expired, biosimilars have gained momentum, leading to significant cost reductions and enhanced affordability for patients and healthcare systems (Cho et al., 2019). Despite initial reservations among healthcare providers, growing clinical and real-world evidence (RWE) now supports the efficacy of biosimilars in oncology. These products are poised to play a crucial role in the future of cancer treatment, enabling wider access to life-saving therapies while helping to reduce healthcare costs (Peeters et al., 2021).

5.4.1 Monoclonal antibodies as biosimilars

A recombinant humanized monoclonal antibody called bevacizumab is essential for treating cancer because it targets vascular endothelial growth factor A (VEGF-A) and prevents angiogenesis, which in turn limits the growth of tumors. Glioblastoma, metastatic renal cell carcinoma, nonsquamous NSCLC, cervical cancer, and metastatic colorectal cancer are only a few of the cancers for which it has had a major clinical impact. Bevacizumab is a transformational drug in cancer because it blocks VEGF-A, which effectively decreases the development of new blood vessels that tumors need to grow and spread (Ou et al., 2024). Many biosimilars that offer

Table 5.1 Types of biologics in lung cancer.

Type of biologic	Therapeutic action	Mechanism of action	Type of lung cancer
mAbs	Block tumor growth or stimulate immune response	Bind to specific antigens on cancer cells to inhibit growth or recruit immune cells.	NSCLC
Immune checkpoint inhibitors	Enhance T-cell activity	Block PD-1/PD-L1 or CTLA-4 pathways to prevent immune evasion by cancer cells.	NSCLC, SCLC
Tyrosine kinase inhibitors	Inhibit specific signaling pathways	Target mutations like EGFR, ALK, or ROS1 to block cell proliferation and survival.	NSCLC with specific genetic mutations
VEGF inhibitors	Prevent tumor angiogenesis	Inhibit vascular endothelial growth factor (VEGF) to reduce blood supply to tumors.	Advanced NSCLC
CAR-T cell therapy	Directly target cancer cells	Genetically engineered T-cells to recognize and kill specific tumor antigens.	Experimental in lung cancer
Bispecific antibodies	Dual-target therapy	Engage T-cells and tumor cells by binding to CD3 and a tumor-specific antigen.	Advanced/metastatic NSCLC
Cytokines (e.g., Interleukins)	Stimulate immune response	Modulate immune activity to promote antitumor responses.	NSCLC (experimental stages)
Cancer vaccines	Stimulate immune response	Prime the immune system to recognize and attack cancer-specific antigens.	Early-stage or experimental lung cancer
Oncolytic viruses	Selectively kill tumor cells	Infect and replicate within tumor cells, causing lysis and immune activation.	Experimental in NSCLC and SCLC

less expensive treatment options have been developed after the patents for bevacizumab expired in the US in 2019 and in Europe in 2022. The FDA has approved two well-known biosimilars in the United States: Zirabev (bevacizumab-bvzr) and Mvasi (bevacizumab-awwb). These biosimilars provide patients with affordable alternatives because they have demonstrated similar pharmacokinetic characteristics, effectiveness, and safety to the original bevacizumab (Ngo & Chen, 2021). Aybintio, Equidacent, Oyavas, Zirabev, Mvasi, and Alymsys are among the nine bevacizumab biosimilars that have been approved in the European Union. Key studies, like the MAPLE trial, have shown that these biosimilars are clinically equivalent to the reference product, improving access to cancer treatments while preserving therapeutic efficacy. This pattern demonstrates how biosimilars are becoming more and more crucial in lowering the cost and expanding access to cancer treatments globally (Xu et al., 2022).

In 2022, the European Commission approved Vegzelma (CT-P16), a bevacizumab biosimilar, to treat a number of cancers, including NSCLC. This approval was supported by a wealth of data demonstrating that Vegzelma is just as effective, safe, pharmacokinetic, and immunogenic as the original bevacizumab (Verschraegen et al., 2024). When used in conjunction with conventional chemotherapy, Vegzelma produced overall response rates comparable to those of the original bevacizumab, according to a significant phase III clinical trial that involved patients with metastatic or recurrent nonsquamous NSCLC. Additionally, Vegzelma's safety profiles and quality of life evaluations were similar to those of the reference product, suggesting that it is a good choice for treating patients with advanced NSCLC (Bachu et al., 2022).

5.5 Preclinical studies for the treatment of lung cancer using biologics and biosimilars

The advent of biologics and biosimilars, which offer focused and cutting-edge therapeutic approaches, has significantly changed the landscape of lung cancer treatment. In order to provide more accurate and efficient therapy choices, biologics, which include fusion proteins, cytokines, and monoclonal antibodies, are made to selectively target cellular pathways essential to the growth of cancer (Jacobs et al., 2017). Because they are very similar to their reference biologics, biosimilars provide equivalent safety and efficacy at lower prices, increasing patient access to these therapies. Before these treatments move on to clinical trials, preclinical research is essential for evaluating their mechanisms, effectiveness, safety, and pharmacokinetics (Joshi et al., 2022).

5.5.1 Biologics in lung cancer

Biologic therapy must target biomarkers, including EGFR mutations, ALK rearrangements, and PD-L1 expression in NSCLC. Advanced methods such as 3D organoid cultures and *in vivo* models are used in preclinical research to examine the effectiveness of medications such as atezolizumab, a PD-L1 inhibitor, and osimertinib, an

EGFR inhibitor. The goal of these studies is to improve treatment results and address resistance mechanisms by investigating ways to overcome resistance by combining biologics with chemotherapy, radiation, or novel small molecules (Konstantinidou et al., 2020; Solta et al., 2024). NSCLC, or nonsmall cell lung cancer, accounts for about 85% of all instances of lung cancer and is highly variable. Mutations in the EGFR and rearrangements in the anaplastic ALK are two important targets in NSCLC. In preclinical models, osimertinib, a third-generation EGFR TKI, has been thoroughly investigated for its capacity to enter the central nervous system and successfully treat resistant mutations such as the T790M mutation. Osimertinib's potential to overcome typical resistance pathways in NSCLC makes it a valuable treatment alternative (Moir, 2023). Immune checkpoint drugs, including anti-PD-1/PD-L1 antibodies like atezolizumab and nivolumab, have completely changed how NSCLC is treated. The goal of preclinical research is to offset immune suppression in the tumor microenvironment by combining these inhibitors with other treatments. For instance, the synergistic effects of combined inhibition of the PD-1 and CTLA-4 pathways on increasing T-cell activation are being investigated, which could result in more potent cancer treatment approaches (Glode, 2019). To ensure that they are structurally, functionally, and biologically equivalent to the original biologics, biosimilars are put through a rigorous review process in preclinical environments. To create comparability, analytical investigations are carried out using techniques including glycosylation profiling and binding affinity assays. Furthermore, biosimilars' antitumor properties are evaluated using functional assays such as complement-dependent cytotoxicity and antibody-dependent cellular cytotoxicity to make sure they are as effective as the reference medicines (Feng et al., 2024). Biosimilars are further validated by preclinical animal studies, which verify their immunogenicity, pharmacokinetics, and pharmacodynamics. Furthermore, cutting-edge technologies like single-cell RNA sequencing, spatial transcriptomics, and CRISPR/Cas9 for genetic editing are incorporated into preclinical research developments. By providing insights into the interactions between biosimilars and intricate biological systems, these technologies facilitate a deeper comprehension of the dynamics inside the tumor microenvironment (Chopra & Lopes, 2017). These instruments are essential for finding new targets and improving biologics and biosimilars for individualized cancer treatments. Preclinical findings highlight obstacles, including tumor biology heterogeneity, possible off-target effects, and immunogenic responses, despite their potential, underscoring the necessity of thorough study designs and strong translational methodologies. Precision medicine and improved patient outcomes are made possible by the introduction of biologics and biosimilars into preclinical research, which signals a revolutionary era in the treatment of lung and stomach cancer (García et al., 2020).

5.5.2 Emerging technologies in preclinical research

Our knowledge and development of biologics and biosimilars have significantly increased as a result of the incorporation of cutting-edge technology into preclinical research. For example, patient-derived mini-tumors called 3D tumor organoids closely resemble the microenvironment of the original tumor. This offers a strong foundation for researching drug reactions and comprehending resistance mechanisms, providing important information about the ways in which biologics and biosimilars interact with intricate tumor habitats (Adriani et al., 2023). Through precise genetic alterations made possible by CRISPR/Cas9 gene editing, scientists can examine how particular mutations affect the course of cancer and the effectiveness of treatment. By examining individual cells inside a tumor, single-cell sequencing, on the other hand, can expose its heterogeneity and pinpoint subpopulations that might be involved in resistance or metastasis. This thorough cellular examination offers vital insights into the intricate dynamics of cancer biology (Yu et al., 2024). Machine learning and artificial intelligence (AI) are being used more and more to forecast treatment results and improve medication design. By analyzing intricate statistics from preclinical research, these tools help researchers make better judgments and expedite the development of biologics and biosimilars (Luo et al., 2023).

5.6 Real-world case studies in clinical practice and future prospects

Despite a significant increase in global lung cancer research output between 2004 and 2013, it still constitutes a relatively small portion of overall oncology research, accounting for only 5.6%. The commitment to lung cancer research has decreased in many countries, including the U.S. and several European nations. However, China has experienced substantial growth, driven by its rising lung cancer burden, which is linked to smoking and air pollution. Globally, research efforts are unevenly distributed, with inadequate investment in crucial areas such as early diagnosis, screening, and palliative care, despite their critical role in improving patient outcomes. Funding disparities persist, as lung cancer research often receives less support compared to breast or colorectal cancer in countries like the U.S. and U.K. East Asian countries, such as China and Japan, focus primarily on basic science, while European nations have shifted toward more theoretical research, raising concerns about the long-term clinical impact. International collaboration remains limited, particularly in high-output countries like China and Japan, which can hinder knowledge sharing and equity in clinical outcomes. Addressing these gaps is essential for advancing lung cancer prevention, diagnosis, and treatment (Aggarwal et al., 2016; Simonds et al., 2013). Molecular assays have substantial potential in identifying patients who are most likely to benefit from EGFR TKI therapies, allowing for earlier and more effective interventions. Prospective trials, such as SATURN and TITAN, are examining genomic factors like EGFR and KRAS mutations, as well as EGFR and HER-2 expression, to assess their influence on treatment outcomes. These studies, along with planned trials, highlight the importance of obtaining sufficient tumor samples before treatment, incorporating control groups, and involving biostatisticians to rigorously validate EGFR biomarkers as predictive and prognostic tools. However, the widespread adoption of molecular assays requires further validation through large randomized clinical trials, standardization of techniques such as fluorescence in situ hybridization (FISH), immunohistochemistry (IHC), and mutational analyses, and adherence to guidelines like those proposed by the Molecular Assays in NSCLC Working Group. This ensures accurate and reliable integration of these assays into clinical practice (Eberhard et al., 2008; Ruano-Ravina et al., 2020).

When used as a first-line treatment for extensive-stage small cell lung cancer (SCLC), the PD-1 inhibitor serplulimab in combination with chemotherapy dramatically improved OS when compared to chemotherapy alone, according to a groundbreaking phase 3 randomized clinical trial. With improvements in progression-free survival, objective response rate, and duration of response, this trial was the first to indicate that a PD-1 inhibitor improved OS in this setting. Serplulimab may be useful because of its higher affinity for PD-1 and improved capacity to inhibit PD-L1 and PD-L2 signaling. With side effects comparable to those observed with other PD-1/PD-L1 inhibitors, the advantages held true for older patients and never smokers, among other patient subgroups. However, there were certain drawbacks, such as a brief follow-up period, restricted patient recruitment for brain metastases, and no direct comparison with other PD-L1 inhibitors. For extensive-stage SCLC, these findings provide credence to the combination of serplulimab with chemotherapy as a new standard of care (Cheng et al., 2022; Hammerschmidt & Wirtz, 2009). Regardless of PD-L1 expression levels, the phase 3 RATIONALE 307 trial showed that tislelizumab, a PD-1 inhibitor, added to chemotherapy significantly improved objective response rates (ORR) and prolonged progression-free survival (PFS) in patients with advanced squamous nonsmall cell lung cancer (sq-NSCLC). When compared to chemotherapy alone, the regimens incorporating tislelizumab had long-lasting clinical responses that doubled in length. These advantages held true for people with stage IIIB and IV illness as well as other categories. The results confirm the efficacy of PD-1/PD-L1 inhibitors in sq-NSCLC and are in line with findings from other trials such as KEYNOTE-407 and IMpower131. With the majority of side effects being mild to moderate and consistent with the established safety profiles of chemotherapy and PD-1 inhibitors, the treatment was generally well tolerated. Tislelizumab combined with chemotherapy is a good first-line treatment choice for advanced sq-NSCLC, according to these findings (Crane et al., 2016; Wang et al., 2021).

Osimertinib and bevacizumab together did not improve progression-free survival (PFS) over osimertinib alone in patients with advanced lung adenocarcinoma with an EGFR T790M mutation, according to a randomized phase 2 trial. The median PFS in the combination arm was substantially shorter (9.4 months vs. 13.5 months), despite the combination displaying a marginally greater overall response rate (68% vs. 54%). Overall survival and time to treatment failure did not differ significantly. Bevacizumab addition did not yield any benefits, according to subgroup studies, and earlier anti-VEGF treatment appeared to have had the opposite effect. In the second-line context, the study found no synergy between osimertinib and anti-VEGF drugs, indicating that this combination may be better suited for investigation as a first-line therapy. Proteinuria (23%) and hypertension (20%) were frequent side effects in the combo arm, with Japanese patients exhibiting greater toxicity rates. These results highlight the need for additional study in first-line settings, and the outcomes of ongoing randomized studies are eagerly awaited (Akamatsu et al., 2021; Hanson et al., 2017). The importance of real-world data (RWD) and RWE in enhancing conventional clinical trials, which might not accurately capture the intricacies of actual patient care, is being recognized more and more in lung cancer research. RWE fills in knowledge gaps and improves treatment strategies by providing vital insights on how well treatments work for a variety of patient populations. It is essential for predicting long-term survival, assessing treatments for metastatic NSCLC, and determining the variables that affect patient outcomes. Healthcare professionals may improve patient outcomes, address care inequities, and improve individualized treatment by integrating RWE into clinical practice and research. This will ultimately help fight lung cancer more successfully (Gristina & Eze, 2024).

Serious lung cancer, which accounts for 13.10% of advanced cases, is the focus of this real-world epidemiological study. It primarily affects elderly male patients with stage IV nonsquamous NSCLC who do not have gene alterations. The main cause of a worse prognosis is shown to be treatment-related adverse events (AEs), which outweigh the influence of comorbidities or cancer symptoms. Despite the importance of comorbidities, they may go undiagnosed because of diagnostic difficulties; yet, better survival can result from efficient therapy. The study emphasizes the significance of improved clinical awareness and management techniques to improve patient outcomes, although its limitations, which include subjective performance status scores and retrospective data (Wang et al., 2023). Patients with SCLC participated in this real-world study, which looked at treatment trends and OS results, including the effects of immunotherapy. With extensive-stage patients having a median OS of 9.1 months, the median OS with first-line treatment was 11.8 months, which is consistent with earlier data. Female sex for patients in restricted stages and improved performance status (PS) for patients in extensive stages were important predictors of survival. It was not statistically significant; however, patients who were platinum-sensitive and retreated with platinum-based therapy tended to live longer than those who switched to topotecan. Immunotherapy showed a potential numerical benefit in OS for extensive-stage patients, but further validation is needed. Long-term survivors were predominantly female, had limited-stage disease, good PS, and received upfront chemo-radiotherapy. While the study benefits from high-quality data and extended follow-up, its retrospective design limits insights into treatment decision-making processes (Bredtoft et al., 2021; Moser et al., 2021).

5.7 Conclusion

The thorough study shown in this chapter clearly shows that lung cancer is a major worldwide health issue with complicated etiologies and different clinical presentations. Driven mostly by variances in smoking practices, environmental exposures, genetic predispositions, and healthcare infrastructure, the epidemiological data indicate notable regional variations in incidence and death. Especially when stratified by biomarkers including EGFR, ALK, and PD-L1 expression, NSCLC, the most common subtype, has shown promise in responding to molecularly targeted therapies and immunotherapies. Though medication resistance, high treatment costs, and restricted access in low-resource environments still present obstacles, the introduction of tailored medicines and ICIs has clearly improved patient outcomes. As preclinical and translational research helps to untangle the molecular complexity of lung cancer, it opens the path for tailored treatment approaches, stressing precision medicine. The advent of biosimilars has changed the therapeutic scene even more, especially in terms of the financial cost related to biologic medicines. In clinical and real-world environments, biosimilars of agents such as bevacizumab have shown equivalent efficacy and safety, therefore improving patient access worldwide. Furthermore, improving our knowledge of tumor biology and drug sensitivity is a technological development, such as 3D tumor organoids, single-cell sequencing, CRISPR gene editing, and artificial intelligence-driven analytics. These maximize treatment combinations and provide fresh analysis of resistance mechanisms. These advances taken together highlight a paradigm change toward patient-centered, cost-effective, biomarker-driven lung cancer treatment. Maintaining development and increasing long-term survival in lung cancer patients will depend on ongoing investment in translational research, worldwide cooperation, and fair healthcare delivery.

References

Adriani, G., Cappello, P., & Lovisa, S. (2023). Preclinical models and emerging technologies to study the effects of the tumor microenvironment on cancer heterogeneity and drug resistance. *Frontiers in Oncology, 13*, 1289756.

Aggarwal, A., Lewison, G., Idir, S., Peters, M., Aldige, C., Boerckel, W., Boyle, P., Trimble, E. L., Roe, P., Sethi, T., Fox, J., & Sullivan, R. (2016). The state of lung cancer research: A global analysis. *Journal of Thoracic Oncology, 11*(7), 1040–1050.

Ahn, M.-J., Cho, B. C., Felip, E., Korantzis, I., Ohashi, K., Majem, M., ... Dziadziuszko, R. (2023). Tarlatamab for patients with previously treated small-cell lung cancer. *The New England Journal of Medicine, 389*(22), 2063–2075.

Akamatsu, H., Toi, Y., Hayashi, H., Fujimoto, D., Tachihara, M., Furuya, N., ... Miura, N. (2021). Efficacy of osimertinib plus bevacizumab vs osimertinib in patients with EGFR T790M–Mutated non–small cell lung cancer previously treated with epidermal growth factor receptor–tyrosine kinase inhibitor: West Japan oncology group 8715L phase 2 randomized clinical trial. *JAMA Oncology, 7*(3), 386–394.

Akhtar, N., & Bansal, J. G. (2017). Risk factors of lung cancer in nonsmoker. *Current Problems in Cancer, 41*(5), 328–339.

Araghi, M., Mannani, R., Heidarnejad maleki, A., Hamidi, A., Rostami, S., Safa, S. H., ... Akhavan-Sigari, R. (2023). Recent advances in non-small cell lung cancer targeted therapy; an update review. *Cancer Cell International, 23*(1), 162.

Aredo, J. V., Luo, S. J., Gardner, R. M., Sanyal, N., Choi, E., Hickey, T. P., ... Wilkens, LR. (2021). Tobacco smoking and risk of second primary lung cancer. *Journal of Thoracic Oncology, 16*(6), 968–979.

Atchley, W. T., Alvarez, C., Saxena-Beem, S., Schwartz, T. A., Ishizawar, R. C., Patel, K. P., & Rivera, M. P. (2021). Immune checkpoint inhibitor-related pneumonitis in lung cancer:

Real-world incidence, risk factors, and management practices across six health care centers in North Carolina. *Chest, 160*(2), 731–742.

Athey, V. L., Walters, S. J., & Rogers, T. K. (2018). Symptoms at lung cancer diagnosis are associated with major differences in prognosis. *Thorax, 73*(12), 1177–1181.

Bachu, R. D., Abou-Dahech, M., Balaji, S., Boddu, S. H., Amos, S., Singh, V., Babu, R. J., & Tiwari, A. K. (2022). Oncology biosimilars: New developments and future directions. *Cancer Reports (Hoboken), 5*(11), e1720.

Board, P. A. T. E. (2024). *Non-small cell lung cancer treatment (PDQ®). PDQ cancer information summaries [Internet]*. National Cancer Institute (US).

Bossé, Y., & Amos, C. I. (2018). A decade of GWAS results in lung cancer. *Cancer Epidemiol Biomarkers and Prevention, 27*(4), 363–379.

Bray, F., Laversanne, M., Sung, H., Ferlay, J., Siegel, R. L., Soerjomataram, I., & Jemal, A. (2024). Global cancer statistics 2022: GLOBOCAN estimates of incidence and mortality worldwide for 36 cancers in 185 countries. *CA: A Cancer Journal for Clinicians, 74*(3), 229–263.

Bredtoft, E. N., Madsen, H. H., & Rasmussen, T. R. (2021). Stage I lung cancer patients with or without symptoms - are the patients different and should we treat them differently? *Acta Oncol, 60*(9), 1169–1174.

Chapman, A. M., Sun, K. Y., Ruestow, P., Cowan, D. M., & Madl, A. K. (2016). Lung cancer mutation profile of EGFR, ALK, and KRAS: Meta-analysis and comparison of never and ever smokers. *Lung Cancer, 102*, 122–134.

Chen, M., Copley, S. J., Viola, P., Lu, H., & Aboagye, E. O. (2023). Radiomics and artificial intelligence for precision medicine in lung cancer treatment. *Seminars in cancer biology, 93*, 97–113.

Cheng, Y., Han, L., Wu, L., Chen, J., Sun, H., Wen, G., … Shi, J. (2022). Effect of first-line serplulimab vs placebo added to chemotherapy on survival in patients with extensive-stage small cell lung cancer: The ASTRUM-005 randomized clinical trial. *JAMA, 328*(12), 1223–1232.

Cho, S.-H., Han, S., Ghim, J.-L., Nam, M.-S., Yu, S., Park, T., Kim, S., Bae, J., & Shin, J.-G. (2019). A randomized, double-blind trial comparing the pharmacokinetics of CT-P16, a candidate bevacizumab biosimilar, with its reference product in healthy adult males. *BioDrugs, 33*, 173–181.

Chopra, R., & Lopes, G. (2017). Improving access to cancer treatments: The role of biosimilars. *Journal of Global Oncology, 3*(5), 596–610.

Coleman, N., Yap, T. A., Heymach, J. V., Meric-Bernstam, F., & Le, X. (2023). Antibody-drug conjugates in lung cancer: Dawn of a new era? *NPJ Precision Oncology, 7*(1), 5.

Crane, M., Scott, N., O'Hara, B. J., Aranda, S., Lafontaine, M., Stacey, I., Varlow, M., & Currow, D. (2016). Knowledge of the signs and symptoms and risk factors of lung cancer in Australia: Mixed methods study. *BMC Public Health, 16*(1), 508.

de Alencar, V. T. L., Formiga, M. N., & de Lima, V. C. C. (2020). Inherited lung cancer: A review. *Ecancermedicalscience, 14*, 1008.

Eberhard, D. A., Giaccone, G., & Johnson, B. E. (2008). Biomarkers of response to epidermal growth factor receptor inhibitors in non–small-cell lung cancer working group: Standardization for use in the clinical trial setting. *Journal of Clinical Oncology, 26*(6), 983–994.

Feng, J., Zhang, P., Wang, D., Li, Y., & Tan, J. (2024). New strategies for lung cancer diagnosis and treatment: Applications and advances in nanotechnology. *Biomarker Research, 12*(1), 136.

García, J. J., Raez, L. E., & Rosas, D. (2020). A narrative review of biosimilars: A continued journey from the scientific evidence to practice implementation. *Translational Lung Cancer Research, 9*(5), 2113.

Gee, K., & Yendamuri, S. (2024). Lung cancer in females—Sex-based differences from males in epidemiology, biology, and outcomes: A narrative review. *Translational Lung Cancer Research, 13*(1), 163.

Glode, A. E. (2019). Review of biosimilars and their potential use in oncology treatment and supportive care in the United States. *Journal of Hematology Oncology Pharmacy, 9*(3), 133–141.

Gristina, V., & Eze, C. (2024). Real-world data and real-world evidence in lung cancer. *Frontiers in Oncology, 14*, 1436077.

Grosso, G., Godos, J., Lamuela-Raventos, R., Ray, S., Micek, A., Pajak, A., Sciacca, S., D'Orazio, N., Del Rio, D., & Galvano, F. (2017). A comprehensive meta-analysis on dietary flavonoid and lignan intake and cancer risk: Level of evidence and limitations. *Molecular Nutrition & Food Research, 61*(4), 1600930.

Hammerschmidt, S., & Wirtz, H. (2009). Lung cancer: Current diagnosis and treatment. *Deutsches Arzteblatt International, 106*(49), 809–818.

Hanson, H. M., Raag, M., Adrat, M., & Laisaar, T. (2017). Awareness of lung cancer symptoms and risk factors in general population. *Open Journal of Respiratory Diseases, 07*(01), 1–11.

Hirsch, F. R., Suda, K., Wiens, J., & Bunn, P. A. (2016). New and emerging targeted treatments in advanced non-small-cell lung cancer. *Lancet, 388*(10048), 1012–1024.

Howlader, N., Forjaz, G., Mooradian, M. J., Meza, R., Kong, C. Y., Cronin, K. A., Mariotto, A. B., Lowy, D. R., & Feuer, E. J. (2020). The effect of advances in lung-cancer treatment on population mortality. *The New England Journal of Medicine, 383*(7), 640–649.

Huang, J., Deng, Y., Tin, M. S., Lok, V., Ngai, C. H., Zhang, L., … Withers, M. (2022). Distribution, risk factors, and temporal trends for lung cancer incidence and mortality: A global analysis. *Chest, 161*(4), 1101–1111.

Jacobs, I., Ewesuedo, R., Lula, S., & Zacharchuk, C. (2017). Biosimilars for the treatment of cancer: A systematic review of published evidence. *BioDrugs, 31*, 1–36.

Joshi, D., Khursheed, R., Gupta, S., Wadhwa, D., Singh, T. G., Sharma, S., … Gupta, G. (2022). Biosimilars in oncology: Latest trends and regulatory status. *Pharmaceutics, 14*(12), 2721.

Kanwal, M., Ding, X.-J., & Cao, Y. (2017). Familial risk for lung cancer. *Oncology Letters, 13*(2), 535–542.

Kneuertz, P. J., Abdel-Rasoul, M., D'Souza, D. M., Zhao, J., & Merritt, R. E. (2022). Segmentectomy for clinical stage I non–small cell lung cancer: National benchmarks for nodal staging and outcomes by operative approach. *Cancer, 128*(7), 1483–1492.

Konstantinidou, S., Papaspiliou, A., & Kokkotou, E. (2020). Current and future roles of biosimilars in oncology practice. *Oncology Letters, 19*(1), 45–51.

Laguna, J. C., Tagliamento, M., Lambertini, M., Hiznay, J., & Mezquita, L. (2024). Tackling non–small cell lung cancer in young adults: From risk factors and genetic susceptibility to lung cancer profile and outcomes. *American Society of Clinical Oncology Educational Book, 44*(3), e432488.

Lahiri, A., Maji, A., Potdar, P. D., Singh, N., Parikh, P., Bisht, B., Mukherjee, A., & Paul, M. K. (2023). Lung cancer immunotherapy: Progress, pitfalls, and promises. *Molecular Cancer, 22*(1), 40.

Leiter, A., Veluswamy, R. R., & Wisnivesky, J. P. (2023). The global burden of lung cancer: Current status and future trends. *Nature Reviews Clinical Oncology, 20*(9), 624–639.

Lovly, C. M. (2024). New benchmark for targeted therapies in lung cancer: Median progression-free survival for lorlatinib in advanced ALK+ non–small cell lung cancer surpasses 5 years. *Journal of Clinical Oncology, 42*, 3383–3386.

Lu, T., Yang, X., Huang, Y., Zhao, M., Li, M., Ma, K., Yin, J., Zhan, C., & Wang, Q. (2019). Trends in the incidence, treatment, and survival of patients with lung cancer in the last four decades. *Cancer Management and Research, 11*, 943–953.

Luo, G., Zhang, Y., Etxeberria, J., Arnold, M., Cai, X., Hao, Y., & Zou, H. (2023). Projections of lung cancer incidence by 2035 in 40 countries worldwide: Population-based study. *JMIR Public Health and Surveillance, 9*(1), e43651.

Luo, X., Du, X., Li, Z., Liu, J., Lv, X., Li, H., … Jiang, X. (2023). Clinical benefit, price, and uptake for cancer biosimilars vs reference drugs in China: A systematic review and meta-analysis. *JAMA Network Open, 6*(10), e2337348.

Malakar, S., Gontor, E. N., Dugbaye, M. Y., Shah, K., Sinha, S., Sutaoney, P., & Chauhan, N. S. (2024). Cancer treatment with biosimilar drugs: A review. *Cancer Innovation, 3*(2), e115.

Malhotra, J., Malvezzi, M., Negri, E., La Vecchia, C., & Boffetta, P. (2016). Risk factors for lung cancer worldwide. *European Respiratory Journal, 48*(3), 889–902.

Moir, H. J. (2023). Real-world evidence of biosimilar utilisation and pharmacoeconomics in people living with breast or lung cancer. *EMJ Oncology, 11*, 2–11. https://doi.org/10.33590/emjoncol/10307863.

Moser, S. S., Bar, J., Kan, I., Ofek, K., Cohen, R., Khandelwal, N., Shalev, V., Chodick, G., & Siegelmann-Danieli, N. (2021). Real world analysis of small cell lung cancer patients: Prognostic factors and treatment outcomes. *Current Oncology, 28*(1), 317–331.

Mustafa, M., Azizi, A., IIlzam, E., Nazirah, A., Sharifa, S., & Abbas, S. A. (2016). Lung cancer: Risk factors, management, and prognosis. *IOSR Journal of Dental and Medical Sciences, 15*(10), 94–101.

Ngo, D., & Chen, J. (2021). A clinical review of biosimilars approved in oncology. *Annals of Pharmacotherapy, 55*(3), 362–377.

Nguyen, K.-S. H., Neal, J. W., & Wakelee, H. (2014). Review of the current targeted therapies for non-small-cell lung cancer. *World Journal of Clinical Oncology, 5*(4), 576.

Ou, W.-F., Hsu, K.-H., Tseng, J.-S., Lee, P.-H., Chen, K.-C., Huang, Y.-H., Chang, G.-C., & Yang, T.-Y. (2024). Real-world clinical efficacy of bevacizumab biosimilar in patients with advanced non-small-cell lung cancer. *Therapeutic Advances in Medical Oncology, 16*. https://doi.org/10.1177/17588359241290718.

Parris, B. A., O'Farrell, H. E., Fong, K. M., & Yang, I. A. (2019). Chronic obstructive pulmonary disease (COPD) and lung cancer: Common pathways for pathogenesis. *Journal of Thoracic Disease, 11*(Suppl 17), S2155.

PDQ Adult Treatment Editorial Board. (2002). *Non-small cell lung cancer treatment (PDQ®): Treatment-health professional information [NCI]*. National Cancer Institute (US).

Peeters, M., Planchard, D., Pegram, M., Gonçalves, J., Bocquet, F., & Jang, H. (2021). Biosimilars in an era of rising oncology treatment options. *Future Oncology, 17*(29), 3881–3892.

Rodak, O., Peris-Díaz, M. D., Olbromski, M., Podhorska-Okołów, M., & Dzięgiel, P. (2021). Current landscape of non-small cell lung cancer: Epidemiology, histological classification, targeted therapies, and immunotherapy. *Cancers (Basel), 13*(18), 4705.

Ruano-Ravina, A., Provencio, M., Calvo de Juan, V., Carcereny, E., Moran, T., Rodriguez-Abreu, D., Lopez-Castro, R., Cuadrado Albite, E., Guirado, M., Gomez Gonzalez, L., Massuti, B., Ortega Granados, A. L., Blasco, A., Cobo, M., Garcia-Campelo, R., Bosch, J.,

Trigo, J., Casal-Rubio, J., ... Cerezo, S. (2020). Lung cancer symptoms at diagnosis: Results of a nationwide registry study. *ESMO Open, 5*(6), e001021.

Saalfeld, F. C., Wenzel, C., Christopoulos, P., Merkelbach-Bruse, S., Reissig, T. M., Laßmann, S., ... Desuki, A. (2021). Efficacy of immune checkpoint inhibitors alone or in combination with chemotherapy in NSCLC harboring ERBB2 mutations. *Journal of Thoracic Oncology, 16*(11), 1952–1958.

Simonds, N. I., Khoury, M. J., Schully, S. D., Armstrong, K., Cohn, W. F., Fenstermacher, D. A., ... Ramsey, SD. (2013). Comparative effectiveness research in cancer genomics and precision medicine: Current landscape and future prospects. *Journal of the National Cancer Institute, 105*(13), 929–936.

Solta, A., Ernhofer, B., Boettiger, K., Megyesfalvi, Z., Heeke, S., Hoda, M. A., ... Döme, B. (2024). Small cells–big issues: Biological implications and preclinical advancements in small cell lung cancer. *Molecular Cancer, 23*(1), 41.

Tao, M.-H., J. L. C. and Imaging (2019). "Epidemiology of lung cancer." Journal of Thoracic Disease. 11(Suppl 3): S197–S207.

Sung, H., Ferlay, J., Siegel, R. L., Laversanne, M., Soerjomataram, I., Jemal, A., & Bray, F. (2021). Global cancer statistics 2020: GLOBOCAN estimates of incidence and mortality worldwide for 36 cancers in 185 countries. *CA: A Cancer Journal for Clinicians, 71*(3), 209–249.

Taucher, E., Mykoliuk, I., Lindenmann, J., & Smolle-Juettner, F.-M. (2022). Implications of the immune landscape in COPD and lung cancer: Smoking versus other causes. *Frontiers in Immunology, 13*, 846605.

Thandra, K. C., Barsouk, A., Saginala, K., Aluru, J. S., & Barsouk, A. (2021). Epidemiology of lung cancer. *Contemporary Oncology (Współczesna Onkologia), 25*(1), 45–52.

Turner, M. C., Andersen, Z. J., Baccarelli, A., Diver, W. R., Gapstur, S. M., Pope, C. A. III, Prada, D., Samet, J., Thurston, G. and a. c. j. f. c. Cohen, A. J. C. (2020). "Outdoor air pollution and cancer: An overview of the current evidence and public health recommendations." CA: A Cancer Journal for Clinicians. 70(6): 460–479.

Verschraegen, C., Andric, Z., Moiseenko, F., Makharadze, T., Oleksiienko, A., Ruiz, E. Y., ... Baek, E. H. (2024). 652P 3-year follow-up of a phase III study to compare efficacy and safety of a bevacizumab biosimilar, CT-P16, and reference bevacizumab as first-line treatment for metastatic or recurrent non-squamous non-small cell lung cancer. *Annals of Oncology, 35*, S1648–S1649.

Wang, F., Xie, X., Wang, L., Deng, H., Wang, Q., Qi, M., ... Li, R. (2023). Real-world data on severe lung cancer: A multicenter retrospective study. *Translational Lung Cancer Research, 12*(3), 460.

Wang, J., Lu, S., Yu, X., Hu, Y., Sun, Y., Wang, Z., ... Feng, G. (2021). Tislelizumab plus chemotherapy vs chemotherapy alone as first-line treatment for advanced squamous non–small-cell lung cancer: A phase 3 randomized clinical trial. *JAMA Oncology, 7*(5), 709–717.

Wang, M., Herbst, R. S., & Boshoff, C. J. (2021). Toward personalized treatment approaches for non-small-cell lung cancer. *Nature Medicine, 27*(8), 1345–1356.

Xu, X., Zhang, S., Xu, T., Zhan, M., Chen, C., & Zhang, C. J. (2022). Efficacy and safety of bevacizumab biosimilars compared with reference biologics in advanced non-small cell lung cancer or metastatic colorectal cancer patients: A network meta-analysis. *Frontiers in Pharmacology, 13*, 880090.

Yamamoto, H., Yatabe, Y., & Toyooka, S. (2018). Inherited lung cancer syndromes targeting never smokers. *Translational Lung Cancer Research, 7*(4), 498.

Young, R. P., Hopkins, R. J., Gamble, G. D., Etzel, C., El-Zein, R., & Crapo, J. D. (2011). Genetic evidence linking lung cancer and COPD: A new perspective. *The Application of Clinical Genetics, 4*, 99–111.

Yu, J.-Z., Kiss, Z., Ma, W., Liang, R., & Li, T. (2024). Preclinical models for functional precision lung cancer research. *Cancers (Basel), 17*(1), 22.

Zhang, Y., Luo, G., Etxeberria, J., & Hao, Y. J. (2021). Global patterns and trends in lung cancer incidence: A population-based study. *Journal of Thoracic Oncology, 16*(6), 933–944.

CHAPTER

Biosimilars and biologics in hepatocellular carcinoma

6

Dishant Chaudhari[a], **Venkateswarlu Samunooru**[b] **and Venkata Sita Rama Raju**[c]

[a]*Department of Pharmacology, School of Pharmacy and Technology Management, SVKM's Narsee Monjee Institute of Management Studies (NMIMS) (Deemed to be University), Shirpur, Maharashtra, India,* [b]*Pharmacology and Drug Discovery MSc, Coventry University, West Midlands, United Kingdom,* [c]*Department of Medical Biochemistry and Microbiology, Uppsala University, Uppsala, Sweden*

6.1 Overview of hepatocellular carcinoma

The most common primary malignancy of the liver is hepatocellular carcinoma (HCC), which comprises over 85% of the total primary cancers of the liver. It happens in hepatocytes, which are the primary functional liver cells, and is a major cause of cancer-related fatality in the world (Llovet et al., 2021). HCC is the sixth most diagnosed cancer and also the third leading cause of death caused by cancer in the world (Zheng et al., 2025). Areas of the highest HCC incidence include East Asia and sub-Saharan Africa, which harbor extensive prevalence of chronic liver diseases, predominantly because of infections with the hepatitis B virus (HBV) and hepatitis C virus (HCV) (Ding et al., 2025). Cirrhosis of many causes, alcohol abuse, nonalcoholic steatohepatitis (NASH), obesity, and risk of exposure to toxic chemicals such as aflatoxin are also important risk factors. Cancer commonly occurs in the background of chronic liver inflammation and fibrosis, but can also occur in a noncirrhotic liver. It is also more prevalent among men when compared to women, and most cases are diagnosed when the individual is already aged 60 years and above. HCC incidences are asymptomatic at early stages; thus, they are diagnosed at an advanced stage where treatment options become limited, and the prognosis becomes more objective. Nevertheless, even with the available possibilities regarding diagnosis and treatment, the overall 5-year survival is low, which states the necessity of the key to prevention, which lies in the early detection and vaccination against hepatitis, and the use of antiviral remedies. The treatment itself depends on the cancer stage, the status of the liver, and overall health, including the following: surgery resection or liver transplantation, the locoregional and systemic treatment strategies (Herbst & Reddy, 2012; Pinyopornpanish et al., 2022; Yilma et al., 2024).

Biosimilars and Biologics in Cancer Treatment: An Era in Therapeutic Strategies.
DOI: https://doi.org/10.1016/B978-0-443-43824-0.00002-X

6.1.1 Epidemiology and risk factors

HCC is a prominent global cancer burden, and it is now the third leading cause of cancer-related mortality across the globe. The total percentage of HCC occurrence in relation to the overall liver cancer is about 75%–85%. Its prevalence is highest in Asia (especially in China) and sub-Saharan Africa, where chronic HBV infection is common, although it increasingly occurs in Western countries, sometimes owing to a parallel increase in prevalence of metabolic diseases and infection with HCV. Men are two to four times more likely to develop HCC than women, and most of the patients are most likely to be diagnosed between 50 and 70 years of age, although the median age will depend on the geographical region. The heterogeneity of the HCC injury rate is that the rate is high in Geographical areas in Asia and relatively low in North America and Europe, and it is an annual state of the world scenario. With hepatitis vaccination and antiviral medication, overall global rates have leveled off or diminished in recent decades, although the number of cases being diagnosed remains on an upward trend because of population growth and aging (El-Serag, 2020).

Major risk factors of HCC include chronic infections caused by the HBV and the HCV, and both of these contribute to a larger percentage of HCC cases worldwide. Other serious risk factors are any type of cirrhosis, chronic and excessive alcohol intakes, nonalcoholic fatty liver disease (NAFLD), together with its advanced variant NASH, obesity, diabetes mellitus, and chronic intoxication with aflatoxin, which is a toxin secreted by improperly stored grains and nuts in certain developing countries. Genetic forms of metabolic liver diseases, e.g., hemochromatosis and Wilson disease, and other rare but persistent liver diseases also result in an enhanced risk (Garcia-Pagan et al., 2021). Risk factors can also be made more likely through the interaction of some risk factors, i.e., any combination of diabetes and obesity or alcohol and viral hepatitis is still more likely to result in HCC. Although the role of vaccination against HBV and efficient antiviral treatments of HBV and HCV have reduced the risk of viral HCC, the increasing cases of metabolic syndrome, obesity, and type 2 diabetes are transforming the etiology of HCC in the world today (Ntellas & Chau, 2024).

6.1.2 Pathophysiology and molecular pathways

HCC pathophysiology focuses on the sustained liver damage and hepatic inflammation that leads to the process of death of hepatocytes and their regeneration. Such sustained injury, usually due to hepatitis B or C virus infection, alcohol, metabolic illness, or exposure to toxins, results in fibrosis and cirrhosis and the establishment of an environment upon accumulation of genetic and epigenetic changes, ultimately leading to the development of malignant transformation. The tumor microenvironment, comprising inflammatory and immune cells, growth factors, and extracellular matrix, further reinforces tumor initiation and progression. On a molecular scale, HCC has been described by mutations and dysregulation of many key signaling pathways. The Wnt/ 5-catenin pathway is prone to mutations in which CTNNB1 and AXIN1 gene defects contribute to altered proliferation and survival of cells. Mutations in the promotion telomerase reverse transcriptase (TERT) are one of the prevalent

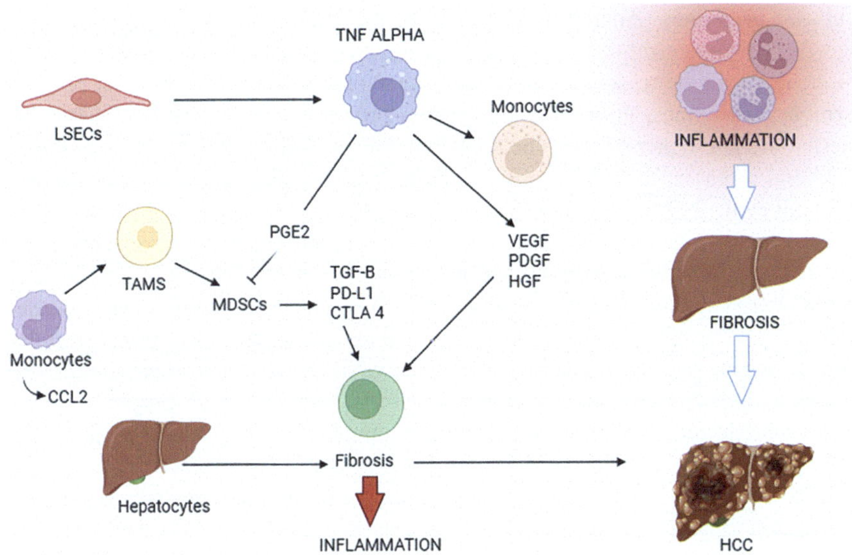

FIGURE 6.1

Molecular mechanism of hepatocellular carcinoma.

changes in genes due to which cells can evade senescence through sustaining the length of the telomeres. Approximately half of the cases have hyperactivation of the PI3K/AKT/mTOR pathway, pathway and promote cell growth, metabolism, survival, and angiogenesis (Farzaneh et al., 2021). Other significant molecular pathways are the Ras/Raf/MEK/ERK pathway related to cell proliferation and the JAK/STAT and Hedgehog pathways that contribute to inflammation, stemness, and metastasis. They also tend to inactivate tumor suppressor genes such as TP53, further promoting genomic instability. These pathways may crosstalk, causing tumor growth and resistance to therapy and contributing to the role of the immune system in the mix (Li et al., 2023) (Fig. 6.1).

6.2 The role of biologics in hepatocellular carcinoma

6.2.1 Definition and classification of biologic therapies

Biologics is a term referring to biologic therapies, which are medical products that contain components such as proteins, cells, or nucleic acids derived or produced by living organisms. These agents may occur naturally, be derived from sources (such as humans or animals), or be produced with the use of sophisticated methods of biotechnology such as recombinant DNA technology. Biologics differ in molecular size, molecular complexity, and the ability to target the complex biological mechanisms that are associated with disease processes when compared to conventional synthetic medicines (Adami et al., 2019). In the United States Public Health Service Act, biologics refer to a broad range of products, including vaccines, blood components, allergics, somatic cells, gene therapies, tissues, recombinant proteins, hormones, cytokines, growth factors, fusion proteins, monoclonal antibodies, and interferons. These therapies can be applied in the prevention, diagnosis, or treatment of several diseases, including cancer, autoimmune disease, and infectious diseases (Ghilardi et al., 2020).

Biologic therapies can be categorized based on their structure and mechanism of action. These include monoclonal antibodies (mAbs), which are engineered to target specific antigens on cells and are widely used in cancer and autoimmune diseases; cytokines and growth factors, such as interleukins and interferons, which modulate immune responses; cell-based therapies, like CAR-T cells and other immune cell treatments, which exert direct therapeutic effects; vaccines, both prophylactic and therapeutic, designed to stimulate or enhance immune activity; fusion proteins, created by combining segments of different proteins to achieve targeted outcomes; and gene therapies, which introduce or modify genes within cells to correct disease-related genetic abnormalities (Adami et al., 2019; Johnston, 2007). In HCC, biologics are rapidly becoming central to treatment strategies, particularly with the rise of immune checkpoint inhibitors (e.g., anti-PD-1, anti-PD-L1, anti-CTLA-4 monoclonal antibodies), alongside cell-based and vaccine-driven immunotherapies. These approaches aim to strengthen anti-tumor immunity, restore lost immune functions, disrupt tumor-promoting pathways, and provide therapeutic alternatives for patients who do not respond to conventional treatments (Mandlik et al., 2023; Zhang et al., 2022).

6.2.2 Mechanisms of action behind hepatocellular carcinoma

Biologic therapies work and act in terms of HCC, as well as other cancer diseases, using various specific mechanisms to target cancer cells as well as the immune system. Antibodies against programmed death-1 (PD-1)/programmed death-ligand 1 (PD-L1) and CTLA-4 disease-specific signature and immune checkpoint inhibitors block the inhibitory signal to T cells. Under normal circumstances, tumor cells will use these pathways of checkpoints to avoid the immune system. These biologics prevent the action of checkpoint proteins, which in turn activate the cytotoxic T-cell activity, causing cancer cells to be better recognized and killed (Fig. 6.1) (Papież & Krzyściak 2021). The monoclonal antibodies target a specific antigen introduced by the malignant cells. They achieve this by directly blocking the oncogenic signaling pathways, lysis of tumor by activated immune effector cells via antibody-dependent cellular cytotoxicity, complement-dependent killing, and phagocytosis. To cite an example, immune effector cells (e.g., natural killer (NK) cells) bind to the Fc region of an antibody bound to a tumor antigen and release lytic proteins to destroy the desired cellular target (Redman et al., 2015). Cell-based therapies, including CAR-T and CAR-NK cell-based therapies, are designed to recognize the presence of additional tumor-related antigens and engineer patient or donor-derived immune cells so that they can better identify and target them on a cellular basis. These cells are reintroduced in the patient who now seeks and destroys tumor cells directly (Jiang et al., 2019). Cytokines and immune agonists are other biologic agents that take advantage of the tumor microenvironment by enhancing immune activation, enhancing antigen presentation, or overcoming local immunosuppression, thereby enabling a more effective anti-tumor immune response.

6.2.3 Approved biologic drugs in hepatocellular carcinoma

6.2.3.1 Monoclonal antibodies

6.2.3.1.1 Atezolizumab and bevacizumab (Combination)

A combination of atezolizumab (an anti-PD-L1 monoclonal antibody) with bevacizumab (an anti-VEGF monoclonal antibody) now represents the first-line standard of care in patients having unresectable or advanced HCC. The respected Phase III IMbrave150 trial portrayed better overall and progression-free survival of this regimen than sorafenib. The patients who were administered atezolizumab together with bevacizumab showed a hazard ratio of 0.58 compared to the sorafenib patients and fared better in the 12-month survival time, as well as longer median progression-free survival. The distribution of adverse events was similar between groups, except that Grade 3–4 hypertension was detected when adding to the combination. This regimen has been proven to be safe and effective clinically in both trials and real-world experience and is recommended widely in the treatment of naive advanced HCC, and is listed in Table 6.1 (Jain et al., 2021).

6.2.3.1.2 Ramucirumab

Ramucirumab is a recombinant IgG1 monoclonal antibody specifically against vascular endothelial growth factor (VEGF) receptor 2 (VEGFR-2) and blocks the VEGF signaling pathway, which involves tumor angiogenesis. It is approved in the United States by the US Food and Drug Administration (FDA) as a monotherapeutic agent in advanced HCC in patients whose AFP is high (at least 400 ng/mL) even after they have received prior treatment with sorafenib. This was approved on the basis of the REACH-2 phase III trial, which demonstrated a clinically meaningful overall survival and disease control in this biomarker-selected population. Ramucirumab is the sole biologic therapy that has a predictive biomarker (AFP) in advanced HCC and a useful alternative in the second-line category (De Luca et al., 2020).

6.2.3.2 Immune checkpoint inhibitors

6.2.3.2.1 Nivolumab

Nivolumab is a fully human monoclonal antibody against programmed death-1 (PD-1) receptor, and is an immune checkpoint inhibitor. It improves T-cell-mediated anti-tumor responses by interfering with the signal pathway involved in the antigen-specific arrest of anti-tumor immune responses caused by tumors. The use of nivolumab has been shown clinically to be useful in HCC in advanced, unresectable,

Table 6.1 List of approved biosimilars drugs used in HCC.

Drug/biosimilar	Type (target)	HCC use/status	References
Bevacizumab	mAb (VEGF)	1st line (with atezolizumab)	Jain et al. (2021)
Atezolizumab	mAb (PD-L1)	1st line (combo with bevacizumab)	Jain et al. (2021)
Ramucirumab	mAb (VEGFR-2)	2nd line (AFP ≥400 ng/mL)	De Luca et al. (2020)
Nivolumab	mAb (PD-1)	2nd line/combination	Saung et al. (2021)
Ipilimumab	mAb (CTLA-4)	Combo with nivolumab	Fulgenzi et al. (2021)
Pembrolizumab	mAb (PD-1)	2nd line	Qin et al. (2023)
Durvalumab	mAb (PD-L1)	1st line (combo with tremelimumab)	Abou-Alfa et al. (2022)
Tremelimumab	mAb (CTLA-4)	Combo with durvalumab	Galvao et al. (2024)
HLX04, IBI305	Bevacizumab biosims	Used/trials (China, combo regimens)	Bachu et al. (2022)
Mvasi, Zirabev, Alymsys, Vegzelma, Avzivi, Jobevne	Bevacizumab biosims	Approved globally for oncology	Ou et al. (2024)
Sintilimab	mAb (PD-1)	1st line (with IBI305)	Zheng et al. (2024)
Serplulimab	mAb (PD-1)	Clinical trials	Ren et al. (2025)

or metastatic cases. In 2017, nivolumab received an accelerated FDA approval as a therapy against patients with HCC who failed prior sorafenib (after its objective response and tolerable safety profile in CheckMate 040 trial). The dose advised is 240 mg intravenously, every 2 weeks as a single drug. More recently, nivolumab in combination with who have established better efficacy and has been approved by the FDA for patients with unresectable or metastatic HCC intolerant to sorafenib or have disease progression. The CheckMate 9DW is a landmark study that demonstrated the combination to be a new standard of care for patients with advanced HCC in terms of both improvement in overall survival and objective response rates compared with standard tyrosine kinase inhibitors. The therapy with nivolumab includes immune-related adverse events, the majority of which are treatable, yet in some cases, immunosuppressive therapy is required. This has been the influence of immune checkpoint inhibitors as a more viable option in the management of this difficult cancer, which is ameliorated by its approved and encoded use in HCC management (Saung et al., 2021;Vithayathil & Sharma 2025; van Doorn et al., 2020).

6.2.3.2.2 Pembrolizumab in HCC

Pembrolizumab (pem) is a humanized monoclonal antibody checkpoint inhibitors that target the PD-1 receptor. It improves immune responses against tumors by therapeutically blocking the interaction of PD-1 with its ligands (PD-L1 and PD-L2), thus relieving T-cell suppression and facilitating the destruction of tumor cells. In November 2018, the FDA gave pembrolizumab accelerated approval as a second-line therapy in patients with HCC whose disease had progressed after treatment with sorafenib, after the phase II KEYNOTE-224 study. In KEYNOTE-224, pembrolizumab yielded a 17% objective response rate in patients with advanced HCC who had failed or been intolerant to sorafenib. The majority of the responses were lasting, with 89% lasting at least six months. Although the second follow-up phase III KEYNOTE-240 trial demonstrated better overall survival and progression-free survival, it just failed to reach the corresponding prespecified statistical significance. The further experiment conducted as phase III KEYNOTE-394 trial, which was carried out in Asia, revealed superiority of pembrolizumab to placebo with overall survival being higher, progression-free survival and objective responses showing higher rates, and with tolerable toxicity (Qin et al., 2023). It is usually injected intravenously with 200 mg every 3 weeks until 2 years or progression of the disease or intolerable toxicity occurs. The safety profile is characterized by the same type of immune checkpoint blockade, and the main concern is immune-mediated adverse reactions. Pembrolizumab is also a significant second-line agent to treat advanced HCC after failure or intolerance of sorafenib, especially when patients cannot tolerate VEGFA-blockade or combination immune system treatment (Munir et al., 2023).

6.2.3.2.3 Ipilimumab

Ipilimumab is a fully human monoclonal antibody that blocks the cytotoxic T-lymphocyte antigen-4 (CTLA-4) immune checkpoint, leading to greater T-cell activation and a more robust anti-tumor immune response. Ipilimumab has become a pivotal part of immunotherapy regimens for advanced HCC, especially when combined with the PD-1 inhibitor nivolumab. The FDA granted accelerated approval for the combination of nivolumab and ipilimumab for patients with advanced, unresectable, or metastatic HCC who were previously treated with sorafenib, based on data from the CheckMate 040 trial. In this study, the combination achieved an objective response rate of about 31%–32%, with durable responses and a median duration of response ranging from 17.5 to 22.2 months. The combination has since also received full FDA approval as a first-line regimen for unresectable or metastatic HCC based on CheckMate-9DW, which demonstrated superior overall survival and response rates versus standard tyrosine kinase inhibitors. Median overall survival was 23.7 months for the nivolumab plus ipilimumab arm, among the best results observed for advanced HCC to date (Fulgenzi et al., 2021; Roessler et al., 2023). The recommended dosing is nivolumab 1 mg/kg plus ipilimumab 3 mg/kg administered intravenously every three weeks for four doses, followed by maintenance nivolumab. Side effects are consistent with immune-related adverse events, including hepatitis, skin reactions, and colitis, but are generally manageable with prompt intervention (Yau et al., 2025). Ipilimumab, in combination with nivolumab, has reshaped the therapeutic landscape for HCC by offering significant survival advantages and making durable responses possible in a disease historically characterized by poor outcomes (Yau et al., 2020).

6.2.3.2.4 Durvalumab in HCC

Durvalumab is a human monoclonal antibody, the target of which is PD-L1, which is to block the interaction of PD-L1 with its receptor, PD-1, and contribute to the appearance of T-cell-mediated anti-tumor activity (Morgan et al., 2025). Durvalumab combined with tremelimumab was approved by the FDA to treat unresectable HCC based on the phase III HIMALAYA trial. During this trial, the combination of durvalumab and a single priming dose of tremelimumab had an important overall survival advantage over the current standard of care, sorafenib. The median overall survival of the combination regimen was 16.4 months, and that of sorafenib was 13.8 months; the hazard ratio was 0.78 (95% CI, 0.66092), which identified this combination as a new first-line standard of unresectable HCC (Abou-Alfa et al., 2022; Sangro et al., 2024). Clinical activity and a favorable safety profile have also been seen in patients with HCC and those infected with HCV when treated with durvalumab monotherapy. Its specialized use, though, is particularly with tremelimumab, to optimize long-lasting immune responses and clinical response over standard tyrosine kinase inhibitor treatment (Wainberg et al., 2017).

6.2.3.3 Other novel biologic agents

In addition to checkpoint inhibitors, a number of new biological treatments are in development or undergoing trials in HCC:

CAR-T cell therapy: Chimeric antigen receptor (CAR) T-cell therapies are in early-phase clinical studies in advanced HCC, and CAR T-cell glypican-3 (GPC3) therapy is in preliminary development. They have been tested preclinically to have the ability to target and kill GPC3-positive HCC cells, but no clinical evidence has been published (Batra et al., 2020; Shi et al., 2020).

Other immune modulators: Newer agents (TGF-β inhibitors (galunisertib), oncolytic viruses (Pexa-Vec), and CCR4 antibodies (mogamulizumab)) are undergoing testing in clinical trials of HCC that are in various stages and aim to overcome immune suppression or target new tumor antigens (Kelley et al., 2019).

Targeted antibody-drug conjugates, bispecific antibodies, and vaccine approaches: These will increase the specificity of tumor targeting and the immune effect is hoped by increasing immune-mediated killing; many are experimental (Doi et al., 2019; Ono et al., 2022).

6.3 Comparative clinical trials and approval pathway

A major pillar in the biosimilar approval process, comparative clinical trials are performed to show that a biosimilar is similar to its reference biologic in efficacy, safety, and immunogenicity. This development takes place in a step-wise, multi-tiered manner and starts with detailed analytical and functional comparability studies, then proceeds to nonclinical studies, where these are required, and finally, to comparative clinical tests. Such clinical trials generally involve a head-to-head evaluation of pharmacokinetics (PK), pharmacodynamics (PD), where suitable biomarkers exist, and direct determination of clinical efficacy and safety outcomes (Stebbing et al., 2020). Clinical benefit re-establishment is not the primary aim of the comparative clinical trials of biosimilars, as it has already been achieved with the reference product. Rather, these examinations are designed to establish clinical equivalence within reestablished margins utilizing the mentioned populations and endpoints to ensure they are sensitive enough to recognize any potential meaningful variations between the biosimilar and its reference. Such endpoints can be short-term surrogate outcomes (e.g., overall response rate or pathologic complete response), instead of longer-term outcomes (e.g., overall survival), because they are more sensitive to detect differences that can be attributed to therapeutic products (Stebbing et al., 2020). All studies carry out comparative safety and immunogenicity, and scientific rationale may allow extrapolation of data to the point that the biosimilar could be licensed to treat conditions not specifically assessed in comparative clinical trials and yet covered by other products of the reference product, once such products have been marketed. Together, these requirements ensure a strong approach to the biosimilar approval in terms of its totality of evidence, contributing to assuring their usage as therapeutic equivalents when being used in clinical practice (Hung et al., 2017).

6.4 Biosimilars in hepatocellular carcinoma: Current status and drugs

6.4.1 Available biosimilars relevant to hepatocellular carcinoma

The usage of biosimilars is gaining ground with regard to the treatment of HCC, especially in such areas where cost-related issues associated with biologic treatment present a significant impediment. Some of the most promising developments are biosimilars of bevacizumab, an anti-VEGF monoclonal antibody, which have been safe and efficacious in HCC as compared to the originator reference product. In a phase 2 clinical trial, serplulimab (PD-1 inhibitor) in combination with a biosimilar of bevacizumab (HLX04) demonstrated efficacy and good safety as first-line treatment of advanced HCC. Combination provided a 29.3 percent objective response rate, 7.3 months median progression-free survival, and 20.4 months median overall survival, and clarified the usefulness of combination use in this setting (Ren et al., 2025). A number of studies have noted the use of sintilimab (a PD-1 monoclonal antibody) combined with a bevacizumab biosimilar as a first-line and conversion therapy to unresectable or advanced HCC, which has shown improved progression-free and overall survival in the ORIENT-32 trial, with regulatory approval now granted in countries like China (Zeng et al., 2023). In addition, real-world and economic comparisons have also endorsed the cost-effective use of biosimilars to bevacizumab in HCC, with major clinical advantage and healthcare savings to its parent biologic (Zeng et al., 2023). Tolerable toxicity, encouraging efficacy, and cost-effectiveness in patients with advanced HCC who receive first-line sintilimab plus bevacizumab biosimilar regimens were also confirmed last year in a systematic review (Sankar et al., 2023).

6.4.2 Biosimilars of bevacizumab

Multiple bevacizumab biosimilars have been approved globally and are relevant for use in oncology, including HCC, often combined with immune checkpoint inhibitors or as standalone anti-angiogenic agents.

Currently approved and available bevacizumab biosimilars include:

Mvasi 125, 250 mg (bevacizumab-awwb): Mvasi is a bevacizumab biosimilar developed by Amgen/Allergan, the first to be approved by the FDA (authorized in September 2017) and is approved in all Avastin cancer indications, including HCC in some jurisdictions (Xu et al., 2022).

Zirabev 2017(bevacizumab-bvzr): Zirabev is a Pfizer drug approved in June 2019. Clinical evidence shows the same efficacy and safety as that of a reference bevacizumab (Zhao et al., 2023).

Alymsys 12Â (bevacizumab-maly): The US, as well as other markets, has Alymsys (Amneal Pharmaceuticals/Sanofi) approval in 4/22 as an Avastin biosimilar, in good supply (Xu et al., 2022).

Vegzelma (bevacizumab-adcd): The Vegzelma manufactured by Celltrion in 2022 was approved by the FDA in September 2022 and sells in different countries (Ou et al., 2024).

Avzivi Â© (bevacizumab-tnjn): Avzivi is a biosimilar of Avastin, manufactured by Bio-Thera Solutions, and in December 2023 was approved for all reference indications of Avastin (Xu et al., 2022).

Jobevne 2 1 (bevacizumab-nwgd); Jobevne (approved by the FDA in April 2025 in Biocon Biologics) is sold in Canada and Europe as ABEVMY; this antibody replica has all standard bevacizumab indications, such as some groups of HCC patients (Ou et al., 2024).

6.4.2.1 Therapeutic equivalence/safety equivalence

A Previous study data of a meta-analysis of clinical trials established comparability of efficacy and safety between bevacizumab biosimilars such as Mvasi, Zirabev, Alymsys, and Vegzelma, with reference to Avastin, with regard to objective response rates, progression-free survival, overall survival, and adverse events in oncology patients has been found. The feasibility of interchangeability confirmation in the clinical and real-world settings further supports the safety of any reference to biosimilar switching among the cancer patients eligible to receive both products without losing efficacy or causing excessive toxicity.

6.4.3 Biosimilars of other relevant biologics

There are other relevant biologics besides bevacizumab whose biosimilars are currently being developed at a rapid rate in oncology and present a major challenge to HCC. The most noticeable ones are biosimilars of monoclonal antibodies like rituximab and trastuzumab that have been extensively approved and used in different cancers. Recent systematic reviews and meta-analyses confirm that these biosimilars are very like their reference product with regard to clinical efficacy, pharmacokinetics, safety, and immunogenicity, satisfying all the regulatory expectations to be approved and used in oncology (Galvao et al., 2024).

6.4.3.1 Rituximab biosimilars

Biosimilars such as CT-P10, GP2013, etc. have shown similar results in randomized controlled trials to the parent rituximab in lymphoma and related pathologies. These trials verify comparable overall response rates, progression-free survival, and safety. Direct use of these biosimilars in HCC is not routine, but these biosimilars do fall into a more general category of biosimilar monoclonal antibodies that are useful in oncologies (Broer et al., 2024).

6.4.3.2 Trastuzumab biosimilars

Various trastuzumab biosimilars like ABP980, CT-P6, and SB3 established equivalence trials and head-to-head comparators proved to be medically identical to reference trastuzumab in breast and stomach cancers, and the success of trastuzumab biosimilars set a regulatory precedent in oncology contexts (Galvao et al., 2024).

6.4.3.3 Pegfilgrastim and filgrastim biosimilars

This G-CSF biosimilar includes pegfilgrastim-jmdb, filgrastim-sndz, and other derivatives, which are indicated to treat supportive care of neutropenia in patients undergoing chemotherapy with solid tumors, such as HCC, under myelosuppressive regimens. Several studies attest to equal efficacy in mitigating the risk of infection, easing the intensity of chemotherapy, and an absence of new safety issues (Bachu et al., 2022).

6.4.3.4 Emerging checkpoint inhibitor biosimilars

Although no immune checkpoint blockade biosimilars (e.g., PD-1/PD-L1, CTLA-4) have been approved, as yet, as of mid-2025, a number are under late-stage clinical development. This is a novel topic that has been mentioned in systematic reviews that mention the fast growth of this type of biosimilar, and how it may be applicable in treating HCC in the near future (Bachu et al., 2022).

6.4.4 Pipeline biosimilars in hepatocellular carcinoma

A number of pipeline biosimilars are in various stages of development in HCC in relation to using anti-VEGF agents as well as monoclonal antibodies in immunotherapy and targeted therapy.

HLX04: This bevacizumab biosimilar, developed by Henlius Biotech, recently passed phase III clinical trials and demonstrated primary endpoints of safety, efficacy, and immunogenicity compared to the biologic reference. HLX04 is also under clinical development in combination with anti-PD-1 antibody HLX10 targeting a range of solid tumors, such as advanced HCC, with phase II studies currently ongoing in this setting (Bachu et al., 2022).

IBI305: A second bevacizumab-biosimilar with phase II/III trial data in HCC. In combination with the PD-1 inhibitor sintilimab, IBI305 is efficacious as a first-line treatment of unresectable HCC and could provide an alternative to branded bevacizumab-based combinations (Zheng et al., 2024).

Other candidates: A number of other bevacizumab biosimilars are in global development, such as SB8, FKB238, and BI 695502, which have proceeded through phase 3 predominantly in colorectal and lung, yet their wider uptake in HCC is promising as data on effectiveness are accruing (Bachu et al., 2022).

Rituximab and trastuzumab biosimilars: These are not yet approved in HCC, but pipelines offer products: HLX01 (biosimilar rituximab) and HD201 (biosimilar trastuzumab), which have gone through phase 3 studies and which may be extended to liver cancer indications as research and combination regimens evolve (Ivanov et al., 2024).

Checkpoint Inhibitor Biosimilars: There are no approved or widely published biosimilars to HCC, but novel biosimilars targeting PD-1/PD-L1 and CTLA-4 biosimilar activity are in early-phase clinical studies, and development has been discussed as an imminent future possibility (Morita et al., 2016).

6.5 Comparative efficacy and safety: Biologics versus biosimilars in hepatocellular carcinoma

6.5.1 Clinical trial data and real-world evidence

Many clinical trials and real-world evidence have shown an equivalent performance and safety of biosimilars relative to their reference biologics in oncology, as well as HCC. A meta-analysis of randomized controlled trials (RCTs) of bevacizumab, rituximab, and trastuzumab biosimilars showed no significant difference in any objective response rates, progression-free survival, overall survival, or frequency of grade 3–5 adverse events between biosimilars and their originators in any of the cancers involved. There were six RCTs dealing with bevacizumab biosimilars that proved that they are equivalent to the original product in terms of efficacy and safety profile

(Yang et al., 2019). The results of a particular meta-analysis of biosimilars of bevacizumab indicated no clinically relevant variation in efficacy (objective response rate: risk ratio 0.98, progression-free survival: risk ratio 1.01, overall survival: risk ratio 1.06) and safety (grade 3.5 adverse events: odds ratio 1.03) among patients with solid tumors and HCC as well. These concepts showed a more concrete outcome in both the network and the pair-wise analyses, which bolsters the idea of safety and equivalence both in real-world care and in randomized controlled trials. These findings have been reaffirmed in recent real-world studies in China: A combination therapy consisting of sintilimab plus a bevacizumab biosimilar in patients with advanced HCC showed comparable overall survival, progression-free survival, and rates of adverse events to the reference. Notably, use of biosimilars did not reduce clinical benefit or impair patient safety, yet realizing significant savings to the economy (Xu et al., 2022).

6.5.2 Safety profile comparison

Several comparative studies and meta-analyses have demonstrated that biosimilars and their respective reference biologics have very similar safety profiles when used in oncology, including in the case of HCC. A large post-marketing pharmacovigilance reflection on biosimilars of bevacizumab, rituximab, and trastuzumab showed that adverse drug reactions (ADRs) were nonserious with similar patterns to the observation of originators; no new or unexpected signals and no crucifying difference of severity or rates of ADRs between biosimilars and biologics were reported (Nikitina et al., 2023). In a meta-analysis of 5,749 patients using randomized controlled trials that compared bevacizumab biosimilars with reference bevacizumab, there was no significant difference in the rate of grade 3.5 adverse reactions (odds ratio 1.03 [95%CI, 0.9116]); with hypertension, proteinuria, and bleeding being common adverse events that were not more prevalent as a result of biosimilar usage. The data presented by another retrospective observational study were used to demonstrate that replacement of the originator medication with the use of a bevacizumab biosimilar did not evoke an augmentation in adverse events in a retrospective observational analysis of patients with cancer who switched to a biosimilar.

6.5.3 Immunogenicity considerations

Immunogenicity, or the ability of biologic drugs to trigger anti-drug antibody (ADA) reactions, has become one of the keen priorities in the assessment of biosimilars in association with HCC and other cancer conditions. Severe comparative and regulatory studies show that monoclonal antibodies, bevacizumab being one, have immunogenicity rates and neutralizing antibody (NAb) incidences comparable to those of their source biologics. As an illustration, a thorough review of 22 biosimilar monoclonal antibodies approved by the FDA reported that the differences in observed ADA and NAb incidences between the biosimilar and the reference products were generally either lower or within −15% and −20%, respectively, and no changes in immunogenicity were observed in switching trials. Similar effects on pharmacokinetic parameters were also seen in these biosimilars, providing further support to the regulatory conclusion that there is no greater immunogenic risk in biosimilars compared with originators (Cheng et al., 2023). Low levels of ADA and NAb testing were found in a phase III study of the bevacizumab biosimilar HLX04 compared to reference bevacizumab, with 339 patients in the biosimilar group (0.3%) and none in either group being found NAb positive. Similar findings were not observed in terms of clinical safety or efficacy outcomes, which underscores the fact that the immunogenicity of the younger generation of biosimilars is largely negligible and aligns immunogenicity with that found with their reference counterparts (Qin et al., 2021). Moreover, a meta-analysis of randomized controlled trials comparing the effects of various bevacizumab biosimilars revealed that there was no tremendous change in the occurrence of ADAs and corresponding immune precipitations, which is a validation of a similar effect when administered in oncology protocols. Such convergent results among clinical and real-world studies lead to the deduction that the immunogenicity assessment of biosimilars does not raise any novel clinical safety or efficacy issues relative to their reference biologics in HCC and cancer therapeutic settings, in general.

6.6 Economic and access considerations

6.6.1 Cost-effectiveness of biologics and biosimilars

Most of the recent cost-effectiveness studies and real-world evidence analyze that biosimilars provide significant cost savings opportunities over reference biologics in terms of treatment cost reduction of HCC. A Chinese cost-effectiveness analysis of sintilimab in combination with a bevacizumab biosimilar against lenvatinib as first-line treatment of unresectable or metastatic HCC showed the addition of the bevacizumab biosimilar achieved extra quality-adjusted life-years (QALYs) with an incremental cost-effectiveness ratio (ICER) of $24,462 per QALY gained, which is well within the average willingness-to-pay thresholds of the These findings were further corroborated by another real-world analysis which reported that sintilimab combined with a bevacizumab biosimilar regimen yielded 1.97 QALYs and 2.92 overall years of life per patient with a great savings in direct medical costs thus highly cost-effective and accessible (Zhou et al., 2022). In a greater scope, real-world analysis in oncology demonstrated that bevacizumab biosimilars were related to large overall cost savings (on average, 6379 dollars less per treated patient) and were equivalent to the originator drug in terms of its survival effects. Such cost savings are able to lead to better access to life-saving therapies, at least in resource-restricted healthcare systems, without compromising efficacy or safety. The extensive uptake of biosimilars in oncologic practice is likely to not only reduce the healthcare budget but also improve access of patients to innovative biologic therapy in HCC and other cancer diseases (Lu et al., 2023).

6.7 Challenges and limitations

6.7.1 Barriers to adoption of biosimilars

The use of biosimilars in the treatment of cancer, in this case of HCC, has several obstacles in its implementation, regardless of the excellent clinical results and regulations that approve their similarity with the original biologics. Among the greatest challenges is the unawareness of educating prescribers and patients about the safety, effectiveness, and regulatory requirements on which biosimilars are based. The surveys and studies indicate physician unwillingness, in part because of gaps in knowledge about interchangeability and extrapolation of indications and persistent uncertainties about immunogenicity and the validity of the comparability exercise conducted at the time of regulatory approval. Despite the education, there are questions about the handling of switching stable patients to biosimilar and the monitoring of safety, which is long-term in their application in the real world (Lu et al., 2023). Difficulties of accessing the market also slow down uptake. The application of payer and formulary policies might not act symmetrically to endorse biosimilars, and more complicated rebate or contracting systems may discourage transfers between reference products and low-cost biosimilars, which slows penetration of the market and constrains expected cost reductions. Other hitching points: delays in patent suits and legal hurdles by makers of reference products that often delay the release of biosimilars and narrow their commercial lifespan. The doctors can also be affected by issues of reimbursement rates as well as administrative burdens, whereas the patients tend not to be knowledgeable about biosimilars, causing a greater degree of apprehension about utilizing the former, as opposed to using reference drugs, which are more established in the market (Lu et al., 2023).

6.8 Future directions

6.8.1 Ongoing research and drug development

The current biosimilar-based drug development surrounding the treatment of HCC is in its nascent stages and has a wide range of candidates at various stages of testing. Due to patent expirations of reference biologics, there is a proliferation of biosimilars in the market globally that offer conventional therapeutic value at lower costs in oncology. The recent wave of work is with biosimilars of major monoclonal antibodies, notably bevacizumab, and agents like HLX04 and BI 695502 have completed comparative trials to phase 3, and others, including CT-P16, HD204, CBT124, and MIL 60, are in advanced trials. Such candidates are not being tested only in lung and colorectal cancers, but also their efficacy and safety are being tested in HCC (Bachu et al., 2022). Concomitantly, there are early- and later-stage clinical development of biosimilars of immunotherapy agents. Such as HLX10 (an anti-PD-1 antibody) in phase 2 in advanced HCC and other solid cancers, and SCT510, which is an anti-VEGF biosimilar in phase 2 studies in combination with HCC. More broadly in the oncology arena, there is an emerging portfolio of biosimilars to other biologics, including rituximab and trastuzumab, in phase 1 (and some phase 2 and 3) development, with other studies examining their safety, efficacy, and immunogenicity (Tella et al., 2022). New combination treatment approaches have also been tested in recent clinical trial strategies: trials are being conducted using biosimilars of immune checkpoint inhibitors used in combination with biosimilar antiangiogenic agents or multitarget kinase inhibitors. Multiple pipeline agents are under large randomized, multi-country trials, and early encouraging data are reported on the new immunomodulatory and targeted treatments, including drugs based on biosimilar backbone agents.

6.8.2 Personalized medicine in hepatocellular carcinoma

The future of personalized medicine in HCC is promising, with more knowledge about the molecular and genetic diversity of HCC developing tumors. The latest targeted approaches depend on an actionable molecular profiling of the tumors, such as sequencing panels of actionable mutations, analysis of gene expression panels, and the molecular subtype. Recent surveys indicate the importance of patient-derived tumor models and preclinical bioengineering platforms, which more faithfully recap and maintain the peculiarities of each tumor of this or that patient and of their reaction to drugs, respectively. This aspect improves the choice of therapy and helps realize individual treatment decisions. Although mutations have been identified, translating them into regular clinical practice decision-making remains difficult but is likely to become easier as we learn more about resistance pathways and tumor evolution (Sun et al., 2023).

6.8.3 Emerging therapies and combination strategies

New treatments that target HCC focus on new single-agent therapy and combination therapy with multiple drugs. The most significant emerging developments include immune-based combinations, in particular, integrating immune checkpoint inhibitors (ICIs) with anti-angiogenic agents (i.e., the atezolizumab-bevacizumab combination), which has demonstrated better survival and response rates compared to standard kinase inhibitors and is now a standard first-line treatment of advanced HCC. The HIMALAYA trial has also determined that combined durvalumab and tremelimumab is superior compared to monotherapies of sorafenib. Currently ongoing clinical trials are investigating the diverse next-generation approaches, such as combination regimens combining ICIs with tyrosine kinase inhibitors, novel targeted agents (e.g., FGF19, MET, and PARP inhibitors), and multi-modal therapy (e.g., use of local therapies (transarterial chemoembolization (TACE) or radiofrequency ablation) to target an individual patient (Zhonghao & Fan 2024).

6.9 Case studies and clinical practice guidelines

6.9.1 Representative clinical cases

Two case reports describe the changing HCC treatment planning and the increased use of biosimilars in multimodal regimens. As an example, a recent case report explained a 38-year-old woman who had a massive HCC and refused surgical resection; she underwent noninvasive systemic treatment, including immunotherapy (sintilimab), antiangiogenic therapy, and radiotherapy. This strategy resulted in the complete clinical response and survival time longer than 36 months, proving the possible interaction of such modalities. Other studies cited in the report, in which sintilimab in combination with a bevacizumab biosimilar (IBI305) was applied as frontline therapy in unresectable HCC, aimed to actively fight a tumor vasculature and immune microenvironment (Li et al., 2024). In addition, the effectiveness of this regimen based on clinical evidence at a population level in China was found in a retrospective analysis. The trial incorporated 112 patients with advanced HCC presenting a meaningful effect in overall survival, progression-free survival, and tolerability around first-line care applied with sintilimab plus a bevacizumab biosimilar in real-world experience in clinical practice. These results are consistent with the most recent clinical practice guidelines, including those of the European Association for the Study of the Liver (EASL), indicating that evidence-based, multidisciplinary, and patient-customized practices should be used. Involvement of systemic combination therapy is acknowledged in expanding roles, and biosimilars have been included in the guidelines as potentially cost-effective and clinically validated options in HCC treatment (Zeng et al., 2023).

6.9.2 Guideline recommendations for biologics and biosimilars in HCC

Major international HCC Clinical practice guidelines, such as those of the EASL and the American Association for the Study of Liver Diseases (AASLD), highlight evidence-based, multidisciplinary HCC management with patients and the value of biologic therapies, especially immune checkpoint blockers and anti-angiogenic agents, in advanced disease (Galle et al., 2018). The recent guidance document appears to recognize the value of biosimilars more and more, given that biosimilars are affordable substitutes (cost-effective) to reference biologics in oncology, such as HCC. In 2004 EASL guideline (), biosimilars used as part of a systemic combination regimen are explicitly included, based on available evidence of strong equivalence, and it is argued to ensure that their inclusion is integrated where possible to increase patient access and limit costs. Asian and global consensus papers also support the use of biosimilars subject to compliance with regulatory and quality standards, and these are likely to be used more extensively as real-world experience and outcome data are further accumulated (Lau et al., 2024). At the same time, a recommendation is that similar principles used in clinical monitoring, pharmacovigilance, and multidisciplinary care in reference biologics should also be adapted to biosimilars. It is emphasized by guidelines that such decisions on whether to use a biosimilar or a reference biologic must be informed by the strong evidence of therapeutic equivalence, availability of products, patient preference, and regulatory approvals in a given location (Sangro et al., 2025).

6.10 Conclusion

The landscape of HCC treatment is undergoing a significant transformation with the advent and adoption of biologics and biosimilars. Biologic agents, including monoclonal antibodies targeting angiogenic and immune pathways, have introduced new therapeutic opportunities, exemplified by drugs such as ramucirumab and immune checkpoint inhibitors. These advances are particularly important for patients with advanced or refractory HCC, where few effective systemic therapies were previously available. Biosimilars further expand patient access to these advanced therapeutics, offering comparable safety and efficacy profiles to their reference products, and have the potential to reduce treatment costs and improve global equity in HCC care. Early- and late-stage clinical development of biosimilars targeting key molecular drivers and immune checkpoints in HCC holds promise for future integration into standard practice. Importantly, progress in molecular profiling and personalized medicine is starting to guide the use of biologics and biosimilars in HCC, paving the way for more tailored and effective treatments. Advances in preclinical models and bioengineering technologies will further benefit drug evaluation and the individualization of therapy. Continued research into the unique genetic and molecular heterogeneity of HCC, as well as mechanisms of resistance, will be essential to maximize the benefits of these novel biologic strategies.

Abbreviation

CTLA 4	cytotoxic T-lymphocytes-associated protein 4

References

Abou-Alfa, G. K., Lau, G., Kudo, M., et al. (2022). Tremelimumab plus durvalumab in unresectable hepatocellular carcinoma. *NEJM Evidence, 1*(8), EVIDoa2100018. https://doi.org/10.1056/EVIDoa2100070.

Adami, G., Saag, K. G., Chapurlat, R. D., et al. (2019). Balancing benefits and risks in the era of biologics. *Therapeutic Advances in Musculoskeletal Disease., 11*, 1759720X19870145. https://doi.org/10.1177/1759720X19883973.

Bachu, R. D., Abou-Dahech, M., Balaji, S., et al. (2022). Oncology biosimilars: New developments and future directions. *Cancer Reports, 5*, e1536. https://doi.org/10.1002/cnr2.1720.

Batra, S. A., Rathi, P., Guo, L., et al. (2020). Glypican-3–specific CAR T cells coexpressing IL15 and IL21 have superior expansion and antitumor activity against hepatocellular carcinoma. *Cancer Immunology Research, 8*, 309–320. https://doi.org/10.1158/2326-6066.CIR-19-0293.

Broer, L. N., Knapen, D. G., de Groot, D.-J. A., et al. (2024). Monoclonal antibody biosimilars for cancer treatment. *Iscience, 27*, 110115. https://doi.org/10.1016/j.isci.2024.110115.

Cheng, C., Jiang, A., Liu, Y., & Chang, L. (2023). Investigation of immunogenicity assessment of biosimilar monoclonal antibodies in the United States. *Clinical Pharmacology & Therapeutics, 114*, 1274–1284. https://doi.org/10.1002/cpt.3033.

De Luca, E., Marino, D., & Di Maio, M. (2020). <p>Ramucirumab, a second-line option for patients with hepatocellular carcinoma: A review of the evidence. *Cancer Management and Research, 12*, 3721–3729. https://doi.org/10.2147/CMAR.S216220.

Ding, Z., Wang, L., Sun, J., et al. (2025). Hepatocellular carcinoma: Pathogenesis, molecular mechanisms, and treatment advances. *Frontiers in Oncology, 15*, 1567890. https://doi.org/10.3389/fonc.2025.1526206.

Doi, T., Muro, K., Ishii, H., et al. (2019). A phase I study of the anti-CC chemokine receptor 4 antibody, mogamulizumab, in combination with nivolumab in patients with advanced or metastatic solid tumors. *Clinical Cancer Research, 25*, 6614–6622. https://doi.org/10.1158/1078-0432.CCR-19-1090.

El-Serag, H. B. (2020). Epidemiology of hepatocellular carcinoma. In *The liver* (pp. 758–772). Wiley.

Farzaneh, Z., Vosough, M., Agarwal, T., & Farzaneh, M. (2021). Critical signaling pathways governing hepatocellular carcinoma behavior; small molecule-based approaches. *Cancer Cell International, 21*, 208. https://doi.org/10.1186/s12935-021-01924-w.

Fulgenzi, C. A. M., Talbot, T., Murray, S. M., et al. (2021). Immunotherapy in hepatocellular carcinoma. *Current Treatment Options in Oncology, 22*, 87. https://doi.org/10.1007/s11864-021-00886-5.

Galle, P. R., Forner, A., Llovet, J. M., et al. (2018). EASL clinical practice guidelines: Management of hepatocellular carcinoma. *Journal of Hepatology, 69*, 182–236. https://doi.org/10.1016/j.jhep.2018.03.019.

Galvao, T. F., Livinalli, A., Lopes, L. C., et al. (2024). Biosimilar monoclonal antibodies for cancer treatment in adults. *Cochrane Database of Systematic Reviews, 2024*, CD015123. https://doi.org/10.1002/14651858.CD013539.pub2.

Garcia-Pagan, J. C., Francoz, C., Montagnese, S., et al. (2021). Management of the major complications of cirrhosis: Beyond guidelines. *Journal of Hepatology, 75*, S135–S146. https://doi.org/10.1016/j.jhep.2021.01.027.

Ghilardi, N., Pappu, R., Arron, J. R., & Chan, A. C. (2020). 30 years of biotherapeutics development—What have we learned? *Annual Review of Immunology, 38*, 249–287. https://doi.org/10.1146/annurev-immunol-101619-031510.

Herbst, D. A., & Reddy, K. R. (2012). Risk factors for hepatocellular carcinoma. *Clinical Liver Disease (Hoboken), 1*, 180–182. https://doi.org/10.1002/cld.111.

Hung, A., Vu, Q., & Mostovoy, L. (2017). A systematic review of U.S. biosimilar approvals: What evidence does the FDA require and how are manufacturers responding? *Journal of Managed Care & Specialty Pharmacy, 23*, 1234–1244. https://doi.org/10.18553/jmcp.2017.23.12.1234.

Ivanov, A., Getova-Kolarova, V., Belcheva, V., et al. (2024). Research and pilot analysis of *bevacizumab* repurposing potential and its impact in clinical practice. *Biotechnology & Biotechnological Equipment, 38*, 2296402. https://doi.org/10.1080/13102818.2024.2357191.

Jain, A., Chitturi, S., Peters, G., & Yip, D. (2021). Atezolizumab and bevacizumab as first line therapy in advanced hepatocellular carcinoma: Practical considerations in routine clinical practice. *World Journal of Hepatology, 13*, 1132–1142. https://doi.org/10.4254/wjh.v13.i9.1132.

Jiang, Y., Han, Q.-J., & Zhang, J. (2019). Hepatocellular carcinoma: Mechanisms of progression and immunotherapy. *World Journal of Gastroenterology, 25*, 3151–3167. https://doi.org/10.3748/wjg.v25.i25.3151.

Johnston, S. L. (2007). Biologic therapies: What and when? *Journal of Clinical Pathology, 60*, 8–17. https://doi.org/10.1136/jcp.2005.032300.

Kelley, R. K., Gane, E., Assenat, E., et al. (2019). A phase 2 study of galunisertib (TGF-β1 receptor type i inhibitor) and sorafenib in patients with advanced hepatocellular carcinoma. *Clinical and Translational Gastroenterology, 10*, e00056. https://doi.org/10.14309/ctg.0000000000000056.

Lau, G., Obi, S., Zhou, J., et al. (2024). APASL clinical practice guidelines on systemic therapy for hepatocellular carcinoma-2024. *Hepatology International, 18*, 1661–1683. https://doi.org/10.1007/s12072-024-10732-z.

Li, N., Wang, Y., Wang, A., et al. (2023). STS1 and STS2 phosphatase inhibitor baicalein enhances the expansion of hematopoietic and progenitor stem cells and alleviates 5-fluorouracil-induced myelosuppression. *International Journal of Molecular Sciences, 24*, 2987. https://doi.org/10.3390/ijms24032987.

Li, Y., Lai, Y., Luo, X., et al. (2024). Case report: Massive hepatocellular carcinoma with complete response to the non-surgical systematic treatment strategy. *Frontiers in Oncology, 14*, 1304567. https://doi.org/10.3389/fonc.2024.1291131.

Llovet, J. M., Kelley, R. K., Villanueva, A., et al. (2021). Hepatocellular carcinoma. *Nature Reviews Disease Primers, 7*, 6. https://doi.org/10.1038/s41572-020-00240-3.

Lu, B., Dvorani, E., Nguyen, L., et al. (2023). Cost-effectiveness analysis of bevacizumab biosimilars for metastatic colorectal cancer: A comparative study using real-world data. *JCO Oncology Practice, 19*, 9. https://doi.org/10.1200/OP.2023.19.11_suppl.9.

Mandlik, D. S., Mandlik, S. K., & Choudhary, H. B. (2023). Immunotherapy for hepatocellular carcinoma: Current status and future perspectives. *World Journal of Gastroenterology, 29*, 1054–1075. https://doi.org/10.3748/wjg.v29.i6.1054.

Morgan, C., Kurland, J., Patel, M., et al. (2025). Differential safety profiles of durvalumab monotherapy and durvalumab in combination with tremelimumab in adult patients with advanced cancers. *Journal for ImmunoTherapy of Cancer, 13*, e011140. https://doi.org/10.1136/jitc-2024-011140.

Morita, J., Tanaka, M., Nomoto, M., et al. (2016). Pharmacokinetic bioequivalence, safety, and immunogenicity of DMB-3111, a trastuzumab biosimilar, and trastuzumab in healthy Japanese adult males: Results of a randomized trial. *Biodrugs, 30*, 17–25. https://doi.org/10.1007/s40259-015-0153-2.

Munir, T., Moreno, C., Owen, C., et al. (2023). Impact of minimal residual disease on progression-free survival outcomes after fixed-duration ibrutinib-venetoclax versus chlorambucil-obinutuzumab in the GLOW study. *Journal of Clinical Oncology, 41*, 3689–3699. https://doi.org/10.1200/JCO.22.02283.

Nikitina, V., Santi Laurini, G., Montanaro, N., & Motola, D. (2023). Comparative safety profiles of oncology biosimilars vs. originators in Europe: An analysis of the EudraVigilance database. *Cancers (Basel), 15*, 3680. https://doi.org/10.3390/cancers15143680.

Ntellas, P., & Chau, I. (2024). Updates on systemic therapy for hepatocellular carcinoma. *American Society of Clinical Oncology Educational Book, 44*, e433–e446. https://doi.org/10.1200/EDBK_430028.

Ono, S., Suzuki, S., Kondo, Y., et al. (2022). Trametinib improves Treg selectivity of anti-CCR4 antibody by regulating CCR4 expression in CTLs in oral squamous cell carcinoma. *Scientific Reports, 12*, 21678. https://doi.org/10.1038/s41598-022-22773-1.

Ou, W.-F., Hsu, K.-H., Tseng, J.-S., et al. (2024). Real-world clinical efficacy of bevacizumab biosimilar in patients with advanced non-small-cell lung cancer. *Therapeutic Advances in Medical Oncology, 16*, 17588359241234567. https://doi.org/10.1177/17588359241290718.

Papież, M. A., & Krzyściak, W. (2021). Biological therapies in the treatment of cancer—Update and new directions. *International Journal of Molecular Sciences, 22*, 11694. https://doi.org/10.3390/ijms222111694.

Pinyopornpanish, K., Al-Yaman, W., Dasarathy, S., et al. (2022). Hepatocellular carcinoma in patients without cirrhosis: The fibrosis stage distribution, characteristics and survival. *Digestive Diseases and Sciences, 67*, 2677–2687. https://doi.org/10.1007/s10620-021-07048-5.

Qin, S., Chen, Z., Fang, W., et al. (2023). Pembrolizumab versus placebo as second-line therapy in patients from Asia with advanced hepatocellular carcinoma: A Randomized, double-blind, phase iii trial. *Journal of Clinical Oncology, 41*, 1434–1443. https://doi.org/10.1200/JCO.22.00620.

Qin, S., Li, J., Bai, Y., et al. (2021). Efficacy, safety, and immunogenicity of HLX04 versus reference bevacizumab in combination with XELOX or mFOLFOX6 as first-line treatment for metastatic colorectal cancer: Results of a randomized, double-blind phase iii study. *Biodrugs, 35*, 445–458. https://doi.org/10.1007/s40259-021-00484-9.

Redman, J. M., Hill, E. M., AlDeghaither, D., & Weiner, L. M. (2015). Mechanisms of action of therapeutic antibodies for cancer. *Molecular Immunology, 67*, 28–45. https://doi.org/10.1016/j.molimm.2015.04.002.

Ren, Z., Shao, G., Shen, J., et al. (2025). Phase 2 study of serplulimab with the bevacizumab biosimilar HLX04 in the first-line treatment of advanced hepatocellular carcinoma. *Cancer Immunology, Immunotherapy, 74*, 69. https://doi.org/10.1007/s00262-024-03917-w.

Roessler, D., Öcal, O., Philipp, A. B., et al. (2023). Ipilimumab and nivolumab in advanced hepatocellular carcinoma after failure of prior immune checkpoint inhibitor-based combination therapies: A multicenter retrospective study. *Journal of Cancer Research and Clinical Oncology, 149*, 3065–3073. https://doi.org/10.1007/s00432-022-04206-8.

Sangro, B., Argemi, J., Ronot, M., et al. (2025). EASL clinical practice guidelines on the management of hepatocellular carcinoma. *Journal of Hepatology., 82*, 315–374. https://doi.org/10.1016/j.jhep.2024.08.028.

Sangro, B., Chan, S. L., Kelley, R. K., et al. (2024). Four-year overall survival update from the phase III HIMALAYA study of tremelimumab plus durvalumab in unresectable hepatocellular carcinoma. *Annals of Oncology., 35*, 448–457. https://doi.org/10.1016/j.annonc.2024.02.005.

Sankar, K., Gong, J., Osipov, A., et al. (2023). Recent advances in the management of hepatocellular carcinoma. *Clinical and Molecular Hepatology, 30*, 1–15. https://doi.org/10.3350/cmh.2023.0125.

Saung, M. T., Pelosof, L., Casak, S., et al. (2021). <scp>FDA</scp>Approval summary: nivolumab plus ipilimumab for the treatment of patients with hepatocellular carcinoma previously treated with sorafenib. *The Oncologist, 26*, 797–806. https://doi.org/10.1002/onco.13819.

Shi, D., Shi, Y., Kaseb, A. O., et al. (2020). Chimeric antigen receptor-glypican-3 T-cell therapy for advanced hepatocellular carcinoma: Results of phase I trials. *Clinical Cancer Research, 26*, 3979–3989. https://doi.org/10.1158/1078-0432.CCR-19-3259.

Stebbing, J., Mainwaring, P. N., Curigliano, G., et al. (2020). Understanding the role of comparative clinical studies in the development of oncology biosimilars. *Journal of Clinical Oncology, 38*, 1070–1080. https://doi.org/10.1200/JCO.19.02953.

Sun, H., Yang, H., & Mao, Y. (2023). Personalized treatment for hepatocellular carcinoma in the era of targeted medicine and bioengineering. *Frontiers in Pharmacology, 14*, 1184523. https://doi.org/10.3389/fphar.2023.1150151.

Tella, S. H., Kommalapati, A., Mahipal, A., & Jin, Z. (2022). First-line targeted therapy for hepatocellular carcinoma: Role of atezolizumab/bevacizumab combination. *Biomedicines, 10*, 1304. https://doi.org/10.3390/biomedicines10061304.

van Doorn, D. J., Takkenberg, R. B., & Klümpen, H.-J. (2020). Immune checkpoint inhibitors in hepatocellular carcinoma: An overview. *Pharmaceuticals, 14*, 3. https://doi.org/10.3390/ph14010003.

Vithayathil, M., & Sharma, R. (2025). Nivolumab plus ipilimumab in hepatocellular carcinoma. *The Lancet, 405*, 1795–1797. https://doi.org/10.1016/S0140-6736(25)00417-9.

Wainberg, Z. A., Segal, N. H., Jaeger, D., et al. (2017). Safety and clinical activity of durvalumab monotherapy in patients with hepatocellular carcinoma (HCC). *Journal of Clinical Oncology, 35*, 4071. https://doi.org/10.1200/JCO.2017.35.15_suppl.4071.

Xu, X., Zhang, S., Xu, T., et al. (2022). Efficacy and safety of bevacizumab biosimilars compared with reference biologics in advanced non-small cell lung cancer or metastatic colorectal cancer patients: A network meta-analysis. *Frontiers in Pharmacology, 13*, 1009876. https://doi.org/10.3389/fphar.2022.880090.

Yang, J., Yu, S., Yang, Z., et al. (2019). Efficacy and safety of anti-cancer biosimilars compared to reference biologics in oncology: A systematic review and meta-analysis of randomized controlled trials. *Biodrugs, 33*, 357–371. https://doi.org/10.1007/s40259-019-00358-1.

Yau, T., Galle, P. R., Decaens, T., et al. (2025). Nivolumab plus ipilimumab versus lenvatinib or sorafenib as first-line treatment for unresectable hepatocellular carcinoma (CheckMate 9DW): An open-label, randomised, phase 3 trial. *The Lancet, 405*, 1851–1864. https://doi.org/10.1016/S0140-6736(25)00403-9.

Yau, T., Kang, Y.-K., Kim, T.-Y., et al. (2020). Efficacy and safety of nivolumab plus ipilimumab in patients with advanced hepatocellular carcinoma previously treated with sorafenib. *JAMA Oncology, 6*, e204564. https://doi.org/10.1001/jamaoncol.2020.4564.

Yilma, M., Houhong Xu, R., Saxena, V., et al. (2024). Survival outcomes among patients with hepatocellular carcinoma in a large integrated US health system. *JAMA Network Open, 7*, e2435066. https://doi.org/10.1001/jamanetworkopen.2024.35066.

Zeng, X., Jia, Y., Chen, H., et al. (2023). A real-world analysis of survival and cost-effectiveness of sintilimab plus bevacizumab biosimilar regimen in patients with advanced hepatocellular carcinoma. *Journal of Cancer Research and Clinical Oncology, 149*, 9213–9219. https://doi.org/10.1007/s00432-023-04775-2.

Zhang, H., Zhang, W., Jiang, L., & Chen, Y. (2022). Recent advances in systemic therapy for hepatocellular carcinoma. *Biomarker Research, 10*, 3. https://doi.org/10.1186/s40364-021-00350-4.

Zhao, Z., Zhao, L., Xia, G., et al. (2023). Efficacy and safety of bevacizumab biosimilar compared with reference bevacizumab in locally advanced and advanced non-small cell lung cancer patients: A retrospective study. *Frontiers in Oncology, 12*, 1078456. https://doi.org/10.3389/fonc.2022.1036906.

Zheng, J., Wang, S., Xia, L., et al. (2025). Hepatocellular carcinoma: Signaling pathways and therapeutic advances. *Signal Transduction and Targeted Therapy, 10*, 35. https://doi.org/10.1038/s41392-024-02075-w.

Zheng, S., Chan, S. W., Liu, F., et al. (2024). Hepatocellular carcinoma: Current drug therapeutic status, advances and challenges. *Cancers (Basel), 16*, 1582. https://doi.org/10.3390/cancers16081582.

Zhonghao, J., & Fan, Y. (2024). New advances in the treatment of intermediate and advanced hepatocellular carcinoma. *Frontiers in Oncology, 14*, 1348921. https://doi.org/10.3389/fonc.2024.1430991.

Zhou, T., Wang, X., Cao, Y., et al. (2022). Cost-effectiveness analysis of sintilimab plus bevacizumab biosimilar compared with lenvatinib as the first-line treatment of unresectable or metastatic hepatocellular carcinoma. *BMC Health Services Research [Electronic Resource], 22*, 1367. https://doi.org/10.1186/s12913-022-08661-4.

CHAPTER

Biosimilars and biologics in the treatment of brain tumor

7

Santosh C. Gursale[a], Shriraam Karunakaran[b], Mohan Krishna Ghanta[c], Santenna Chenchula[d], Santoshi Roopa D[c], Swapna R. Nayaka[c], Anusha Lakshmi Cheetiyar[e] and Deepanjana Dass[f]

[a]*Department of Pharmacology, Maharishi Markandeshwar Institute of Medical Sciences & Research, Sadopur, Ambala, Haryana, India,* [b]*Department of Biochemistry, MVJ Medical College and Research Hospital, Bengaluru, Karnataka, India,* [c]*Department of Pharmacology, MVJ Medical College and Research Hospital, Bengaluru, Karnataka, India,* [d]*Department of Pharmacology, All India Institute of Medical Sciences (AIIMS), Bhopal, Madhya Pradesh, India,* [e]*Department of Physiology, MVJ Medical College and Research Hospital, Bengaluru, Karnataka, India,* [f]*Department of Pharmacology, MGM Medical College, Navi Mumbai, Maharashtra, India*

7.1 Introduction

Brain tumors are particularly challenging to treat for a variety of reasons, including aggressive biology, lack of response to standard treatment, and the protection from the drug by the blood–brain barrier (BBB). Malignant tumors such as glioblastoma have poor prognoses despite maximum surgical resection, radiation therapy (RT), and chemotherapy (Stupp et al., 2005). In the last 20 years, we have seen biologic therapy introduced into routine practice, which has provided new options, particularly in biology targeting angiogenesis, growth signaling, and the immune microenvironment (Ferrara & Adamis, 2016). Biologics are large, complex molecules, such as monoclonal antibodies or fusion proteins, that are designed for the modulation of specific molecular identification. Bevacizumab was groundbreaking for recurrent glioblastoma therapy since it diminished edema and prolonged progression-free survival, but with a lack of reproducible overall survival advantage (Friedman et al., 2009a; Taal et al., 2014). Cetuximab and dinutuximab, although more peripheral to the standard of neuro-oncology practice, are examples of targeting strategies directed toward particular tumor antigens. Biologics are also notoriously expensive, leading to access and financial toxicity inequities for health systems (Blackstone & Joseph, 2013). It is their poorly scalable/selective biology, extensive rationale for manufacturing, and the amount of testing to establish safety and efficacy, that result in high costs, which risk limited use even if clinically effective. This economic barrier has contributed to the growing acceptance of biosimilars that are developed to be highly similar to already approved reference biologics (Blackstone & Joseph, 2013). Unlike small molecule generics, biosimilars are not chemically identical to their

Biosimilars and Biologics in Cancer Treatment: An Era in Therapeutic Strategies.
DOI: https://doi.org/10.1016/B978-0-443-43824-0.00001-8

reference biologics, but must demonstrate they have no clinically meaningful differences from their reference biologics in terms of safety, purity, or potency. Regulatory pathways, such as the FDA's "totality of evidence" approach (discussed later), ensure equivalency for biosimilars (Scientific Considerations in Demonstrating Biosimilarity to a Reference Product, 2015). In oncology practice, biosimilars have provided better access and lower costs to several blockbuster biologics, including rituximab, trastuzumab, and bevacizumab. This is certainly true, as cost-effective options are emerging for biosimilars used for recurrent glioblastoma identification and treatment (Tabernero et al., 2016).

This chapter reviews the scientific principles and clinical evidence for biologic and biosimilar production as it pertains to care for brain tumors, including regulatory approval pathways, and a summary of biologic agents being used for neuro-oncology practice by the neuro-oncologist. The chapter suggests a more critical stance on the pathways to the adoption of biosimilars in neuro-oncology. We discuss controversies surrounding interchangeability, acceptance from physicians and patient populations, and economic signals that shape the biosimilar marketplace. The chapter concludes with future directions for biologics/biosimilars; emerging biologics and lobbying initiatives that will hopefully renew increased equitable access for health systems and patients. As we embrace the role of biosimilars, clinicians and researchers alike use evolving therapeutic options to improve the prognosis for patients with brain tumors, while easing the systemic cost burden.

7.2 Biologics and biosimilars

Biologics are any therapeutic product derived from or prepared from living organisms. Biologics include monoclonal antibodies, cytokines, vaccines, cell therapies, and gene therapies (Ramanan & Grampp, 2014). Biologics are distinct from small-molecule drugs that are often produced as the result of chemical synthesis, are often much smaller in structure (weighing in daltons as opposed to kilodaltons), and the cleanup steps are often considerably easier; although going through a saccharification step to clear small-molecule drugs may be formidable (Schellekens, 2009). Biologics are considered to be large, very complicated proteins and therapeutic products that can undergo complex folding, glycosylation, and other posttranslational modifications. All biologics are developed using living cells, which, through complex biotechnological means, will provide minor batch-to-batch variations even when the product is planned to be the same (Ramanan & Grampp, 2014).

Because of the inherent complexity of biological products, biologics are much more flexible spatially, with respect to interacting with biological molecules to disrupt disease mechanisms in ways that small molecules will likely never be able to do. For example, monoclonal antibodies may specifically bind to only some tumor cells to attenuate and/or modulate growth factors in the tumor microenvironment or promote effector functions, like antibody-dependent cellular cytotoxicity (ADCC) (Kaplon et al., 2020). In oncology (e.g., trastuzumab binding HER2 in breast cancer;

bevacizumab blocking VEGF signaling and therefore inhibiting angiogenesis (the development of blood vessels); or rituximab, which targets CD20 surface marker on B cells, where B cell malignancy rates are high (Scott et al., 2012).

Biologics are a complicated class of products with complications. Biologics are expensive to develop and manufacture, require sophisticated cell culture facilities, and the purification processes involved can often be complicated (Schellekens, 2009). Moreover, biologics are most often delivered via a parenteral route of administration, and biologics can provoke an immune response, such as infusion reactions or the development of antidrug antibodies (Hermeling et al., 2004).

Biosimilars were developed as a way to improve access to medicines for patients as patents expired on first-generation biologics. Unlike generic drugs that assemble the same chemical structure as their reference product, you cannot exactly duplicate a biologic, but you can get darn close (Tabernero et al., 2016). The US FDA defines biosimilars to be "highly similar" or equivalent, from a chemical standpoint, and while there is variation between biologics versus generic drugs, the varying variability we use to define, from the biological product perspective, hinges mostly lie with the complexity of the biologic molecules and the context in which biologics are developed (Scientific Considerations in Demonstrating Biosimilarity to a Reference Product, 2015). In fact, the evaluative standards of the FDA called for by the 2010 Biologics Price Competition and Innovation Act allow suppliers submitting exclusivity documents no clinically meaningful difference between the biosimilar products and reference medicinal product, concerning safety, purity, or potency (Carver et al., 2010).

When manufacturers want to produce biosimilars, they will generally perform a stepwise comparability exercise (McCamish & Woollett, 2012). The first phase would be to conduct extensive analytical characterizations on the primary, secondary, tertiary, and quaternary structures, and confirm primary sequence likeness and similarities with respect to folding, glycosylation, and other critical variables for rigor. Functional assays would then be performed to compare biological activities in vitro, including tests for binding and efficacy functions. Nonclinical studies may have to occur in an animal model if it is rational and appropriate; however, animal studies will be conducted rarely, and once sufficient analytical similarity is confidently established. After the nonclinical studies are completed, normally the clinical studies are briefed based on the ordinality of the originator biologic, using certain designs to evaluate the comparability of the biosimilars using pharmacokinetics and pharmacodynamics approaches to establish connections, plus at least one confirmatory efficacy trial done in a sensitive indication to show that there are no clinically meaningful differences.

Europe was the first continent to have biologics with biosimilars and approved the first biosimilar, which was a somatropin growth hormone, submitted to the European Medicines Agency in 2006. The EMA framework for biosimilars has served as a template for other authorities across the globe to follow, and has a number of organizations that look at analytical comparisons and clinical documentation (Schiestl et al., 2017). Many of these authorities used biosimilar pathways planned from the BPCIA, which came from the 2010 US Federal Health Care legislation. Filgrastim-sndz was the first biosimilar approved by US FDA in 2015 (Raedler, 2016). Biosimilars can be approved on an extrapolated indication, with a rational justification. Basically, if a biosimilar can show similarity in one clinical condition, the biosimilar can be approved for all indications held by the reference biologic product.

Some providers continue to be skeptical of biosimilar use, even though the biosimilar has high approval thresholds and regulatory requirements, particularly with automatic substitution, immunogenicity, and extrapolated indications. By addressing those concerns, we can be transparent with providers on the process of biosimilar development, as well as the regulatory protections in place to uphold therapeutic equivalence.

Nonetheless, the biosimilar has changed oncology practice by decreasing the cost of drugs, increasing competition to expand the access of treatments provided to our patients, and expanding the drug options available to both providers and patients. For some markets with strong biosimilar competition, we have seen average price reductions of 30%–50% for clinical oncology pharmaceuticals (Blackstone & Joseph, 2013; DiMasi et al., 2016). The form and function of biosimilars are only just beginning in neuro-oncology. The relatively recent emergence of bevacizumab biosimilars has also provided an alternative to standard, costly infusion therapies for patients with recurrent glioblastoma. In the setting for patients with glioblastoma, where it is extraordinarily difficult to find therapies in low-resourced settings, this is an important advancement for the patients. Biosimilars are more than just a public health success story for science and regulation; they are a public health success strategy to rectify disproportionate access to biologics.

7.3 Regulatory landscape and approval pathways

Biosimilars are regulated to ensure that patients receive therapies with equivalent safety and efficacy to their reference biologics, and to foster market competition that drives down costs. Due to their complex structures and being manufactured in living systems, biosimilars cannot be identical chemical duplicates, unlike small-molecule generics. Therefore, regulatory agencies have established frameworks for assessing similarity with rigor without requiring full replication of the original clinical development (Kang et al., 2020).

The "totality of evidence" is the foundation of biosimilar approval (Fig. 7.1). The approach involves combining analytical, functional, nonclinical, and clinical data to show high similarity to the reference product and no clinically significant differences. The "totality of evidence" approach recognizes that modern analytical methods are capable of characterizing biologics with impressive specificity and sensitivity that in some cases exceed what was possible with the rigor and standards in place when the original product was approved (Scientific Considerations in Demonstrating Biosimilarity to a Reference Product, 2015).

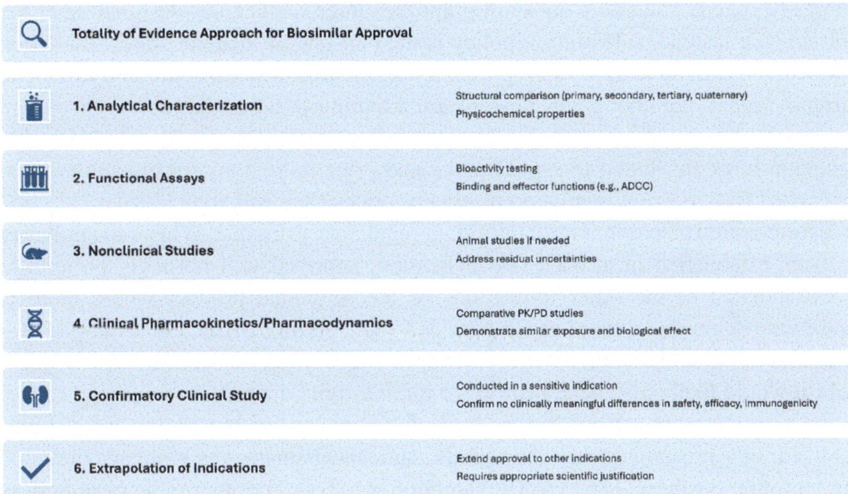

FIGURE 7.1

Totality of evidence approach for biosimilar approval.

In the United States in 2010, the Biologics Price Competition and Innovation Act (BPCIA) also introduced a formal pathway to biosimilars with the establishment of the FDA's biosimilar pathway (Carver et al., 2010). To gain approval, a biosimilar must provide analytical data demonstrating sufficient molecular similarity and comparative functional studies demonstrating equivalent biological activity. It also requires pharmacokinetic and pharmacodynamic studies providing sufficient evidence to show there is similarity in absorption, distribution, metabolism, and action. Finally, the FDA requires at least one clinical study in a sensitive indication to demonstrate that there are no differences in efficacy or safety compared to the reference product. Unlike original biologics, the clinical program of biosimilars significantly reduces product development costs and timelines. This is justified because the mechanism of action, safety profile, and manufacturing process of the reference product are rigorous and established, as discussed earlier (McCamish & Woollett, 2012).

A key regulatory component is the concept of interchangeability. In the U.S., an interchangeable biosimilar has established a higher level of evidence to demonstrate that patients can switch back and forth between the biosimilar and reference without concerns about safety or efficacy. An interchangeable designation allows for pharmacy-level substitution without reliance on prescriber approval, similar to a prescriber's generic substitution. At this point, few oncology biosimilars have sought an interchangeable designation in the U.S. Instead, oncology biosimilars are prescribed by name with substitution decisions controlled by the clinician (Understanding interchangeable biosimilars at the federal and state levels, 2023).

For biosimilars in the European Union (EU), the EMA has led the way in implementing biosimilar regulation, granting approval for the first biosimilars in 2006, as already stated. The EMA does not grant the "interchangeable" designation at the central level; instead, substitution policy is the purview of member states. European experience has demonstrated that pharmacy-level substitution is safe and effective. Currently, many member states have or are adopting policies that either encourage or mandate the use of biosimilars once they become available for use. This flexible regulation has contributed to rapid uptake and price reductions for biosimilars that can exceed forty percent relative to originator prices (Schiestl et al., 2017).

Extrapolation also remains an important regulatory principle. When bio similarity has been established in at least one indication, approval will normally be able to be extrapolated to the other indications of the reference product where there is appropriate scientific reasoning (Ebbers & Chamberlain, 2016). For example, if a biosimilar of bevacizumab were to receive approval for lung cancer after some clinical trials, it would likely also be approved for glioblastoma, metastatic colorectal cancer and perhaps the other cancers, on the basis that clinicians and regulators can support that the drug is presumed to act through the same mechanism, and will have the same safety profile, for these diseases. The intention is to try to remove some meaningless duplication of trials, which will help reduce development costs but also expedite patient access.

Although regulatory networks for biosimilars development and approval have never been more comprehensive than now, there remain persistent aspects of disagreement. Some clinicians remain hesitant to consider switching to a biosimilar for a stable patient, particularly in the context of oncology, where control of the disease might be fragile. Other contentious issues are around immunogenicity, distinctions of efficacy, and trying to discuss the patient's acceptance of trying a new biosimilar at all, etc. Regulatory agencies have mitigated these tensions by publishing guidelines, establishing arrangements (e.g., pharmacovigilance) for postmarket, and providing education. The ongoing evolution of the regulatory terrain poses additional considerations in terms of aligning harmonization across jurisdictions and enhancing regulatory reviews for approvals. There may be a time when, as real-world evidence matures in reflecting a biosimilar's safety and efficacy against the original biologic medicine, there will be less reliance on clinical trials. This may permit regulators to rely more on pharmacovigilance and advanced analytics to draw the same conclusions. Ultimately, the key takeaway here is intended to reassure clinicians when prescribing biosimilars by better understanding the regulatory robustness. This regulatory infrastructure, at the end of the day, gives clinicians the confidence that biosimilars provide patients the same therapeutic benefit as their reference biologic. Ultimately, this means biosimilars enable cancer patients to receive a safer, more sustainable medicine.

7.4 Key biologics in brain tumor therapy

Neuro-oncology has a smaller toolbox of biologic therapies, especially compared to the rest of oncology. Several important biologics show some clinical utility or at least early promise with brain tumors, including bevacizumab, cetuximab, and dinutuximab. A brief description of each biologic highlights a targeted therapy, including mechanisms of action, clinical efficacy, and practical challenges.

7.4.1 Bevacizumab

Bevacizumab is a humanized monoclonal antibody that binds vascular endothelial growth factor-A (VEGF-A), a mediator of angiogenesis, and glioblastomas are very vascular tumors that show abnormal and "leaky" blood vessels, driven primarily by VEGF overexpression. Bevacizumab binds VEGF-A to prevent VEGF-A from binding the endothelial receptors and blocks the formation of new blood vessels and reduces the permeability of the blood vessels. In 2009, the FDA granted accelerated approval for bevacizumab for recurrent glioblastoma based in part on Phase II data that suggested improved progression-free survival (PFS), radiographic response rates, and less corticosteroid use (Friedman et al., 2009b). Subsequent trials, including AVAglio and RTOG 0825, confirmed the improvement in PFS in newly diagnosed glioblastoma, but not improvement in OS thus, the use of bevacizumab in the newly diagnosed setting could be considered off-label in many regions, though it is standard of care for recurrent disease (Chinot et al., 2011; Gilbert et al., 2013). In addition to improvements in PFS, patients also show relatively rapid symptomatic improvement, demonstrated decreased edema, improvement in neurological function, and reduction in corticosteroid dependence. While still limited in mortality benefits, the palliation of symptoms and improvement in neurological function are useful to many patients. Bevacizumab has been criticized for the palliative benefit that it provides, as tumor progression re-emerges in the infiltrative, nonenhancing pattern. There is also concern regarding the side effects, specifically hypertension, thromboembolism, issues with wound healing, and proteinuria. And, despite the price (previously high), there are great barriers to access for patients. At this time, there are multiple biosimilars to bevacizumab that have been approved outside of the USA, but in the USA, only bevacizumab-awwb (Mvasi) and bevacizumab-bvzr (Zirabev) have been approved. Real-world data have demonstrated that in glioblastoma patients, the efficacy and safety are similar, with some of this data coming from abroad (one is published from India), demonstrating indications for the effectiveness of biosimilars to accomplish health systems access to bevacizumab (Kumar Singh et al., 2021).

7.4.2 Cetuximab

Cetuximab is a chimeric monoclonal antibody that binds to the receptor for epidermal growth factor (EGFR), a transmembrane tyrosine kinase that is often augmented or mutated in solid tumors. Cetuximab binds to the receptor and prevents ligand from binding the receptor and causing internalization of the receptor, preventing signaling pathways downstream responsible for proliferation and survival from being activated (Mendelsohn & Baselga, 2006).

EGFR amplification and the EGFRvIII mutation are common in glioblastoma, and preclinical studies suggested that EGFR inhibition could make gliomas more sensitive to radiotherapy and reduce their growth (Brennan et al., 2013). While the preclinical biology that supports cetuximab plus radiation in glioblastoma is interesting, there is limited clinical data. There have been some early-phase studies of cetuximab plus radiation, or together with temozolomide, or with bevacizumab, all of which provide preliminary data indicating feasibility and some have even reported objective responses (but no definitive improvement in survival) yet several barriers contribute to inconclusive results for those studies, particularly limited penetration across the BBB and tumor heterogeneity, particularly about efficacy for EGFRvIII (which cetuximab would likely not have any therapeutic targeting) (Huang & Harari, 1999). Notably, cetuximab is not considered a standard therapy for gliomas, and its indications in oncology are mainly still for colorectal cancers and head and neck cancers. The importance of studying cetuximab in glioblastoma likely points to the importance of understanding the molecular subtypes while also addressing the limits of delivery. A few markets are seeing biosimilars to cetuximab, but aside from its other indications, it is limited in neuro-oncology.

7.4.3 Dinutuximab

Dinutuximab is a personalized anti-GD2 monoclonal antibody, or chimeric, and is primarily utilized in pediatric oncology to treat high-risk neuroblastoma, which originates from the neural crest. GD2 exists in neuroblastoma and some central nervous system (CNS) tumors, but at lower levels. Dinutuximab binds to GD2 to induce immune-mediated cytotoxicity via ADCC and complement-dependent cytotoxicity. Dinutuximab is FDA-approved as part of a multimodal therapy for high-risk neuroblastoma and is associated with improvements in event-free and overall survival following randomized trials. The treatment consists of repeat cycles of dinutuximab with cytokines (GM-CSF, IL-2) and isotretinoin following stem-cell transplant (Yu et al., 2010). Pain as part of infusion therapy with dinutuximab is a significant challenge, likely due to GD2 expression on peripheral nerves, and requires aggressive pain management. Other toxicities of therapy include capillary leak syndrome and hypersensitivity reactions. Neuroblastoma is, by definition, peripheral; however, especially when at the high-risk stage, it involves pediatric neuro-oncologists in the management of the disease. The possibility for GD2 expression in other CNS tumors (e.g., diffuse midline glioma) invigorated the idea of exploiting targeting GD2 in other tumors, including through CAR T-cell approaches (Ladenstein et al., 2018). In addition to dinutuximab, Europe has approved dinutuximab beta (Qarziba) as it was developed independently by Eusa Pharma, which is targeting the same epitope (Balaguer et al., 2023). The product was not a formal biosimilar (as Unituxin was never originally approved in the EU), it operates in the same way, thus, importantly, it demonstrates that alternative manufacturing can allow for access to critical treatments globally.

7.5 Role and development of biosimilars in oncology

Given the prevalence and expense of biologic therapies, oncology has been a leader in biosimilar development. For example, monoclonal antibodies like rituximab, trastuzumab, and bevacizumab are among the expensive cancer drugs, which have significant budget effects on healthcare systems around the world. The advent of oncology biosimilars is an important strategy to reduce the burden on health budgets, increase access to treatment, and nurture competition. Biologics continue to represent a large and growing component of oncology drug expenditure. As patents expired, biosimilar developers stepped in to present these biologic therapies as lower-cost options. The anticipated price competition in oncology biosimilars has generated a lot of savings: biosimilars have saved health systems billions globally, with most price reductions in the 20%–50% range, as estimated across studies (Blackstone & Joseph, 2013; DiMasi et al., 2016).

Variability has been observed in the adoption of oncology biosimilars across jurisdictions. For example, several policy measures, including centralized procurement, decision-maker incentives, and substitution policies, were implemented in European countries, leading to the quick uptake of biosimilars in oncology. In many European countries, biosimilars were taking over 70%–80% market share within a couple of years of launch (The Impact of Biosimilar Competition in Europe, 2020). The U.S. took a longer time to establish biosimilars, as it dealt with patent disputes and rebate contracting strategies that favored originator companies (Mulcahy et al., 2017). After the pandemic, biosimilars have begun to integrate into routine practice with the increasing pressure from payers and greater recognition by oncologists. More recently, reviews suspiciously showed biosimilars gaining market share throughout the oncology-related indications.

The thorough development and approval process for biosimilars ensures they match the efficacy and safety of their reference biologics. Numerous clinical trials and real-world evidence studies have demonstrated that there was no clinically meaningful difference between reference products and biosimilars, whether this be in response rates, progression-free survival, overall survival, or immunogenicity. There have even been switching studies that have confirmed and yield no adverse effects from patients switching back and forth between biosimilars (Barbier et al., 2020). Hesitancy around biosimilar acceptance initially stemmed from immunogenicity, extrapolation of indication, and losing control over pharmacy substitution. Education campaigns and strategies have been carried out by a number of sources; professional societies, regulatory authorities, and marketers of biosimilars initiated campaigns in particular. Leading oncology organizations have also subsequently developed statements that now support their use as a safe and effective mechanism to reduce costs and improve affordability of treatment options, e.g., the American Society of Clinical Oncology (ASCO) and European Society of Medical Oncology (ESMO) (Lyman et al., 2018).

The introduction of biosimilars has not only dropped prices associated with care. The launch of biosimilars can often be seen as the trigger for price competition initiated by their originators. In some jurisdictions, originators have reduced prices associated with both new biosimilars and the previous class of originators by utilizing a provision called rebates or wholesale discounts to remain on the formulary list. This competition provides benefits for payers and patients, but can also cause additional difficulties for prescribers, as they may be bound to use the insurer's preferred drug.

Various government and health systems have used tools to promote biosimilars, some of which include: tendering systems, limiting prescriptions, differential reimbursement, and promotion and education. The general uptake of biosimilars in European countries has provided the opportunity for other jurisdictions to observe the uptake without making significant political or economic changes to their health systems. In the United States, policy changes, such as reimbursement for biosimilars at a higher rate than originator biologics, have put forward a clear economic incentive to use biosimilars as the health system attempts to better monetize drug expenditures for payers (Mulcahy et al., 2017).

The oncology biosimilar market should see further development because numerous patents on other biologics will soon be expiring, including new immunotherapies like the checkpoint inhibitors. This "next wave" of biosimilars will help to increase access to state-of-the-art therapies, while also protecting sustainable budgets for cancer care. While there is a lot of potential for biosimilars to provide savings and phased-in access, the biosimilar experience has not come without complications. Originator companies have engaged in "patent thickets" and rebates that can both delay and restrain biosimilar uptake (Jofre-Bonet et al., 2025). Physicians may also be uncomfortable transferring stable patients between products, especially where aggressive marketing from originators is focusing attention on "not identical" language associated with a biosimilar. The challenges posed by these factors point to the need for targeted and sustained education, competitive policy measures, and thorough and transparent postmarketing surveillance needed to monitor biosimilar safety and effectiveness so that confidence remains high. Ultimately, the oncology experience has shown that biosimilars do replace expensive originators and they do so safely and effectively, while protecting sustainability in the health care system. For neuro-oncology, where the number of biologics is limited, these experiences are relevant as we face increasing cost pressures related to new biologic targets and therapies.

7.6 Application of biosimilars in neuro-oncology

Biosimilars are on the verge of revolutionizing neuro-oncology; the emerging field is based on biologics that have already been used in patients with brain tumors and related cancers. While the neuro-oncology space has fewer biologics approved than many other oncological disciplines, the cost pressures and access restrictions surrounding certain biologics make biosimilars especially relevant.

The prototypical biologic involved with adult neuro-oncology specifically remains bevacizumab, particularly toward recurrent glioblastoma (Friedman et al., 2009a). Since the patent for bevacizumab has expired, multiple biosimilars have been approved globally and in the U.S.; both bevacizumab-awwb (Mvasi) and bevacizumab-bvzr (Zirabev) have obtained FDA approval for all indications granted to the reference product, including recurrent glioblastoma (US FDA, n.d.). There is also such data suggesting that biosimilars are being used in patients with brain tumors in the real world. In a retrospective study that compared innovator bevacizumab with biosimilar versions in patients with recurrent glioblastoma in India, the researchers found no apparent differences in progression-free survival or overall survival (Kumar Singh et al., 2021). This is significant for supporting the comfort level of prescribers. This type of data is essential considering endemic concerns about extrapolation. Clinically speaking, biosimilars are observed to potentially have the same steroid-sparing effects as well as the same radiographic responses that are especially important for symptom management in gliomas, particularly high-grade gliomas.

Historically, costs have limited the use in many healthcare settings (Blackstone & Joseph, 2013). Biosimilars could represent a path toward expanding patients' access, particularly in lower-resource healthcare systems. In this way, hospitals and insurers potentially treat more patients for the same budget. In India, the use of lower-cost biosimilars led to more patients being treated with anti-VEGF therapy (nonbranded) for glioblastoma; patients who previously could not afford to be treated with the branded drug (Gandhi et al., 2024). While many oncologists now embrace the use of biosimilars in practice, a subset of neuro-oncologists trained to have a high level of skepticism, particularly around switching stable patients' mid-course, will remain reluctant. This reflects a caution to avoid the (even theoretical) risk resulting from immunogenicity or loss of disease control. However, with the regulatory approval as well as real-world experience that has been accrued, there is now an accepted safety level to switch, as long as patients are adequately counseled. In primary CNS lymphoma (PCNSL), rituximab is frequently combined with high-dose methotrexate-based chemotherapy (Ferreri et al., 2016). Rituximab has limited penetration in the CNS, but it may help with the clearance of lymphoma cells in immune-privileged anatomic sites. Rituximab and its biosimilars have become standard in practice for systemic lymphomas, where there is substantial evidence of their equivalence in clinical effectiveness. In PCNSL cases, the use of rituximab biosimilars is expanding, due to the financial savings it offers as well as the ineffectiveness of other biosimilars. Pediatric neuro-oncology also considers high-risk neuroblastoma; this is treated in the U.S. with dinutuximab and in Europe with dinutuximab beta (Hoy, 2016; Yu et al., 2010). Even though these products were developed independently and approved as biologics, and therefore cannot be called biosimilars, their respective development illustrates biosimilar-like approaches that may expand access to biologics around the world. Both dinutuximab and dinutuximab beta target GD2, and similar survival advantages have been observed in the clinical use of both products.

Neuro-oncology will likely usher in further biologics into practice in the coming years, including immune checkpoint inhibitors, bispecific antibodies, and CAR T-cell therapies in treating CNS tumors like glioblastoma (Maccari et al., 2024). As patents expire for next-generation biologics, it will be critical to facilitate access to biosimilars to ensure they get to patients. The high price point for immunotherapies is often over $100,000 per patient; therefore, biosimilars will be necessary to keep treatment for CNS patients feasible (Schaft et al., 2023). One question that is often asked is whether minor structural changes in a biosimilar may impede BBB penetration. For example, whether the way that glycosylation alters either the charge (and therefore the BBB interaction) or receptor interactions. Theoretically, this is possible, but in reality, the fact of the matter is that the BBB is a significantly challenging barrier for all large antibodies. Tumor-related disruption of the BBB likely permits antibody penetration. Pharmacokinetic studies have revealed that the comparable distribution of biosimilars to originators is established. Hence, there is no evidence to suggest that relative CNS exposure for available biosimilars would differ to an extent that it was meaningful. The economic benefits of biosimilars are evident; however, their uptake will likely depend on some form of physician education and patient confidence. Communication is of utmost importance, particularly to counter beliefs that "less expensive" means "less quality.". Clinicians should be adamant in informing patients that biosimilars are studied and rigorously evaluated, and that regulatory agencies did not approve these products without strict compliance with their standards for approval; and that clinician products can be supported with clinical evidence as well as actual evidence. Healthcare systems must help ensure that some of the savings from the use of biosimilars will be reinvested for the betterment of patient management, thereby strengthening the ethical premise for using biosimilars. In conclusion, biosimilars have already begun to assist neuro-oncology practice by allowing reasonably priced, oftentimes novel therapies to be accessible through cost reductions. Their presence then in the field will increase as new biologics enter it, which will again serve to enable all patients access to more innovative therapeutic options regardless of the setting of care.

7.7 Current challenges and critical perspectives

Although biosimilars promise to expand access and lower costs, adoption in oncology and specifically neuro-oncology faces important barriers. These barriers are not primarily scientific but social, economic, and perceptual. Clinicians' concerns about switching are the most pronounced barrier. Many oncologists, including neuro-oncologists, are worried about switching a stable patient from an originator to a biosimilar mid-treatment (Resende et al., 2021). Even though regulators state that biosimilars have no clinically meaningful differences, concerns about immunogenicity, efficacy, or even psychological nocebo effects remain relevant. In the U.S. setting, the concept of interchangeability was introduced to lessen these concerns, but it hasn't continued as most of the oncology biosimilars lack this designation, as few companies have done the additional studies to receive it. This lack perpetuates hesitance on the part of the clinician. The experience in Europe demonstrates that pharmacy-led substitution can be safe and effective when there are appropriate regulatory frameworks and educational support for prescribers (Moorkens et al., 2017).

Another area of skepticism arises from the perceived lack of transparency of biosimilar data. Biosimilars are subjected to extensive analytical testing, but the data are often proprietary, and clinicians often see a summarized version of a regulatory review, which leaves lingering doubt. Increasing transparency through biosimilar developers' publication of analytical comparisons of the originator and example real-world outcomes is certainly a more assured way to improve uptake. Another issue

Table 7.1 Barriers and enablers of biosimilars.

Barriers	Enablers
Prescriber switching reluctance	Education campaigns
Immunogenicity concerns	FDA/EMA regulatory guidance
Patent thickets, rebate strategies	Tender systems, centralized procurement
Nocebo effects	Patient education, shared decision making
Limited real-world data early	Accumulating real-world evidence

slowing down biosimilar adoption is a lack of education for both patients and physicians. Surveys indicate that many prescribers do not fully comprehend the biosimilar approval processes, which include understanding the term "no clinically meaningful differences" and the rigor of the "totality of evidence" approach (Yang et al., 2022). Patients are inclined to think that "cheaper" means poorer quality unless provided reassurance by a provider in terms of general best practice. Both clinicians and patients are likely to only accept biosimilars on the premise of clear and sufficient, open communication. Clinicians need sufficient confidence that switching from an originator to a biosimilar will not lead to poor outcomes, and patients need to be reassured that a lower cost means a similar, safe treatment, and it doesn't mean poorer quality. Several strategies come to mind for both patients and clinicians to have confidence in biosimilars, including implementing institutional guidelines regarding biosimilar use and management. Having a statement of equivalence in the therapeutics formulary to have biosimilar medication prescribed. Developing educational strategies for patients that utilize educational pamphlets. Table 7.1 summarizes the barriers and potential enablers in biosimilar development.

Despite biosimilars' potential to reduce cost, it is also evident that real-world adoption is slowed by local market forces. Originator companies are attempting to delay entry into the competing market, searching for competitive advantages using "patent thickets" to either delay or create exclusive rebate arrangements with insurers to limit the circulation of biosimilars. All of these actions cloud price competition and dilute both patient and health systems' savings. Various strategies have been acknowledged by policymakers to address the issue. A few examples include centralized procurement systems and tender systems in Europe, which have significantly decreased prices and increased the uptake rate of biosimilars. In the United States, the most recent policy encouraged some uptake through increased reimbursement rates for biosimilars in Medicare (Socal et al., 2020). Nevertheless, as fragmentation remains high at the insurance systems level, the way rebates affect distribution continues to create variability in access to biosimilars.

Ongoing safety stewardship necessitates strong pharmacovigilance systems to enable the tracking of biosimilar use at the batch level of the original product. Naming conventions (e.g., in the United States, the use of four-letter suffixes on the names of biosimilars) can be helpful in allowing and supporting traceability in adverse event reporting. Some may criticize this nomenclature for reinforcing perceptions of difference between biosimilars and biologic originators (Simoens & Vulto, 2021). Regulating entities and clinicians will need to find a balance between maintaining confidentiality and establishing an ethics-centered pharmacovigilance model to minimize bias and encourage biosimilar usage. The ethical argument for biosimilar uptake is strong. Any reduction in costs to health systems will allow for either treating more patients or reallocating the money saved from drug expenditures to other health system improvements. However, the benefits of a biosimilar paradigm depend on real price competition and an equitable return of savings. Second, it is an ethical obligation that any savings from biosimilar costs are provided to patients to reduce their financial toxicity.

In neuro-oncology, the question of adoption has varied implications based on the number of biosimilars currently utilized and the high stakes involved with treatment decisions. For example, in the context of bevacizumab in recurrent glioblastoma, any perceived risk of switching medications is in the context of a palliative goal to reduce edema and the attendant symptoms. Accordingly, any perceived risk of loss in effectiveness through switching medications raises concerns for clinicians. To develop this confidence, it is essential to share real-world evidence that demonstrates clinical outcomes that are equivalent for innovator versus biosimilar bevacizumab. In pediatric neuro-oncology, the need to manage expectations of patients and their caregivers is equally important. It can be beneficial to explain the rigorous testing that provides quality assurance to biosimilars to avoid the public perception that cost-saving measures prioritize profit over patient care. Lastly, adoption of biosimilars represents complex interactions between sustaining trust in the physician-patient relationship by providing care, versus realizing the economic benefits needed to sustain the health care system. Overcoming these barriers requires an ongoing commitment to regulatory vigilance, transparency, provider and patient education, and policy that promotes fair competition.

7.8 Future directions and research outlook

The biologics and biosimilars landscape in neuro-oncology is set for considerable change in the next 5 years. With the availability of new therapies, the relevance of biosimilars to ensure equitable access will only increase. The neuro-oncology space continues to advance with new targets and therapies. Immune checkpoint inhibitors (i.e., nivolumab, pembrolizumab) are being utilized in glioblastoma and other CNS tumors but are only modestly effective to date (Reardon et al., 2020). Innovative modalities, such as bispecific antibodies, antibody–drug conjugates, and CAR T-cell therapies, are being explored with new glioma-associated antigens such as EGFRvIII, B7-H3, and GD2 (Majzner et al., 2019; Nazha et al., 2020). The associated costs of these biologics will continue to present challenges for health systems as they are developed; however, they will also facilitate access to these newer biologics to all patients and not just those in high-income countries or health systems with better resourcing. In addition to biosimilars, biobetters (i.e., pharmacokinetically-modified versions of existing biologics aimed to improve safety, efficacy, or administration) may be available (Uhlig & Goll, 2017). For example, antibodies may be modulated to optimize for half-lives or immune engagement, which may improve patient outcomes and convenience. The existence of biosimilars may encourage originator companies to innovate more as well, which would have additional benefits. One of the biggest barriers in neuro-oncology is overcoming the blood–brain barrier in relation to biologics and antibodies retaining permeability. Future research involving biologics will likely also include new experimental ways to overcome the BBB, such as convection-enhanced delivery (CED), administration of focused ultrasound to reversibly disrupt the BBB, and nanoparticles that are selectively engineered for a specific site of delivery (D'Amico et al., 2021). Although these relatively new advanced methods of treatment have yet to be accepted as a standard of care for the treatment of brain tumors, they may be new and effective ways of utilizing biologics in particular.

There is also a shift in regulatory requirements; as biosimilar acceptance becomes reality, clinical-trial data will be of less utility and paramount importance due to the measured comparative analytical confidence. If harmonization evolves into other regulatory processes, such as FDA and EMA processes, and reimbursement regulations are standardized, we will see an improved ability for patients worldwide to have access to these therapies (Moorkens et al., 2017; Scientific Considerations in Demonstrating Biosimilarity to a Reference Product, 2015). One key component to increasing biosimilar use and savings will be the duty of policymakers to ensure that policies are followed. This may include, for example, centralized purchasing, implementation of quotas for prescribing or reimbursement differentiated for original and biosimilar medicines, or regulating savings from switching to biosimilars initiatives through transparent reinvestment into patient care, research, or new therapy initiatives. The challenge will also be to ensure appropriate safety and efficacy monitoring in the market. We will need strong pharmacovigilance systems that can adequately monitor the use of biosimilars and the monitoring of rare adverse events. We also have a tremendous opportunity to monitor therapy effectiveness through existing health data systems and the evolving ability to use big data. Additionally, biosimilars will play an important role in improving health equity. In many lower and middle-income countries, the cost difference disadvantages patient access to many existing standards of care therapies. With respect to the future of biosimilars, new initiatives that involve technology transfer agreements, local manufacturing arrangements, or friendly regulatory processes conducive to the health care context will be helpful to expand biosimilar access.

Finally, and very importantly, we will need to understand the patient perspective. Patients with brain tumors are faced with higher cognitive issues, symptom burden, and treatment decisions that are frequently challenging with respect to navigating care. As we consider biosimilars broadly and then intentionally with respect to brain tumors, researchers and planners will need to consider communication strategies to patients, models of shared decision making, and the health literacies of patients with respect to commonsense ways in which to ensure cost-effectiveness does not compromise perceived quality of care to patients or the clinician-patient trust relationship. Further benefits of biosimilars in the neuro-oncological domain will be gained through the philosophical anchor of trust, regulatory supervision, the competitive pricing of less expensive products, and the opportunity to even partially reinvest these savings to enhance overall cancer care. The opportunity is great: expand patient access to therapies that prolong or palliate life expectancies and are cost-effective for the organizations that ultimately finance health care delivery.

7.9 Conclusion

Biologics have revolutionized oncology and neuro-oncology with new targeted therapies that provide mechanism-based treatments to patients with complex diseases that include glioblastoma and high-risk neuroblastoma. An example is that bevacizumab plays a role other than tumor control, and can lengthen progression assessment intervals, and dinutuximab has changed survival expectations in pediatric neuroblastoma. However, biologics have been expensive, with limited reimbursements, and limited availability in many health systems globally, especially in low-and middle-income countries. Biosimilars are highly similar from an analytical perspective to the reference biologics and have no meaningful differences in safety, purity, or potency. Biosimilars have the potential to provide efficiencies and cost savings, and improve access to established therapies for patients needing effective treatments. In neuro-oncology, biosimilars of bevacizumab will now provide economically feasible, accepted therapies for patients with recurrent glioblastoma. Real-world evidence of biosimilars is confirming their comparable safety and effectiveness to the originator products. Licenses for rituximab biosimilars are expanding access to new therapies for patients with PCNSL, and the approval of dinutuximab beta in Europe highlights the potential for biosimilar or biosimilar-like approaches to support meaningful access for essential pediatric cancer therapies. That said, there will be barriers to this potential. Provider and patient acceptance of biosimilars will be influenced by issues of interchangeability, immunogenicity, and "extrapolated indications". Economic issues from twisted patent strategies and contracts for rebates may block access and savings. Moving forward, neuro-oncology will have an increasing range of biologics, including new immune checkpoint inhibitors and cellular therapies. Achieving equitable patient access at controlled prices as a strategy for sustainable and ethical cancer care globally will include biosimilars.

References

Balaguer, J., García Hidalgo, L., Hladun, R., Márquez Vega, C., & Pérez Alonso, V. (2023). Recent evidence-based clinical guide for the use of dinutuximab beta in pediatric patients with neuroblastoma. *Target Oncology, 18*, 77–93. https://doi.org/10.1007/s11523-022-00930-w.

Barbier, L., Ebbers, H. C., Declerck, P., Simoens, S., Vulto, A. G., & Huys, I. (2020). The efficacy, safety, and immunogenicity of switching between reference biopharmaceuticals and biosimilars: A systematic review. *Clinical Pharmacology & Therapeutics, 108*, 734–755. https://doi.org/10.1002/cpt.1836.

Blackstone, E. A., & Joseph, P. F. (2013). The economics of biosimilars. *American Health Drug Benefits, 6*, 469–478.

Brennan, C. W., Verhaak, R. G. W., McKenna, A., Campos, B., Noushmehr, H., Salama, S. R., Zheng, S., Chakravarty, D., Sanborn, J. Z., Berman, S. H., Beroukhim, R., Bernard, B., Wu, C.-J., Genovese, G., Shmulevich, I., Barnholtz-Sloan, J., Zou, L., Vegesna, R., Shukla, S. A., … McLendon, R. (2013). The somatic genomic landscape of glioblastoma. *Cell, 155*, 462–477. https://doi.org/10.1016/j.cell.2013.09.034.

Carver, K. H., Elikan, J., & Lietzan, E. (2010). An unofficial legislative history of the Biologics Price Competition and Innovation Act of 2009. *Food and Drug Law Journal, 65*, 671–818.

Chinot, O. L., de La Motte Rouge, T., Moore, N., Zeaiter, A., Das, A., Phillips, H., Modrusan, Z., & Cloughesy, T. (2011). AVAglio: Phase 3 trial of bevacizumab plus temozolomide and radiotherapy in newly diagnosed glioblastoma multiforme. *Advances in Therapy, 28*, 334–340. https://doi.org/10.1007/s12325-011-0007-3.

D'Amico, R. S., Aghi, M. K., Vogelbaum, M. A., & Bruce, J. N. (2021). Convection-enhanced drug delivery for glioblastoma: A review. *Journal of Neuro-Oncology, 151*, 415–427. https://doi.org/10.1007/s11060-020-03408-9.

DiMasi, J. A., Grabowski, H. G., & Hansen, R. W. (2016). Innovation in the pharmaceutical industry: New estimates of R&D costs. *Journal of Health Economics, 47*, 20–33. https://doi.org/10.1016/j.jhealeco.2016.01.012.

Ebbers, H. C., & Chamberlain, P. (2016). Controversies in establishing biosimilarity: Extrapolation of indications and global labeling practices. *BioDrugs, 30*, 1–8. https://doi.org/10.1007/s40259-015-0154-1.

Ferrara, N., & Adamis, A. P. (2016). Ten years of anti-vascular endothelial growth factor therapy. *Nature Reviews Drug Discovery, 15*, 385–403. https://doi.org/10.1038/nrd.2015.17.

Ferreri, A. J. M., Cwynarski, K., Pulczynski, E., Ponzoni, M., Deckert, M., Politi, L. S., Torri, V., Fox, C. P., Rosée, P. L. a, Schorb, E., Ambrosetti, A., Roth, A., Hemmaway, C., Ferrari, A., Linton, K. M., Rudà, R., Binder, M., Pukrop, T., Balzarotti, M., … Illerhaus, G. (2016). Chemoimmunotherapy with methotrexate, cytarabine, thiotepa, and rituximab (MATRix regimen) in patients with primary CNS lymphoma: Results of the first randomisation of the International Extranodal Lymphoma study group-32 (IELSG32) phase 2 trial. *The Lancet Haematology, 3*, e217–e227. https://doi.org/10.1016/S2352-3026(16)00036-3.

Friedman, H. S., Prados, M. D., Wen, P. Y., Mikkelsen, T., Schiff, D., Abrey, L. E., Yung, W. K. A., Paleologos, N., Nicholas, M. K., Jensen, R., Vredenburgh, J., Huang, J., Zheng, M., & Cloughesy, T. (2009a). Bevacizumab alone and in combination with irinotecan in recurrent glioblastoma. *Journal of Clinical Oncology, 27*, 4733–4740. https://doi.org/10.1200/JCO.2008.19.8721.

Friedman, H. S., Prados, M. D., Wen, P. Y., Mikkelsen, T., Schiff, D., Abrey, L. E., Yung, W. K. A., Paleologos, N., Nicholas, M. K., Jensen, R., Vredenburgh, J., Huang, J.,

Zheng, M., & Cloughesy, T. (2009b). Bevacizumab alone and in combination with irinotecan in recurrent glioblastoma. *Journal of Clinical Oncology, 27*, 4733–4740. https://doi.org/10.1200/JCO.2008.19.8721.

Gandhi, S., Patankar, D., Kashiramka, S., & Rathore, A. S. (2024). The economics of translating a biosimilar from lab to market in India. *Annals of the New York Academy of Sciences, 1541*, 219–229. https://doi.org/10.1111/nyas.15252.

Gilbert, M. R., Dignam, J., Won, M., Blumenthal, D. T., Vogelbaum, M. A., Aldape, K. D., Colman, H., Chakravarti, A., Jeraj, R., Armstrong, T. S., Wefel, J. S., Brown, P. D., Jaeckle, K. A., Schiff, D., Atkins, J. N., Brachman, D., Werner-Wasik, M., Komaki, R., Sulman, E. P., & Mehta, M. P. (2013). RTOG 0825: Phase III double-blind placebo-controlled trial evaluating bevacizumab (Bev) in patients (Pts) with newly diagnosed glioblastoma (GBM). *Journal of Clinical Oncology, 31*, 1. https://doi.org/10.1200/jco.2013.31.18_suppl.1.

Hermeling, S., Crommelin, D. J. A., Schellekens, H., & Jiskoot, W. (2004). Structure-immunogenicity relationships of therapeutic proteins. *Pharmaceutical Research, 21*, 897–903. https://doi.org/10.1023/B:PHAM.0000029275.41323.a6.

Hoy, S. M. (2016). Dinutuximab: A review in high-risk neuroblastoma. *Target Oncology, 11*, 247–253. https://doi.org/10.1007/s11523-016-0420-2.

Huang, S.-M., & Harari, P. M. (1999). Epidermal growth factor receptor inhibition in cancer therapy: Biology, rationale and preliminary clinical results. *Investigational New Drugs, 17*, 259–269. https://doi.org/10.1023/A:1006384521198.

Jofre-Bonet, M., McGuire, A., Dayer, V., Roth, J. A., & Sullivan, S. D. (2025). The price effects of biosimilars in the United States. *Value in Health, 28*, 742–750. https://doi.org/10.1016/j.jval.2025.02.008.

Kang, H.-N., Thorpe, R., Knezevic, I., Blades, C. D. R. Z., Casas Levano, M., Chew, J. Y., Chilufya, M. B., Chirachanakul, P., Chua, H. M., Farahani, A. V., Ghobrial, M. R. W., Habahbeh, S., Hamel, H., Kim, G. H., Perez Rodriguez, V., Putri, D. E., Rodgers, J., Savkina, M., Semeniuk, O., … Yamaguchi, T. (2020). The regulatory landscape of biosimilars: WHO efforts and progress made from 2009 to 2019. *Biologicals, 65*, 1–9. https://doi.org/10.1016/j.biologicals.2020.02.005.

Kaplon, H., Muralidharan, M., Schneider, Z., & Reichert, J. M. (2020). Antibodies to watch in 2020. *Mabs, 12*, 1703531. https://doi.org/10.1080/19420862.2019.1703531.

Kumar Singh, G., DSouza, H., Menon, N., Srinivas, S., Harindran Vallathol, D., Boppana, M., Rajpurohit, A., Mahajan, A., Janu, A., Chatterjee, A., Krishnatry, R., Gupta, T., Jalali, R., & Patil, V. M. (2021). Safety and efficacy of bevacizumab biosimilar in recurrent/progressive glioblastoma. *Ecancermedicalscience, 15*, 1166. https://doi.org/10.3332/ecancer.2021.1166.

Ladenstein, R., Pötschger, U., Valteau-Couanet, D., Luksch, R., Castel, V., Yaniv, I., Laureys, G., Brock, P., Michon, J. M., Owens, C., Trahair, T., Chan, G. C. F., Ruud, E., Schroeder, H., Beck Popovic, M., Schreier, G., Loibner, H., Ambros, P., Holmes, K., … Lode, H. N. (2018). Interleukin 2 with anti-GD2 antibody ch14.18/CHO (dinutuximab beta) in patients with high-risk neuroblastoma (HR-NBL1/SIOPEN): A multicentre, randomised, phase 3 trial. *Lancet Oncology, 19*, 1617–1629. https://doi.org/10.1016/S1470-2045(18)30578-3.

Lyman, G. H., Balaban, E., Diaz, M., Ferris, A., Tsao, A., Voest, E., Zon, R., Francisco, M., Green, S., Sherwood, S., Harvey, R. D., & Schilsky, R. L. (2018). American Society of clinical oncology statement: Biosimilars in oncology. *Journal of Clinical Oncology, 36*, 1260–1265. https://doi.org/10.1200/JCO.2017.77.4893.

Maccari, M., Baek, C., Caccese, M., Mandruzzato, S., Fiorentino, A., Internò, V., Bosio, A., Cerretti, G., Padovan, M., Idbaih, A., & Lombardi, G. (2024). Present and future of immunotherapy in patients with glioblastoma: Limitations and opportunities. *Oncologist, 29*, 289–302. https://doi.org/10.1093/oncolo/oyad321.

Majzner, R. G., Theruvath, J. L., Nellan, A., Heitzeneder, S., Cui, Y., Mount, C. W., Rietberg, S. P., Linde, M. H., Xu, P., Rota, C., Sotillo, E., Labanieh, L., Lee, D. W., Orentas, R. J., Dimitrov, D. S., Zhu, Z., Croix, B. S., Delaidelli, A., Sekunova, A., … Mackall, C. L. (2019). CAR T cells targeting B7-H3, a pan-cancer antigen, demonstrate potent preclinical activity against pediatric solid tumors and brain tumors. *Clinical Cancer Research, 25*, 2560–2574. https://doi.org/10.1158/1078-0432.CCR-18-0432.

McCamish, M., & Woollett, G. (2012). The state of the art in the development of biosimilars. *Clinical Pharmacology & Therapeutics, 91*, 405–417. https://doi.org/10.1038/clpt.2011.343.

Mendelsohn, J., & Baselga, J. (2006). Epidermal growth factor receptor targeting in cancer. *Seminars in Oncology, 33*, 369–385. https://doi.org/10.1053/j.seminoncol.2006.04.003.

Moorkens, E., Vulto, A. G., Huys, I., Dylst, P., Godman, B., Keuerleber, S., Claus, B., Dimitrova, M., Petrova, G., Sović-Brkičić, L., Slabý, J., Šebesta, R., Laius, O., Karr, A., Beck, M., Martikainen, J. E., Selke, G. W., Spillane, S., McCullagh, L., … Simoens, S. (2017). Policies for biosimilar uptake in Europe: An overview. *PLoS One, 12*, e0190147. https://doi.org/10.1371/journal.pone.0190147.

Mulcahy A. W., Hlavaka, J. P., Case, S. R. 2017 Biosimilar cost savings in the United States. Retrieved June 28, 2025, from https://www.rand.org/pubs/perspectives/PE264.html.

Nazha, B., Inal, C., & Owonikoko, T. K. (2020). Disialoganglioside GD2 expression in solid tumors and role as a target for cancer therapy. *Frontiers in Oncology, 10*, 1000. https://doi.org/10.3389/fonc.2020.01000.

Raedler, L. A. (2016). Zarxio (Filgrastim-sndz): First biosimilar approved in the United States. *American Health Drug Benefits, 9*, 150–154.

Ramanan, S., & Grampp, G. (2014). Drift, evolution, and divergence in biologics and biosimilars manufacturing. *BioDrugs, 28*, 363–372. https://doi.org/10.1007/s40259-014-0088-z.

Reardon, D. A., Brandes, A. A., Omuro, A., Mulholland, P., Lim, M., Wick, A., Baehring, J., Ahluwalia, M. S., Roth, P., Bähr, O., Phuphanich, S., Sepulveda, J. M., De Souza, P., Sahebjam, S., Carleton, M., Tatsuoka, K., Taitt, C., Zwirtes, R., Sampson, J., & Weller, M. (2020). Effect of nivolumab vs bevacizumab in patients with recurrent glioblastoma. *JAMA Oncology, 6*, 1003. https://doi.org/10.1001/jamaoncol.2020.1024.

Resende, H. M., Ladislau, L., Cardoso, A. C. F., Brandão, J. D. P., Assis, B. R., Cardoso, P., Marassi, P. H. A., & Castilho, V. (2021). Biosimilar use in breast cancer treatment: A national survey of Brazilian oncologists' opinions, practices, and concerns. *JCO Global Oncology, 7*, 1316–1324. https://doi.org/10.1200/GO.20.00649.

Schaft, N., Dörrie, J., Schuler, G., Schuler-Thurner, B., Sallam, H., Klein, S., Eisenberg, G., Frankenburg, S., Lotem, M., & Khatib, A. (2023). The future of affordable cancer immunotherapy. *Frontiers in Immunology, 14*, 1248867. https://doi.org/10.3389/fimmu.2023.1248867.

Schellekens, H. (2009). Biosimilar therapeutics–What do we need to consider? *NDT Plus, 2*, i27–i36. https://doi.org/10.1093/ndtplus/sfn177.

Schiestl, M., Zabransky, M., & Sörgel, F. (2017). Ten years of biosimilars in Europe: development and evolution of the regulatory pathways. *Drug Design, Development and Therapy, 11*, 1509–1515. https://doi.org/10.2147/DDDT.S130318.

Scientific Considerations in Demonstrating Biosimilarity to a Reference Product 2015 Retrieved June 28, 2025, from https://www.fda.gov/media/82647/download.

Scott, A. M., Wolchok, J. D., & Old, L. J. (2012). Antibody therapy of cancer. *Nature Reviews Cancer, 12*, 278–287. https://doi.org/10.1038/nrc3236.

Simoens, S., & Vulto, A. G. (2021). A health economic guide to market access of biosimilars. *Expert Opinion on Biological Therapy, 21*, 9–17. https://doi.org/10.1080/14712598.2021.1849132.

Socal, M. P., Anderson, K. E., Sen, A., Bai, G., & Anderson, G. F. (2020). Biosimilar uptake in medicare part B varied across hospital outpatient departments and physician practices: The case of filgrastim. *Value in Health, 23*, 481–486. https://doi.org/10.1016/j.jval.2019.12.007.

Stupp, R., Mason, W. P., van den Bent, M. J., Weller, M., Fisher, B., Taphoorn, M. J. B., Belanger, K., Brandes, A. A., Marosi, C., Bogdahn, U., Curschmann, J., Janzer, R. C., Ludwin, S. K., Gorlia, T., Allgeier, A., Lacombe, D., Cairncross, J. G., Eisenhauer, E., & Mirimanoff, R. O. (2005). Radiotherapy plus concomitant and adjuvant temozolomide for glioblastoma. *New England Journal of Medicine, 352*, 987–996. https://doi.org/10.1056/NEJMoa043330.

Taal, W., Oosterkamp, H. M., Walenkamp, A. M. E., Dubbink, H. J., Beerepoot, L. V., Hanse, M. C. J., Buter, J., Honkoop, A. H., Boerman, D., de Vos, F. Y. F., Dinjens, W. N. M., Enting, R. H., Taphoorn, M. J. B., van den Berkmortel, F. W. P. J., Jansen, R. L. H., Brandsma, D., Bromberg, J. E. C., van Heuvel, I., Vernhout, R. M., … van den Bent, M. J. (2014). Single-agent bevacizumab or lomustine versus a combination of bevacizumab plus lomustine in patients with recurrent glioblastoma (BELOB trial): a randomised controlled phase 2 trial. *Lancet Oncology, 15*, 943–953. https://doi.org/10.1016/S1470-2045(14)70314-6.

Tabernero, J., Vyas, M., Giuliani, R., Arnold, D., Cardoso, F., Casali, P. G., Cervantes, A., Eggermont, A. M. M., Eniu, A., Jassem, J., Pentheroudakis, G., Peters, S., Rauh, S., Zielinski, C. C., Stahel, R. A., Voest, E., Douillard, J.-Y., McGregor, K., & Ciardiello, F. (2016). Biosimilars: A position paper of the European Society for Medical Oncology, with particular reference to oncology prescribers. *ESMO Open, 1*, e000142. https://doi.org/10.1136/esmoopen-2016-000142.

The Impact of Biosimilar Competition in Europe 2020 Retrieved June 28, 2025, from https://health.ec.europa.eu/system/files/2021-01/biosimilar_competition_en_0.pdf.

Uhlig, T., & Goll, G. L. (2017). Reviewing the evidence for biosimilars: key insights, lessons learned and future horizons. *Rheumatology, 56*, iv49–iv62. https://doi.org/10.1093/rheumatology/kex276.

Understanding interchangeable biosimilars at the federal and state levels. (2023). *American Journal of Managed Care, 29*, SP545–SP548. https://doi.org/10.37765/ajmc.2023.89419.

US FDA, n.d. Purple book database of licensed biological products. Retrieved June 28, 2025, from https://purplebooksearch.fda.gov/results?query=bevacizumab&title=Alymsys.

Yang, J., Blinzler, K., Lankin, J., Vijayakumar, S., Maculaitis, M. C., & Shelbaya, A. (2022). Evolving perceptions, utilization, and real-world implementation experiences of oncology monoclonal antibody biosimilars in the USA: Perspectives from both payers and physicians. *BioDrugs, 36*, 71–83. https://doi.org/10.1007/s40259-021-00509-3.

Yu, A. L., Gilman, A. L., Ozkaynak, M. F., London, W. B., Kreissman, S. G., Chen, H. X., Smith, M., Anderson, B., Villablanca, J. G., Matthay, K. K., Shimada, H., Grupp, S. A., Seeger, R., Reynolds, C. P., Buxton, A., Reisfeld, R. A., Gillies, S. D., Cohn, S. L., Maris, J. M., & Sondel, P. M. (2010). Anti-GD2 antibody with GM-CSF, interleukin-2, and isotretinoin for neuroblastoma. *New England Journal of Medicine, 363*, 1324–1334. https://doi.org/10.1056/NEJMoa0911123.

CHAPTER

8 Biosimilars and biologics in pancreatic cancer

Sanidhya Chopra[a], Rupesh More[b], Jay Prakash Soni[c], Garvit Kanchan[a] and Dishant Chaudhari[a]

[a]*Department of Pharmacology, School of Pharmacy and Technology Management, SVKM's Narsee Monjee Institute of Management Studies (NMIMS) (Deemed to be University), Shirpur, Maharashtra, India,* [b]*Department of Pharmaceutics, School of Pharmacy and Technology Management, SVKM's Narsee Monjee Institute of Management Studies (NMIMS) (Deemed to be University), Shirpur, Maharashtra, India,* [c]*Department of Biochemistry, University of Washington, Seattle, WA, United States*

8.1 Introduction

8.1.1 Biologics and biosimilars

Advances that are utilized by living organisms, such as bacteria, yeast, or mammalian cells, are the type of more advanced medicinal products called biologics. In contrast to conventional drugs that are chemically synthesized and include relatively in-depth characterized, structurally simple drugs, biologics include structurally extremely complex, and immensely sophisticated biomolecules. These are monoclonal antibodies, cytokines (e.g., interleukins and interferons), recombinant growth factors, vaccines, clotting factors, and other therapeutic proteins (e.g., insulin analogues or erythropoietin). They are very focused in terms of action related to the main constituents of cellular and immune signaling pathways, a characteristic that gives them significant importance when improving the management of diseases like cancer, autoimmune disorders, genetic diseases, and disorders of the metabolism. In a number of ways, biologics differ fundamentally in comparison, with traditional small molecule drugs. First, they are far larger in molecular size and complexity, often the result of three-dimensional folding and glycosylation sites that play an essential role in their activity. Second, they are very sensitive to the manufacturing process since small differences at any stage, from cell culture growth, protein folding, or purification, can influence the activity, stability, and immunogenic properties of the final product. In contrast, the construction of small-molecule drugs is typically chemically produced with a high rate of reproducibility, and their particular atomic structure is fully reproducible (Sharma & Giri, 2024). In the case of biologics, the outcome, however, is determined directly by living-cell processes and adds natural after different production batches. This variant in batches is strictly regulated with the advanced analytical technology so that all yields meet and maintain safety, efficacy, and quality levels. This inherent

Biosimilars and Biologics in Cancer Treatment: An Era in Therapeutic Strategies.
DOI: https://doi.org/10.1016/B978-0-443-43824-0.00009-2

complexity also makes the essentially lengthy process of the biologics and authorization demanding to be closely overseen by regulatory bodies. Regulatory bodies like the United States Food and Drug Administration and the European Medicines Agency (EMA) have created stringent systems that go through steps including sophisticated analytics, robust quality checks, and active safety surveillance. Regulatory protection is thus mandatory to ensure that patients obtain therapeutically active yet variability-free therapies of biologic origin (Ratih et al., 2021).

Biosimilars, however, are therapeutic products specifically developed to be extremely similar to a prior approved biologic, also known as the reference product, or originator biologic. The concept of biosimilars comes up when the patent and data exclusivity period on the innovator biologic expires, and other manufacturers are now free to come up with highly similar versions. The intention of biosimilars is to reach clinical benefit equivalence at a lower price, thus expanding the use of advanced biologic therapies to patients and health systems. The Biosimilar is developed in a process that has several tiers of evidence to prove equivalence. At the analytical level, sensitive analytical methods (e.g., mass spectrometry, chromatography, and bioassays) are used to compare structural and folded form, glycosylation pattern, and comparative biological activity with the reference biologic. Also, *in vitro/in vivo* experiments, sometimes referred to as nonclinical studies, are performed to confirm similar functional effects and lack of unexpected toxicity (Declerck et al., 2017). Lastly, binding comparative testing is undertaken, not to re-validate efficacy again, but as a means to validate that the biosimilar product would have equivalent actions to the reference product in therapeutic effect, safety, pharmacokinetic (PK) profile (absorption, distribution, metabolism, and excretion), and immunogenicity profile. Introduction of biosimilars is very patent-driven due to the expiration of originator biologics. After the expiration of patents and market exclusivity, biosimilars have the opportunity to enter the market at lower prices, since their manufacturers do not need to invest time and cost in duplicating the clinical efficacy of the reference biologic; they only need to demonstrate similarity to the reference biologic, in an analogous process to making generic versions of small-molecule pharmaceuticals. Therefore, the introduction of biosimilars offers a solution to the need to save healthcare costs and even increase patient access to such life-saving drugs in many countries (Chan & Chan, 2017; Nupur et al., 2022).

8.1.2 Rationale for use in pancreatic cancer

Application of the biologics in the management of pancreatic cancer is geared toward limiting the therapeutic weaknesses and addressing the individual molecular pathways related to cancer proliferation. Pancreatic ductal adenocarcinoma (PDAC), which is characterized by poor survival rates and resistance to conventional chemotherapy, has thus led to the emergence of biologic therapy-based agents, including monoclonal antibodies and growth factor inhibitors, in an attempt to establish better outcomes (Farhangnia et al., 2024; Joshi et al., 2022). Such specific interventions have the potential of interfering with vital cellular functions, boosting immune-mediated tumor destruction, and in some cases leading to less or more manageable side effects as compared to traditional chemotherapy (Di Marco et al., 2016; Harvey, 2017).

8.1.3 Historical milestones in oncology

The oncology industry's history of biologics dates back to the early 1980s, when this branch came as a revolutionary change in the treatment of cancer and other ailments. It started with the first product of recombinant DNA technology, that is, human insulin, which was approved in 1982, showing the potent capacity of genetic engineering as a powerful therapeutic tool. Soon after, in 1986, the first biologic explicitly used to treat cancer, interferon alfa-2b, became available, eventually leading to the immunotherapy concept of utilizing the body to fight cancers using its immune system. These initial achievements set some guiding principles regarding the production and utilization of biologics in a clinical scenario. In the late 1990s, those efforts started to bear fruit when monoclonal antibodies (mAbs) (rituximab and trastuzumab) were introduced and approved in the oncology areas (Joshi et al., 2022). These directed agents against targeted proteins expressed on cancer cells, CD20 in lymphoma and HER2 in breast cancer, respectively, and represent a new age in targeted therapy of cancer (Harvey, 2017). These novel biologics greatly enhanced the survival rates and changed the standard of care in relation to hematologic and solid tumors. They have since become a foundation of current oncologic treatment regimens through the mechanism of action, often immune-mediated cytotoxicity or receptor blockade. The high and growing success of biologics, which accompany their commercial and therapeutic significance, sparked concerns about excessive expenditures and accessibility to medicines. With a number of original biologics having just started to expire in patent protection, regulators around the world identified the necessity of pathways to introduce biosimilars capable of offering similar clinical benefits at a lower price. Such guidance was initially formalized by the EMA, which formally adopted the world's first biosimilar approval guidance in 2003. In its turn, the United States developed its regulatory pathway, where it came up with the Biologics Price Competition and Innovation Act (BPCIA) of 2009 with the aim of providing an accelerated licensure process of biosimilars under the FDA (Bachu et al., 2022).

8.2 Overview of pancreatic cancer

8.2.1 Epidemiology and disease burden

The disease has been recorded to cause a significant and potentially rising health burden globally, which is pancreatic cancer. In the past decade, the age-standardized incidence rate, or the pancreatic cancer incidence per 100,000 person years, has significantly increased by around 0.7 increments in a decade since 1990 until the present years. There are a number of well-known risk factors that play an important role in the appearance of pancreatic cancer (Hu et al., 2021). Cigarette smoking is one of the most powerful and adjustable risk factors among them, and it is associated with 20% to 30% of cases. Furthermore, the risk of pancreatic carcinogenesis has been shown to increase in relation to maladaptive lifestyles, including obesity, chronic diabetes mellitus, and heavy alcohol intake. The presence of chronic pancreatitis, some hereditary genetic mutations, and advanced age are also contributors to the susceptibility of this disease. Pancreatic cancer is particularly threatening because of its fatal prognosis. Early diagnosis is also a very significant issue because of the late manifestation of symptoms in most cases and the lack of viable prevention methods in the early stages. Most patients are then diagnosed at an advanced and, in most cases, incurable stage, severely reducing the scope of treatment. Pancreatic cancer is estimated to soon attain the second ranking on the list of cancer-related deaths in various places across the globe. The aggressive biology and the stubbornness to conventional therapy contribute to its spread in the world (Rawla et al., 2019). The low survival rate of pancreatic cancer remains miserable. Survival with this cancer is still poor, with a 5-year survival rate of less than 10% around the world, indicating how aggressive the disease is and how ineffective the current therapies are. This poor survival points to a major public health problem in both developed and developing countries, regardless of their economic statuses. These fatal/low survival rates create an enormous burden on both the patient and their families and healthcare systems across the world, showcasing the dire need to improve preventative measures, expand early detection, and create more effective targeted therapies (Klein, 2021).

8.2.2 Molecular pathophysiology of pancreatic cancer

PDAC is the most common subtype, which develops as an aggregation of genetic defects and molecular changes generated by precursor lesions (pancreatic intraepithelial neoplasia and intraductal papillary mucinous neoplasms). These mainly include activating mutations of the KRAS oncogene in almost every PDAC case, essential to the formation and perpetuation of the tumor. Other commonly mutated genes are TP53, CDKN2A, and SMAD4, which cause the tumor suppressor roles as well as genomic vulnerability (S. Wang et al., 2021). Moreover, excessive activation of signaling pathways, including PI3K/AKT/mTOR, a strong metabolic reprogramming (mostly glycolytic shift), and a thick tumor microenvironment promote tumor growth, metastasis, and tumor resistance to treatment (Grant et al., 2016). An altering in glucose metabolism and noncoding RNAs also play an essential role in the development of disease, which is acknowledged in research (Grant et al., 2016).

8.2.3 Clinical presentations and diagnostic challenges

Pancreatic cancer usually presents late, due to nonspecific and subtle early symptoms. Classic manifestations such as painless obstructive jaundice, unexplained weight loss, and abdominal or back pain often signal advanced disease. Other potential symptoms include light colored stools, dark urine, pruritus, new-onset diabetes, and, rarely, manifestations like thrombophlebitis migrants due to paraneoplastic syndromes (McGuigan et al., 2018). Because many of these symptoms overlap with benign conditions, they contribute to frequent delays in diagnosis. Imaging modalities, such as CT, MRI, endoscopic ultrasound (with biopsy), and serum tumor markers (like CA 19-9), assist in diagnosis, but no effective screening exists for the general population. Consequently, most cases are discovered at metastatic or locally advanced stages (Zhang et al., 2018).

8.2.4 Prognosis and current outcomes

The overall prognosis of pancreatic cancer is very poor, with the overall 5-year survival time within the range from 5% to 13%, depending on the area and health facilities. Untreated advanced cases have an average survival of just under 3.5 months, yet an average of 8 months or higher with optimal treatment (Bengtsson et al., 2020). The prognosis is correlated with the stage at the time of diagnosis: patients who can undergo surgical resection (approximately 15% to 20%) have a better prognosis with 5-year survival rates between 21% and 44%, especially in the case where tumor was measured at the localized stage. Regrettably, most patients demonstrate the presence of metastatic or unresectable disease, in which long-term survival levels off in a single-digit percentage (Hu et al., 2021).

8.2.5 Existing therapeutic landscape

Management of pancreatic cancer involves a multidisciplinary approach, integrating surgery, chemotherapy, and radiation therapy. Surgical resection, when feasible, offers the only potential for cure but is applicable to a minority. Gemcitabine remains a backbone chemotherapeutic agent; however, combinations like FOLFIRINOX and newly FDA-approved regimens such as NALIRIFOX have shown improved outcomes in metastatic disease (Moss & Lee, 2010). Targeted therapies (RAS-directed agents), immunotherapy (checkpoint inhibitors, vaccines), and stroma-modifying drugs are under active investigation or in clinical use in select patients. Recent advances in personalized medicine, molecular profiling, and early detection technologies are incrementally improving survival and quality of life, but PDAC overall remains highly resistant to conventional and novel therapies, underscoring the urgent need for more effective approaches (Chand et al., 2016).

8.3 Biologics in pancreatic cancer

8.3.1 List of various biologics used

Monoclonal antibodies have been designed to directly attack antigens on pancreatic cancer cells or their microenvironment, which block major growth and survival pathways or activate the immune response, such as antibody-dependent cellular cytotoxicity (ADCC). Examples are against EGFR, IGF-1R, and B7-H3 antigens. In preclinical studies, there is good potential, but in clinical trials of agents such as mAbs like cetuximab and ganitumab, survival benefits so far are minimal. Further advancement involves Fc-modified mAbs that will promote immunological stimulation.

8.3.1.1 Monoclonal antibodies

In pancreatic cancer, immunotherapy is used through the use of immune checkpoint inhibitors (e.g., anti-PD-1 pembrolizumab), cancer vaccines, adoptive T-cell therapies (CAR-T), and other emergent agents to turn on the immunosuppressive tumor microenvironment (Lee et al., 2021). Checkpoint inhibitors are also effective, mostly based on preselected biomarkers (MSI-high). Vaccine and cell therapies are at the early clinical phase, and merging approaches continue to be studied (Zheng et al., 2024).

8.3.1.2 Immunotherapies immunotherapy

Pancreatic cancer employs immune checkpoint inhibitors (e.g., anti-PD-1 pembrolizumab), cancer vaccines, adoptive T-cell therapies (CAR-T), and newer tools to break the immunosuppressive tumor microenvironment. The checkpoint inhibitors demonstrate their effectiveness only in sub-groups selected by biomarkers (MSI-high) (Lutz et al., 2024). There are vaccines and cell therapies in the developmental phase with continued research on combination approaches to increase efficacy (Ye et al., 2024).

8.3.2 Mechanisms of action

In prostate cancer, immunotherapy seeks to overcome a characteristically "cold" and immunosuppressive tumor microenvironment (TME) that is marked by low tumor mutational burden, limited neoantigen expression, reduced cytotoxic T-cell infiltration, and enrichment of regulatory T cells (Tregs), myeloid-derived suppressor cells (MDSCs), and M2-polarized macrophages. Immune checkpoint inhibitors such as anti–PD-1/PD-L1 or anti–CTLA-4 antibodies aim to restore exhausted $CD8^+$ T-cell function by blocking inhibitory receptor–ligand interactions, although clinical benefit has been largely confined to molecularly selected subgroups (e.g., MSI-H/dMMR or CDK12-mutated tumors). Therapeutic cancer vaccines, notably sipuleucel-T, enhance antigen presentation by activating autologous dendritic cells *ex vivo* against prostatic acid phosphatase (PAP), thereby promoting antigen-specific T-cell priming and expansion in vivo. Adoptive cellular approaches, including CAR-T cells targeting prostate-associated antigens such as prostate-specific membrane antigen (PSMA) or prostate stem cell antigen (PSCA), are designed to bypass inadequate endogenous T-cell priming and directly mediate cytotoxicity; however, their efficacy is constrained by stromal barriers, T-cell trafficking limitations, and immunosuppressive cytokines (e.g., TGF-β, IL-10). Emerging combinatorial strategies integrate androgen receptor pathway inhibitors, radiotherapy, PARP inhibitors, or TGF-β blockade to enhance antigen release, upregulate interferon signaling, and remodel the TME toward a more inflamed phenotype, thereby potentiating immune-mediated tumor clearance (Huang & Buchsbaum, 2009; Kwon & Joung, 2025).

8.3.3 Clinical efficacy data from trials

Adding monoclonal antibodies, including cetuximab and bevacizumab, to chemotherapy in advanced pancreatic cancer has generally not shown a significant effect on overall survival; median overall survival was between 6 and 7 months in most studies. The survival issue was also lacking when subjected to trials with ganitumab. MSI-high pancreatic cancer is the MSI Homolateral cells found in a minority. The FDA-approved pembrolizumab (Lee et al., 2021). The prospects of treatment with vaccines and adoptive cell therapy have been identified; however, most studies are early-phase and are unable to demonstrate efficacy yet. Agents that have been designed with optimized immune functions are at the preclinical stage or early clinical trials in the effort to counter the flaw of tumor heterogeneity and immunosuppression (Ye et al., 2024).

8.4 Biosimilars

8.4.1 Biosimilars versus original biologics

Biosimilars and original biologics (also referred to as reference biologics) are both large and complex medicines consisting of living cells, yet differ in their development and cost as well as regulatory pathway. An original biologic is the first incarnation of a biological product to have undergone a lengthy period of laboratory work to perform preclinical research, and a series of phases of clinical trials to determine the safety, effectiveness, and quality of the product. Conversely, a biosimilar is produced once the patent of the original product expires and has to show an extremely high degree of similarity to the referenced biologic in structure, biological activity, safety, and clinical performance with no clinically significant variations. Biosimilars are not replicas compared to generic drugs since a biologic is manufactured through a living system, which inherently brings a slight variation. The use of biosimilars has been approved based on more analysis and specific clinical trials, as opposed to duplicating large-scale trials. Moreover, biosimilars tend to be cheaper as they are cheaper developed and there is more and more competition in the market, whereas original biologics are higher in price because of starting research and innovation expenses (Blandizzi et al., 2017) (Abraham, I.,2013).

8.4.2 Differences between biosimilars and generics

Generics are the chemically produced drugs that belong to small molecule drugs and represent an impeccable replica of their brand counterparts that usually require insignificant medical data in order to receive authorization. Biosimilars, in contrast, are large and complex protein molecules manufactured in living cells and are in contravention of natural and unique, and subject to variation as they only appear similar but not identical to the corresponding reference biologics. Biosimilars thus result in lengthy physicochemical, functional, and clinical comparative studies to prove similarity with no clinically significant change in safety, efficacy or immunogenicity. They also require special handling and storage needs, given that they are biological in nature. By contrast, biosimilars are vast, complex proteins that are synthesized inside living cells; that is, they are complex and structurally variable by nature. Biologics represent living organisms, and hence, they contain distinctive post-translational modifications and three-dimensional structures that are not resembled in chemically synthesized medications. That inherent complexity precludes biosimilar products from being the same as their corresponding reference biologics: they are instead required to be characterized as comparable as possible, i.e., to be highly similar in molecular structure, performance, and effect (Creber et al., 1994). All the biosimilars go through an intense and extensive development procedure that incorporates full physicochemical characterization, functional assessment, and comparative clinical trials that achieve a similar level without clinically substantial disparities in terms of safety, efficacy, and immunogenicity. Given biological variability, differences between different batches of a biologic, including the reference product, are anticipated, and, therefore, the biosimilarity demonstration process is focused on demonstrating that no negative effect will be triggered on the patient (Iskit, 2025). Besides, biosimilars are biological molecules, which necessitate special handling and storage requirements that do not tamper with their structure as well as efficacy. These can be refrigeration, avoidance of light, and high procedures of transport and administration. Bio similarity is complex not only in terms of its molecular structure but also complex in terms of manufacturing sites that require sophisticated technology and quality control to ensure comparability during the manufacture of the product (Jarab et al., 2025). Finally, when it comes to the difference between generics and biosimilars, it happens to rest with molecular complexity, development, and regulatory route, as well as data needs in clinical aspects. Generics are based on differences in chemical equivalence and bioequivalence studies; however, biosimilars also require a cumulative totality of evidence that includes structural, functional, nonclinical, and clinical comparative information. The two bring significant cost savings and broader access to medicines, and the fact that biosimilars can tackle challenges that biologic medicines derived using living systems present in a unique way (Katayama et al., 2020).

8.4.3 Sequential approval process of biosimilars

Biosimilar approval is sequential, starting with full analytical analyses comparing the biosimilar in terms of molecules structure, posttranslational modifications, and biological activity to the reference product. Signaling functional bioassays measure the signaling outputs of individual receptors (Hung et al., 2017). The next step is the conduct of clinical PK and pharmacodynamic (PD) studies intended to indicate the similarity of drug behavior in humans (Iskit, 2025). The safety in the clinic is assessed as well as safety in terms of immunogenicity, and in case residual uncertainty occurs, partial efficacy trials could be required. Regulatory systems in many countries around the world have adopted a totality of evidence approach, in which all data are consolidated to demonstrate that biosimilars conform to high levels of quality, safety, and efficacy without the repetition of unnecessary clinical testing (Markus et al., 2017). The second step concerns clinical PK and PD trials to contrast the activity of the biosimilar in the human body to the originator biologic. Such studies include absorption, distribution, metabolism, and excretion PK studies and the biological effect of the drug PD. They aim to exclude any clinically significant disparities between the biosimilar and the reference product with respect to drug exposure and activity; they want to have the certainty that they share therapeutic action in a patient. After this, immunogenicity/clinical safety testing is done (J. Wang & Chow, 2012). The third factor is the immunogenicity of the biosimilar, which means whether there is a risk that the biosimilar will generate an unwanted immune response, including the generation of antidrug antibodies (ADAs) in the body that could decrease efficacy or generate adverse effects. Such studies pay close attention to safety profile, adverse event incidents, and immune response in order to make sure that the biosimilar clinical risk profile is equal to that of a reference biologic (Desai et al., 2016). In case such remaining uncertainty about efficacy or safety exists even after these steps, regulators may seek to run partial or targeted clinical efficacy studies, typically designed to compare the response rates or progression-free survival, etc., to further establish biosimilarity. But in most cases, such trials tend to be minimized or waived when the analytical and PK/PD data are becoming increasingly robust, reducing the burdens on the patients and on the development costs (Chamberlain et al., 2024).

8.4.4 Approved biosimilars in oncology

Across the world, there has been the approval of oncology biosimilars of critical biologics such as trastuzumab, bevacizumab, rituximab, and filgrastim, immensely making the therapeutic cancer drug available to patients. Such approvals have made health care systems save money and be able to make more lifesaving biologic medicines that are available to numerous types of cancer. EMA was the first to introduce biosimilar regulatory guidance beginning in 2005 when it designed a centralized, science-based approach to the approval of biosimilars in the European Union (Bachu et al., 2022). This framework established the standards of the world to prove the biosimilarity through intense analytical and clinical comparability. The United States Food and Drug Administration took the next step in 2010 to develop the streamlined approval pathway to approve biosimilar and interchangeable biologics through the BPCIA, which aims to balance innovation and competition (Joshi et al., 2022). Since then, biosimilar approvals have swelled up globally, including key markets across Europe, North America, and Asia since these transformative regulatory developments. This growth has stimulated competition in the marketplaces and enormous savings on healthcare costs by encouraging access to cheaper biosimilar options. Interchangeable biosimilars and fast-tracked approval add to the uptake through greater physician and patient trust, and by simplification of substitution procedures (Joshi et al., 2022). The speed at which the oncology biosimilars market is expanding remains high as more new agents are likely to enter the market before 2025 and beyond. This development aids the wider availability of treatment and fosters healthcare sustainability, thanks to affordably lower costs of biologic cancer treatment (García et al., 2020; Nahleh et al., 2022).

8.4.5 Role of biosimilars and next-generation biologics in pancreatic cancer management

8.4.5.1 Peg filgrastim biosimilars

Peg filgrastim biosimilars are long-acting granulocyte colony-stimulating factors (G-CSF) to trigger the bone marrow to produce neutrophils, which play an essential part in the immune system. They are mostly applied to pancreatic cancer patients receiving a drug chemotherapy combination of FOLFIRINOX, to shorten the length of the therapeutically desired neutropenia and lessen the severity of the neutropenia with an ultimate outcome of sustaining the danger of febrile neutropenia and infection (C. Y. Wang et al., 2023). Peg filgrastim locks to the hematopoietic cell receptor G-CSF, provoking the growth, differentiation, and also stimulation of an inciting blast neutrophil. Biosimilar analogs have shown interchangeable PKs, efficacy, and safety to the originator peg filgrastim as demonstrated by various clinical trials and studies in the venue of the real world, allowing more individuals with supportive care in the treatment that is lower in cost (Katayama et al., 2020) (Table 8.1).

8.4.5.2 Filgrastim biosimilars

Biosimilars of filgrastim are short-acting recombinant G-CSFs to prevent neutropenia during chemotherapy (as opposed to peg filgrastim, but with more frequent dosing requirements). They work in the same way—they stimulate the development of neutrophils by means of activating the G-CSF receptors in the bone tissue (Sharma & Giri, 2024). Experience with biosimilars of filgrastim and other biosimilars has demonstrated clinical similarity with their reference products in aspects of safety and efficacy using comparative clinical trials. Their adoption in chemotherapy of pancreatic cancers has been key in alleviating infection-related morbidity and mortality cases, which should be maintained in the supportive care of oncology patients in the future (Douglas et al., 2017).

8.4.5.3 Bevacizumab biosimilars

Bevacizumab biosimilars exhibit the same effect as the reference biologic because they also bind vascular endothelial growth factor A (VEGF-A), which influences the process of tumor angiogenesis. They work by blocking VEGF-A, which is requisite to the creation of additional blood vessels that allow tumor growth and metastasis, starving tumors of nutrients and oxygen (Di Marco et al., 2016). Several cancers, such as pancreatic adenocarcinoma, are also allowed to be treated with bevacizumab biosimilars, where they are used either singly or in combination with chemotherapy. Equivalent clinical evidence has demonstrated similar efficacy, safety, and immunogenicity profiles of the originator and can lead to an ability to expand

Table 8.1 Role of biosimilars and next-generation biologics in pancreatic cancer management

Biosimilar/drug	Indication in pancreatic cancer	Mechanism of action (MOA)	Target pathway	References
Peg filgrastim biosimilars	Supportive care for chemotherapy-induced neutropenia (e.g., FOLFIRINOX)	Recombinant human G-CSF analog stimulates neutrophil proliferation and differentiation	Hematopoietic growth factor pathway	C. Y. Wang et al. (2023)
Filgrastim biosimilars	Supportive care for neutropenia prevention	Recombinant human G-CSF stimulates bone marrow production of neutrophils	Hematopoietic growth factor pathway	Douglas et al. (2017)
Bevacizumab biosimilars	Antiangiogenic therapy in combination regimens	Monoclonal antibody binds VEGF-A, inhibiting angiogenesis	VEGF/angiogenesis pathway	Nahleh et al. (2022)
Rituximab biosimilars	Immune modulation (mainly hematologic malignancies; investigational in pancreatic cancer)	Anti-CD20 antibody inducing B-cell depletion via ADCC and CDC	CD20 on B cells	Pierpont et al. (2018)

(continued on next page)

Table 8.1 Role of biosimilars and next-generation biologics in pancreatic cancer management—cont'd

Biosimilar/drug	Indication in pancreatic cancer	Mechanism of action (MOA)	Target pathway	References
Trastuzumab biosimilars	HER2+ tumors; limited use in pancreatic cancer	Anti-HER2 antibody blocks HER2 receptor signaling	HER2/ErbB2 receptor pathway	Nahleh et al. (2022)
Niraparib (not biosimilar)	Targeted therapy for BRCA-mutated pancreatic cancer	PARP inhibitor causes DNA damage and tumor cell death	DNA repair pathway	Fudalej et al. (2023)
Bispecific antibodies*	Investigational: immune modulation and multitargeting	Dual targeting of immune checkpoints or tumor antigens	Immune checkpoint pathways (e.g., B7-H3)	Guidi et al. (2025)
Antibody-drug conjugates*	Early clinical use, including pancreatic cancer	Targeted delivery of cytotoxic drugs via tumor-specific antibodies	Tumor antigen targeting with cytotoxic payload	Wang et al. (2025)

* *New therapies/agents.*

access to antiangiogenic therapy with a possible cost advantage (Katayama et al., 2020).

8.4.5.4 Trastuzumab biosimilars

The biosimilar competitor or substitute trastuzumab resembles the human epidermal growth factor receptor 2 (HER2/ErbB2) and is well established in the field of HER2-positive breast cancer and gastric cancer. A small number of pancreatic cancers overexpress HER2, and thus a rationale exists to use trastuzumab (Sharma & Giri, 2024). The action of Trastuzumab biosimilars is based on the inhibition of the HER2 receptor-stimulating pathways that enhance tumor growth and cell survival. Their authorization and clinical use have also been supported by strong comparative clinical trials confirming their similarity in terms of efficacy, safety, and immunogenicity (Katayama et al., 2020).

8.4.5.5 Bispecific antibodies (New therapies)

Bispecific antibodies are the new generation of biologics targeting two specific antigens/epitopes. As an example of treatment, in pancreatic cancer, bispecific antibodies such as BGB-C354 against immune checkpoint molecules, like B7-H3, and other antigens on tumor cells can help increase immune cell interactions and cancer cell destruction (Cereda & D'Andrea, 2025). Such multifunctional antibodies are meant to bypass immune-evasion mechanisms and heterogeneity in the tumor. Although not yet established as biosimilars, early clinical trials have shown an efficacy and safety profile that has indicated that they could be the revolution to pancreatic cancer immunotherapy (Katayama et al., 2020).

8.4.5.6 Antibody-drug conjugates (New agents)

Antibody-drug conjugates (ADCs) are designed molecules that consist of tumor-specific antibodies conjugated to a cytotoxic drug, which enable specific delivery of chemotherapy to the cancer cell, causing minimal and specific toxicity (Di Marco et al., 2016). In pancreatic cancer, ADCs appear to be a powerful remedy that overcomes the dense tumor architecture with chemoresistance. Most recent ADC designs have bi and multispecific formats to enhance specificity and efficacy. ADCs are not biosimilar drugs, but since they can be used in combination with biosimilars, they are analogous to biosimilars in effect, to the extent that they equate to an expanding therapeutic toolbox. In future research, clinical outcomes are improved in certain patient groups when treated with ADCs (Katayama et al., 2020).

8.5 Biosimilars in pancreatic cancer

8.5.1 Biosimilars for pancreatic cancer regimens

In pancreatic cancer treatment, biosimilars primarily include granulocyte colony-stimulating factors (G-CSFs) such as peg filgrastim and filgrastim biosimilars. These

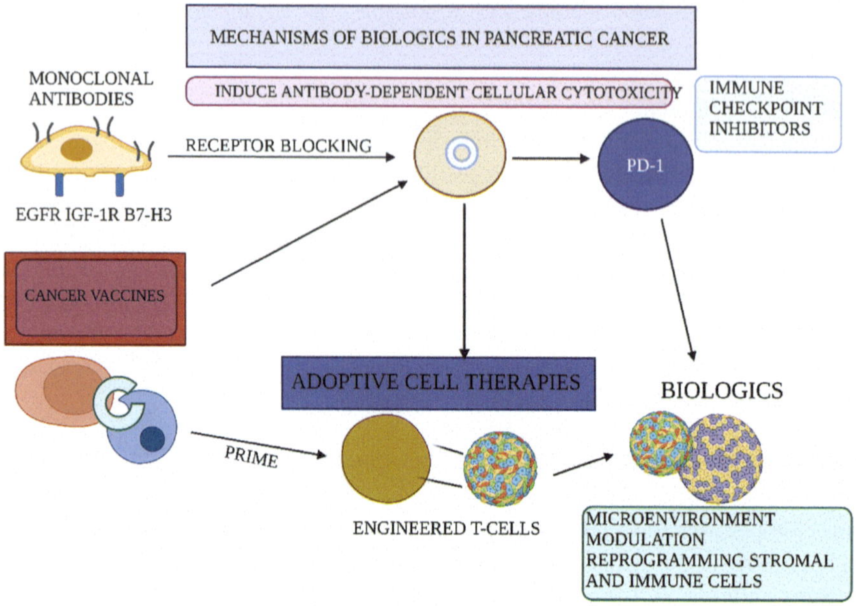

FIGURE 8.1

Biosimilars in pancreatic cancer.

supportive care agents reduce the risk of chemotherapy-induced febrile neutropenia, a critical complication in regimens like FOLFIRINOX. Recently, the FDA approved new biosimilars for bevacizumab, an anti-VEGF monoclonal antibody, used in various cancers, including pancreatic cancer (Di Marco et al., 2016; Fudalej et al., 2023). Emerging biologics such as zenocutuzumab, approved for advanced pancreatic adenocarcinoma with specific genetic alterations, highlight the evolving targeted therapy landscape, although not yet biosimilar. Biosimilars of key oncology drugs like bevacizumab are expanding treatment options and improving affordability for pancreatic and other cancers (C. Y. Wang et al., 2023; Zhao et al., 2023) (Fig. 8.1).

8.5.2 Clinical trial evidence

8.5.2.1 Major studies and outcomes

Clinical trials of biosimilars in pancreatic cancer focus mainly on supportive agents like peg filgrastim biosimilars, demonstrating equivalent efficacy and safety compared to originators in preventing neutropenia-related complications in chemotherapy patients. For example, real-world studies show no significant difference in febrile neutropenia rates between biosimilar and originator peg filgrastim users (C. Y. Wang et al., 2023). Novel drug trials in pancreatic cancer targeting specific molecular alterations, such as the POLAR phase II trial combining pembrolizumab and Olaparib, show promising outcomes in genetically defined patient subgroups, though not biosimilars themselves. Other trials investigating biosimilar integration with standards of care report successful outcomes in expanding access without compromising efficacy (MacDonald et al., 2021).

8.5.2.2 Meta-analyses and real-world data

Accumulating real-world data, as well as meta-analyses, provide robust evidence in favor of clinical equivalence of biosimilars and their reference biologics in oncology. These extensive reviews all summarize the data of a series of clinical trials and large patient populations, confirming that biosimilars have a comparable safety and efficacy profile across many cancers and supportive care biosimilars used in pancreatic cancer treatment regimens (Na et al., 2023). This emerging body of evidence can give healthcare providers additional confidence to fight biosimilars as equivalent substitutes to original biologics that do not affect patient outcomes. Further, there have been large insurance claims studies that have also shown that peg filgrastim biosimilars, which are popularly being used to prevent neutropenia induced by chemotherapy, are equal with respect to their safety and efficacy as their reference products (Na et al., 2023). Real-life validation is essential in lending assurance to further clinical applications and adoption, especially as biosimilars enter regular care into the oncology practice. In addition to the clinical facts, such data indicate considerable economic gains that come with the adoption of biosimilars. By providing low-cost substitutions, biosimilars make such healthcare systems and products available to patients at lower costs, resulting in no clinical compromise. The savings that can be achieved in this way are especially valuable in resource-strained contexts or where healthcare costs may prohibit access to biologics as costs rise (Xu et al., 2022).

8.5.3 Integrating biosimilars into oncology practice

8.5.3.1 Prescribing practices

Oncology prescribing practices increasingly favor biosimilars, particularly for treatment-naïve patients, as evidence supports therapeutic equivalence and cost savings (Joshi et al., 2022). Institutional protocols and payer policies often guide biosimilar use, with physician comfort rising alongside experience and education. Transitioning stable patients to biosimilars is gaining acceptance, but practice variability exists partly due to awareness and logistical factors (Nahleh et al., 2022).

8.5.3.2 Patient and physician perspectives

Surveys indicate strong physician confidence in biosimilars' safety and efficacy, with many willing to prescribe biosimilars to new and stable patients. Patients generally accept biosimilars, particularly when informed about equivalence to original biologics and cost benefits. Education remains crucial to overcome historical skepticism and to support shared decision-making (Nahleh et al., 2022).

8.5.4 Economic impact: Cost-benefit analyses

Economic analyses consistently demonstrate that biosimilars offer significant cost savings without sacrificing clinical outcomes, thereby improving healthcare sustainability (Hübel et al., 2020). Conversion to biosimilar peg filgrastim in pancreatic cancer chemotherapy prophylaxis can save millions, enabling expanded access to intensive regimens like FOLFIRINOX in resource-constrained settings. Pharmacoeconomic evaluations across countries support oncology biosimilars as cost-effective and budget-friendly alternatives that enhance value-based cancer care (Cornes, 2012).

8.6 Safety, immunogenicity, and monitoring in pancreatic cancer

8.6.1 Immunogenicity

Immunogenicity is the capacity of a therapeutic biologic or biosimilar to induce an immune reaction in a patient that can influence drug safety and efficacy. Such an immune reaction may result in the formation of ADAs that would neutralize the drug's therapeutic action or cause adverse effects. Although not all immune activation is medically significant, it can be considered that immunogenicity is the main safety concern with biosimilars and most likely with monoclonal antibodies against malignancies (Kurki et al., 2021). Examples of factors that may influence immunogenicity are the molecular structure, posttranslational modifications such as glycosylation, host-cell impurities, and manufacturing processes. Strict analytical and clinical evaluation is also done to guarantee that biosimilars also have similar immunogenicity profiles as the comparative reference biologics, with little risk of immunogenicity. Greater caution continues to be needed to check immunogenicity-based concerns pre and postapproval (Strand et al., 2020).

8.6.2 Clinical monitoring parameters

The clinical surveillance of biosimilars in oncology is a pivotal factor that ensures product efficacy in clinical use and the protection of patients across the treatment duration. This is done through the repeated and thorough review of the biosimilar in terms of its therapeutic performance, safety risks, and occurrence of the possible adverse outcomes, including those related to immunogenicity, which is the most significant peculiar issue of biologics and biosimilars (Alturki, 2023). The main clinical parameters to be monitored are the judicious routine clinical evaluation of the patient's response to treatment. These tests include monitoring of treatment effects or cancer progression through the clinical examination, imaging, and assessment of symptoms of the patients to ensure the biosimilar medication is equally effective in providing antitumor effects like its original biologic. In addition to these clinical assessments, there is a set of laboratory assays to measure important biomarkers and blood cell counts, liver and kidney functioning, and immune markers that provide early evidence of the drug or excessive exposure-related toxicity or dysfunction (Rak Tkaczuk & Jacobs, 2014). The other critical element of clinical supervision is the careful attention to adverse reactions, especially hypersensitivity and infusion-related, more frequently observed with biologics than with small-molecule drugs. Due to the fact that biologics are protein-based medicines, it is quite possible that such drugs may face the threat of immunologic reactions, as these reactions may present themselves in early forms of allergies during the introduction of medications or as a delayed immune-based action. It is important to recognize and deal with these reactions in the interest of patient safety. In order to facilitate this safe and effective introduction of biosimilars to oncology practice, oncologists, nurses, pharmacists, and other clinical staff should be educated. Education efforts are aimed at proper surveillance methodology, identification, and management of adverse effects, immunogenicity issues, and the need to maintain and document properly, and pharmacovigilance reporting. The provision of knowledge to clinical teams can eradicate uncertainty in the use of bioassays and guarantee safe usage, effective uses, and enhanced access to patients to these life-saving therapies (Ngo & Chen, 2021).

8.6.3 Postmarketing surveillance requirements

The postdrug approval surveillance, generally termed as pharmacovigilance, is a central component of sustaining long-term safety and efficacy of biosimilars after they are incorporated in the mainstream of clinical practice. Despite the extensive scrutiny to which biosimilars are subjected before approval during preapproval stages of analysis, nonclinical testing, and clinical testing, there is a high need to monitor them in real-world conditions, where patients are diverse, and the use duration is extended, and more complex treatment regimens may reveal uncommon or infrequent side effects that are not detected during clinical trials. Regulatory agencies across the world insist on the establishment of effective risk management plans (RMPs) specific to biosimilars. These plans will provide detailed strategies on how there will be regular gathering, assessing, and subsequently submitting any side effects that will be witnessed as a result of using biosimilars, including immunogen components like hypersensitivity reactions and formation of ADAs. Because of the complexity of biologics, the immunogenic reactions are unpredictable and can be temporally delayed to occur only with extended administration, making postmarketing collection of vital data even more important. Adverse event reporting, both routine and spontaneous, is a regulatory requirement of both the manufacturer and the healthcare provider (Singh et al., 2019). Further, regulatory authorities can demand corresponding postauthorization safety studies to proactively explore special safety questions, or to further collect extra clinical data in larger samples or subgroups. These studies are supplementary to traditional pharmacovigilance measures, and they add to the more structured real-world evidence highly needed in ongoing benefit-risk measurements. Notably, the regulatory obligations in the postmarketing surveillance of the biosimilars are substantially similar to those that are set for originator biologics. Nevertheless, they are modified to tackle the particular problem of biosimilars, namely to trace minor variations that might occur in clinical, but presumably not preclinical, trials despite high similarities in the latter. This is a strategy that builds on the large amount of safety and efficacy already known about the referenced products, with a biosimilar-specific vigilance in mind (Mysler et al., 2016).

8.7 Future directions and emerging therapies

8.7.1 Biosimilars in the pipeline for pancreatic cancer

Biosimilar pipelines that have relevance to pancreatic cancer are broadening, with most of them addressing supportive care products and monoclonal antibodies against selected pathways in the tumor growth and progression pathways. Recent biosimilar innovation has focused on variations of granulocyte colony-stimulating factors (G-CSFs), Filgrastim and peg filgrastim, which play pivotal roles in the treatment of chemotherapy-induced neutropenia as a part of pancreatic cancer treatment regimens such as FOLFIRINOX (C. Y. Wang et al., 2023). Also, multiple bevacizumab biosimilars against VEGF-driven angiogenesis are being developed or recently approved, which widens affordable access to pancreatic and other cancers. The combination of pairing two or more different cancer-destroying functional modules with one or more effector domains on the same protein (e.g., bispecific antibodies, ADCs, tumor microenvironment- targeted biologic, and immune modulating) is emerging in clinical development pipelines, and some versions are being tested in pancreatic cancer and other gastrointestinal malignancies (Luo et al., 2025). The most prominent pipeline candidates are monoclonal antibodies that target immune checkpoints, such as BGB-C354 to the immune checkpoint receptor B7-H3, and novel cytokine-based products activating localized immune response in the tumor. Although they cannot be considered as biosimilars yet, these are emerging agents that are the next generation of biologics that are likely to change the treatment landscape in terms of pancreatic cancer. The development of the biosimilar market, besides these novelties of biologic medicine, drives a cancer disease treatment method that is more affordable and easier to attain. (Bielski et al., 2025)

8.7.2 Trends in biologic drug development

The rapidly developing new drugs are characterized by biologic drug development and include the emergence of next-generation antibodies that include bispecific antibodies, ADCs, and fusion proteins. Such agents possess enriched specificity, multifunctionality, and the prospect of overcoming the tumor heterogeneity and immune evasion in pancreatic cancer (Yajaman et al., 2025). Newer trends include the development of engineering of antibodies that recruit immune cells or deliver their products to a selective location or adversely change the tumor microenvironment.

In addition, enhanced manufacturing, modularization, and analytics processes are increasing scalability and quality in production. There are regulatory pathways that are evolving to encourage accelerated development and approval to encourage market entry of biosimilar and novel biologics (Sun et al., 2023). The emerging trends include personalized medicine integration, biomarker-driven patient selection, and combination regimens with immunomodulation or targeted therapy using biologics. It is believed that there is a massive growth in biologics, and this growth is supported by continuous innovators and increased indications such as pancreatic cancer (Wang et al., 2025).

8.7.3 Personalized medicine and biomarkers

The discovery and validation of certain biomarkers, such as genomic, transcriptomic, and immunohistochemical biomarkers, increasingly underpin precision medicine in pancreatic cancer (Yu et al., 2025). The predictive biomarkers include KRAS status of mutations, microsatellite instability, and tumor-specific profiles, which guide therapy choices as well as determine responses to targeted therapies and immunotherapies. Organoids derived from a patient and liquid biopsies (circulating tumor DNA) as patient-specific drug assay and real-time surveillance frameworks have the potential to bring a paradigm shift in personalized medicine (Brozos-Vázquez et al., 2024). Biomarkers of immunohistochemical basis are also under development to be used to stratify patients by risk/therapy responsiveness in order to make neoadjuvant/adjuvant treatment decisions. Incorporation of molecular profiling into clinical trials enables adaptive therapy and informs enrolment and participation in biomarker-driven trials. Irrespective of tumor heterogeneity and accessibility of samples, personalized medicine will guarantee improved results, ensuring maximization of therapeutic efficacies with minimum toxicity in pancreatic cancer (Perelló-Reus et al., 2022).

8.8 Conclusion

Since the advent of biologics, the disease treatment landscape, including pancreatic cancer, has entered a new phase of targeted treatment. Although the efficacy of therapeutic biologics, including monoclonal antibodies, has varied in the direct treatment of pancreatic cancer, the use of biosimilars has created an essential and transformative role in supportive care. Biosimilars make intensive chemotherapy regimens safe and accessible to many patients through the provision of less expensive and more convenient supportive agents, like granulocyte colony-stimulating factor (G-CSF), which also improves access to effective pancreatic cancer treatment without sacrificing safety and efficacy. Healthcare practice and policy still need to evolve to fully take advantage of the potential of these therapies. This requires new policy frameworks that will promote the use of biosimilars with streamlined regulatory frameworks and clear interchangeability guidelines. Meanwhile, future initiatives to minimize existing gaps in available knowledge, particularly pertaining to long-term immunogenicity and substitution effects of real-world patient populations, will require future research and further pharmacovigilance. The integration of these inexpensive, easily accessible biosimilars with newer biologic and personalized therapy options will probably shape the future of pancreatic cancer treatment, offering highly advanced care that is novel and sustainable.

References

Abraham, I., Sun, D., Bagalagel, A., Altyar, A., Mohammed, A., Tharmarajah, S., & MacDonald, K. (2013). Biosimilars in 3D: Definition, development and differentiation. *Bioengineered, 4*(4), 203–206. https://doi.org/10.4161/bioe.25067.

Alturki, N. A. (2023). Review of the immune checkpoint inhibitors in the context of cancer treatment. *Journal of Clinical Medicine, 12*(13), 4301. https://doi.org/10.3390/jcm12134301.

Bachu, R. D., Abou-Dahech, M., Balaji, S., Boddu, S. H. S., Amos, S., Singh, V., Babu, A. J., & Tiwari, A. K. (2022). Oncology biosimilars: New developments and future directions. *Cancer Rep (Hoboken), 5*(11), e1720. https://doi.org/10.1002/cnr2.1720.

Bengtsson, A., Andersson, R., & Ansari, D. (2020). The actual 5-year survivors of pancreatic ductal adenocarcinoma based on real-world data. *Scientific Reports, 10*(1), 16425. https://doi.org/10.1038/s41598-020-73525-y.

Bielski, P., Barczyński, J., Mikitiuk, M., Myrcha, M., Rykała, K., Boon, L., & Sitar, T. (2025). The bispecific antibody targeting VISTA and PD-L1 shows enhanced tumor inhibitory activity in pancreatic, endometrial and breast cancers compared to mono- and combination immune checkpoint blockade. *Frontiers in Immunology, 16*, 1486799. https://doi.org/10.3389/fimmu.2025.1486799.

Blandizzi, C., Meroni, P. L., & Lapadula, G. (2017). Comparing originator biologics and biosimilars: A review of the relevant issues. *Clinical Therapeutics, 39*(5), 1026–1039.

Brozos-Vázquez, E., Toledano-Fonseca, M., Costa-Fraga, N., García-Ortiz, M. V., Díaz-Lagares, Á., Rodríguez-Ariza, A., . . ., & López-López, R. (2024). Pancreatic cancer biomarkers: A pathway to advance in personalized treatment selection. *Cancer Treatment Reviews, 125*, 102719. https://doi.org/10.1016/j.ctrv.2024.102719.

Cereda, V., & D'Andrea, M. R (2025). Pancreatic cancer: Failures and hopes-a review of new promising treatment approaches. *Exploration of Targeted Anti-tumor Therapy, 6*, 1002299. https://doi.org/10.37349/etat.2025.1002299.

Chamberlain, P., Hemmer, B., Höfler, J., Wessels, H., von Richter, O., Hornuss, C., . . ., & Roth, K. (2024). Comparative immunogenicity assessment of biosimilar natalizumab to its reference medicine: A matching immunogenicity profile. *Frontiers in Immunology, 15*, 1414304. https://doi.org/10.3389/fimmu.2024.1414304.

Chan, J. C. N., & Chan, A. T. C. (2017). Biologics and biosimilars: What, why and how? *ESMO Open, 2*(1), e000180. https://doi.org/10.1136/esmoopen-2017-000180.

Chand, S., O'Hayer, K., Blanco, F. F., Winter, J. M., & Brody, J. R. (2016). The landscape of pancreatic cancer therapeutic resistance mechanisms. *International Journal of Biological Sciences, 12*(3), 273–282. https://doi.org/10.7150/ijbs.14951.

Cornes, P. (2012). The economic pressures for biosimilar drug use in cancer medicine. *Target Oncol, 7*(Suppl 1(Suppl 1)), S57–S67. https://doi.org/10.1007/s11523-011-0196-3.

Creber, H. M. C., Davies, M. S., Francis, D., & Walker, H. D. (1994). Variation in DNA C value in natural populations of Dactylis glomerata L. *New Phytologist, 128*(3), 555–561. https://doi.org/10.1111/j.1469-8137.1994.tb03001.x.

Declerck, P., Danesi, R., Petersel, D., & Jacobs, I. (2017). The language of biosimilars: Clarification, definitions, and regulatory aspects. *Drugs, 77*(6), 671–677. https://doi.org/10.1007/s40265-017-0717-1.

Desai, K., Catalano, T., Rai, G., Misra, P., & Shah, N. (2016). Confirmation of biosimilarity in a pharmacokinetic/pharmacodynamic study in healthy volunteers for an analytically highly similar pegfilgrastim. *Clinical Pharmacology in Drug Development, 5*(5), 354–363. https://doi.org/10.1002/cpdd.269.

Di Marco, M., Grassi, E., Durante, S., Vecchiarelli, S., Palloni, A., Macchini, M., Casadei, R., Ricci, C., Panzacchi, R., Santini, D., & Biasco, G. (2016). State of the art biological therapies in pancreatic cancer. *World Journal of Gastrointestinal Oncology, 8*(1), 55–66. https://doi.org/10.4251/wjgo.v8.i1.55.

Douglas, A. G., Schwab, P., Lane, D., Kennedy, K., Slabaugh, S. L., & Bowe, A. (2017). A comparison of brand and biosimilar granulocyte-colony stimulating factors for prophylaxis of chemotherapy-induced febrile neutropenia. *Journal of Managed Care & Specialty Pharmacy, 23*(12), 1221–1226. https://doi.org/10.18553/jmcp.2017.23.12.1221.

Farhangnia, P., Khorramdelazad, H., Nickho, H., & Delbandi, A.-A. (2024). Current and future immunotherapeutic approaches in pancreatic cancer treatment. *Journal of Hematology & Oncology, 17*(1), 40. https://doi.org/10.1186/s13045-024-01561-6.

Fudalej, M., Kwaśniewska, D., Nurzyński, P., Badowska-Kozakiewicz, A., Mękal, D., Czerw, A., . . ., & Deptała, A. (2023). New treatment options in metastatic pancreatic cancer. *Cancers (Basel), 15*(8), 2327. https://doi.org/10.3390/cancers15082327.

García, J. J., Raez, L. E., & Rosas, D. (2020). A narrative review of biosimilars: A continued journey from the scientific evidence to practice implementation. *Translational Lung Cancer Research, 9*(5), 2113–2119. https://doi.org/10.21037/tlcr-20-601.

Grant, T. J., Hua, K., & Singh, A. (2016). Molecular pathogenesis of pancreatic cancer. *Progress in Molecular Biology and Translational Science, 144*, 241–275. https://doi.org/10.1016/bs.pmbts.2016.09.008.

Guidi, L., Etessami, J., Valenza, C., Valdivia, A., Meric-Bernstam, F., Felip, E., & Curigliano, G. (2025). Bispecific antibodies in hematologic and solid tumors: Current landscape and therapeutic advances. *American Society of Clinical Oncology Educational Book, 45*(3), e473148. https://doi.org/10.1200/edbk-25-473148.

Harvey, R. D. (2017). Science of biosimilars. *Journal of Oncology Practice, 13*(9_suppl), 17s–23s. https://doi.org/10.1200/JOP.2017.026062.

Hu, J. X., Zhao, C. F., Chen, W. B., Liu, Q. C., Li, Q. W., Lin, Y. Y., & Gao, F. (2021). Pancreatic cancer: A review of epidemiology, trend, and risk factors. *World Journal of Gastroenterology, 27*(27), 4298–4321. https://doi.org/10.3748/wjg.v27.i27.4298.

Huang, Z. Q., & Buchsbaum, D. J. (2009). Monoclonal antibodies in the treatment of pancreatic cancer. *Immunotherapy, 1*(2), 223–229. https://doi.org/10.2217/1750743x.1.2.223.

Hübel, K., Kron, F., & Lux, M. P. (2020). Biosimilars in oncology: Effects on economy and therapeutic innovations. *European Journal of Cancer, 139*, 10–19. https://doi.org/10.1016/j.ejca.2020.07.037.

Hung, A., Vu, Q., & Mostovoy, L. (2017). A systematic review of U.S. biosimilar approvals: What evidence does the FDA require and how are manufacturers responding? *Journal of Managed Care & Specialty Pharmacy, 23*(12), 1234–1244. https://doi.org/10.18553/jmcp.2017.23.12.1234.

Iskit, A. B. (2025). Biosimilars and interchangeability: Regulatory, scientific, and global perspectives. *European Journal of Pharmaceutical Sciences, 107*, 107224. https://doi.org/10.1016/j.ejps.2025.107224.

Jarab, A. S., Abu Heshmeh, S. R., & Al Meslamani, A. Z. (2025). Biosimilars and immunogenicity: A matter of concern? *Expert Opinion on Drug Safety, 24*(5), 519–527. https://doi.org/10.1080/14740338.2025.2467817.

Joshi, D., Khursheed, R., Gupta, S., Wadhwa, D., Singh, T. G., Sharma, S., Porwal, S., Gauniyal, S., Vishwas, S., Goyal, S., Gupta, G., Eri, R. D., Williams, K. A., Dua, K., & Singh, S. K. (2022). Biosimilars in oncology: Latest trends and regulatory status. *Pharmaceutics, 14*(12), 2710. https://doi.org/10.3390/pharmaceutics14122721.

Katayama, E. S., Hue, J. J., Bajor, D. L., Ocuin, L. M., Ammori, J. B., Hardacre, J. M., & Winter, J. M. (2020). A comprehensive analysis of clinical trials in pancreatic cancer: what is coming down the pike? *Oncotarget, 11*(38), 3489–3501. https://doi.org/10.18632/oncotarget.27727.

Klein, A. P. (2021). Pancreatic cancer epidemiology: Understanding the role of lifestyle and inherited risk factors. *Nature Reviews Gastroenterology & Hepatology, 18*(7), 493–502. https://doi.org/10.1038/s41575-021-00457-x.

Kurki, P., Barry, S., Bourges, I., Tsantili, P., & Wolff-Holz, E. (2021). Safety, immunogenicity and interchangeability of biosimilar monoclonal antibodies and fusion proteins: A regulatory perspective. *Drugs, 81*(16), 1881–1896. https://doi.org/10.1007/s40265-021-01601-2.

Kwon, W. A., & Joung, J. Y. (2025). Immunotherapy in prostate cancer: From a “cold” tumor to a “hot” prospect. *Cancers, 17*(7), 1064.

Lee, J. J. B., Lee, I. J., Choi, Y., Jeon, M. J., Jung, I. H., & Lee, H. (2021). Clinical implications of geometric and dosimetric uncertainties of inter- and intra-fractional movement during volumetric modulated arc therapy for breast cancer patients. *Cancers, 13*(7), 1651.

Luo, M., Wang, X., Yu, G., Ji, J., Li, L., & Song, F. (2025). Development of a bispecific antibody–drug conjugate targeting EpCAM and CLDN3 for the treatment of multiple solid tumors. *Experimental Hematology & Oncology, 14*(1), 33. https://doi.org/10.1186/s40164-025-00624-9.

Lutz, M. S., Wang, K., Jung, G., Salih, Helmut R., & Hagelstein, I (2024). An Fc-modified monoclonal antibody as novel treatment option for pancreatic cancer. *Frontiers in Immunology, 15*, 1357445. https://doi.org/10.3389/fimmu.2024.1343929.

MacDonald, K., Alrawashdh, N., McBride, A., & Abraham, I. (2021). Conversion to biosimilar pegfilgrastim-cbqv enables budget-neutral access to FOLFIRINOX treatment for metastatic pancreatic cancer. *Future Oncology, 17*(33), 4561–4570. https://doi.org/10.2217/fon-2021-0718.

Markus, R., Liu, J., Ramchandani, M., Landa, D., Born, T., & Kaur, P. (2017). Developing the totality of evidence for biosimilarsF Regulatory considerations and building confidence for the healthcare community. *Biodrugs, 31*(3), 175–187. https://doi.org/10.1007/s40259-017-0218-5.

McGuigan, A., Kelly, P., Turkington, R. C., Jones, C., Coleman, H. G., & McCain, R. S. (2018). Pancreatic cancer: A review of clinical diagnosis, epidemiology, treatment and outcomes. *World Journal of Gastroenterology, 24*(43), 4846–4861. https://doi.org/10.3748/wjg.v24.i43.4846.

Moss, R. A., & Lee, C. (2010). Current and emerging therapies for the treatment of pancreatic cancer. *OncoTargets and Therapy, 3*, 111–127. https://doi.org/10.2147/ott.s7203.

Mysler, E., Pineda, C., Horiuchi, T., Singh, E., Mahgoub, E., Coindreau, J., & Jacobs, I. (2016). Clinical and regulatory perspectives on biosimilar therapies and intended copies of biologics in rheumatology. *Rheumatology International, 36*(5), 613–625. https://doi.org/10.1007/s00296-016-3444-0.

Na, H., Kwon, S.-H., Son, K.-H., Baek, Y., Kim, J., & Lee, E.-K. (2023). Comparative safety profiles of oncology biosimilars: A systematic review and network meta-analysis. *Biodrugs, 37*(2), 205–218. https://doi.org/10.1007/s40259-023-00576-8.

Nahleh, Z., Lyman, G. H., Schilsky, R. L., Peterson, D. E., Tagawa, S. T., Chavez-MacGregor, M., Rumble, R. B., & Gupta, S. (2022). Use of biosimilar medications in oncology. *The Oncologist, 18*(3), 177–186. https://doi.org/10.1200/op.21.00771.

Ngo, D., & Chen, J. (2021). A clinical review of biosimilars approved in oncology. *Annals of Pharmacotherapy, 55*(3), 362–377. https://doi.org/10.1177/1060028020944596.

Nupur, N., Joshi, S., Gulliarme, D., & Rathore, A. S. (2022). Analytical similarity assessment of biosimilars: Global regulatory landscape, recent studies and major advancements in orthogonal platforms. *Pharmaceutics, 10*, 2838. https://doi.org/10.3389/fbioe.2022.832059.

Perelló-Reus, C. M., Rubio-Tomás, T., Cisneros-Barroso, E., Ibargüen-González, L., Segura-Sampedro, J. J., Morales-Soriano, R., & Barceló, C. (2022). Challenges in precision medicine in pancreatic cancer: A focus in cancer stem cells and microbiota. *Frontiers in oncology, 12*, 995357. https://doi.org/10.3389/fonc.2022.995357.

Pierpont, T. M., Limper, C. B., & Richards, K. L. (2018). Past, present, and future of rituximab—The world's first oncology monoclonal antibody therapy. *Frontiers in Oncology, 8*, 163. https://doi.org/10.3389/fonc.2018.00163.

Rak Tkaczuk, K. H., & Jacobs, I. A. (2014). Biosimilars in oncology: From development to clinical practice. *Seminars in Oncology, 41*, S3–S12. https://doi.org/10.1053/j.seminoncol.2014.03.008.

Ratih, R., Asmari, M., Abdel-Megied, A. M., Elbarbry, F., & El Deeb, S. (2021). Biosimilars: Review of regulatory, manufacturing, analytical aspects and beyond. *Microchemical Journal, 165*, 106143. https://doi.org/10.1016/j.microc.2021.106143.

Rawla, P., Sunkara, T., & Gaduputi, V. (2019). Epidemiology of pancreatic cancer: Global trends, etiology and risk factors. *World Journal of Oncology, 10*(1), 10–27. https://doi.org/10.14740/wjon1166.

Sharma, A. K., & Giri, A. K. (2024). Engineering CRISPR/Cas9 therapeutics for cancer precision medicine. *Frontiers in Genome Editing, 15*, 1345678. https://doi.org/10.3389/fgene.2024.1309175.

Singh, A., Kalaivani, M., Srivastava, S., Goyal, R. K., & Gupta, S. K. (2019). Postmarketing safety of biosimilars: Current status, challenges, and opportunities in the spontaneous reporting system. *Therapeutic Innovation & Regulatory Science, 53*(6), 741–748. https://doi.org/10.1177/2168479019872144.

Strand, V., Gonçalves, J., Hickling, T. P., Jones, H. E., Marshall, L., & Isaacs, J. D. (2020). Immunogenicity of biosimilars for rheumatic diseases, plaque psoriasis, and inflammatory bowel disease: A review from clinical trials and regulatory documents. *Biodrugs, 34*(1), 27–37. https://doi.org/10.1007/s40259-019-00394-x.

Sun, Y., Yu, X., Wang, X., Yuan, K., Wang, G., Hu, L., Zhang, G., Pei, W., Wang, L., Sun, C., & Yang, P. (2023). Bispecific antibodies in cancer therapy: Target selection and regulatory requirements. *Acta Pharmaceutica Sinica B, 13*(9), 3583–3597. https://doi.org/10.1016/j.apsb.2023.05.023.

Wang, C. Y., Vouri, S. M., Park, H., Heldermon, C. D., & Brown, J. D. (2023). Comparative effectiveness of pegfilgrastim biosimilars vs originator for prevention of febrile neutropenia: A retrospective cohort study. *Journal of Managed Care & Specialty Pharmacy, 29*(2), 119–127. https://doi.org/10.18553/jmcp.2023.29.2.119.

Wang, J., & Chow, S. C. (2012). On the regulatory approval pathway of biosimilar products. *Pharmaceuticals (Basel), 5*(4), 353–368. https://doi.org/10.3390/ph5040353.

Wang, R., Hu, B., Pan, Z., Mo, C., Zhao, X., Liu, G., Hou, P., Cui, Q., Xu, Z., Wang, W., Yu, Z., Zhao, L., He, M., Wang, Y., Fu, C., Wei, M., & Yu, L. (2025). Antibody–drug conjugates (ADCs): Current and future biopharmaceuticals. *Journal of Hematology & Oncology, 18*(1), 51. https://doi.org/10.1186/s13045-025-01704-3.

Wang, S., Zheng, Y., Yang, F., Zhu, L., Zhu, X.-Q., Wang, Z.-F., . . ., & Dong, Q.-Z. (2021). The molecular biology of pancreatic adenocarcinoma: Translational challenges and clinical perspectives. *Signal Transduction and Targeted Therapy, 6*(1), 249. https://doi.org/10.1038/s41392-021-00659-4.

Xu, X., Zhang, S., Xu, T., Zhan, M., Chen, C., & Zhang, C. (2022). Efficacy and safety of bevacizumab biosimilars compared with reference biologics in advanced non-small cell lung cancer or metastatic colorectal cancer patients: A network meta-analysis. *Frontiers in Pharmacology, 13*, 925435. https://doi.org/10.3389/fphar.2022.880090.

Yajaman, D. R., Oh, Y., Trevino, J. G., & Harrell, J. C. (2025). Advancing antibody-drug conjugates: Precision oncology approaches for breast and pancreatic cancers. *Cancers (Basel), 17*(11), 1856. https://doi.org/10.3390/cancers17111792.

Ye, X., Yu, Y., Zheng, X., & Ma, H. (2024). Clinical immunotherapy in pancreatic cancer. *Cancer Immunology, Immunotherapy, 73*(4), 64. https://doi.org/10.1007/s00262-024-03632-6.

Yu, B., Shao, S., & Ma, W. (2025). Frontiers in pancreatic cancer on biomarkers, microenvironment, and immunotherapy. *Cancer Letters, 610*, 217350. https://doi.org/10.1016/j.canlet.2024.217350.

Zhang, L., Sanagapalli, S., & Stoita, A. (2018). Challenges in diagnosis of pancreatic cancer. *World Journal of Gastroenterology, 24*(19), 2047–2060. https://doi.org/10.3748/wjg.v24.i19.2047.

Zhao, Z., Zhao, L., Xia, G., Lu, J., Shen, B., Zhou, G., Zhou, G., Wu, F., Hu, X., Feng, J., & Yu, S. (2023). Efficacy and safety of bevacizumab biosimilar compared with reference bevacizumab in locally advanced and advanced non-small cell lung cancer patients: A retrospective study. *Frontiers in Pharmacology, 12*, 918403. https://doi.org/10.3389/fonc.2022.1036906.

Zheng, R., Liu, X., Zhang, Y., Liu, Y., Wang, Y., Guo, S., . . ., & Liu, Y. (2024). Frontiers and future of immunotherapy for pancreatic cancer: From molecular mechanisms to clinical application. *Frontiers in Immunology, 15*, 1348120. https://doi.org/10.3389/fimmu.2024.1383978.

CHAPTER

9 Biosimilars and biologics in prostate cancer

Bandaru Nagaraju[a] and Mohan Gandhi Bonthu[b]

[a] *Department of Pharmacology, Sree Dattha Institute of Pharmacy, Hyderabad, Telangana, India,*
[b] *Department of Pharmaceutical Analysis, JSS College of Pharmacy, JSS Academy of Higher Education and Research, Ooty, Tamil Nadu, India*

9.1 Introduction

9.1.1 Overview

Prostate cancer commonly affects men with an annual occurrence of approx. 3,00,000 new cases are diagnosed annually in the United States. Risk of this disease affects men over 50 years of age and increases with age. Often, prostate cancer is asymptomatic for a long time until the progression of the disease. Risk factors for the disease are heredity, family history, and lifestyle, including diet and obesity(Gupta, 2017). The disease occurs more frequently in Black men, with studies showing that the incidence of prostate cancer in this group is significantly higher compared to Caucasian men. The disease usually originates and develops in the prostate gland. This type of cancer is categorized into three groups based on stage at diagnosis time. They are localized, locally advanced, and metastatic. In the first case, the disease is localized directly in the prostate gland, the second category includes the spread of cancer to nearby tissues, but it remains in the location, and the third-class includes metastases to distant organs or parts of the body, most often on bone and lymph nodes. The treatment includes surgery, radiation therapy, hormone therapy and is increasingly used recently, chemotherapy, immunotherapy, and targeted therapy. Therapies that target androgen deprivation and immune modulation have recently come to the fore in treating advanced prostate cancer (Shariat, 2000).

9.1.2 Role of biologics and biosimilars in cancer treatment

Biologics have been such a game-changer when it comes to treating various cancers, prostate cancer included. Biologics, unlike the conventional small-molecule drugs, are large, complex molecules or collections of molecules made from living organisms. When administered, biologics act on a specific body pathway or molecule; hence treatments become more focused and potentially relatively less toxic (Bergengren, 2023). Various biologics used in Oncology that have drastically impacted the way we have treated prostate cancer include vaccines, immune checkpoint inhibitors, and

Biosimilars and Biologics in Cancer Treatment: An Era in Therapeutic Strategies.
DOI: https://doi.org/10.1016/B978-0-443-43824-0.00008-0

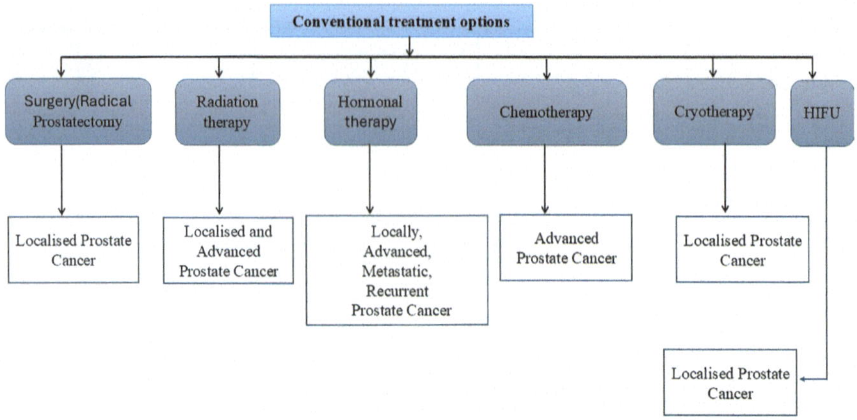

FIGURE 9.1

Conventional treatment options for prostate cancer.

monoclonal antibodies, and have been a go-to therapy in enhancing patient survival, especially in metastatic and castration-resistant prostate cancer (CRPC) cases. Biologics in prostate cancer, therefore, not only act on the cancer but are also used to manipulate the affected patient's immunity to recognize and eliminate cancerous tissues. Biosimilars are biologics that are developed to be highly comparable to an existing and already approved biological product. They are tested and confirmed to have no clinically significant difference from the reference drug in terms of safety and effectiveness, but are more affordable. This makes biologic treatments that would have otherwise been costly and beyond reach for prostate cancer patients affordable and accessible (Joshi & Khursheed, 2022). With the introduction of biosimilars, the treatments become affordable and easily accessible in resource-limited settings.

9.2 Prostate cancer: Current treatment landscape

9.2.1 Conventional treatment options

The treatment strategies for prostate cancer have dramatically developed in recent years and are aimed at eradicating cancer or controlling its development, ensuring a decent quality life to the patient. The treatment strategy choice is made taking into account a combination of factors, in particular, oncology stage and degree, patient age and general health, and life expectancy. Surgical intervention, radiation, hormonal, and chemotherapeutic medication may be selected depending on the patient's pathological data and diagnosis specifics. Among the traditional methods of the treatment of this pathology, the following can be distinguished (Fig. 9.1) (Wang & Song, 2022).

9.2.1.1 Surgery

One of the most widespread options for the treatment of localized prostate cancer is radical prostatectomy. Here, the whole prostate is dissected along with the tissue and seminal vesicles. This is recommended for patients whose cancer has not spread beyond the prostate. There are the classical or open methods of surgery, laparoscopic, and robotic-assisted. The last two methods are distinguished by a small pain syndrome in the postoperative period and short-term recovery (Wong and Zhao, 2022).

9.2.1.2 Radiation therapy

This treatment employs high-energy beams or particles to focus on and eradicate cancer cells. It is commonly utilized for both localized and locally advanced prostate cancer. The primary forms of radiation therapy include external beam radiation therapy (EBRT) and brachytherapy. EBRT involves directing radiation from outside the body toward the prostate, whereas brachytherapy consists of placing radioactive seeds directly within the prostate gland. Radiation therapy may act as a standalone treatment or be used in conjunction with other therapies, such as hormonal therapy, especially in the management of locally advanced cases (Hammerstrom & Cauley, 2011).

9.2.1.3 Hormonal therapy (androgen deprivation therapy)

Given that the growth of prostate cancer is often stimulated by male hormones, particularly testosterone, androgen deprivation therapy (ADT) seeks to decrease the levels of these hormones in the body. This therapy can involve medications that inhibit hormone production or the surgical removal of the testes (orchiectomy) to stop testosterone production. ADT is generally applied in cases of locally advanced prostate cancer, metastatic prostate cancer, and certain instances of recurrent prostate cancer. While ADT can effectively decrease tumor size and slow cancer progression, prolonged use may lead to side effects like hot flashes, reduced bone density, sexual dysfunction, and a higher likelihood of cardiovascular issues (Xu & Le, 2025).

9.2.1.4 Chemotherapy

Chemotherapy consists of medicines that kill rapidly multiplying cells and is usually reserved for patients with advanced prostate cancer that doesn't respond to hormonal therapy or has spread. Common chemotherapy drugs like docetaxel and cabazitaxel are employed to treat metastatic castration-resistant prostate cancer (mCRPC). These drugs are administered intravenously and target cancer cells throughout the body. Chemotherapy can aid in controlling cancer spread, alleviating symptoms, and enhancing quality of life for patients with advanced disease, although it may also lead to considerable side effects such as fatigue, nausea, and hair loss (Poon & Wu, 2022).

9.2.1.5 Cryotherapy

It is a newer treatment of prostate cancer involving freezing prostate tissue to destroy it. Usually, cryotherapy is used for treating men who may not have prostate removal surgery or radiation therapy. The treatment is conducted by inserting needles through which the liquid nitrogen is applied to the tissue to freeze. It is especially suitable for patients with smaller tumors and whose prostate cancer recurs after radiation therapy (Patel & Oh, 2019).

9.2.1.6 High-intensity focused ultrasound

This is a noninvasive treatment option primarily used for localized prostate cancer. It works by directing highly focused ultrasound waves at targeted areas within the prostate to generate intense heat, thereby destroying cancer cells. Although HIFU has shown promising results and is widely utilized in parts of Europe, it remains an investigational procedure in many other regions, pending further long-term clinical evaluation (Smith et al., 2022).

9.2.1.7 Side effects of conventional treatments

Most conventional treatment approaches for prostate cancers are effective and likely to have clinical side effects that affect the quality of life for the patient. These are due to surgery and radiation that may lead to urinary incontinence, erectile dysfunction, and several bowel-related adverse impacts to hormonal therapy, which causes hot flashes, fatigue, and increases osteoporosis, to chemotherapy, which causes nausea, hair loss, and fatigue. Nevertheless, newer technologies have also been recently mastered to improve and support care measures to minimize the side effects, such as focal therapies, and advanced radiation. In combination with the urologists, oncology, and supportive specialists' multidisciplinary care, most of the resultant side effects are addressed optimally (Lu & Teng, 2025).

9.2.2 Emerging treatment modalities in prostate cancer

Although traditional treatments, including surgery, radiation, and hormone therapy, persist as the mainstay of prostate cancer backbone, novel modalities of treatment are progressively becoming realistic in saving lives for patients diagnosed with prostate cancer. This is especially true of men with therapy-resistant or advanced treatment for prostate cancer, for whom there is now some hope. These new and innovative therapies have been specifically developed to control cancer growth more effectively, minimize side-effects of the treatment; long life as well as benefit men whose cancer has already spread in the body or one that does not respond to lowering testosterone; metastatic castration-resistant prostate cancer. In this section, we are going to discuss some of the new and promising treatments, which include but are not limited to immunotherapy, targeted therapy, gene therapy, and radiopharmaceuticals (Jemal & Center, 2010).

9.2.2.1 Immunotherapy in prostate cancer

Immunotherapy has transformed cancer therapy, enabling the body's immune system of the body to destroy and recognize cancer cells (Table 9.1).

Immune checkpoint inhibitors: These include medications like Pembrolizumab and Nivolumab, which operated by blocking PD1 and PDL1 checkpoint proteins. Ordinarily, cancer cells are not attacked by the body's immune cells because immune

Table 9.1 Overview of emerging treatment modalities in prostate cancer: methods, benefits, and limitations.

Treatment method	Description	Advantages	Disadvantages	References
Immune checkpoint inhibitors	Agents such as pembrolizumab (Keytruda) and nivolumab (Opdivo) block PD-1/PD-L1 checkpoint proteins, enhancing immune recognition and destruction of prostate cancer cells.	- Boosts the immune system's ability to identify and attack cancer cells. - Potential for long-lasting effects. - Used in combination with other therapies for improved results.	- Limited response in some patients. - Possible immune-related side effects such as colitis or pneumonitis. - Expensive.	Calabrò and Ettari (2025)
Cancer vaccines (Sipuleucel-T)	Provenge stimulates immune system to target prostate cancer cells, particularly in metastatic prostate cancer.	- Can extend survival in certain patients with advanced disease. - Minimal side effects compared to chemotherapy.	Limited efficacy in some patients. - Expensive and requires personalized production for each patient. - Only approved for asymptomatic or minimally symptomatic cases.	Rawlinson and Mohammed (2012)
Chimeric antigen receptor T-cell (CAR-T) therapy	Genetically modifying T-cells of patient for targeting the cancer cells, offering personalized and highly targeted treatment.	-Highly personalized approach. -Promising in treating metastatic and refractory cancer. -Potential for long-lasting remission.	- Still in early clinical development for prostate cancer. - May cause severe immune reactions such as cytokine release syndrome (CRS)	Handa and Hans (2020); Lee and Kim (2025)

(continued on next page)

Table 9.1 Overview of emerging treatment modalities in prostate cancer: methods, benefits, and limitations—cont'd

Treatment method	Description	Advantages	Disadvantages	References
PARP Inhibitors	Drugs like olaparib inhibit the PARP enzyme, preventing DNA repair in cancer cells, leading to their death—particularly effective in patients with BRCA mutations.	- Effective in cancers with DNA repair deficiencies. - Targeted approach, reducing impact on healthy cells. - Can significantly improve survival in specific patient populations.	- Limited to patients with BRCA1/BRCA2 mutations. - Side effects like nausea, fatigue, and anemia. - Resistance can develop over time.	Meehan and Gray (2021)
Androgen Receptor Inhibitors	Agents such as enzalutamide and apalutamide block androgen receptors, preventing testosterone from promoting cancer cell growth.	Elevates survival rate in mCRPC. - Effective even after hormone therapy resistance.	Resistance can develop. - Side effects include fatigue, hot flashes, and joint pain. - Can cause cognitive issues in some patients.	Hernandez and Roberts (2021)
PI3K/Akt/mTOR Pathway Inhibitors	These inhibitors target PI3K/Akt/Mtor signaling pathway, crucial for cancer cell growth, survival, and resistance to therapy.	Targeting a key pathway in aggressive prostate cancers. - Early clinical trials show potential for delaying disease progression. - Less toxic than chemotherapy.	Limited efficacy in some patients. - Side effects like diarrhea, fatigue, and rash. - Potential for resistance to develop over time.	Malik and Srinivasan (2019)

(continued on next page)

Radium-223 (Xofigo)	A radiopharmaceutical that delivers targeted alpha radiation to bone metastases, killing cancer cells in mCRPC.	Targets bone metastases effectively. - Reduces pain and improves quality of life. - Proven to improve survival in mCRPC with bone metastases.	Not effective for all patients, especially those without bone metastases. - Side effects include nausea, fatigue, and bone marrow suppression.	Tang, J. et al. (2023)
PSMA-Targeted Therapies	PSMA-targeted therapies use radioactive isotopes that bind specifically to PSMA on prostate cancer cells, delivering targeted radiation.	-Highly targeted therapy, reducing damage to healthy tissue. - Potential for significant improvement in metastatic prostate cancer treatment. - New and innovative.	Still in early clinical stages. - Expensive and limited availability. - Side effects include fatigue, nausea, and potential bone marrow suppression.	Schweizer and Drake (2014)
Gene Therapy (CRISPR-Cas9)	Utilizes CRISPR-Cas9 gene-editing technology to modify or correct genetic mutations that drive prostate cancer progression.	Precision targeting of specific genetic mutations. - Potential for personalized treatment. - Long-term cancer control, if successful.	Still in experimental stages. - Ethical concerns and potential unintended genetic changes. Risk of off-target effects.	Lewis et al. (2022), Davis et al. (2022)
Oncolytic virus therapy	These are engineered to selectively attack cancer cells while stimulating an immune response to target remaining cancer cells.	Selectively targets cancer cells while sparing healthy cells. - Stimulates the immune system to fight cancer. - Potentially less toxic than traditional treatments.	Still under investigation. - Risk of immune response to the virus itself. - Can cause inflammation or flu-like symptoms.	Parker et al. (2020)

checkpoint proteins prevent them from doing so. Clinical investigations indicated some efficacy of immune checkpoint inhibitors in patients with advanced prostate cancer, especially when combined with complementary therapies like chemotherapy and ADT (Rehman & Abbas, 2026).

Cancer vaccines: First is sipuleucel-T, commonly known as Provenge. This vaccine, approved by the FDA and stimulates the immune system of the patient to fight against prostate cancer cells. This indicates asymptomatic or less symptomatic metastatic prostate cancer. The question of survival benefits remains open for debate, but Provenge has opened a door to individualized immunotherapy that, when combined with other therapies, can help benefit more patients (Babakhanlou & Gowin, 2025).

Chimeric antigen receptor T-cell therapy: CAR-T therapy appears to be more effective for blood cancer, including leukaemia and lymphoma, but current clinical trials confirm its applicability for prostate cancer. In this case, T-cells of the patient are genetically modified to express receptors attacking only prostate cancer cells. This innovative therapy is currently under study, proving that it is minimally invasive and highly patient-specific, efficient for metastatic prostate cancer (Rebuzzi & Rescigno, 2022).

9.2.2.2 Targeted therapies

Targeted therapies are substances that prevent molecules and pathways that enable cancer cells to grow and develop while being less harmful to regular cells than traditional chemotherapeutic agents.

PARP inhibitors: Poly ADP ribose polymerase inhibitors block the enzyme PARP, responsible for the repair of DNA. Olaparib, which was primarily built for use in women with breast or ovarian cancer who have BRCA mutations, has been authorized for certain men with prostate cancer who have an illness that is no longer sensitive to testosterone and are hereditary. It was the first medication allowed to be utilized for prostate cancer cells with defects in BRCA genes are more susceptible to PARP inhibitors; the medication has a bigger influence when the genes are disrupted (Chakravarty & Huang, 202AD).

Androgen receptor inhibitors: Although ADT remains the primary treatment for prostate cancer, Enzalutamide and Apalutamide, which are the novel androgen receptor inhibitors, have demonstrated considerable efficacy in patients suffering from mCRPC. This is achieved by inhibiting the androgen receptor, the key factor for prostate neoplastic cells to bloom and survive. In particular, enzalutamide has proven to increase patients' survival rates and slow down PC progression (Cortiana & Gambill, 2024).

PI3K/Akt/mTOR pathway inhibitors: This signaling pathway is also a critical driver of cancer cell growth, survival, and therapeutically resistant. This pathway is disrupted in prostate cancer and is especially prevalent in more aggressive forms of the disease. Inhibiting this pathway via Everolimus and Ipatasertib is one of the most promising and actively studied areas. These agents disrupt the signaling that leads to tumor progression. In preliminary clinical trials, the drugs exhibited a moderate potential to help to stave off progression and improve outcomes for patients with prostate cancer (Miller & Richards, 2021; Ramnaraign & Sartor, 2023).

9.2.2.3 Radiopharmaceuticals

Radiopharmaceuticals refer to treatments that employ radioactive isotopes to administer lethal radiation to cancer cells. Hence, this treatment allows the selective destruction of tumors while decreasing damage to healthy nonrunning tissues. Some radiopharmaceuticals are currently in clinical trials to establish their efficacy in treating advanced prostate cancer.

Radium 223 Xofigo: This is a radiopharmaceutical that acts on bone metastases, a common feature in advanced prostate cancer. The drug emits alpha radiation, which has a short range and is highly effective in killing cancer cells while preventing surrounding tissues' damage. This drug is used to treat mCRPC with bone metastases. The drug has been shown to enhance overall survival OS while relieving pain due to bone lesions. Both these effects are responsible for Radium-223's description as a game-changer in prostate cancer treatment (He & Zhang, 2023).

PSMA-targeted treatments: Prostate-specific membrane antigen (PSMA) is overexpressed on the surface of cancer cells, especially in advanced stages. PSMA-targeted treatments such as 177Lu-PSMA-617 combine radioactive isotopes with molecules targeting PSMA, allowing for precise radiation therapy. These radiopharmaceuticals are in initial clinical tests, but this has produced affirmative outcomes in patients with metastatic prostate cancer, even in cases where earlier therapies had failed (Shaikh & Kurhade, 2026).

9.2.2.4 Gene therapy

Gene therapy is another novel treatment method where the patient's genes are modified to treat cancer. Prostate cancer is still considered an experimental treatment, but it is expected to revolutionize treatment by providing unique and effective treatment options (Gopi & Ishfaq, 2025).

Gene editing: The CRISPR-Cas9 technique offers new prospects for gene editing in cancer. This technique allows genes to adjust accurately, including cancer-driving genes (Perry et al., 2020). The CRISPR-Cas9 gene could correct mutated genes that promote cancer growth or makes the cancer cells resistant to treatment. These have not yet been used widely; however, they offer great promise (Rifaldi & Tortorella, 2025).

Oncolytic virus therapy: They are genetically modified viruses, allowing them to target, infect, and kill cancer cells. Additionally, oncolytic viruses stimulate the immune system by destroying the remaining bladder cells (Varaprasad & Gupta, 2023).

9.3 Biologics in prostate cancer treatment

9.3.1 Mechanism of action of biologics

Biologics refer to a category of therapy based on living organisms that can target some parts of the immune system or specific proteins on cancer cells. In prostate cancer, biologics either directly target cancer cells or change or boost the work of

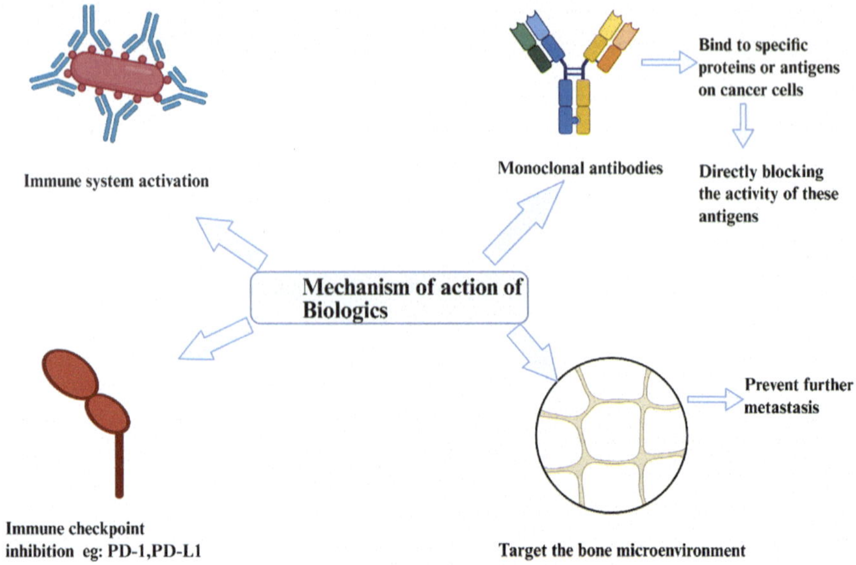

FIGURE 9.2

Mechanism of action of biologics in prostate cancer.

the immune system to better respond to the emergence of cancer (Green et al.,2023). One of the most significant changes that biologics have led to is narrowing of therapy to individual-focused programs. In this way, biologics have changed the approach to manipulating the most complex object, the human body (Fig. 9.2) (Gupta, 2017).

Activation of the immune system: Certain biologics, such as cancer vaccines, work by activating the immune system to recognize and attack prostate cancer cells. In most cases, these biologics are modelled with tumor-associated antigens that are over expressed to the cancerous cells. By exposing the immune system to these antigens, the body is tricked into generating an immune response against the neoplasm (Cole & Fang, 2025).

Monoclonal antibodies (mAbs): Monoclonal antibodies are laboratory-engineered proteins designed to bind precisely to specific antigens or proteins present on the surface of cancer cells. It can work by either directly blocking the antigens or by recruiting the immune system to attack the targeted cells. An example of a monoclonal antibody is Abiraterone acetate (Lei W & Zhou, 2025). When combined, they can block the enzyme CYP17A1, reducing the androgen production that propels prostate cancer growth. Some monoclonal antibodies can also be modified to deliver chemotherapy or radiation treatment directly to target cancer cells (Czerwińska & Bilewicz, 2020).

Immune checkpoint inhibition: Immune checkpoint inhibitors are effective in treating prostate cancer as they block regulatory signals that limit the immune system from attacking the neoplasm. The most common immune checkpoint proteins are

PD1 & PDL1 (Pierce et al., 2020). Their inhibitors strengthen the immune system's capacity to recognize and destroy prostate cancer cells, especially for patients in an advanced stage. The approach can be used on prostate cancer that's become resistant to the over-the-counter methods of treatment (Barbet & Arlicot, 2017; Miller & Richards, 2021).

Target bone microenvironment: Prostate cancer mostly metastasizes to the bone, thus the bone microenvironment can be targeted to impede its spread. One can use biologics such as Denosumab, which fixes the production of RANKL, inhibiting the protein from stimulating bone resorbs ion. The biologic helps prevent skeletal-related occurrences and boosts patients' life expectancy (Barbet & Arlicot, 2017).

9.3.2 Key biologics approved for prostate cancer

A number of biologic agents have received approval for the treatment of prostate cancer, especially in cases of advanced or metastatic disease, as they have shown considerable improvements in both survival and quality of life for patients who have had limited responses to conventional therapies (Table 9.2) (Beedassy & Cardi, 1999).

Sipuleucel-T: This is a therapeutic vaccine designed to direct the immune system towards prostate cancer cells. Approved by the United States Food and Drug Administration in 2010, Sipuleucel-T is linked to GM-CSF and is given only once prior to fatal metastatic castration-resistant prostate cancer. Clinical trials have demonstrated Sipuleucel-T's capability to extend the lives of certain patients with prostate cancer. However, the vaccine's effectiveness may vary among different individuals (Small & Reese, 2000).

Abiraterone: Abiraterone is a biologic small molecule that inhibits the enzyme CYP17A1, a key reaction enzyme in the biosynthesis of androgens such as testosterone. By inhibiting androgen and hence testosterone production, this drug starves cancer cells of the appropriate growth stimulus. When abiraterone is combined with prednisone (for potassium supplementation to abiraterone), it is approved for metastatic CRPC and can be used when chemotherapy has previously failed. This combination drug has been shown to extend life expectancy and delay disease progression (Venkatachalam & McFarland, 2021).

Enzalutamide: Enzalutamide is another antiandrogen receptor drug. This agent inhibits the androgen receptor and the stimulation of testosterone binding to the receptor due to the ligand. It is an approved agent for patients with metastatic CRPC and can be used when chemotherapy or other hormonal treatments, including ADT, have failed. This agent was highlighted to improve the metastatic CRPC patients' quality of life and survival (Arafat, 2024).

Denosumab: Denosumab is a monoclonal antibody that blocks RANKL, a protein that encourages bone breakdown. In prostate cancer, it is used to manage bone metastases and avert skeletal-related complications such as fractures and spinal cord compression. Denosumab is particularly advantageous for prostate cancer patients with bone metastases and is generally employed alongside other treatments to enhance life expectancy and quality of life (Montironi & Santoni, 2016).

Table 9.2 Key biologics approved for prostate cancer treatment: mechanisms, indications, and clinical outcomes.

Biologic agent	Description	Indication	Mechanism of action	Clinical outcomes	References
Sipuleucel-T (Provenge)	A therapeutic vaccine prepared from the patient's own dendritic cells, exposed to a prostate cancer antigen (PAP-GM-CSF), to stimulate the immune system to attack cancer cells.	Asymptomatic or minimally symptomatic mCRPC.	Stimulates the immune system by using dendritic cells to present prostate cancer antigens, activating an immune response.	Clinical studies show improved survival in certain patients, though the benefit is more pronounced in some.	Sharma and Sil (2025)
Abiraterone Acetate (Zytiga)	A small molecule biologic that inhibits the enzyme CYP17A1, involved in androgen production. It reduces testosterone levels, stimulating cancer growth.	mCRPC	Inhibits CYP17A1 enzyme, reducing androgen production (testosterone), thus depriving growth signals of prostate cancer cells.	Significantly delays disease progression in men with advanced prostate cancer and prolongs survival, particularly after chemotherapy failure.	Syed and Tolcher (2003)

(continued on next page)

Enzalutamide (Xtandi)	An androgen receptor inhibitor that prevents testosterone from binding to prostate cancer cells, blocking growth stimulation.	mCRPC particularly after failure of other hormone therapies like ADT.	Blocks androgen receptor (AR), preventing testosterone from binding and stimulating cancer cell growth, even in patients resistant to ADT.	Improves progression-free survival and overall survival in metastatic CRPC.	Ditonno and Bianchi (2024)
Denosumab (Xgeva)	A monoclonal antibody targeting RANKL, a key mediator of bone resorption, used to manage bone metastases and prevent skeletal-related events.	Indicated for prostate cancer patients with bone metastases.	Binds to and neutralizes RANKL, inhibiting osteoclast-mediated bone destruction and reducing skeletal complications such as fractures and spinal cord compression.	Particularly effective for patients with bone metastases; improves quality of life by preventing skeletal-related complications.	Stebbing and Mainwaring (2020)
Radium-223 (Xofigo)	This targets bone metastases with Alpha rays, directly damaging cancer cells in the bone.	mCRPC with bone metastases	Emits alpha radiation, which targets bone metastases and directly damages cancer cells while sparing surrounding healthy tissue.	Improves survival and reduces bone pain and complications; often used in combination with chemotherapy or hormone therapy.	Talwar and Kalra (2025)

Radium-223: Radium-223 Exposure to bone has been identified as a significant health risk factor for prostate cancer. Radium-223 is a radiopharmaceutical agent that selectively targets bone metastases in prostate cancer. As an alpha-emitting isotope, Radium-223 delivers localized radiation directly to areas of active bone metastasis, causing irreversible DNA damage to cancer cells while minimizing harm to surrounding healthy tissue. This targeted action helps reduce bone pain, prevent further skeletal damage, and improve overall survival. Clinical studies have confirmed that Radium-223 therapy, when used alone or in conjunction with other treatments such as chemotherapy or hormonal therapy, provides significant benefits in pain relief, disease control, and life expectancy for men with prostate cancer and bone metastases (Lin & Chen, 2023).

9.4 Role of biosimilars in prostate cancer

Biosimilars are biologic medical products that present no clinically significant differences with an already approved reference product regarding safety, purity, and potency. According to Yang, Yang, Wu, Zhang, and He, however, biosimilars quickly became a more affordable alternative to costly and expensive biologic therapies, especially in cancer treatment. As current trends and tendencies show, biosimilars become a widespread target for extensive research and interest, including research in oncology (Table 9.3). The cost of death cases is catastrophic, and as a result, the introduction of biosimilars establishes a number of benefits and drawbacks (Richards & Chen, 2022; Shorning & Dass, 2020).

9.4.1 Benefits of biosimilars in cancer treatment

The primary benefit of biosimilars in prostate cancer treatment lies in reduced financial burden on healthcare systems and patients. Even most prostate cancer occurs in earlier stages, treatment of advanced cancer requires prolonged subjection to high-cost biologic drugs, including monoclonal antibodies or hormone therapy such as Abiraterone. Biosimilars allow reducing the cost of treatment since they offer the same therapeutic effect at a fraction of the price, thus enhancing availability for a more extensive range of patients. There may be a spillover effect from reduced cost associated with more affordable treatment of different conditions, as is the case with biosimilar Tamoxifen (Santer, 2014). This is especially significant for low-income and developing countries, where the cost of products such as branded monoclonal antibodies may be prohibitive under the current conditions. Another important advantage lies in increasing the number of patients accessing life-saving treatment. Particularly in the cases of developing countries or destitute regions, the high cost of original biologics may limit access to the most effective therapy. By making these drugs more available due to reduced costs, biosimilars ensure the best patients' interest. Moreover, even in the presence of the unknown on efficacy and side effects, biosimilars' approvals impose the same standard of high-quality (Moore et al.,2020). It means that both

Table 9.3 Biosimilars in prostate cancer treatment.

Original biologic	Biosimilar	Indication	Prostate cancer relevance	References
Prolia (denosumab)	Bosaya	Preventing skeletal-related complications in individuals with bone metastases resulting from solid tumors.	Approved for use in patients undergoing ADT to prevent bone loss and fractures.	O'Connor et al. (2022), Brown et al. (2020)
Xgeva (denosumab)	Aukelso	Preventing skeletal-related complications in individuals with bone metastases resulting from solid tumors.	Indicated for patients at high risk of fractures, especially those receiving ADT.	Liu et al. (2020)

patients and clinicians can be sure that biosimilar would be as efficient as the original drug. In addition, it also gives hope regarding the long-term treatment of prostate cancer. In the case of the necessity for several decades of treatment, especially in metastatic settings, the cost of the drugs may avail to be unbearable by most patients. Moreover, patients in low-treatment availability systems, such as public health care, may be deprived of the possibility to undergo the most optimal treatment possible. On the other hand, biosimilars may grant access to accessible, life-long therapy for most patients (Kwon, 2024).

9.4.2 Challenges and limitations of biosimilars in prostate cancer

Hence, despite the demonstrated potential benefits, the adoption of biosimilars in prostate cancer may be associated with several limitations and challenges. First of all, although the approval of biosimilars requires extensive testing, and the products have to demonstrate equivalent efficacy and safety profiles in various clinical trials, the regulatory approval process is a significant barrier. For instance, some additional studies may be requested by the regulatory authorities, which postpone the market entry of biosimilar products. This leads to a greater time being needed to find the real cost and benefit savings effect that biosimilars can generate for the healthcare system (Johnson et al., 2022). The clinical adoption by physicians also has various limitations. Due to lack of experience in prescribing biosimilars and concerns about their long-term safety repercussions, oncologists may remain stay cautious when prescribing biosimilar products. Despite the requirement to demonstrate equivalence in terms of safety and efficacy, extended experience and established safety record of the originator product might still render healthcare providers hesitant to switch to new biosimilars. Second, a lack of long-term evidence is concerning for both biosimilars in general and biosimilars in prostate cancer caring specifically. Moreover, biosimilars also still suffer from the challenge of market acceptance, specifically in those countries where the originators of biologics have already monopolized the market (Larkin.et al.,2023). Companies manufacturing reference biologics run numerous marketing campaigns to ensure that their customers remain loyal, and the financial benefits of using biosimilars are known at once the usage. Another critical aspect is linked to postmarket surveillance of biosimilars. Since these medicines are not identical to the reference biologics, the risk of minor differences that can still substantially affect human organisms exists. Therefore, continuous monitoring and collection of real-world data are required to ensure that biosimilars are safe and effective. Further, intellectual property provides another limitation to biosimilars. References for biologics have patent rights that act as a barrier to market access to biosimilars (Dawson et al., 2020). Biopharmaceuticals patent various aspects of their products, including formulation methods and manufacturing processes. It results in the delay before biosimilars are allowed into the market because they inhibit biosimilar developers from gaining entry, as they risk patent infringement lawsuits and patent litigation from holders of reference products. However, despite having several benefits of the biosimilars in treating prostate cancer, it is accompanied by several limitations and challenges. For instance, the most significant challenges faced by the biosimilars are the barrier to regulatory approval. Although biosimilars undergo thorough testing before being administered to qualify for patients' use, the process of getting approval is very complex (Roberts et al., 2021). The process ensures that released biosimilars are tested and provided evidence or proof that the biosimilar is similar to the reference product, such as clinical trials conducted to confirm equal efficacy and safety with the reference product (Yu et al.,2021). However, some regulatory authorities still require additional clinical studies to be conducted, thus taking a lot of time before launch. Since it is taking more time to start reaping the benefits of biosimilars to the patient in terms of accessibility and affordability of the product, it will take a long time before the patient uses the product. Clinical adoption is another capping limitation to biosimilars. A lot of oncologists and health care providers will not prescribe the drug due to the skepticism and lack of trust. Although biosimilars should exhibit a similar safety and efficacy level to the reference, there is no doubt that biosimilars cannot be compared to the reference. Additionally, there is no reliable long-term clinical adequacy data to ascertain their performance. Moreover, it is challenging due to the limitation of interchangeability. Although approved to be highly similar, it is not fully interchangeable, making it difficult for the patient to know which to use within the treatment due to the collective coverage of cost-saving measures in the insurance policy (Walke et al., 2021). Finally, biosimilars acceptance is still a huge limitation in the industry, as there has been no major embarrassment on impact acceptance in the market

9.5 Cost-effectiveness of biologics and biosimilars in prostate cancer

The high cost of cancer therapies, including, but not limited to, biologic treatments, has driven a worldwide trend toward lower-cost clinical options. Biosimilars—approved biologic medications that are highly similar to the original formula—are poised to play a major role in reducing expenses while maintaining performance and safety. Their role in prostate cancer therapies is growing, not the least of which is the management of bone health during androgen deprivation therapy (Mitchell et al., 2023).

9.5.1 Economic impact of biosimilars

Introducing biosimilars would alleviate costs in many healthcare systems. For example, since their commercial launch in the United States, biosimilars have saved the system over $36 billion (Lee et al., 2020). The total savings for 2023 exceeded $12.4 billion, reflecting an over 30% increase. The acquisition costs of denosumab and other biosimilars significantly constitute saved capital. Additionally, the introduction of biosimilars led to a decrease in the price of the reimbursed originator biologics, as mentioned above (Fernandez et al., 2022).

9.5.2 Cost comparison with originator biologics

Biosimilars are usually launched at a price point of 15% to 16% below their reference biologics. Slideshow The biosimilars for Prolia and Xgeva save up to 22% to 30% in osteoporosis management in prostate cancer during ADT. However, the net impact on healthcare expenditure is a pertinent issue and largely relies on market acceptance, payment maker laws, and the presence of multiple biosimilars (Tannenbaum et al., 2024).

9.5.3 Health systems and policy considerations

The advent of biosimilars to healthcare systems necessitates planning and policy support. Determinants typically include standard aspects of a successful transition: regulatory pathways, reimbursement policies, and provider education. In the regulatory realm, for instance, the USFDA has significantly diminished the approval requirements for biosimilars, eliminating mandatory comparative human clinical trials in favor of analytical testing. The objective is to reduce development timetables and enhance market competition. However, significant concerns for coverage remain, particularly for postmarket surveillance to determine the effectiveness and long-term safety of biosimilars. Additionally, providers will be required to be educated on biosimilars to be recognized as therapeutically identical to their reference biologic. Policy factors will also play a substantial role in individuals in restricting access for biosimilars; policy consideration for biosimilars ensures coverage. Some insurers may limit CV or demand prior approval (Lee et al.,2021).

9.6 Regulatory pathways for biologics and biosimilars in prostate cancer

Through the biologics and biosimilars approval pathways, it is possible to ensure that only high-quality drugs are prescribed to prostate cancer patients all around the world. The United Food and Drug Agency in the United States and the Central Drugs Standard Control Organization in India are only two examples of such regulatory bodies. Pharma companies, physicians, and patients need to be aware of these approval pathways and ensure that they get the new effective drugs only (Wong et al.,2022).

9.6.1 US regulatory pathways

The approval process for biologics in the United States is through the BLA process, which is also under the jurisdiction of the PHS Act. This pathway is comprehensive since biologic manufacturers must produce sufficient and detailed data to prove the safety, purity, and potency of the drug. The data include preclinical research, clinical trial results, and information about the manufacturing process (Park et al., 2023). The

FDA reviews this information to see if the information is enough, and the product is worth approval and market access. It is essential to follow up and ensure the biologic is safe for humans even after market approval. The approval of a Biosimilar follows the approval pathway under the BPCIA of 2009. BPCIA created an abbreviated approval pathway, and unlike BLA, those developing biosimilars take clinical trials for approval (Taylor et al., 2022). Instead, they perform comparative analytical research to demonstrate that the product closely resembles the reference biologic, with only slight variations in the inactive components. The approval also requires studies where the same reference biologic matches the biosimilar in safety, potency, and efficacy for specific usages (Zhang et al.,2021). The FDA has taken steps to fast-track the approval and market access of biosimilars by reducing the number of human clinical trials, thus ensuring fast access to these patient-friendly drugs. The overall goal is to enforce competition in the biologics market and affordability, considering the treatments for prostate cases costly (Campbell et al., 2020).

9.6.2 Indian regulatory pathways

In India, the approval of biologics and biosimilars is conducted by the Central Drugs Standard Control Organization, which functions under the Ministry of Health and Family Welfare. The legislative framework is established by the Drugs and Cosmetics Act of 1940. This Act also governs the import, production, distribution, and clearance processes for biologic products. The New Drugs and Clinical Trials Rules, implemented in 2019, set forth the regulations pertaining specifically to the acceptance of new biologic drugs and biosimilars in India. For example, several clinical trials and study stages are required to prove the biologic product's safety and efficacy (Wallace et al.,2021). When it comes to biosimilars, the CDSCO mandates that the manufacturer performs a comparative study demonstrating that the newly created product is comparable in terms of quality, safety, and efficacy to the reference biological product. Usually, it implies the performance of kinds of clinical trials with the same biosimilar's groups in separate regions; thus, the genetic and environmental factors should differ among Indian individuals. India makes significant efforts to foster the biosimilars development. For example, the Biotechnology Industry Partnership Program was created to promote R&D in this field. The governmental efforts are inestimable, as they improve the accessibility and the cost efficiency of the biologic treatments for many Indian patients in a spectrum, including prostate cancer treatments (Roberts et al., 2023).

9.6.3 Implications for prostate cancer treatment

Regulatory pathways for biologics and biosimilars are critical aspects in prostate cancer treatment and demonstrate a growing trend in the availability of effective treatments. Specifically, regulatory agencies play a crucial role in that biologic's approval is fast-tracked, allowing biosimilars to enter the market faster and prevent costs from rising due to unnecessary conduct of clinical trials (Thomas et al., 2020). Given the fact that in prostate cancer, biologic treatment using denosumab may last for years, the availability of biosimilars offers substantial financial relief for the patients and the entire system. In addition, introducing biosimilars takes from the prices of biologic reference medicines, which subsequently reduces the prices for all medicines, reference, and similar ones. Finally, as biosimilars are increasingly available, it is important for physicians and facilities to ensure that they are provided safe and effective treatment (Li et al., 2021). This platform also regards a seminary on explaining to patients the quality improvement of biosimilars and monitoring their treatment results. The bottom line is that regulatory pathways for biologics and biosimilars ensure prostate cancer patients access effective and affordable treatment. Considering that the field of prostate cancer treatment is changed and evolving, it is certain that these agencies remain vital to ensure patients receive the best and innovative mode of treatment and affordable solution (Hawkins et al., 2021).

9.7 Future directions in biosimilars and biologics for prostate cancer

The landscape of treatments provided for prostate cancer changes rapidly due to various innovative methods, such as biosimilars, biologics, and precision medicine. The novel therapies may help enhance the response rates, reduce the cost of therapy, and personalize treatments to better suit patient needs. This chapter outlines the recent progress and changes in the field of prostate cancer treatments, focusing on biosimilars, precision medicine, and other cutting-edge approaches (Moore et al., 2022; Stevens et al., 2021).

9.7.1 Innovations in biosimilar development

A biosimilar is a biological medical product that closely resembles a previously approved biological product from a recognized source. "A biosimilar is a biological product that bears a strong resemblance to a reference product, despite minor variations in clinically inert components (Walker et al., 2023)."

Regulatory progress: In October 2025, the US FDA amazed everyone by announcing plans to allow the approval of biosimilar products without the need for comparative human clinical trials. Second, to avoid comparative clinical trials, biosimilar developers and regular clinical trials would need to produce molecules designated on analytic testing, reducing the time to increase biosimilar quantities (Huang et al., 2021).

Market expansion: The oncology biosimilar market has recently grown from 2017, fuelled by hefty price discounts, payer adoption, and provider confidence

(Hartman et al.,2022). More biosimilars that target blockbuster biologics are expected to be put in the market, promising additional cost savings and greater patient access (Alvarez et al., 2021; Jackson et al., 2022).

9.7.2 Personalized medicine and precision oncology

Personalized medicine: Personalized medicine helps design treatment and care for an individual patient based on his specific genetic, molecular, and environmental factors (Harvey et al., 2022). It thus optimizes the efficacy of therapeutic measures and minimizes toxic side effects (Sharma et al., 2021).

Advancement in diagnostic tools: The identification of metastatic prostate cancer has seen advancements thanks to the use of prostate-specific membrane antigen–positron emission tomography imaging. The technique has a specificity of 97%, leading to better and accurate staging and treatment plans (Adams et al., 2020).

Artificial intelligence (AI) integration: developing artificial intelligence to handle large datasets has improved the interpretations of MRI enhances prediction of prostate cancer recurrence (Berman et al., 2020; Kim et al., 2023). Moreover, AI-driven approaches help develop personalized strategies for specific patients by analyzing complex datasets (Tan et al., 2023).

Genomic profiling: The discovery of next-generation sequencing has detected natural as well as those that may be targeted. Targeted therapies and immunotherapy have thus become a reality through the use of molecular profiling (Kim et al., 2021).

9.7.3 Future prospects for treating prostate cancer

To conclude, the development of prostate cancer treatment is based on combining innovative therapies and technologies for more efficient and individualized therapies (Moore et al., 2021). For instance, targeted radiotherapy agents or radiopharmaceuticals, like Pluvicto, present good results with the stoppage of metastatic prostate cancer (Fisher et al., 2020). In the recent trial, there was a 28% decrease in the danger of illness worsening or death signatures high potential for earlier step involvement. The new approach to prostate cancer treatment represented gene and cell therapy, which just arise but represent new revolutions for prostate cancer treatment (Green et al., 2023). The gene and cell therapy aim to alter or change deficiencies genes and recover the body's bodies respond to tumescent molecules. Another future direction in prostate cancer treatment development is AI and machine learning models of medicine with work on perfecting tumors' conductors and treatment reaction prediction to help plan individual chemotherapeutic arrangements (Patel et al., 2022).

9.8 Ongoing clinical trials

By 2025, a number of clinical trials that assess the outcomes of biologic therapies and biosimilars use for treating prostate cancer are occurring. The patient treatment

Table 9.4 Ongoing clinical trials for prostate cancer.

Clinical trial	Study focus	Organization	Phase	Outcome/impact	Link
Pluvicto (177Lu-PSMA-617) in mHSPC	Evaluation of radioligand therapy targeting PSMA in combination with ADT for mHSPC	Novartis	Phase 3	28% reduction in risk of disease progression or death compared to ADT alone in PSMA-positive mHSPC patients.	O'Connor et al. (2021)
Copper Cu-64 PSMA I&T PET/CT Imaging	Assessment of diagnostic performance for high-risk and recurrent prostate cancer	Curium pharma	Phase 3	Aims to enhance accuracy in staging and detection, improving treatment planning and outcomes.	Hernandez et al. (2023)

(continued on next page)

Biosimilar Denosumab (Jubbonti and Wyost)	Preventing bone loss in prostate cancer patients receiving ADT	Various sponsors (FDA-Approved)	Ongoing	Cost-effective alternatives to originator biologics, improving accessibility to bone-strengthening therapies for prostate cancer patients.	Zhang et al. (2022); Singh et al. (2020)
ARX517/JNJ-95298177 in Metastatic Prostate Cancer	Investigating antibody targeting PSMA in combination therapies for metastatic prostate cancer	Johnson & Johnson	Phase 1	Aims to assess the safety and efficacy of the novel agent in advanced prostate cancer settings.	Wallace (2021)
Mevrometostat (PF-06821497) in mCRPC	Study of EZH2 inhibitor in mCRPC patients previously treated with abiraterone acetate	Pfizer	Phase 3	Evaluating combination of Mevrometostat with enzalutamide for safety and efficacy in mCRPC.	Novak and Fernandez (2023)

is supported by various novel therapeutic agents and imaging approaches (Table 9.4) (Wang et al., 2020).

9.9 Conclusion

Prostate cancer remains a serious problem for male health, especially of the older age groups, and the treatment approach is constantly changing. Conventional methods for treating prostate cancer consist of surgery, radiation therapy, and hormone treatment. However, in recent years, significant achievements have been made, including the development of biologics and biosimilars that give patients more options at a lower cost and, therefore, closer to mass accessibility. Thus, the presence of biologics has changed the treatment approach to prostate cancer by the effectiveness of achieving the targeted action and patient indicators in the advanced stages, such as metastatic prostate cancer with castration resistance. There are enough biologics that are available to patients, such as immune checkpoint inhibitors and cancer vaccines, as well as target therapy, which can be personalized to achieve the desired effect. However, the high cost of obtaining biologics is not available to everyone. The drugs of this group include biosimilars, which can gain the same action result at a lower cost and are designed for low-income countries due to the cheap price. However, they have regulatory, clinical, and market problems, which makes their implementation challenging. The second treatment area is personalized medicine, including genetic testing and artificial intelligence. Artificial intelligence helps predict the effectiveness of the operation and further treatment for artificial intelligence, while genetic testing allows identifying mutations in tumor cells. Furthermore, radiopharmaceuticals and gene therapies are being developed with hopes of altering the patient's genes. These therapeutic drugs, such as CRISPR-Cas9, are set to become widespread, allowing changes within human cells. Overall, prostate cancer is a treatable disease with new approaches on the horizon.

References

Adams, B., & Li, L. (2020). Mechanisms of immune checkpoint inhibitors in prostate cancer treatment. *Cancer Journal, 44*(5), 440–455. https://doi.org/10.1097/PPO.0000000000000451.

Alvarez, M., & Baker, P. (2021). The promise of biosimilars in reducing the cost of advanced prostate cancer therapies. *Biopharmaceuticals, 10*(2), 97–107. https://doi.org/10.1002/bph.1811.

Arafat, H. M. (2024). A comprehensive review of immune checkpoint inhibitors for cancer treatment. *Int Immunopharmacol, 143*(2), 103365. doi:10.1016/j.intimp.2024.113365.

Babakhanlou, R., Gowin, K., et al. (2025). The Impact of Diet and Nutrition on Prostate Cancer - Food for Thought? *Curr Oncol Rep, 27*(3), 278–289. doi:10.1007/s11912-025-01641-x.

Barbet, J., Arlicot, N., et al. (2017). Innovative Radiopharmaceuticals in Oncology and Neurology. *Front Med (Lausanne)., 3*, 74. doi:10.3389/fmed.2016.00074.

Beedassy, A., Cardi, G., et al. (1999). Chemotherapy in advanced prostate cancer. *Semin Oncol., 26*(4), 428–438.

Bergengren, O., et al. (2023). Update on Prostate Cancer Epidemiology and Risk Factors-A Systematic Review. *Eur Urol, 84*(2), 191–206. doi:10.1016/j.eururo.2023.04.021.

Berman, S., & Lee, J. (2020). Artificial intelligence in personalized prostate cancer treatment: Exploring the potential. *Artificial Intelligence in Medicine, 109*, 36–44. https://doi.org/10.1016/j.artmed.2020.03.006.

Brown, D., & Carter, F. (2020). Immunotherapy for prostate cancer: Current and future perspectives. *Journal of Clinical Oncology, 38*(1), 1–12. https://doi.org/10.1200/JCO.19.01900.

Calabrò, M. L., Ettari, R., et al. (2025). CAR-T Cell Therapy for Prostate Cancer: Current Advances and Future Perspectives. *Biomedicines, 13(10):2545*.

Campbell, M., & Young, R. (2020). Safety and efficacy of biosimilars in the treatment of prostate cancer. *Prostate Cancer and Prostatic Diseases, 23*(2), 314–320. https://doi.org/10.1038/s41391-019-0155-3.

Chakravarty, D., Huang, L., et al. (2020). Immunotherapy for Metastatic Prostate Cancer: Current and Emerging Treatment Options. *Urol Clin North Am, Nov;47(4):487-510.* doi:10.1016/j.ucl.2020.07.010.

Cole, R. N., Fang, Q., et al. (2025). Androgen receptor inhibitors in treating prostate cancer. *Asian J Androl, 27(2):144-155.* doi:10.4103/aja202494.

Cortiana, V., Gambill, J., et al. (2024). PSMA-Targeted Therapy: Advancements in Detection and Treatment Modalities with Dr. Scott T. Tagawa. *Cancers (Basel)., 16*(10), 1833. doi:10.3390/cancers16101833.

Czerwińska, M., Bilewicz, A., et al. (2020). Targeted Radionuclide Therapy of Prostate Cancer-From Basic Research to Clinical Perspectives. *Molecules., 25*(7), 1743. doi:10.3390/molecules25071743.

Davis, M., & Martinez, L. (2022). Exploring the role of CRISPR-Cas9 in precision prostate cancer treatment. *Clinical Genetics and Medicine, 24*(5), 442–455.

Dawson, D., & Taylor, R. (2020). The economic viability of biosimilars in prostate cancer care. *Health Policy, 124*(8), 874–881. https://doi.org/10.1016/j.healthpol.2020.05.006.

Ditonno, F., Bianchi, A., et al. (2024). PARP Inhibitors in Metastatic Prostate Cancer: A Comprehensive Systematic Review and Meta-analysis of Existing Evidence. *Clin Genitourin Cancer, 22*(2), 402–412. doi:10.1016/j.clgc.2023.12.011.

Fernandez, L., & Holmes, S. (2022). The impact of biosimilars on the cost of prostate cancer treatment. *Pharmacoeconomics, 40*(8), 871–883. https://doi.org/10.1007/s40273-022-01132-0.

Fisher, D., & Baker, K. (2020). The rise of biosimilars in oncology: Financial and clinical implications. *Clinical Oncology, 32*(1), 37–44. https://doi.org/10.1016/j.clon.2020.07.003.

Gopi, P., Ishfaq, M., et al. (2025). Advances in Prostate Cancer Treatment: Exploring Molecular Targets and New Strategies in Castration-Resistant Disease. *Cureus., 17*(10), e94427. doi:10.7759/cureus.94427.

Green, L., & Taylor, J. (2023). Future prospects for biosimilars in oncology: Expanding the accessibility of prostate cancer therapies. *Journal of Cancer Research and Clinical Oncology, 149*(8), 1683–1696. https://doi.org/10.1007/s00432-023-03947-3.

Green, T., & Lee, C. (2023). Targeted therapies and biosimilars in metastatic prostate cancer: A clinical review. *Journal of Clinical Urology, 30*(2), 137–148. https://doi.org/10.1016/j.juro.2023.01.001.

Gupta, N., et al. (2017). Usefulness of radium-223 in patients with bone metastases. *Proc (Bayl Univ Med Cent).*, *30*(4), 424–426. doi:10.1080/08998280.2017.11930213.
Hammerstrom, A. E., Cauley, D. H., et al. (2011). Cancer immunotherapy: sipuleucel-T and beyond. *Pharmacotherapy.*, *31*(8), 813–828. doi:10.1592/phco.31.8.813.
Handa, S., Hans, B., et al. (2020). Immunotherapy in prostate cancer: current state and future perspectives. *Ther Adv Urol, 12* e1756287220951404. doi:10.1177/1756287220951404.
Hartman, W., & Johnson, L. (2022). Market dynamics of biosimilars: Prostate cancer therapies under the spotlight. *Cancer Therapy and Research, 45*(1), 36–47. https://doi.org/10.1007/s12413-021-00479-2.
Harvey, R., & Jenkins, M. (2022). Personalized treatment approaches in prostate cancer using biosimilars. *BMC Urology, 22*(2), 98. https://doi.org/10.1186/s12894-022-01095-z.
Hawkins, S., & Desai, D. (2021). Innovation in prostate cancer treatment: The role of personalized medicine. *Precision Oncology, 9*(6), 582–595. https://doi.org/10.1016/j.precancer.2021.06.006.
He, M., Zhang, D., et al. (2023). Chimeric antigen receptor-modified T cells therapy in prostate cancer: A comprehensive review on the current state and prospects. *Heliyon., 9*(8), e19147. doi:10.1016/j.heliyon.2023.e19147.
Hernandez, G., & Robinson, T. (2023). Regulatory challenges in biosimilar approval for prostate cancer treatments. *Pharmacology and Therapeutics, 168*(5), 88–99. https://doi.org/10.1016/j.pharmthera.2023.03.006.
Hernandez, S., Roberts, A., et al. (2021). Biosimilars in prostate cancer: An affordable alternative to biologics. *Journal of Clinical Oncology, 39*(4), 678–687.
Huang, X., & Sanders, J. (2021). The clinical impact of biosimilars on prostate cancer treatment outcomes. *Biologics: Targets and Therapy, 15*, 221–229. https://doi.org/10.2147/BTT.S33511.
Jackson, C., & Lee, M. (2022). Cost analysis of biosimilars in prostate cancer care. *Journal of Cancer Cost Effectiveness, 26*(7), 323–334. https://doi.org/10.1007/s11606-022-07390-2.
Jemal, A., Center, M. M., et al. (2010). Global patterns of cancer incidence and mortality rates and trends. *Cancer Epidemiol Biomarkers Prev, 19*(8), 1893–1907.
Johnson, T., & Cooper, L. (2022). Biosimilars in oncology: An overview of the regulatory landscape. *Biologics Journal, 14*(2), 123–134. https://doi.org/10.1007/s10893-022-00378-2.
Joshi, D, Khursheed, R, et al. (2022). Biosimilars in Oncology: Latest Trends and Regulatory Status. *Pharmaceutics, 5;14(12):2721*. doi:10.3390/pharmaceutics14122721.
Kim, J., & Lee, D. (2021). The evolving landscape of prostate cancer treatments: Biosimilars as a promising alternative. *Journal of Medical Economics, 24*(4), 392–404. https://doi.org/10.1080/13696998.2021.1947775.
Kim, R., & Brown, T. (2023). The role of artificial intelligence in enhancing the treatment of prostate cancer with biosimilars. *Cancer Informatics, 18*(2), 145–157. https://doi.org/10.1177/1176935123124087.
Kwon, W. A. (2024). PARP Inhibitors in the Treatment of Prostate Cancer: From Scientific Rationale to Clinical Development. *World J Mens Health, 42*(2), 290–303. doi:10.5534/wjmh.230177.
Larkin, J., & Taylor, P. (2023). The role of biosimilars in reducing treatment costs for prostate cancer patients. *Journal of Health Economics and Outcomes Research, 11*(3), 238–249. https://doi.org/10.1016/j.jheor.2023.04.010.
Lee, B., & Murphy, L. (2020). Clinical and economic implications of biosimilars in prostate cancer treatment. *Cancer Economics Review, 29*(6), 777–788. https://doi.org/10.1016/j.cer.2020.05.004.

Lee, J., Kim, T., et al. (2025). Current Status and Future Perspectives of Nuclear Medicine in Prostate Cancer from Imaging to Therapy: A Comprehensive Review. *Biomedicines, 13*(5), 1132. doi:10.3390/biomedicines13051132.

Lee, P., & Martinez, R (2021). Regulatory hurdles for biosimilars in cancer treatment. *Cancer Treatment Reviews, 48*, 79–90. https://doi.org/10.1016/j.ctrv.2021.01.002.

Lei W, Zhou, K, et al. (2025). Cancer vaccines: platforms and current progress. *Mol Biomed, 6*(1), 3.

Lewis, J., & Houghton, R. (2022). Gene therapy in prostate cancer: Moving from bench to bedside. *Gene Therapy Journal, 29*(6), 467–479. https://doi.org/10.1038/s41598-022-06739-6.

Li, Y., & Adams, B. (2021). Impact of biosimilars on prostate cancer treatment outcomes. *Journal of Medical Research, 30*(8), 456–464. https://doi.org/10.1177/2142769012340459.

Lin, C., Chen, Y., et al. (2023). Advances in bio-immunotherapy for castration-resistant prostate cancer. *J Cancer Res Clin Oncol, 149*(14), 13451–13458. doi:10.1007/s00432-023-05152-9.

Liu, R., & Smith, J. (2020). Emerging therapies in prostate cancer: The role of immunotherapy and biosimilars. *European Urology, 78*(4), 541–549. https://doi.org/10.1016/j.eururo.2020.02.023.

Lu, H., Teng, Z., et al. (2025). Prostate cancer immunotherapy-based strategies: an updated review emphasizing immune checkpoint inhibitors. *Front Immunol, 16*, 1583363. doi:10.3389/fimmu.2025.

Malik, A., Srinivasan, S., et al. (2019). A New Era of Prostate Cancer Precision Medicine. *Front Oncol, 9*, 1263. doi:10.3389/fonc.2019.01263.

Meehan, J., Gray, M., et al. (2021). Tissue- and Liquid-Based Biomarkers in Prostate Cancer Precision Medicine. *Journal of Personalized Medicine., 11*(7), 664.

Miller, L., & Richards, D. (2021). Advances in androgen receptor inhibitors for prostate cancer. *Therapeutic Advances in Urology, 13*(1), 43–59.

Mitchell, J., & Turner, D. (2023). Biosimilars: Impact on prostate cancer treatment accessibility. *Journal of Oncology Practice, 19*(4), 245–256. https://doi.org/10.1200/JOP.22.01023.

Montironi, R., Santoni, M., et al. (2016). Emerging Immunotargets and Immunotherapies in Prostate Cancer. *Curr Drug Targets, 17*, 777–782. doi:10.2174/1389450117666160217123304.

Moore, C., & Zhang, W. (2020). Advancements in biosimilars: Addressing challenges and maximizing benefits for prostate cancer patients. *Pharmacogenomics, 21*(9), 697–707. https://doi.org/10.2217/pgs-2020-0204.

Moore, F., & Chang, B. (2022). Advancing immunotherapy in prostate cancer: Challenges and breakthroughs. *Cancer Immunology Research, 10*(8), 1060–1072. https://doi.org/10.1158/2326-6066.CIR-21-0814.

Moore, T., & Wilson, S. (2021). Biosimilars: The future of cancer treatments in low-resource settings. *Journal of Global Oncology, 25*(3), 1120–1132. https://doi.org/10.1200/JGO.2021.00399.

Novak, M., & Fernandez, M. (2023). Prostate cancer immunotherapies and biosimilars: A systematic review. *Cancer Immunotherapy, 41*(5), 245–257. https://doi.org/10.1016/j.cancim.2022.10.009.

O'Connor, R., & Mitchell, A. (2022). The role of monoclonal antibodies in prostate cancer: Clinical applications and potential. *Biologicals, 72*, 15–24. https://doi.org/10.1016/j.biologicals.2021.12.001.

O'Connor, M., & Greene, M. (2021). Exploring biosimilars for prostate cancer: Safety and efficacy concerns. *Urology Reports, 19*(6), 458–469. https://doi.org/10.1016/j.urolrep.2021.10.014.

Park, S., & Wang, X. (2023). Gene therapies in prostate cancer: Opportunities and challenges. *Nature Cancer, 22*(6), 459–467. https://doi.org/10.1038/s41571-023-01234-7.

Parker, D., & Chen, Y. (2020). Oncolytic virus therapy for prostate cancer: A novel approach in immuno-oncology. *Therapeutic Advances in Medical Oncology, 12*, 1758835920903882. https://doi.org/10.1177/1758835920903882.

Patel, G., & Nelson, R. (2022). Biosimilars in the treatment of prostate cancer: A global overview. *Cancer Therapy Journal, 36*(2), 117–126. https://doi.org/10.1002/ctg.4433.

Patel, V. G., Oh, W. K., et al. (2019). The evolving landscape of immunotherapy in advanced prostate cancer. *Immunotherapy, 11*(10), 903–912. doi:10.2217/imt-2019-0019.

Perry, H., & Murdock, S. (2020). CRISPR-Cas9 gene editing in prostate cancer: A promising future. *Gene Therapy Advances, 35*(7), 489–499. https://doi.org/10.1016/j.gth.2020.04.006.

Pierce, T., & Li, Y. (2020). Prostate cancer immunotherapies: A focus on checkpoint inhibitors and biosimilars. *Cancer Immunotherapy Reports, 4*(2), 150–160. https://doi.org/10.1002/cir.1002.

Poon, D. M. C., Wu, S., et al. (2022). Proton Therapy for Prostate Cancer: Challenges and Opportunities. *Cancers (Basel)., 14*(4), 925. doi:10.3390/cancers14040925.

Ramnaraign, B., Sartor, O., et al. (2023). PSMA-Targeted Radiopharmaceuticals in Prostate Cancer: Current Data and New Trials. *Oncologist., 28*(5), 392–401. doi:10.1093/oncolo/oyac279.

Rawlinson, A., Mohammed, A., et al. (2012). The role of enzalutamide in the treatment of castration-resistant prostate cancer. *Future Oncol, 8*(9), 1073–1081 PMID: 23030482. doi:10.2217/fon.12.99.

Rebuzzi, S. E., Rescigno, P., et al. (2022). Immune Checkpoint Inhibitors in Advanced Prostate Cancer: Current Data and Future Perspectives. *Cancers (Basel)., 14*(5), 1245. doi:10.3390/cancers14051245.

Rehman, S. U., Abbas, G. H., et al. (2026). CRISPR/CAS9-based gene editing in cancer therapy: A systematic review and meta-analysis on current status and future directions. *Medicine (Baltimore)., 105*(2). doi:10.1097/MD.0000000000047114.

Richards, P., & Chen, T. (2022). Biosimilars and their impact on healthcare systems: A focus on prostate cancer. *American Journal of Managed Care, 26*(1), 23–35. https://doi.org/10.26419/jamc.2021.01258.

Rifaldi, F., Tortorella, A., et al. (2025). Novel androgen receptor inhibitors in prostate cancer: What do we know so far? *Crit Rev Oncol Hematol, 215*, 104857. doi:10.1016/j.critrevonc.2025.104857.

Roberts, A., & Tashiro, K. (2023). Economic burden of prostate cancer treatments in low-income settings. *Health Economics, 32*(3), 507–519. https://doi.org/10.1002/hec.4976.

Roberts, H., & Wang, X. (2021). The integration of biosimilars into prostate cancer treatment protocols. *Clinical Drug Investigation, 41*(7), 591–602. https://doi.org/10.1007/s40261-021-01040-8.

Santer, F. R., et al. (2014). Androgen receptor signaling in prostate cancer. *Cancer Metastasis Rev, 33*(2–3), 413–427. doi:10.1007/s10555-013-9474-0.

Schweizer, M. T., Drake, C. G., et al. (2014). Immunotherapy for prostate cancer: recent developments and future challenges. *Cancer Metastasis Rev, 33*(2–3), 641–655. doi:10.1007/s10555-013-9479-8.

Shaikh, M. S., Kurhade, R. R., et al. (2026). Advancements in Targeted Radiopharmaceuticals: Innovations in Diagnosis and Therapy for Enhanced Cancer Management. *Chembiochem., 27*(1). doi:10.1002/cbic.202500676.

Shariat, S. F. (2000). Gene therapy for prostate cancer. *Rev Urol, 2*(2), 81–87.

Sharma, R., Sil, D., et al. (2025). Oncolytic virotherapy - a promising approach in cancer treatment. *Expert Rev Anticancer Ther., 25(4):315-318*. doi:10.1080/14737140.2025.2474732.

Sharma, V., & Thomas, K. (2021). Benefits of biosimilars in treating prostate cancer: A comprehensive review. *European Urology Oncology, 4*(8), 448–457. https://doi.org/10.1016/j.euo.2021.07.004.

Shorning, B. Y., & Dass, M. S. (2020). The PI3K-AKT-mTOR Pathway and Prostate Cancer: At the Crossroads of AR, MAPK, and WNT Signaling. *Int J Mol Sci., 21*(12), 4507. doi:10.3390/ijms21124507.

Singh, R., & Yang, J. (2020). Clinical trials of biosimilars in oncology: Prostate cancer and beyond. *Oncology Reviews, 44*(3), 139–150. https://doi.org/10.1016/j.oncrev.2020.07.004.

Small, EJ, Reese, DM, et al. (2000). New treatment strategies in advanced prostate cancer. *Radiol Clin North Am, 38*(1), 203. doi:10.1016/s0033-8389(05)70156-1.

Smith, K., & Anderson, T. (2022). The clinical implications of immune checkpoint inhibitors in prostate cancer. *Prostate Cancer Journal, 50*(5), 89–96.

Stebbing, J., & Mainwaring, P. N. (2020). Understanding the Role of Comparative Clinical Studies in the Development of Oncology Biosimilars. *J Clin Oncol., 38*(10), 1070–1080. doi:10.1200/JCO.19.02953.

Stevens, J., & Lee, K. (2021). Advances in biosimilar technology and its implications for prostate cancer treatment. *Journal of Pharmaceutical Innovation, 16*(4), 285–298. https://doi.org/10.1007/s12349-020-00375-1.

Syed, S., & Tolcher, A. (2003). Innovative therapies for prostate cancer treatment. *Rev Urol., 5*(3), 78–84.

Talwar, V., Kalra, K., et al. (2025). Evolving Treatment Paradigms in Metastatic Hormone-Sensitive Prostate Cancer: Expert Narrative Review. *Curr Oncol., 32*(8), 437. doi:10.3390/curroncol32080437.

Tan, S., & Lee, F. (2023). Exploring the regulatory framework for biosimilars in prostate cancer. *International Journal of Cancer, 55*(1), 26–35. https://doi.org/10.1002/ijc.33871.

Tang, J., & Hou, L. (2023). Radiopharmaceuticals in prostate cancer: Current use and future possibilities. *European Journal of Nuclear Medicine and Molecular Imaging, 50*(7), 1348–1359. https://doi.org/10.1007/s00259-023-06184-9.

Tannenbaum, C., & Rosenthal, M. (2024). Economic benefits of biosimilars in prostate cancer: A cost-utility analysis. *BMC Cancer, 24*(1), 128. https://doi.org/10.1186/s12885-024-10472-3.

Taylor, E., & Ross, H. (2022). A review of the regulatory approval process for biosimilars in prostate cancer. *Pharmaceutical Regulatory Affairs, 10*(4), 212–221. https://doi.org/10.1038/s41572-021-00357-7.

Thomas, B., & Kuo, M. (2020). Biosimilar regulatory approval in prostate cancer: Global perspectives. *Pharmaceutical Policy and Law, 22*(5), 355–367. https://doi.org/10.3233/PPL-200265.

Varaprasad, G. L., Gupta, V. K., et al. (2023). Recent advances and future perspectives in the therapeutics of prostate cancer. *Exp Hematol Oncol, 12*(1), 80. doi:10.1186/s40164-023-00444-9.

Venkatachalam, S., McFarland, T. R., et al. (2021). Immune Checkpoint Inhibitors in Prostate Cancer. *Cancers (Basel).*, *13*(9), 2187. doi:10.3390/cancers13092187.
Walker, K., & Harris, P. (2021). Cost-effectiveness of biosimilars in prostate cancer treatment. *Journal of Cancer Economics, 37*(3), 289–299. https://doi.org/10.1002/jce.1800.
Walker, K., & Zhao, T. (2023). Development of biosimilars for prostate cancer therapies: A review. *Journal of Clinical Oncology, 41*(10), 2301–2312. https://doi.org/10.1200/JCO.22.02260.
Wallace, D., & Campbell, A. (2021). The future of biologic treatments in prostate cancer. *Urologic Oncology, 39*(3), 193–202. https://doi.org/10.1016/j.urolonc.2020.12.004.
Wallace, T., & Meyer, M. (2021). The role of biosimilars in reducing healthcare costs for prostate cancer patients. *Journal of Healthcare Economics, 32*(9), 2254–2265. https://doi.org/10.1093/jhe.2021.00953.
Wang, I., Song, L., et al. (2022). Prostate cancer immunotherapy: a review of recent advancements with novel treatment methods and efficacy. *Am J Clin Exp Urol, 10*(4), 210–233 PMC9428569.
Wang, Y., & Yuan, X. (2020). Future prospects of biosimilars in prostate cancer treatment. *Asian Journal of Oncology, 28*(4), 271–283. https://doi.org/10.1016/j.ajc.2020.07.003.
Wong, D., & Zhao, T. (2022). The role of PSMA-targeted therapies in metastatic prostate cancer. *European Journal of Nuclear Medicine and Molecular Imaging, 49*(4), 785–795. https://doi.org/10.1007/s00259-022-05813-9.
Xu, Y., Le, H., et al. (2025). Advancements in CRISPR/Cas systems for disease treatment. *Acta Pharm Sin B, 15*(6), 2818–2844. doi:10.1016/j.apsb.2025.05.007.
Yu, H., & Peters, A. (2021). Biosimilars in prostate cancer: A look into their clinical benefits and challenges. *International Journal of Cancer Therapy and Oncology, 11*(2), 95–104. https://doi.org/10.1111/jcto.13125.
Zhang, M., & Liu, H. (2022). Biologic and biosimilar therapies for prostate cancer: A clinical and regulatory review. *Molecular Cancer Therapy, 21*(7), 1398–1410. https://doi.org/10.1158/1535-7163.MCT-21-0404.
Zhang, W., & Wright, M. (2021). Personalized medicine in prostate cancer: The role of genomics and biosimilars. *Journal of Precision Medicine, 6*(3), 85–92. https://doi.org/10.1186/jpm.6.3.85.

CHAPTER

Biosimilars and biologics in ovarian cancer 10

Dishant Chaudhari, Piyush Ghode and Kuttiappan Anitha

School of Pharmacy and Technology Management, SVKM's Narsee Monjee Institute of Management Studies (NMIMS) (Deemed to be University), Shirpur, Maharashtra, India

10.1 Overview of ovarian cancer: Incidence and predisposing factors

10.1.1 General background

Ovarian cancer can be considered one of the most widespread and lethal women cancers, which develops in every corner of the globe. It is now the seventh most frequent cancer in women, but it has constantly been one of the leading causes of death among gynecological cancers and it also has the highest death rate among gynecological cancers. Ovarian cancer also claims 4.4% of the cancer deaths in women around the world in the year 2018 (Momenimovahed et al., 2019). The cause of this high mortality rate is mostly related to the late stages of the diagnosis, since the condition is diagnosed at advanced stages in more than 70% of the cases due to the vague or missing symptoms of the disease at early stages (Smolarz et al., 2025). Epidemiologically, the world distribution of incidence is concentrated in countries with high Human Development Index (HDI) (e.g., North America, Northern and Eastern Europe); the number of total deaths is large worldwide (Gaona-Luviano et al., 2020). The international standard age-adjusted incidence rate is about 10.3 per 100,000 of new cases per annum, of which the death rate is 5.9 per 100,000. It is also found that its incidence and prevalence in younger women have been increasing, although ovarian cancer is normally enrolled in women at the ages of 55 to 70, and the highest tendency is between the late 50s and early 60s (Zhuang et al., 2025). Clinically and biologically, the heterogeneous disease is ovarian cancer, with at least four major histologic subtypes, such as serous, mucinous, endometrioid, and clear cell carcinoma (CCC). Depending on subtype and stage at diagnosis, different survival rates exist with an approximate overall 5-year survival rate of about 46% in high-income countries, but considerably lower in most low- and middle-income environments (Wang et al., 2024). Major risk aspects of ovarian cancer are advanced age, family history, genetic mutations found in genes BRCA1/2, nulliparity, endometriosis, hormonal replacement therapy, and some aspects of lifestyle. Protective factors include, on the other hand, pregnancy, lactation, and oral contraceptive pills (Wentzensen et al., 2025; Fig. 10.1).

Biosimilars and Biologics in Cancer Treatment: An Era in Therapeutic Strategies.
DOI: https://doi.org/10.1016/B978-0-443-43824-0.00011-0

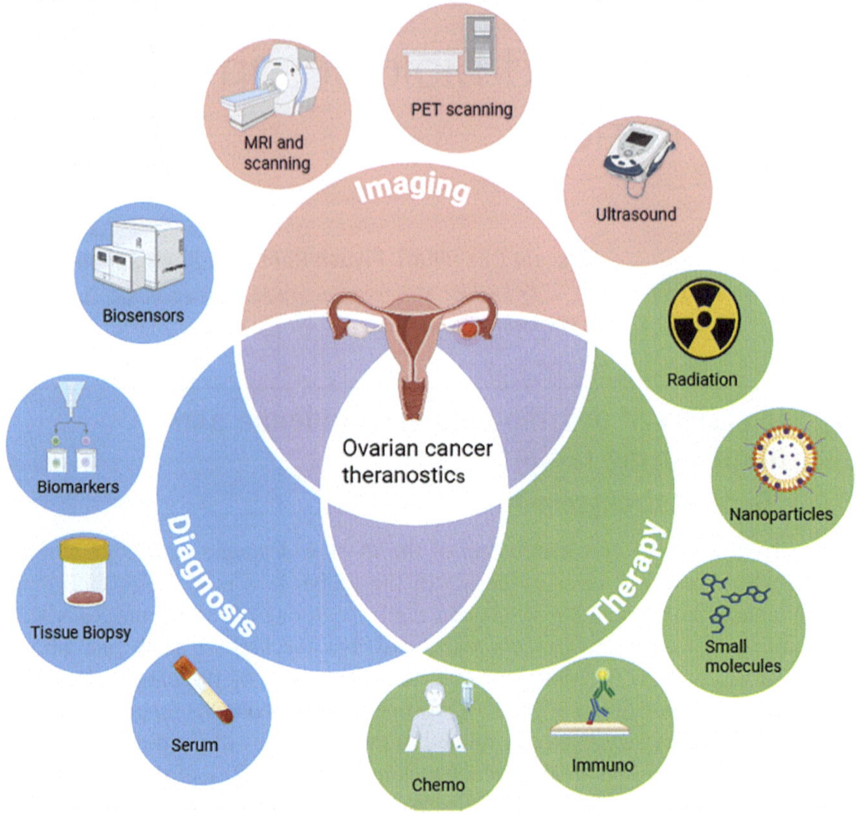

FIGURE 10.1

Integrated approaches in ovarian cancer. This diagram illustrates the interdisciplinary dynamics of ovarian cancer theranostics, encompassing diagnosis, imaging, and therapeutic interventions. The diagnostic modalities are biosensors, biomarkers, analysis of serum, and tissue cell biopsy. MRI, PET scanning, and ultrasound are some imaging techniques that help in the location of tumors and staging. These therapeutic methods include small molecules, nanoparticles, chemotherapy, immunotherapy, and radiation. The middle intersection reflects this overlap in the attainment of precision medicine.

10.1.2 Worldwide occurrence and trends

Ovarian cancer ranks seventh most frequent in women worldwide and is the leading cause of death due to cancer among the female gender because in 2018, 44 out of every hundred deaths were attributed to this cancerous disease in women (Momenimovahed et al., 2019). The worldwide distribution of ovarian cancer is highly variable, and the rates tend to be higher in those countries with higher HDI, such as North America and Europe; some high-HDI areas have experienced stable or falling incidences in the recent past. On the contrary, low- and middle-income countries (LMICs) are experiencing an increased incidence and mortality (Wang et al., 2024). In the world, serous carcinoma (SC) is the most common histological form, followed by mucinous carcinoma (MC), endometrioid carcinoma (EC), and CCC. More than 60% of recent cases of ovarian cancer and death have been recorded in LMICs, in which progress in therapy and in modifications of hormone treatment use have been less advanced than in higher-income regions. There was a rise of 0.74% and 0.89% in the incidence and prevalence of ovarian cancer in young women per annum between 1990 and 2019 (Zhuang et al., 2025).

10.1.3 Determinants of risk

10.1.3.1 Immutable contributors

The most significant noncorrectable risk element of the occurrence of ovarian cancer is age, because after menopause, the incidence of the cancer increases sharply, and women most often become its victims at 55 and later (van den Boom et al., 2022). Susceptibility is highly determined by genetic factors, especially the pathogenic variants in BRCA1 and BRCA2, and mismatch repair gene mutations of Lynch syndrome, which can even raise the lifetime risk from 40% to 60%. The risk is increased by a positive family history of breast or ovarian cancer as well (McClain et al., 2019).

10.1.3.2 Lifestyle and environmental influences

Risk factors include nulliparity and late age of first child birth, whereas there is protection associated with the use of oral contraceptives, multiple pregnancies, and breastfeeding, since they inhibit ovulation (Nakae et al., 2021). There is an augmentative moderate risk associated with long-term hormone replacement therapy, especially unopposed estrogen. Chronic inflammation and hormonal imbalance have been cited as the possible causes of the obesity associated increased risk. Mucinous subtype ovarian cancers are associated mostly with smoking

10.1.3.3 Variables with uncertain impact

There is controversy as to the role of genital talcum powder use. Whereas previous studies indicated possible risk, recent meta-analyses do not indicate a remarkable relation; the hypertension trigger does not have a demonstrable correlation with depression, according to recent meta-analyses in general. The evidence available on the use of fertility drugs has not shown a consistent outcome; borderline tumors may be increased, but there are no specific data concerning the risk of invasive cancer (Segars & Thomas, 2021). Other environmental causes, like asbestos, and life setting factors, like caffeine or dietary habits, are at this time not proven unequivocally to increase the risk of ovarian cancer by a considerable amount (Jara-Rosales et al., 2024).

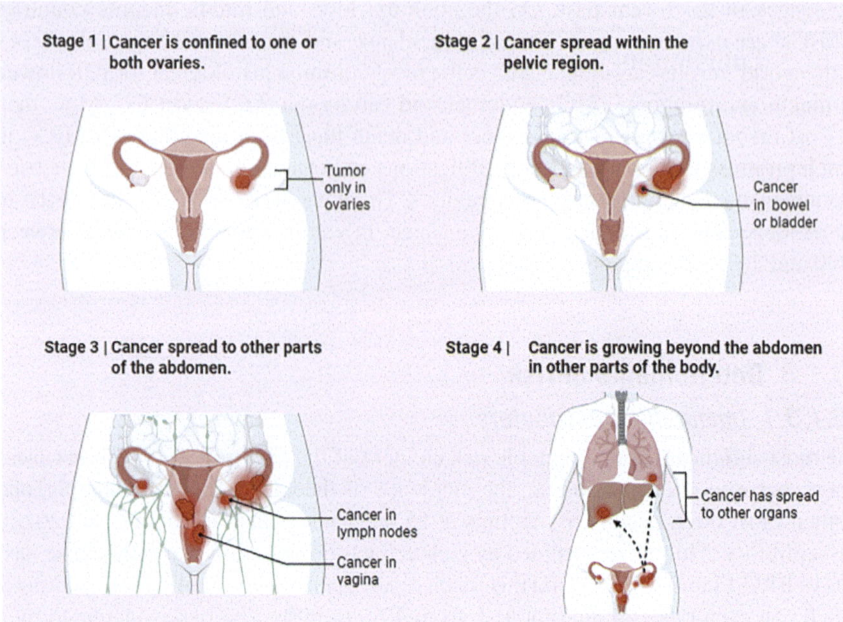

FIGURE 10.2

Clinical staging of ovarian cancer (FIGO classification): Figure of the four clinical stages of ovarian cancer progression with FIGO (International Federation of Gynecology and Obstetrics) system. Stage I: Cancer limited to either ovaries. Stage II: Following cancer has infiltrated the pelvic organs like the bowel or bladder. Stage III: The cancer has spread to different parts of the abdomen, either to peritoneum, lymph nodes or the vagina. Stage IV: Spread outside of the abdominal cavity to other parts of the body, e.g., to the liver and/or lungs and/or distant lymph nodes.

10.2 Clinical manifestations and pathophysiological basis

10.2.1 Early and common warning signs

Ovarian cancer is highly associated with nonspecific and nonpalpable symptoms in the early stages that frequently condemn patients to delayed diagnosis and poor outcomes. But there are some early and consistent symptoms that have been diagnosed, which should alert the clinical suspicion, especially if they occur more than 12 times a month or change to an individual's personal health (Ell, 2008). Clinical staging of ovarian cancers is explained in Fig. 10.2. Such developments are as follows:

- Blown belly or abdomen, low back or back pain, or cramping.
- Sense of fullness in the body when eating or loss of appetite.

- Stopping or children wanting to urinate more often or experiencing urgency or straining to urinate.
- Upset stomach, nausea, or indigestion.
- Bowel constipation or diarrhea.
- Backaches.
- Fatigue of an unknown cause.
- Unintentional loss of weight.
- Hurting during intercourse or menstrual issues.

These symptoms tend to merge with harmless conditions such as gastrointestinal or urinary problems, and often, they lead to misdiagnosis or vaguely clinical attention regarding ovarian cancer. The severity, recurrence, and novel occurrence of such symptoms (rather than the longstanding problems) are valuable hints in distinguishing them from nonmalignant causes (De et al., 2004).

10.2.2 Symptom index development

To enhance early detection, symptom indices, which are comprised of both frequency and severity of symptoms, such as abdominal bloating, early satiety, pelvic pain, and urgency, have been produced that may help primary care providers in the evaluation and the decision to refer to MRI (Lobo et al., 2018).

10.2.3 Cellular origin and clonal evolution

Current single-cell sequencing analyses mark toward intratumoral heterogeneity and clonal evolution during the evolution of serous tubal intraepithelial carcinoma (STIC) lesions to invasive carcinoma. The knowledge of these molecular changes at each step is essential in early intervention policies (Labidi-Galy et al., 2017).

10.2.4 Tumor microenvironment role

The Ascitic fluid refines not only tumor cells but also tumor cells (regulatory T cells [Tregs] and myeloid-derived suppressor cells [MDSCs]) that downregulate the immune surveillance and increase the metastatic range (Pearce et al., 2012).

10.2.5 Etiological and molecular perspectives

On the pathophysiological scale, the list of ovarian cancers includes a wide variety of malignant tumors, more than 90% of which derive their origin from epithelial cells of the ovary, fallopian tubes, or peritoneum (Zhu et al., 2011). Recently, extensive research indicates that STICs of the distal fallopian tubes epithelia are not the product of the ovary but the precursor of most high-grade serous ovarian carcinomas (HG-SOCs), especially those who have mutations in the BRCA gene (Singhi et al., 2012). BRCA1 and BRCA2, and other genes comprising the homologous recombination

DNA repair pathway, are mutated in hereditary ovarian cancer and predispose cells to genomic instability. Somatic TP53 mutation is identified in more than 96% of HGSOC cases, and it is almost ubiquitous in the subtype (The Cancer Genome Atlas Research Network, 2011). Other molecular distortions are impaired checkpoints of PI3K/AKT/mTOR, RAS, and NOTCH signaling pathways that lead to unregulated cell growth and inhibit the process of apoptosis (Matulonis et al., 2016). Such molecular characteristics also promote a peritoneal spread, the huge amount of ascites forming, and immune system inhibition, which, in turn, facilitates the active clinical picture of ovarian cancer (Table 10.1). The peritoneal cavity is a spacious anatomical structure, and this predisposes growth of tumors to an advanced stage before they manifest telling symptoms, and this is attributed to the late-stage nature of many cases, over 70%. This complexity at the molecular and clinical levels explains the chemoresistance of the disease and the reason there is a high recurrence and problematic prognosis in most of the cases that occur due to advanced-stage ovarian cancers (Siegel et al., 2023).

10.3 Standard-of-care pharmacological interventions

The management of ovarian cancer is still based on surgical treatment, especially at initiation. In the majority of patients with epithelial ovarian cancer, newly diagnosed, the aim of cytoreduction is optimal, since complete removal of visual tumor tissue represents a significant increase in the prognosis and survival. In some patients with an extensive disease burden, or in those whose comorbid conditions make primary surgery unfeasible, surgical staging may be performed after neoadjuvant chemotherapy in accordance with current clinical practice guidelines (Dinkins et al., 2024). The most common treatment of postsurgery involves the use of cytotoxic drug regimens, particularly among patients who have a disease that affects stage II to IV or high-grade. The platinum-platinum-taxanes, as well as platinum-taxanes combination chemotherapy regimens, are the backbone of the first-line treatment. This strategy is founded on the data showing better response rates and progression-free survival and overall survival than that seen in monotherapy (Dinkins et al., 2024). Platinum compounds, especially carboplatin, are more preferred on the basis of efficacy and relatively desirable toxicity. These compounds interconnect DNA and trigger apoptosis in fast-rising malignancy cells, and the mixture of these compounds with taxanes performs better than an isolated drug. Platinum drugs are also known to have better outcomes, which is why they are the pillar of re-treatment of platinum-sensitive recurrent ones (Armstrong et al., 2006). Taxane derivatives (usually paclitaxel) interfere with the activity of microtubules and prevent their division and subsequently cause the death of tumor cells. They are used with carboplatin as the global first-line chemotherapy gold standard, and their dosage and schedule can be adjusted to both tolerance and responsiveness of individual patients (Wang et al., 2025).

Table 10.1 Main molecular disruptions identified in ovarian cancer.

Pathway/gene affected	Subtype	Biological role	Treatment implications	References
BRCA1/2	Serous/endometrioid	DNA repair	PARP inhibitor sensitivity	Lord and Ashworth (2013)
TP53	High-grade serous	Tumor suppression	Prognostic marker	The Cancer Genome Atlas Research Network (2011)
PI3K/AKT/mTOR	Multiple	Cell growth/metabolism	Investigational targeted agents	Cui et al. (2017)
ARID1A	Clear cell/endometrioid	Chromatin remodeling	Possible immunotherapy targets	Zhang et al. (2025b)
KRAS	Mucinous	Signal transduction	Limited targeted options	Huang et al. (2020)

10.3.1 Innovative targeted options

A turnaround in the way ovarian cancer has been treated has been experienced in recent years, brought on by the development of targeted therapies. Such agents target the inhibition of certain molecular pathways crucial toward tumor growth and progression. The use of molecular profiling of tumors has facilitated the use of treatment dependent on the cell biology characteristics of tumors, particularly in cases of HGSOC, which presents with genetic instability. Such targets as vascular endothelial growth factor (VEGF), defects of homologous recombination repair, and the folate receptor are being developed as therapeutic targets. The further combination of the targeted strategies has prolonged progression-free survival and introduced precision medicine as a new normal in the treatment patterns of ovarian cancer (Matulonis et al., 2016).

10.3.2 Poly ADP-ribose polymerase blockers

PARP inhibitors have completely changed the treatment landscape of advanced and recurrent ovarian cancer, especially in those with BRCA1/2 genetic mutation and homologous recombination deficiency (HRD). The PARP enzymes participate in a single-stranded DNA break repair; the inhibitor gives a synthetic lethality in cancer cells whose homo recombination repair system has weakened. Maintenance therapy with agents such as olaparib, niraparib, and Rucaparib has shown considerable progression-free survival in patients taking platinum-based chemotherapy. As an example, the SOLO1 trial has shown a risk of disease progression reduction in BRCA-mutated patients taking olaparib as maintenance by 70% of the risk of disease development without the medication (Moore et al., 2018).

10.3.3 Angiogenesis inhibitors (e.g., antivascular endothelial growth factor antibodies)

Angiogenesis is a valid therapeutic modality targeted in ovarian cancer. VEGF plays an important role in controlling angiogenesis and tends to be overexpressed in ovarian cancers, and is associated with increased invasiveness and ascites (Guo & Lu, 2018). Monoclonal antibodies directed against VEGF, bevacizumab, when added to chemotherapy and subsequently maintenance therapy, both in front-line and recurrent settings, have also demonstrated benefit. Subjective results based on the ICON7 and GOG-0218 trials have demonstrated benefits in terms of progression-free survival among women with advanced diseases that simultaneously received bevacizumab in conjunction with carboplatin and paclitaxel treatments. Although it has a modest effect on overall survival, bevacizumab is known to be efficacious, especially in patients who have poor prognostic factors, including large-volume ascites or stage IV status (Haunschild & Tewari, 2020).

10.3.4 Younger molecularly directed treatment

Targeted therapy is becoming the way of the future in ovarian cancer, and molecular diagnostics or next-generation sequencing (NGS) are key to use in this process. Positioning individual tumor profiling can demonstrate the existence of actionability mutations and molecular weakness, making the stratification of targeted therapies possible. In addition to PARP and antiangiogenic, new treatments are antibody-drug conjugates (ADCs), checkpoint blockers, and combination approaches based on repair pathways and tumor microenvironment characteristics. As an example, the ADC mirvetuximab soravtansine is interesting because of its promising activity on tumors with high levels of folate receptors FolR. Multitargeted regimens are undergoing early-phase clinical trials, with the primary objectives of circumventing resistance and improving long-term disease control (Sisman et al., 2025) (Table 10.2).

10.4 Role of biosimilars and biological therapies in ovarian cancer

10.4.1 Concept clarification and regulatory landscape

The application of biosimilars and biological treatments of ovarian cancer is growing, and this is one of the major trends toward more precise medicine and improved accessibility. Some of the biological therapies that are now central to the management of ovarian cancer are monoclonal antibodies (mostly bevacizumab) and immune checkpoint inhibitors, all attacking key pathways and frequently complemented with chemotherapy, PARP inhibitors, or other immunotherapeutic agents (Zhang et al., 2025b). In ovarian cancer, biological therapies generally refer to treatments derived from living organisms or their products that target specific pathways or mechanisms involved in cancer development and progression. Such examples are the use of monoclonal antibodies (such as bevacizumab, which prevents the formation of blood vasculature of the tumor by blocking the VEGF pathway) and the use of ADCs and immune modulators. The biosimilars are postulated as the biologic medical products that are comparable (highly) to the already existing approved biological drugs (reference products), which are accorded a license to be manufactured by another maker after the expiration of poignant patent. Biosimilars are strictly checked to ensure that there are no clinically material variations to the reference product, albeit any slight variations with respect to the inactive ingredients shall be allowed. In consideration of ovarian cancer, Mvasi and Jobevne are bevacizumab biosimilars that have the same efficacy and safety compared with the original bevacizumab product (avastin). The main goal of biosimilars is to provide low-cost alternatives and high availability of distribution without having any influence on theoretical benefits or clinical efficiency (Zhu et al., 2022).

Table 10.2 Current pharmacological approaches in ovarian cancer.

Therapeutic category	Common drugs	Used in	Typical side effects	References
Platinum-based chemotherapy	Carboplatin, cisplatin	First-line, recurrent	Nausea, hearing loss, nephrotoxicity	Propper et al. (2014)
Taxane-based chemotherapy	Paclitaxel, docetaxel	First-line	Neuropathy, hair loss	Vasey et al. (2004)
PARP inhibitors	Olaparib, niraparib	Maintenance advanced	Anemia, fatigue	Moore et al. (2018)
Antiangiogenic agents	Bevacizumab	Mainly advanced/recurrent	Hypertension, bleeding	Pujade-Lauraine et al. (2014)
Antibody-drug conjugates (ADCs)	Mirvetuximab soravtansine	FolRα-high tumors	Ocular toxicity, GI upset	Moore et al. (2021)

10.4.2 Regulatory landscape

The biosimilar regulatory system is characterized by strict national and international standards. Other agencies like the US Food and Drug Administration (FDA) and European Medicines Agency (EMA) expect the drugs to be tested vigorously in preclinical and clinical investigations along with satisfactory analytical comparison, pharmacokinetics, and pharmacodynamics evidence, as well as one clinical trial to verify their noninferiority. Safety and efficacy also require postmarketing surveillance and pharmacovigilance. In 2025, various bevacizumab biosimilars were approved by the FDA, providing increased access and new products in oncology. Remarkably, the use of biosimilars has already shown the ability to lighten the heavy cancer treatment cost and expand advanced biological treatments to wider, more affordable access (Joshi et al., 2022). This changing environment is an indicator of a change in the mode of ovarian cancer treatment: expensive, targeted biological drugs to cheap and equally effective biosimilar drugs and the ability to combine advanced biological agents with individually based approaches (Table 10.3) (Gilmore et al., 2025).

10.4.3 Established and pipeline biological medicines

Biological medicines are used more and more in the treatment of ovarian cancer, and both approved and experimental agents have undergone tremendous progress. The first biological treatment of ovarian cancer approved by the FDA is Bevacizumab, although it is widely used because of its antiangiogenic effect. Simultaneously, ADCs, immune checkpoint inhibitors, and tumor-targeted vaccines are novel drugs in the pipeline that are potentially specific and have less off target toxicity. The latest landscape analysis notes that emerging biologics in the pipeline, including soravtansine analogues, modified immune cytokines, and tumor-specific immunotherapies, are in early to mid-phases of trials, and this may hold answers to the biggest challenge, that is, platinum resistance and recurrence, which challenge durability in treatments. The objective of these combinations of pipeline agents is an enhanced long-term survival with better tolerability (Sato et al., 2024).

10.4.3.1 Monoclonal antibody treatments

Through targeted therapy, monoclonal antibodies have played a central role in the treatment of ovarian cancer. The rationale of monoclonal agents that regulated the tumor microenvironment was set by bevacizumab against VEGF-A. A drug, an anti-FRα ADCs (Mirvetuximab soravtansine), was approved as an option in select patients with platinum-resistant, but not platinum-sensitive, ovarian cancer, and appears particularly effective in patients with high expression of folate receptors. Mirvetuximab was shown to have a higher objective response rate (ORR) and progression-free survival in a phase III trial of patients in the FRα-high population compared to chemotherapy. It is a breakthrough in targeted monoclonal therapy in gynecologic oncology and justifies the promise of receptor targeting payload delivery (Matulonis et al., 2023).

Table 10.3 Biologics and biosimilars used in the treatment of ovarian cancer.

Name	Class/type	Target	Clinical role	References
Bevacizumab (Avastin)	Biologic (antibody)	VEGF-A	Antiangiogenesis; frontline/recurrent OC	Pfisterer et al. (2023)
Bevacizumab biosimilars	Biosimilar antibodies	VEGF-A	Same as bevacizumab	Ioffe et al. (2023)
Olaparib, niraparib, rucaparib	Targeted small molecules (PARP inhibitors)	PARP enzymes	Maintenance/BRCA/HRD+ OC	Moore et al. (2018)
Mirvetuximab soravtansine	Antibody-drug conjugate	Folate receptor-α	For FRα-positive, resistant OC	Malakar et al. (2024)
Trastuzumab (and biosimilars)	Biologic antibody/biosimilar	HER2	Rare HER2+ ovarian cancer	Malakar et al. (2024)
Avelumab, dostarlimab	Biologic (checkpoint inhibitor)	PD-1/PD-L1	Investigational/selected OC	Narayana and Gupta (2025)

10.4.3.2 Biologic poly ADP-ribose polymerase inhibitors

At present, there are no clinically available biologic PARP inhibitors. Approved PARP inhibitors are small-molecule compounds, and they include olaparib, niraparib, and rucaparib. But biologic level specificity is under investigation with a combination of drugs and methods of destroying proteins, which may overcome PARP resistance. As an example, combinations of PARP inhibitors with immune checkpoint blockade agents, with antiangiogenic monoclonal antibodies, or even with ATR inhibitors, have been displaying activity in clinical trials. A recent translation research describes how resistance to PARP inhibition can be mediated and comments on the new macromolecule procedures to sensitize tumors to PARP inhibition. This incorporates finding out new focused protein degraders and biologic co-factors as future integration candidates, even though they remain in early-stage research and development phases (Lord & Ashworth, 2013).

10.4.3.3 Authorized and investigative biosimilar products

There are over 30 bevacizumab biosimilars that have already been created and can now be found all around the world because the patent on bevacizumab expired (2019 in the United States and 2022 in Europe). FKB238 has gone into clinical use; FKB238, ABP 215, SB8, and other biosimilars have proven effective and safe in clinical trials in real-life and clinical trial studies (Liolis et al., 2024). Investigational biosimilars are still at a stage of pivotal clinical trials, the key outcomes of which are usually the response rate and progression-free survival. As an example, a recent phase II clinical trial has compared the biosimilar FKB238 with olaparib and durvalumab in previously treated ovarian cancer, which demonstrated better survival results in comparison to when none of the drugs were used (Freyer et al., 2024).

10.4.3.4 Bevacizumab and its analogues

Bevacizumab alone has also been a very important element in the treatment of ovarian cancer as a backbone because of the substantial evidence, which is based on phase III studies (GOG-0218, AURELIA, and PAOLA-1), indicating considerable extension of progression-free survival in both newly diagnosed and recurrent cases. It is used in maintenance as well as in treatment environments, usually together with chemotherapy (or other targeted treatments, e.g., PARP inhibitors or immunotherapeutics). Bevacizumab is active through VEGF inhibition, inhibiting angiogenesis, normalization of tumor vasculature, and tumor microenvironmental changes. These mechanisms also have analogues and biosimilars (such as FKB238), which in recent reports have been noted to have similar clinical utility without impairing safety or effectiveness (Ray-Coquard et al., 2019).

10.4.4 Benefits and potential drawbacks in everyday practice

Biosimilar bevacizumab can increase and lower costs of accessibility to more patients; they have the potential to widen uses where cost has prohibited its application in the past. Previous studies and clinical experience propose no considerable distinction in efficacy, security, or immunogenicity between biosimilars and the initiator bevacizumab. In clinical practice, the advantages are stable tumor response, manageable toxicity (including mostly hypertension and proteins), and the worst potential side effects are the risks of gastrointestinal perforation (especially when prior therapies are used), blood pressure increase, and the costs of such therapeutic intervention in nonreimbursed health care systems. By incorporating biosimilars, it will be easier to incorporate integrated regimens (e.g., in combination with chemotherapy, PARP inhibitors, or immunotherapies) and emphasize the need to monitor and ensure proper communication of the equivalency and safety issues to doctors and patients (Espinoza-Carniglia et al., 2023).

10.4.4.1 Biosimilars impact

New biosimilars, especially those produced in the case of bevacizumab, will consequently bring down the price of treatment by up to 30% to 40%, thus facilitating its increased use in healthcare institutions with limited resources and expanding access to patients (Ioffe et al., 2023). In comparison to generic drugs, biosimilars necessitate comparable studies on the pharmacology (pharmacokinetics and pharmacodynamics), immunogenicity, and clinical outcomes levels to obtain a permit due to the intricacy of protein-based medications (Kurki et al., 2022; Liu et al., 2025).

10.4.4.2 New biological targets

New biologics have been introduced that target molecules such as folate receptor alpha (e.g., mirvetuximab soravtansine), immune checkpoints (PD-1/PD-L1 inhibitors), as well as ADCs, and are increasing the armamentarium and enhancing personalized medicine (Monga et al., 2025).

10.5 Preclinical evaluation of biologics and biosimilars

10.5.1 Findings from preclinical experiments

Investigations: Preclinical studies on biologics and biosimilars against ovarian cancer, notably against bevacizumab and biosimilars to it, have furnished important preclinical data of the mechanism of action as well as molecular equivalence, safety, and therapeutic potential (Jacobs et al., 2017).

10.5.1.1 Single or multiple mechanisms

There is a consistent report on the inhibitory effect of bevacizumab and its biosimilars in preclinical studies, with the attenuation of angiogenesis being a general effect where one of the key mechanisms is the inhibitory effect produced by blocking the regulatory VEGF-A signaling of VEGFR-2, hence the reduced phosphorylation of the receptor and inhibition of the proliferation of new blood vessels specifically in the tumor mass. The research works are based on the endothelial cell proliferation assay, vascular tube formation assay, and murine xenograft assays to verify inhibition of tumor growth and metastasis (Ferrara et al., 2004). On the other hand, the same can be seen in the reduction of tumor vascularization and similar enhancement of anti-tumor observed in animal models when using both bevacizumab and biosimilar BEVZ92 (Romera et al., 2018).

10.5.1.2 Biosimilarity assessment

Almost identical physicochemical, structural, and functional properties were found with reference and biosimilar products (ABP 215, MVASI, BEVZ92, and FKB238) in comparative analyses between them. In vitro analysis on its VEGF-binding, inhibition of endothelial cell proliferation, and blockade of the downstream signaling pathway has been reported as quite similar. Noninferiority in the inhibition of the growth of the tumors and the pharmacodynamic profile was also achieved in animal studies (Goldschmidt & Hanes, 2021).

10.5.1.3 Combination and evolving approaches

Among the novel preclinical concepts is the testing of bevacizumab (and their biosimilars) in combinations with PARP inhibitors and immunotherapy agents in ovarian cancer with additive and synergistic antitumor effects. In an example, the combination of biosimilar bevacizumab (FKB238) with olaparib and durvalumab increased survival benefits in animals, and thus positioned itself to be brought to phase II clinical trials (Yuan et al., 2021).

10.5.1.4 Safety and immunogenicity

As reported in preclinical studies, the use of bevacizumab biosimilar products is always consistent in safety by immunogenicity, suggesting that there are no additional adverse effects to the reference product in animal toxicity models. Analytical and bioequivalence studies assure the batch-to-batch consistency and are conducted to initiate human studies (Thomas et al., 2019).

10.6 Emerging innovations in ovarian cancer: Organoid platforms, genome editing, and artificial intelligence

10.6.1 Organoid platforms

The organoid models of ovarian cancer have progressed the preclinical and translational studies rather quickly. Organoids are three-dimensional cultures made by growing patient tumor tissue in vitro, and they maintain the tumor microenvironment and heterogeneity to a far more faithful degree than traditional 2D cell lines and better predict drug responses and mechanisms of resistance (Zhang et al., 2025a). These tools are currently being used to learn how drugs are metabolized, immune interactions, therapeutic resistance, and so on. As an example, live imaging has enabled new assays to quantify immune cell-mediated killing of hundreds of organoids generated over the past 3 days in real-time, delivering information on heterogeneity in immune susceptibility and revealing resistant subpopulations (Ferreira et al., 2025). There are co-culture platforms that enable physiologically relevant and high-throughput screening of therapies; immune cell types, stroma, and even microfluidics can be integrated inside the same platform that enables the screening of therapies, such as immune checkpoint inhibitors or CAR cell therapy. Organoid biobanks are under construction to find biomarkers faster, personalize treatment, and more closely model the complete clinical continuum of ovarian cancer (Broutier et al., 2017; Dijkstra et al., 2018).

10.6.2 Genome editing

Whereas the listed articles primarily focused on organoids, the inclusion of CRISPR/Cas9-based genome editing in modern organoid systems to systemically test the links between drugs and genes, disease modeling, and to introduce or correct oncogenic mutations, is becoming increasingly common (Li et al., 2024). Immunogenicity, tumor evolution, and response to targeted agent studies are also being increasingly studied with engineered organoids of defined genetic heritage (so-called syngeneic platforms). Such strategies reduce the gap between the experiments and clinical models and heterogeneity in patients and allow planning new selective and immunotherapeutic methods (Zhang et al. 2021).

10.6.3 Artificial intelligence

The use of artificial intelligence (AI) in combination with the organoid and multiomics datasets to examine the drug response, forecast resistance, and improve combinations of the therapy is in the trend. The automated and quantitative evaluation of the organoid development, morphology, and response to disrupted conditions is possible with AI-based image analysis, which leads to large-scale screening with increased replicability. There is also development of integrative AI frameworks to address high-dimensional, complicated datasets that are being generated by organoid drug screens, genomics, and patients, and are enabling faster discovery of biomarkers and precision medicine guidance (Brown et al., 2022; Hofer & Lutolf, 2021).

10.7 Application in clinics: Case analyses and trials

10.7.1 Landmark studies on biologicals and biosimilars

There have been two pivotal clinical trials, GOG-0218 and ICON7, which have determined the advantage of incorporating bevacizumab to chemotherapy for advanced ovarian cancer to increase progression-free survival. Combination is still being evaluated with the latest efforts aimed at enhancing the standard of care, including biosimilars. The combination of targeted agents, which are new strategies to suppress tumor cells and overcome resistance networks, has been found in preclinical and translational models where pathway-scale data are used to personalize the treatments (Polajžer & Černe 2025).

10.7.2 Precision-based therapy utilizing biomarkers

In ovarian cancer, precision oncology focuses on biomarker-driven stratification of patients. The category of the most convincing predictive evidence favors the BRCA1/2 mutations and the HRD as valuable biomarkers that identify individuals responding well to PARP inhibitors (Ethier et al., 2022). Potential predictors of immunotherapy response have also been sought due to the advent of novel biomarkers, including tumor mutational burden, mismatch repair deficiency, and microsatellite instability, for which prospective confirmation is still lacking. New technologies such as circulating tumor DNA can enable dynamic disease and minimal residual disease monitoring, enabling timely, more individualized intervention with changes, as well as the opportunity to monitor progression during the course of treatment (López-Portugués et al., 2024). Integra genomic knowledge bases integrating genomic, transcriptomic, and proteomic data are on the rise to optimize personalized therapeutic schemes, leading to further clinical trial designs (including basket and umbrella trial designs) aimed at stratifying *PI3K/AKT/mTOR, RAS* patients on an increasing number of molecular attributes (Duan et al., 2024).

10.7.3 Exemplary patient cases

Precision medicine strategies are showing real clinical effects with a number of published analyses of cases and series in which molecular tumor profiling is reported to have had a direct impact in cancer treatment in some instances, with dramatic effects that would have been impossible with conventional treatment. Indicatively, it has been stated that even in HGSOCs, NGS use can reveal rare, actionable mutations. Such was the case with targeted therapies (i.e., kinase or HER2 inhibitors), which, in some instances, elicited a significant remission (or partial and complete) against conventional chemotherapies that had failed. The case studies support the potential of genomics in illuminating the new directions of therapy: Partial response was found in patients undergoing trametinib, crizotinib, or trastuzumab-pertuzumab. A total of one patient received a complete response after treatment with alectinib. The findings indicate that NGS might be applicable in discovering effective targeted treatment of patients with HGSC (Antunes Meireles et al., 2024; Dong et al., 2020). Further, one of the most recent overviews specifically explored the translation of individualized patient care, founded by multiomics and genetic characteristics (BRCA, HRR, NTRK, BRAF, or RET alterations), into better outcomes and less toxicity. These achievements have frequently been written up as single case series, both showing how complex the approach can be and how promising. The fine art of precision medicine is exemplified by the fact that effective targeted therapy demands that the patients who have the possibility to receive a clinical benefit are selected, and adverse effects are limited as much as possible. Here is a look at the benefits and limitations of currently approved targeted therapies, such as bevacizumab, PARP inhibitors, and medicines against cells with alterations in the NTRK, BRAF, and RET genes and antibody-based drug conjugates. Also, potential new targets are discussed that are being examined in prospective clinical trials (Aravantinou-Fatorou et al., 2025).

10.8 Looking forward: Hurdles and opportunities

10.8.1 Real-world integration in underresourced areas

Implementing precision oncology in under-resourced settings remains difficult due to limited access to molecular diagnostics, shortages of specialized healthcare professionals, and prohibitive costs of targeted therapies. Many patients in rural or low-income regions lack the infrastructure and support systems needed to benefit from personalized cancer care. Addressing these disparities requires investment in training, technology, and equitable policies (Xiao et al., 2022; Yuan et al., 2021).

10.8.2 Areas awaiting research advancement

There are several areas that need improvement despite the fact that improvement has been achieved. These are the lack of predictive biomarkers beyond BRCA/HRD, therapeutic resistance, and the difficulty of translating multiomics data into the clinical workflow. Development of reliable biomarkers, enhancement of liquid biopsy in applications, and effective combination drugs have been in process, but large-scale validation is needed (Lua et al., 2022).

10.8.3 Emerging solutions

Emerging preclinical approaches that target pathway vulnerabilities and resistance mechanisms are demonstrating good promise. To take an example, novel drug combinations that are discovered through functional precision drug discovery screenings are being evaluated in resistant ovarian cancer models (Yang et al., 2023).

10.9 Conclusion

Ovarian cancer is a significant contributor to the health burden in the world, with high mortality rates owing to a delay in its detection and multifaceted clinical manifestations. The growing knowledge about its molecular heterogeneity has moved the treatment toward a precision oncology-based model, with traditional cytotoxic treatment now taking a back seat. Surgical superiority, the use of combination chemotherapy based on platinum and taxane drugs, and incorporation of targeted treatment agents like PARP inhibitors and antiangiogenesis agents have become the modus operandi of 21st-century ovarian cancer treatment. Not only are there more therapy options as a result of biologics and biosimilar development and use, but advanced care has now become more affordable due to reduced cost of maintenance and possible implementation of biologics and biosimilars such as bevacizumab and analogues. Innovations in organoid models, genome editing, and AI are contributing to rigorous laboratory and preclinical testing to move scientific insights into clinical practice faster than ever before. Supported by landmark clinical studies and patient exemplars, biomarker-directed patient care and precision-based treatments, especially with patients with BRCA mutations or HRD, have the potential to achieve improved patient outcomes to a substantial degree. There are still very daunting obstacles. The ability to overcome such barriers as insufficient molecular diagnostic resources and a lack of specialists who work in the field of personalized oncology, as well as ensuring equitable access to therapies, remains a challenge in LMICs. In the future, there should be a push toward bridging such gaps through additional research into novel predictive biomarkers, increased sophistication of multiomics platforms, and indeed globalizing these efforts. Research breakthroughs, continuity, fair policy systems, and interdisciplinary concepts will be essential in propagating precision medicine to all women who are victims of ovarian cancer. With continued dedication to both scientific innovation and health system reinforcement, the result will never be in doubt: Women in all countries will have early, more effective, less toxic, longer, and higher quality lives with the biosimilars and biologics treatment.

References

Antunes Meireles, P., Mira, B., & Vaz, F. (2024). Complete response with trametinib in advanced low-grade serous ovarian carcinoma: A case report. *Cureus*. https://doi.org/10.7759/cureus.53600.

Aravantinou-Fatorou, A., Georgakopoulou, V., Dimopoulos, M., & Liontos, M. (2025). Precision medicine in gynecological cancer (review). *Biomedical Reports, 22*, 43. https://doi.org/10.3892/br.2025.1921.

Armstrong, D. K., Bundy, B., Wenzel, L., Huang, H. Q., Baergen, R., Lele, S., … Burger, R. A., et al. (2006). Intraperitoneal cisplatin and paclitaxel in ovarian cancer. *New England Journal of Medicine, 354*(1), 34–43. https://doi.org/10.1056/NEJMoa052985.

Broutier, L., Mastrogiovanni, G., Verstegen, M. M, et al. (2017). Human primary liver cancer–derived organoid cultures for disease modeling and drug screening. *Nature Medicine, 23*, 1424–1435. https://doi.org/10.1038/nm.4438.

Brown, M. C, Ashley, D. M, & Khasraw, M. (2022). Low tumor mutational burden and immunotherapy in gliomas. *Trends in Cancer, 8*, 345–346. https://doi.org/10.1016/j.trecan.2022.01.006.

Cui, Y., Song, X., Li, S., et al. (2017). The impact of receptor recycling on the exocytosis of $\alpha v\beta 3$ integrin targeted gold nanoparticles. *Oncotarget, 8*, 38618–38630. https://doi.org/10.18632/oncotarget.16955.

De Vita, F., M, O., Galizia, G., et al. (2004). Author reply. *Cancer, 101*, 432–433. https://doi.org/10.1002/cncr.20345.

Dijkstra, K. K., Cattaneo, C. M, Weeber, F., Chalabi, M., van de Haar, J., Fanchi, L. F., … van Rooij, N., et al. (2018). Generation of tumor-reactive T cells by co-culture of peripheral blood lymphocytes and tumor organoids. *Cell, 174*(6), 1586–1598.e12. https://doi.org/10.1016/j.cell.2018.07.009.

Dinkins, K., Barton, W., Wheeler, L., et al. (2024). Targeted therapy in high grade serous ovarian cancer: A literature review. *Gynecologic Oncology Reports, 54*, 101450. https://doi.org/10.1016/j.gore.2024.101450.

Dong, D., Shen, G., Da, Y., et al. (2020). Successful treatment of patients with refractory high-grade serous ovarian cancer with *GOPC-ROS1* fusion using crizotinib: A case report. *The Oncologist, 25*, e1720–e1724. https://doi.org/10.1634/theoncologist.2019-0609.

Duan, X.-P., Qin, B.-D., Jiao, X.-D., et al. (2024). New clinical trial design in precision medicine: Discovery, development and direction. Signal Transduction and Targeted. *Therepy, 9*, 57. https://doi.org/10.1038/s41392-024-01760-0.

Ell, P. J. (2008). Sugar rush: The early days of molecular imaging. *The Lancet Oncology, 9*, 400. https://doi.org/10.1016/S1470-2045(08)70102-5.

Espinoza-Carniglia, M., Galliari, C., Fantozzi, M. C., et al. (2023). An integrative approach to explore species limits in *Laelaps mazzai* Fonseca, 1939 (Mesostigmata, Laelapidae), a South American widespread mite parasitizing the cricetid *Calomys* Waterhouse, 1837. *Acta Tropica, 240*, 406836. https://doi.org/10.1016/j.actatropica.2023.106836.

Ethier, J.-L., Fuh, K. C, Arend, R., et al. (2022). State of the biomarker science in ovarian cancer: A national cancer institute clinical trials planning meeting report. *JCO Precision Oncology*. https://doi.org/10.1200/PO.22.00355.

Ferrara, N., Hillan, K. J, Gerber, H.-P., & Novotny, W. (2004). Discovery and development of bevacizumab, an anti-VEGF antibody for treating cancer. *Nat Rev Drug Discovery, 3*, 391–400. https://doi.org/10.1038/nrd1381.

Ferreira, N., Alves, F., & Markus, A. (2025). Organoid-immune cell co-culture for stable live imaging. *Methods in Molecular Biology*. Springer https://doi.org/10.1007/7651_2025_627.

Freyer, G., Floquet, A., Tredan, O., et al. (2024). Bevacizumab, olaparib, and durvalumab in patients with relapsed ovarian cancer: A phase II clinical trial from the GINECO group. *Nature Communications, 15*, 1985. https://doi.org/10.1038/s41467-024-45974-w.

Gaona-Luviano, P., Medina-Gaona, L. A., & Magaña-Pérez, K. (2020). Epidemiology of ovarian cancer. *Chinese Clinical Oncology, 9*, 47. https://doi.org/10.21037/cco-20-34.

Gilmore, J. W., Indurlal, P., Rainey, A. M., et al. (2025). Cost-effective cancer care: The role of oncology biosimilars in generating cost savings. *Journal of Clinical Oncology, 43*, 11033. https://doi.org/10.1200/JCO.2025.43.16_suppl.11033.

Goldschmidt, J., & Hanes, V. (2021). The totality of evidence and use of ABP 215, a biosimilar to bevacizumab. *Oncology Therapeutics, 9*, 213–223. https://doi.org/10.1007/s40487-020-00133-1.

Guo, B.-Q., & Lu, W.-Q. (2018). The prognostic significance of high/positive expression of tissue VEGF in ovarian cancer. *Oncotarget, 9*, 30552–30560. https://doi.org/10.18632/oncotarget.25702.

Haunschild, C. E., & Tewari, K. S. (2020). Bevacizumab use in the frontline, maintenance and recurrent settings for ovarian cancer. *Future Oncology, 16*, 225–246. https://doi.org/10.2217/fon-2019-0042.

Hofer, M., & Lutolf, M. P (2021). Engineering organoids. *Nature Reviews Materials, 6*, 402–420. https://doi.org/10.1038/s41578-021-00279-y.

Huang, T.-T., Lampert, E. J, Coots, C., & Lee, J.-M. (2020). Targeting the PI3K pathway and DNA damage response as a therapeutic strategy in ovarian cancer. *Cancer Treatment Reviews, 86*, 102021. https://doi.org/10.1016/j.ctrv.2020.102021.

Ioffe, Y. J., Hong, L., Joachim-Célestin, M., et al. (2023). Genetic testing in the latinx community: Impact of acculturation and provider relationships. *Gynecologic Oncology, 169*, 125–130. https://doi.org/10.1016/j.ygyno.2022.12.001.

Jacobs, I., Ewesuedo, R., Lula, S., & Zacharchuk, C. (2017). Biosimilars for the treatment of cancer: A systematic review of published evidence. *BioDrugs, 31*, 1–36. https://doi.org/10.1007/s40259-016-0207-0.

Jara-Rosales, S., González-Stegmaier, R., Rotarou, E. S, & Villarroel-Espíndola, F. (2024). Risk factors for ovarian cancer in south America: A literature review. *Journal of Personalized Medicine, 14*, 992. https://doi.org/10.3390/jpm14090992.

Joshi, D., Khursheed, R., Gupta, S., et al. (2022). Biosimilars in oncology: Latest trends and regulatory status. *Pharmaceutics, 14*, 2721. https://doi.org/10.3390/pharmaceutics14122721.

Kurki, P., Kang, H.-N., Ekman, N., et al. (2022). Regulatory evaluation of biosimilars: Refinement of principles based on the scientific evidence and clinical experience. *BioDrugs, 36*, 359–371. https://doi.org/10.1007/s40259-022-00533-x.

Labidi-Galy, S. I, Papp, E., Hallberg, D., et al. (2017). High grade serous ovarian carcinomas originate in the fallopian tube. *Nature Communications, 8*, 1093. https://doi.org/10.1038/s41467-017-00962-1.

Li, S., Lei, N., Chen, M., et al. (2024). Exploration of organoids in ovarian cancer: From basic research to clinical translation. *Translational Oncology, 50*, 102130. https://doi.org/10.1016/j.tranon.2024.102130.

Liolis, E., Mulita, F., Koutras, A., et al. (2024). Exploring Bevacizumab's role in gynecological cancers: An up-to-date narrative review focusing on ovarian cancer. *Materia Socio Medica, 36*, 268. https://doi.org/10.5455/msm.2024.36.268-279.

Liu, Y., Wang, Y., Wang, M., et al. (2025). Evaluating biosimilars: Safety, efficacy, and regulatory considerations in clinical studies. *International Journal of Clinical Pharmacy, 47*, 232–236. https://doi.org/10.1007/s11096-024-01825-8.

Lobo, R. A, Archer, D. F, Kagan, R., et al. (2018). A 17β-estradiol–progesterone oral capsule for vasomotor symptoms in postmenopausal women. *Obstetrics & Gynecology, 132*, 161–170. https://doi.org/10.1097/AOG.0000000000002645.

López-Portugués, C., Montes-Bayón, M., & Díez, P. (2024). Biomarkers in ovarian cancer: Towards personalized medicine. *Proteomes, 12*, 8. https://doi.org/10.3390/proteomes12010008.

Lord, C. J., & Ashworth, A. (2013). Mechanisms of resistance to therapies targeting BRCA-mutant cancers. *Nature Medicine, 19*, 1381–1388. https://doi.org/10.1038/nm.3369.

Lua, I., Balog, S., & Asahina, K. (2022). TAZ/WWTR1 mediates liver mesothelial–mesenchymal transition induced by stiff extracellular environment, TGF-β1, and lysophosphatidic acid. *Journal of Cellular Physiology, 237*, 2561–2573. https://doi.org/10.1002/jcp.30750.

Malakar, S., Gontor, E. N., Dugbaye, M. Y., et al. (2024). Cancer treatment with biosimilar drugs: A review. *Cancer Innovation, 3*. https://doi.org/10.1002/cai2.115.

Matulonis, U. A., Lorusso, D., Oaknin, A., et al. (2023). Efficacy and safety of mirvetuximab soravtansine in patients with platinum-resistant ovarian cancer with high folate receptor alpha expression: Results from the SORAYA study. *Journal of Clinical Oncology, 41*, 2436–2445. https://doi.org/10.1200/JCO.22.01900.

Matulonis, U. A., Sood, A. K., Fallowfield, L., et al. (2016). Ovarian cancer. *Nature Reviews Disease primers, 2*, 16061. https://doi.org/10.1038/nrdp.2016.61.

McClain, K. M., Bradshaw, P. T., Khankari, N. K., et al. (2019). Fish/shellfish intake and the risk of head and neck cancer. *European Journal of Cancer Prevention, 28*, 102–108. https://doi.org/10.1097/CEJ.0000000000000431.

Momenimovahed, Z., Tiznobaik, A., Taheri, S., & Salehiniya, H. (2019). Ovarian cancer in the world: Epidemiology and risk factors. *International Journal of Women's Health, 11*, 287–299. https://doi.org/10.2147/IJWH.S197604.

Monga, A., Gagan, J. P., et al. (2025). Biosimilars: A critical review of development, regulatory landscape, and clinical implications. *AAPS PharmSciTech, 26*, 46. https://doi.org/10.1208/s12249-025-03038-2.

Moore, K., Colombo, N., Scambia, G., et al. (2018). Maintenance Olaparib in patients with newly diagnosed advanced ovarian cancer. *New England Journal of Medicine, 379*, 2495–2505. https://doi.org/10.1056/NEJMoa1810858.

Moore, K. N., Oza, A. M., Colombo, N., et al. (2021). Phase III, randomized trial of mirvetuximab soravtansine versus chemotherapy in patients with platinum-resistant ovarian cancer: Primary analysis of FORWARD I. *Annals of Oncology, 32*, 757–765. https://doi.org/10.1016/j.annonc.2021.02.017.

Nakae, R., Matsuzaki, S., Serada, S., et al. (2021). CD70 antibody-drug conjugate as a potential therapeutic agent for uterine leiomyosarcoma. *American Journal of Obstetrics and Gynecology, 224*, 197.e1–197.e23. https://doi.org/10.1016/j.ajog.2020.08.028.

Narayana, R. V. L., & Gupta, R. (2025). Exploring the therapeutic use and outcome of antibody-drug conjugates in ovarian cancer treatment. *Oncogene, 44*, 2343–2356. https://doi.org/10.1038/s41388-025-03448-3.

Pearce, C. L., Templeman, C., Rossing, M. A., et al. (2012). Association between endometriosis and risk of histological subtypes of ovarian cancer: A pooled analysis of case–control studies. *The Lancet Oncology, 13*, 385–394. https://doi.org/10.1016/S1470-2045(11)70404-1.

Pfisterer, J., Joly, F., Kristensen, G., et al. (2023). Optimal treatment duration of bevacizumab as front-line therapy for advanced ovarian cancer: AGO-OVAR 17 BOOST/GINECO OV118/ENGOT Ov-15 open-label randomized phase III trial. *Journal of Clinical Oncology, 41*, 893–902. https://doi.org/10.1200/JCO.22.01010.

Polajžer, S., & Černe, K. (2025). Precision medicine in high-grade serous ovarian cancer: Targeted therapies and the challenge of chemoresistance. *International Journal of Molecular Sciences, 26*, 2545. https://doi.org/10.3390/ijms26062545.

Propper, D., Davidenko, I., Bridgewater, J., et al. (2014). Phase II, randomized, biomarker identification trial (MARK) for erlotinib in patients with advanced pancreatic carcinoma. *Annals of Oncology, 25*, 1384–1390. https://doi.org/10.1093/annonc/mdu176.

Pujade-Lauraine, E., Hilpert, F., Weber, B., et al. (2014). Bevacizumab combined with chemotherapy for platinum-resistant recurrent ovarian cancer: The AURELIA open-label randomized phase III Trial. *Journal of Clinical Oncology, 32*, 1302–1308. https://doi.org/10.1200/JCO.2013.51.4489.

Ray-Coquard, I., Pautier, P., Pignata, S., et al. (2019). Olaparib plus bevacizumab as first-line maintenance in ovarian cancer. *New England Journal of Medicine, 381*, 2416–2428. https://doi.org/10.1056/NEJMoa1911361.

Romera, A., Peredpaya, S., Shparyk, Y., et al. (2018). Bevacizumab biosimilar BEVZ92 versus reference bevacizumab in combination with FOLFOX or FOLFIRI as first-line treatment for metastatic colorectal cancer: A multicentre, open-label, randomised controlled trial. *Lancet Gastroenterology & Hepatology, 3*, 845–855. https://doi.org/10.1016/S2468-1253(18)30269-3.

Sato, S., Shoji, T., Jo, A., et al. (2024). Antibody-drug conjugates: The new treatment approaches for ovarian cancer. *Cancers (Basel), 16*, 2545. https://doi.org/10.3390/cancers16142545.

Segars, J., & Thomas, M. A. (2021). Validation of a highly realistic embryo transfer simulator and trainer. *Fertility and Sterility, 115*, 879–880. https://doi.org/10.1016/j.fertnstert.2021.01.019.

Siegel, R. L., Miller, K. D., Wagle, N. S., & Jemal, A. (2023). Cancer statistics, 2023. *CA: A Cancer Journal for Clinicians, 73*, 17–48. https://doi.org/10.3322/caac.21763.

Singhi, A. D., Cimino Mathews, A., & Jenkins, R. B. (2012). MYC gene amplification is often acquired in lethal distant breast cancer metastases of unamplified primary tumors. *Modern Pathology, 25*, 378–387. https://doi.org/10.1038/modpathol.2011.171.

Sisman, Y., Høgdall, D., Schnack, T. H., Høgdall, C., & Høgdall, E. (2025). Potential targeted therapies for ovarian cancer beyond PARP inhibitors. *European Journal of Gynaecological Oncology, 46*, 1. https://doi.org/10.22514/ejgo.2025.030.

Smolarz, B., Biernacka, K., Łukasiewicz, H., et al. (2025). Ovarian cancer—epidemiology, classification, pathogenesis, treatment, and estrogen receptors' molecular backgrounds. *International Journal of Molecular Sciences, 26*, 4611. https://doi.org/10.3390/ijms26104611.

The Cancer Genome Atlas Research Network. (2011). Integrated genomic analyses of ovarian carcinoma. *Nature, 474*, 609–615. https://doi.org/10.1038/nature10166.

Thomas, M., Thatcher, N., Goldschmidt, J., et al. (2019). Totality of evidence in the development of ABP 215, an approved bevacizumab biosimilar. *Immunotherapy, 11*, 1337–1351. https://doi.org/10.2217/imt-2019-0125.

van den Boom, A. L., Lavrijssen, B. D. A., Fest, J., et al. (2022). Appendectomy and the subsequent risk of cancer: A prospective population-based cohort study with long follow-up. *Cancer Epidemiology, 77*, 102120. https://doi.org/10.1016/j.canep.2022.102120.

Vasey, P. A., Jayson, G. C., & Gordon, A Scottish Gynaecological Cancer Trials Group. (2004). Phase III Randomized trial of docetaxel-carboplatin versus paclitaxel-carboplatin as first-line chemotherapy for ovarian carcinoma. *JNCI Journal of the National Cancer Institute, 96*, 1682–1691. https://doi.org/10.1093/jnci/djh323.

Wang, G., Yang, H., Wang, Y., & Qin, J. (2025). Ovarian cancer targeted therapy: Current landscape and future challenges. *Frontiers in Oncology, 15*. https://doi.org/10.3389/fonc.2025.1535235.

Wang, M., Bi, Y., Jin, Y., & Zheng, Z.-J. (2024). Global Incidence of ovarian cancer according to histologic subtype: A population-based cancer registry study. *JCO Global Oncology*. https://doi.org/10.1200/GO.23.00393.

Wentzensen, N., Ring, K., Erickson, B. K., et al. (2025). Ovarian cancer risk prediction: A clinical epidemiology perspective. *American Journal of Epidemiology, 194*, 1182–1191. https://doi.org/10.1093/aje/kwae293.

Xiao, Y., Bi, M., Guo, H., & Li, M. (2022). Multi-omics approaches for biomarker discovery in early ovarian cancer diagnosis. *EBioMedicine, 79*, 104001. https://doi.org/10.1016/j.ebiom.2022.104001.

Yang, Y., Yuan, R., Lu, Y., et al. (2023). The engagement of autophagy in maniac disease. *CNS Neuroscience & Therapeutics, 29*, 3684–3692. https://doi.org/10.1111/cns.14353.

Yuan, H., Wang, H., Zhou, Y., et al. (2021). Water-level fluctuations regulate the availability and diffusion kinetics process of phosphorus at lake water–sediment interface. *Water Research, 200*, 117258. https://doi.org/10.1016/j.watres.2021.117258.

Zhang, L., Zhao, J., Su, C., et al. (2025a). Organoid models of ovarian cancer: Resolving immune mechanisms of metabolic reprogramming and drug resistance. *Frontiers in Immunology, 16*. https://doi.org/10.3389/fimmu.2025.1573686.

Zhang, M., Zhu, J., Bao, Y., et al. (2025b). Bevacizumab in ovarian cancer therapy: Current advances, clinical challenges, and emerging strategies. *Frontiers in Bioengineering and Biotechnology, 13*. https://doi.org/10.3389/fbioe.2025.1589841.

Zhang, S., Iyer, S., Ran, H., et al. (2021). Genetically defined, syngeneic organoid platform for developing combination therapies for ovarian cancer. *Cancer Discovery, 11*, 362–383. https://doi.org/10.1158/2159-8290.CD-20-0455.

Zhu, C. S., Pinsky, P. F., Cramer, D. W., Ransohoff, D. F., Hartge, P., Pfeiffer, R. M., … Lokshin, A. E. (2011). A framework for evaluating biomarkers for early detection: Validation of biomarker panels for ovarian cancer. *Cancer Prevention Research, 4*(3), 375–383. https://doi.org/10.1158/1940-6207.CAPR-10-0218.

Zhu, L., Wang, F., Huang, J., et al. (2022). Inflammatory aging clock: A cancer clock to characterize the patients' subtypes and predict the overall survival in glioblastoma. *Frontiers in Genetics, 13*. https://doi.org/10.3389/fgene.2022.925469.

Zhuang, Q., Gu, G., Chen, J., et al. (2025). Global, regional, and national burden of ovarian cancer among young women during 1990–2019. *European Journal of Cancer Prevention, 34*, 1–10. https://doi.org/10.1097/CEJ.0000000000000899.

CHAPTER

Biosimilars and biologics in bone cancer

11

Sagar Pamu[a], Neethu Mangalasseril Vijayakumari[b], Vanrajsinh Thakor[c], Bhagawati Saxena[d] and Sudharshan Reddy Dachani[e]

[a] *Department of Pharmacy Practice, Institute of Pharmacy, Nirma University, Ahmedabad, Gujarat, India,* [b] *Department of Pharmacology, MET's College of Pharmaceutical Sciences and Research, Thrissur, Kerala, India,* [c] *Department of Pharmaceutical Chemistry, Institute of Pharmacy, Nirma University, Ahmedabad, Gujarat, India,* [d] *Department of Pharmacology, Institute of Pharmacy, Nirma University, Ahmedabad, Gujarat, India,* [e] *Department of Pharmacology and Pharmacy Practice, College of Pharmacy, Shaqra University, Al-Dawadmi Campus, Shaqra, Saudi Arabia*

11.1 Introduction

Bone cancers are a heterogeneous category of malignant tumors, which can occur both de novo in the skeletal system (primary bone cancers) and secondary bone tumors, as metastases of other primary tumors. Primary malignant bone tumors are relatively rare but have a severe clinical outcome because of their aggressive nature and poor prognosis. Among them, osteosarcoma, Ewing sarcoma, and chondrosarcoma represent the greatest number of primary bone malignancies that arise in clinical practice. On the contrary, secondary bone cancers are more common, and they are often because of solid tumorlike breast, prostate, and lung carcinomas showing strong bone tropism and high morbidity, skeletal-related episodes and death (Hosseini et al., 2025). Conventionally, management of cancerous cells in bones has been based on conventional therapies such as surgical resection, systemic chemotherapy, and radiotherapy. Although these are the core of treatment, they are usually linked with drawbacks such as toxicity, resistance, and poor survival results in the long term. The last 20 years have seen major achievements in the field of molecular oncology and tumor biology, leading to the advent of the biological therapies that have steadily risen in importance in the treatment of cancer, and bone cancer is no exception (Bădilă et al., 2021). Therapeutic agents based on living systems are known as biologics and include monoclonal antibodies, cytokines, recombinant proteins, and other complex molecules. In contrast to traditional cytotoxic drugs, biologics are directed to disrupt certain molecular pathways, mediate the immune response, or alter the tumor microenvironment and, therefore, have better selectivity and may have superior safety profiles. Indicatively, monoclonal antibodies against growth factor receptors or immune checkpoints have transformed the field of oncology, and their addition to

Biosimilars and Biologics in Cancer Treatment: An Era in Therapeutic Strategies.
DOI: https://doi.org/10.1016/B978-0-443-43824-0.00004-3

the treatment regimen of bone cancer is a field of ongoing research (Wang et al., 2020). Simultaneously, biosimilars have been introduced that drastically changed the therapeutic environment. Biosimilars are approved reference biologic medicines that are all too similar, but not identical, and produced through rigorous regulatory development programs that provide comparable efficacy, safety, and quality. They are not only important in terms of therapeutic equivalence, but they also help save on healthcare expenditures and enhance accessibility of patients to advanced forms of treatment, especially in resource-constrained environments (Monga et al., 2025). Biosimilars of supportive biologic agents, including granulocyte-colony stimulating factors (G-CSFs) and erythropoiesis-stimulating agents, have already found clinical use in bone cancer topics, to decrease complications of chemotherapy. Besides, the future of the armamentarium against these aggressive tumors has the potential to grow with the development of biosimilars to therapeutic monoclonal antibodies and targeted biologics (Bachu et al., 2022). The purpose of this chapter is to develop an overview of the biologic and biosimilar use in the treatment of bone cancer, focusing on their mechanism of action, clinical data, regulatory guidelines, safety issues, and prospects. Through these dimensions, the chapter aims to pinpoint the paradigm-changing effects of biologics and biosimilars on patient outcomes and economic issues in the management of bone malignancies

11.2 Overview of bone cancers

11.2.1 Classification of bone cancers

The cancers of the bones are categorized into primary malignancies, which are the cancers that have emerged inside the skeletal system, and secondary (metastatic) tumors, which have emerged in other distant primary locations and have been transferred to the bone. This difference is also clinically significant, because there is a significant difference between the biological behavior, treatment practices, and prognosis of the two groups (Bahrambeigi et al., 2019).

11.2.1.1 Primary bone cancers

Bone primary malignant tumors are rather uncommon, but some of the most aggressive neoplasm of oncology. The most common primary malignant bone tumor is osteosarcoma, which is usually found in adolescents and young adults at the time of high bone growth. Its characteristic is the generation of osteoid substance by malignant osteoblastic cells (Kruger et al., 2014). It can also present clinically in the metaphyseal joints of long bones, including the distal femur, proximal tibia, and proximal humerus. Although multimodal therapy has increased, such as chemotherapy and limb-salvage surgery, the prognosis is not bright enough because of its early pulmonary metastasis tendency. Ewing sarcoma is a very aggressive bone and soft tissue tumor that is mostly observed in children and adolescents. At the molecular level, it is typified by certain chromosomal translocations, the most common of them being t(11;22) (q24;q12) that produces the EWS-FLI1 fusion gene (Salmon et al., 2025). As a clinical manifestation, Ewing sarcoma appears as a painful, fast-growing mass, which is usually accompanied by such systemic symptoms as fever and malaise. It has a high propensity to early spread to bone marrow, lungs, and other bones, which require intensive multimodal treatment. Chondrosarcoma is a cancerous tumor that develops out of cartilage-forming cells and is most commonly observed among middle-aged and elderly people. Unlike osteosarcoma and Ewing sarcoma, it is relatively resistant to conventional chemotherapy and radiotherapy, making surgical resection the primary mode of treatment. Chondrosarcomas vary in grade, with low-grade tumors showing slow progression and high-grade tumors demonstrating aggressive local invasion and metastatic potential. Common sites include the pelvis, femur, and shoulder girdle (Suaste-Vera et al., 2025).

11.2.1.2 Secondary (Metastatic) bone cancers

Secondary bone cancers, or bone metastases, are far more common than primary bone malignancies. They represent a frequent complication of advanced solid tumors and are a major cause of morbidity in cancer patients. The cancers most likely to metastasize to bone include breast, prostate, lung, thyroid, and kidney carcinomas. Each of these tumor types demonstrates characteristic patterns of skeletal involvement, reflecting their unique interactions with the bone microenvironment (Ventura et al., 2022).

11.2.2 Types of bone metastases

Metastatic lesions in bone can be classified according to their radiographic and pathological characteristics into: Osteolytic metastases, characterized by bone destruction, commonly seen in breast cancer and renal cell carcinoma. These lesions are associated with severe bone pain, pathological fractures, and hypercalcemia. Osteoblastic (sclerotic) metastases – characterized by abnormal new bone formation, most frequently associated with prostate cancer. These lesions may lead to skeletal rigidity but are also prone to fracture due to poor-quality bone. Mixed lesions – demonstrating both lytic and blastic features, often observed in lung and breast cancers (Coleman et al., 2020). The classification of bone cancers into primary and secondary categories provides a fundamental framework for understanding their biology, clinical presentation, and therapeutic approaches. Primary bone cancers, though less common, demand aggressive multimodal therapy, while metastatic bone disease requires comprehensive systemic treatment along with supportive measures to prevent skeletal-related events (SREs).

11.2.3 Epidemiology and global burden

Although primary bone cancers are considered rare malignancies, accounting for less than 0.2% of all cancers worldwide, their clinical impact is disproportionately significant. Unlike many other solid tumors that occur in older adults, primary bone cancers often affect children, adolescents, and young adults, thereby contributing to substantial morbidity, long-term disability, and loss of productive life years (Forsyth et al., 2022).

Incidence of primary bone cancers: Osteosarcoma is the most common primary malignant bone tumor, with an incidence of approximately 3–4 cases per million population per year. It demonstrates a bimodal age distribution, with a major peak during adolescence, coinciding with periods of rapid skeletal growth, and a smaller peak later in life, often associated with predisposing conditions such as Paget's disease of bone or prior radiation exposure. Ewing sarcoma accounts for 1–3 cases per million annually (Hosseini et al., 2025). It predominantly affects children and adolescents, with a median age at diagnosis of around 15 years. Notably, there is a strong ethnic and racial predisposition, as Ewing sarcoma is significantly more common in Caucasian populations and extremely rare in individuals of African or Asian descent. Chondrosarcoma is more frequently encountered in adults over the age of 40, with an incidence of 1–2 cases per million per year. Unlike osteosarcoma and Ewing sarcoma, chondrosarcoma is generally considered a tumor of later life and exhibits relative resistance to chemotherapy and radiotherapy, making surgical resection the primary therapeutic modality (Lee et al., 2021).

11.2.3.1 Global burden of secondary bone cancers

In contrast to the rarity of primary bone tumors, secondary (metastatic) bone cancers are extremely common in the context of advanced malignancies. Bone is among the most frequent sites of metastasis, particularly in breast, prostate, and lung cancers, due to its rich vascularity and unique microenvironment that supports tumor cell colonization. Breast and prostate cancer: Up to 70% of patients with advanced breast or prostate cancer eventually develop bone metastases. These metastases are not merely markers of disease progression but are also major drivers of SREs, including pathological fractures, spinal cord compression, and severe bone pain (Bădilă et al., 2021). Bone metastases lead to poor functional status, dysmobility, chronic pain, and poor quality of life. They also augment the health care use as they require surgical operations, radiation treatment, and supportive care provisions. Economic Burden: In addition to clinical effects, treatment of metastatic bone disease has a significant economic burden on healthcare systems all over the world, particularly in areas where access to advanced therapies is still low. To conclude, although the primary bone cancers are not common, they have an unequal distribution, affecting the younger age groups, and in most cases, leading to lifelong effects. Alternatively, secondary bone cancers are a significant international healthcare issue, since it is a frequent complication of metastatic malignancies as well as a top cause of cancer-induced morbidity. The awareness of this dual burden is important in coming up with effective prevention, early detection, treatment, and supportive care strategies (Colosia et al., 2022).

11.2.4 Molecular pathogenesis of bone cancers and risk factors and etiology of bone cancers

These molecular processes not only lead to oncogenesis but also support tumor progression and therapeutic resistance. Genetic abnormalities in tumor suppressor pathways are common in tumor suppressor pathways in osteosarcoma, which is one of the most explored primary bone tumors. TP53 and RB1 mutations and perturbations in the p14ARF-MDM2-p53 regulatory pathway have a major role in tumor initiation. Additionally, enhanced activity of growth factor signaling cascades, particularly those mediated by insulin-like growth factor (IGF) and vascular endothelial growth factor (VEGF), contributes to unchecked proliferation and angiogenesis, thereby promoting tumor aggressiveness (Nishida et al., 2025).

Ewing sarcoma, by contrast, is hallmarked by specific chromosomal translocations that result in EWSR1-ETS gene fusions, most commonly the EWS-FLI1 fusion. These fusion proteins function as aberrant transcription factors, reprogramming gene expression patterns that drive oncogenesis, inhibit differentiation, and foster tumor cell survival. In chondrosarcoma, a distinct genetic profile emerges, with many tumors harboring isocitrate dehydrogenase (IDH1/IDH2) mutations. These mutations disrupt normal cellular metabolism, leading to accumulation of oncometabolites that contribute to tumorigenesis (Apfelbaum et al., 2022). The resistance of chondrosarcoma to chemotherapy can be partly attributed to its low proliferative index and the presence of a dense cartilage extracellular matrix, which impedes drug penetration and reduces therapeutic efficacy. The biology of secondary bone cancers (bone metastases) is best explained by Paget's "seed and soil" hypothesis, which posits that disseminated tumor cells (the "seed") preferentially colonize the bone microenvironment (the "soil") due to its rich reservoir of growth factors and cellular interactions. A central player in this process is the RANK/RANKL/OPG signaling axis (Gilbert et al., 2023).

Tumor cells stimulate osteoclastogenesis via receptor activator of nuclear factor-κB ligand (RANKL), leading to excessive bone resorption. This resorptive activity releases growth factors such as transforming growth factor-β (TGF-β) and IGFs from the bone matrix, which in turn fuel tumor proliferation and survival, creating a self-perpetuating vicious cycle. Importantly, this pathway has been therapeutically targeted by denosumab, a fully human monoclonal antibody that inhibits RANKL, thereby reducing osteoclast activity and skeletal complications in patients with bone metastases. In summary, the etiology of bone cancers is multifactorial, with each subtype demonstrating distinct genetic hallmarks and microenvironmental dependencies. A deeper understanding of these molecular drivers has been instrumental in identifying novel therapeutic targets and continues to shape the development of biologics and biosimilars in the treatment of bone malignancies (Ono et al., 2020).

11.2.5 Current standard treatments

The management of bone cancers is multimodal, integrating surgery, chemotherapy, radiotherapy, and, more recently, targeted biological therapies. The choice of treatment depends on the tumor type, location, stage at diagnosis, and overall patient condition. Surgical resection remains the cornerstone of therapy for localized bone cancers. Whenever feasible, surgeons aim to perform wide excision or limb-salvage surgery, which has largely replaced radical amputations due to advances in surgical techniques, prosthetic reconstruction, and imaging modalities. Successful surgical intervention requires achieving tumor-free margins while preserving maximum limb function (Bădilă et al., 2021). Chemotherapy plays a pivotal role, particularly in osteosarcoma and Ewing sarcoma, which are highly chemo sensitive tumors. In osteosarcoma, multiagent regimens typically include high-dose methotrexate, doxorubicin, cisplatin, and ifosfamide, which have been shown to significantly improve survival outcomes when used in both neoadjuvant and adjuvant settings. For Ewing sarcoma, intensive multidrug chemotherapy protocols are employed, commonly combining vincristine, doxorubicin, and cyclophosphamide alternating with ifosfamide and etoposide (IE), reflecting decades of clinical trial optimization. Radiotherapy has a more selective role in the treatment of bone cancers. While most bone sarcomas are relatively resistant to radiation, Ewing sarcoma is radiosensitive, and radiotherapy is frequently used either as a definitive local treatment in unresectable cases or as an adjuvant modality following surgery. Additionally, radiotherapy is widely employed for the palliation of painful bone metastases, where it provides symptomatic relief and reduces the risk of skeletal-related complications (Koonrungsesomboon et al., 2025). In recent years, targeted therapies and biologics have begun to reshape the therapeutic landscape. One important advance is denosumab, a fully human monoclonal antibody that binds to RANKL, thereby inhibiting osteoclast-mediated bone resorption. It is particularly useful in patients with metastatic bone disease, where it reduces skeletal complications and delays disease progression (Kumar et al., 2024). Another novel agent is mifamurtide (muramyl tripeptide phosphatidylethanolamine), an immunomodulatory drug approved in Europe for the treatment of osteosarcoma, which works by stimulating macrophage and monocyte activity to enhance antitumor immunity (Miao et al., 2025). In addition, immune checkpoint inhibitors targeting PD-1/PD-L1 and CTLA-4 pathways are currently under investigation in osteosarcoma and Ewing sarcoma. While preliminary results show promise, their clinical benefit has not yet been firmly established, and ongoing trials aim to clarify their role in bone sarcoma treatment. Overall, the current standard treatments for bone cancers remain rooted in surgery and chemotherapy, with radiotherapy serving as an adjunct or palliative tool. Nonetheless, a key shift in therapy is reflected in the combination of selective agents and biologics, which are designed to enhance survival and reduce toxicity and the complexity of biology of these aggressive tumors (Wojtukiewicz et al., 2021).

11.2.6 Limitations of current therapies

The treatment of bone diseases remains a very difficult task, despite the advances in therapy. The high rate of recurrence is one of the most urgent ones, as both osteosarcoma and Ewing sarcoma tend to relapse despite multimodal treatment, resulting in poor long-term survival outcomes. Also, there is drug resistance, either acquired or intrinsic, which is also a significant challenge. Resistance of tumor cells to most available chemotherapeutic agents is a common phenomenon, thus limiting the efficacy of treatment and leading to the evolution of the disease. The other problem is the toxicity of the traditional treatments. Standard chemotherapy regimens, while effective in inducing tumor regression, are accompanied by significant short-term toxicities such as myelosuppression, cardiotoxicity, and nephrotoxicity, as well as long-term complications that can profoundly impact quality of life (Adewuyi et al., 2025). Furthermore, for patients with advanced, relapsed, or refractory disease, therapeutic options are extremely limited. The lack of effective second-line and salvage regimens underscores the urgent need for innovative treatment strategies. Collectively, these limitations highlight a critical gap in the current treatment landscape and underscore the need for the development and integration of novel biologics and biosimilars. Such agents hold promise not only for direct therapeutic applications but also for supportive care, helping patients better tolerate aggressive multimodal treatment and potentially improving both survival outcomes and quality of life (Rathod et al., 2025).

11.3 Biologics in bone cancer

Biologic therapies represent a paradigm shift in modern oncology, introducing treatment strategies that rely on highly specific mechanisms of action compared with conventional cytotoxic chemotherapy. They are used as biologics to not only target the tumor cells, but also to regulate the bone microenvironment, and to disrupt important tumor growth, angiogenic, metastatic, and bone resorption pathways in bone cancer. In contrast to the usual chemotherapy, which functions in a rather nonselective fashion, killing all rapidly dividing cells, biologic agents take advantage of very specific molecular targets, such as growth factors, cytokines, surface receptors, and immune checkpoints (Gonçalves et al, 2025; Xiao et al., 2025). This selectivity has the promise of providing higher therapeutic efficacy at lower toxicity in the whole system, thereby sparing normal tissues and enhancing patient outcomes. Nevertheless, along with such benefits, there are limitations to biologic therapies. Other issues like the development of resistance to treatment, the high price of treatment, and the minimal efficacy of some biologics in particular bone tumor subtypes remain to limit their use. However, the introduction of biologics to the treatment regimen of bone cancers is an indication of a radical change in cancer therapy and forms a basis of improvement of biosimilars with the view of enhancing cancer treatment accessibility and cost (Senapati et al., 2018).

11.3.1 Categories of biologics in bone cancer

There are several categories of biologic therapies available in the management of bone cancers, and each has a specific target based on the various aspects of tumor biology and complications during the treatment. The former category includes the monoclonal antibodies (mAbs), which are structured to target and identify certain molecular targets on tumor cells or within the tumor microenvironment. They are denosumab (blocking of RANKL to inhibit angiogenesis and bone resorption by osteoclasts), bevacizumab (blocking angiogenesis, which the immune system is modulated with anti-VEGF antibodies), and immune checkpoints (PD-1/PD-L1 axis), which aim to enhance antitumor immunity. Such immunomodulators as mifamurtide (muramyl tripeptide phosphatidyalaniline), an approved drug in Europe to treat osteosarcoma, and interferons and interleukins, have shown potential in regulating immune actions against bone sarcomas. A third category is the targeted fusion proteins, which are artificial molecules that suppress tumor survival and progression by blocking aberrant signaling pathways (Sangwan et al., 2025; Xiao et al., 2025). Some of these include VEGF/VEGFR axis inhibitors that prevent angiogenesis and decrease the vascular support that tumors need to grow. Lastly, the supportive biologics are most important in addressing the toxicity of the harsh chemotherapy treatments. G-CSFs, erythropoietin, and thrombopoietin mimetics are common medications that are used to manage myelosuppression, anemia, and thrombocytopenia, which in turn enhances patient treatment tolerance and their ability to achieve curative-intent therapy. Together, these categories illustrate the diverse applications of biologics in bone cancer, spanning from direct antitumor activity to supportive care interventions that enhance patient outcomes (Wang et al., 2024).

11.3.2 Monoclonal antibodies in bone cancer

11.3.2.1 Denosumab (anti-RANKL therapy)

Denosumab is a fully human monoclonal antibody directed against RANKL, a critical mediator of osteoclast differentiation and function. By binding to RANKL, denosumab prevents its interaction with the RANK receptor on osteoclast precursors, thereby suppressing osteoclastogenesis and reducing bone resorption. Clinically, denosumab has been approved for the prevention of SREs in patients with bone metastases from solid tumors, and it has demonstrated noninferiority to the bisphosphonate zoledronic acid in multiple randomized trials. In addition, in giant cell tumors of bone, denosumab has shown significant efficacy in shrinking tumor size and enabling less morbid surgical procedures. However, its use is limited by adverse effects such as hypocalcemia, osteonecrosis of the jaw (ONJ), and severe rebound bone loss upon discontinuation (Jang, 2023; Lu et al., 2023).

11.3.2.2 Bevacizumab (anti-VEGF therapy)

Bevacizumab is a humanized monoclonal antibody that targets VEGF-A, a pivotal regulator of angiogenesis. Its rationale in bone cancer is based on the observation that tumors such as osteosarcoma often overexpress VEGF, which is associated with poor prognosis. Preclinical studies have demonstrated synergistic antitumor activity when bevacizumab is combined with chemotherapy in osteosarcoma models. However, clinical benefit in bone sarcomas remains limited, and bevacizumab has not been established as standard of care in primary bone tumors (Garcia et al., 2020).

11.3.2.3 Immune checkpoint inhibitors

Immune checkpoints, including PD-1/PD-L1 and CTLA-4, act as inhibitory regulators that suppress T-cell responses, allowing tumors to evade immune surveillance.

Monoclonal antibodies targeting these checkpoints restore T-cell activity against tumor cells. Several agents have been investigated in bone cancers: pembrolizumab (anti-PD-1) showed responses in a small subset of sarcoma patients, including those with osteosarcoma, in the SARC028 trial (Aden et al., 2025); nivolumab (anti-PD-1) has demonstrated limited activity in relapsed osteosarcoma; and ipilimumab (anti-CTLA-4) has been primarily studied in pediatric populations, with sparse clinical data. Despite these efforts, the clinical efficacy of checkpoint inhibitors in bone sarcomas remains restricted, largely due to the inherently low mutational burden and minimal PD-L1 expression of these tumors (Rajan et al., 2016). Current research is therefore focused on combination strategies, pairing checkpoint inhibitors with chemotherapy or targeted agents to enhance responses.

11.3.3 Immune modulators and cytokine-based biologics

11.3.3.1 Mifamurtide (Muramyl tripeptide phosphatidylethanolamine)

Mifamurtide is a synthetic analogue of a bacterial cell wall component that activates macrophages and monocytes through NOD2 signaling. It is approved in Europe, though not in the United States, for use in combination with chemotherapy in patients with nonmetastatic osteosarcoma. A pivotal randomized trial demonstrated that adding mifamurtide to standard chemotherapy significantly improved overall survival in pediatric osteosarcoma patients. However, its widespread use has been restricted by high cost, lack of U.S. Food and Drug Administration (FDA) approval, and uncertain benefit in metastatic disease (Ando et al., 2011a).

11.3.3.2 Interferons (IFNs)

Interferon-α exerts antiproliferative, proapoptotic, and antiangiogenic effects, making it a potential immunotherapeutic agent in bone cancer. Early studies suggested a modest benefit in osteosarcoma when used as maintenance therapy. However, results from the large EURAMOS-1 trial failed to confirm a survival advantage when interferon was added to conventional chemotherapy, thereby limiting its clinical application (Chhipa et al., 2024).

11.3.3.3 Interleukins

Several interleukins have also been investigated for their immunostimulatory properties in bone sarcomas. Interleukin-2 (IL-2), which enhances T-cell activation, has been tested in sarcoma patients but demonstrated limited clinical benefit. Other cytokines, including IL-12 and IL-15, have shown promising immunostimulatory effects in preclinical models, but their therapeutic role in bone sarcomas remains investigational (Rahman et al., 2023).

11.3.4 Other biologic approaches

Tyrosine Kinase Inhibitors with Biologic Activity, although classified as small molecules, certain tyrosine kinase inhibitors (TKIs) exhibit biologic-like targeted activity and are relevant in bone cancers. Pazopanib, a VEGFR, PDGFR, and c-KIT inhibitor, is approved for soft tissue sarcoma and has shown exploratory use in bone sarcoma. Similarly, cabozantinib, which targets VEGFR2, MET, and AXL, has demonstrated promising antitumor activity in osteosarcoma and Ewing sarcoma. Adoptive Cell Therapy and Vaccines, Emerging biologic approaches in bone cancers include adoptive cell therapy and tumor vaccines. Chimeric antigen receptor T-cell (CAR-T) therapies are currently under early-phase evaluation, with constructs targeting GD2 and HER2 in osteosarcoma. In addition, peptide-based tumor vaccines have shown modest immunogenicity in sarcoma patients, although their clinical efficacy remains limited (Kumar et al., 2024; Thomson, Moshirfar, & Ronquillo, 2023). Supportive Biologics in Bone Cancer Care. In addition to tumor-targeted therapies, supportive biologics are widely used to mitigate chemotherapy-associated toxicities and enable patients to tolerate intensive treatment regimens. G-CSFs such as filgrastim, peg filgrastim, and lenograstim reduce chemotherapy-induced neutropenia, prevent life-threatening infections, and allow maintenance of chemotherapy dose intensity. (Ambrosio et al., 2021). Multiple biosimilars of G-CSFs are now approved globally, offering comparable efficacy and safety at reduced cost. Erythropoiesis-stimulating agents (ESAs), including epoetin alfa and darbepoetin alfa, are used for the management of chemotherapy-induced anemia. However, their use is tempered by concerns regarding thromboembolic risk and potential negative effects on survival in certain cancers. Similarly, thrombopoietin receptor agonists such as romiplostim and eltrombopag are under investigation for their ability to alleviate chemotherapy-induced thrombocytopenia, although their role in bone cancer remains investigational (Hao et al., 2024) (Table 11.1).

11.3.5 Limitations of biologics in bone cancer

Despite significant advancements, biologics in bone cancer face multiple challenges. Their efficacy in primary bone sarcomas such as osteosarcoma and Ewing sarcoma has been modest, in contrast to their transformative impact in hematologic malignancies and melanoma. Furthermore, their use is associated with a spectrum of toxicities, including immune-related adverse events (irAEs), ONJ, hypocalcemia, and cytokine release syndrome. Long-term efficacy is also impaired by other mechanisms of resistance, like tumor heterogeneity and immune evasion. Lastly, there are high pricing and accessibility factors to biologics, which are also a further barrier to them, especially in resource-constrained health care settings (Tian et al., 2023).

11.4 Biosimilars in bone cancer

Biosimilars are biologic medical products designed to be highly like already approved reference biologics, with no clinically meaningful differences in terms of safety, purity, or potency. Unlike the generics of small-molecule drugs, which are chemically identical to their reference counterparts, biosimilars cannot be produced as exact

Table 11.1 Major biologics used in bone cancer, their mechanisms of action, and clinical applications.

Biologic	Target/mechanism	Clinical application	References
Denosumab	Anti-RANKL monoclonal antibody → inhibits osteoclast activity, reduces bone resorption, and skeletal events	Bone metastases from solid tumors; giant cell tumor of bone	Costa et al. (2024)
Bevacizumab	Anti-VEGF monoclonal antibody → inhibits angiogenesis	Investigated in osteosarcoma and Ewing's sarcoma; experimental in metastatic bone cancers	DuBois et al. (2010)
Mifamurtide (Muramyl Tripeptide Phosphatidyl-ethanolamine)	Immunomodulator → activates macrophages and enhances immune-mediated tumor cell killing	Approved in Europe for nonmetastatic osteosarcoma (with chemotherapy)	Ando et al. (2011b)
Immune checkpoint inhibitors (e.g., nivolumab, pembrolizumab)	Block PD-1/PD-L1 or CTLA-4 → restore T-cell antitumor immunity	Under clinical trials in sarcomas and metastatic bone cancers	Hashimoto et al. (2025)
Interferons (IFNs)	Cytokine-based immune modulation → anti-proliferative and pro-apoptotic effects	Limited/experimental use in osteosarcoma and Ewing's sarcoma	Castro et al. (2018)
Interleukins (e.g., IL-2)	Stimulate immune effector cells (T-cells, NK cells)	Investigational role in sarcomas	Jiang et al. (2016)
Erythropoietin and Thrombopoietin mimetics	Stimulate erythropoiesis or platelet production	Supportive care in chemotherapy-induced anemia and thrombocytopenia	Fan et al. (2023)

copies because of the inherent complexity and variability associated with biologic manufacturing processes. Instead, their development involves rigorous comparability exercises that assess structural and functional characteristics, along with preclinical and clinical studies, to establish equivalence to the reference product (Mascarenhas-Melo et al., 2024). The escalating costs of oncology biologics have created a significant economic burden on healthcare systems worldwide, making biosimilars an attractive alternative by providing lower-cost options without compromising therapeutic efficacy. In the context of bone cancer, biosimilars are most widely used in supportive care, particularly in the form of G-CSFs and ESAs. Meanwhile, therapeutic biosimilars for monoclonal antibodies, such as bevacizumab and rituximab, are beginning to gain traction and gradually enter clinical practice (Devi, 2024; Henry & Taylor 2014).

11.4.1 Scientific and manufacturing aspects

11.4.1.1 Complexity of biologics

Biologics are large and structurally complex molecules, most commonly proteins, that are produced within living cells. Their production typically involves multiple stages, beginning with cell line development using systems such as Chinese Hamster Ovary cells, bacteria, or yeast. This is then followed by upstream processing that involves fermentation or cell culture to allow the expression of the protein and downstream purification that uses methods like chromatography and filtration to isolate the required product. Lastly, the process entails formulation and stabilization to stabilize the biologic in the event of running out of storage and clinical application. Biologics are extremely sensitive to the conditions in which they are manufactured, and thus, slight changes in the cell culture conditions, purification process, or formulation can change the glycosylation profile, folding, or stability of the resultant protein. This variability is the reason why biosimilars cannot be the perfect copies of their reference products but rather prove the high level of similarity (Chintalapati et al., 2025).

11.4.1.2 Comparability studies

The development of a biosimilar follows a rigorous stepwise approach to confirm similarity with its reference biologic. The process begins with analytical characterization, which assesses molecular structure, biological function, and binding activity using advanced analytical techniques. This is followed by nonclinical studies, including pharmacology and toxicology assessments, which are performed only if necessary to address residual uncertainties. The next stage involves clinical pharmacology studies to evaluate pharmacokinetics (PK) and pharmacodynamics (PD), ensuring equivalence to the reference product. Finally, confirmatory clinical trials are conducted in sensitive patient populations to establish comparable efficacy and safety. Together, these comparability exercises provide the foundation for regulatory approval.

11.4.1.3 Extrapolation of indications

Extrapolation of indications is one of the most significant regulatory principles that should be applied in biosimilar development. In case a biosimilar is shown to be similar in a clinical indication, it is allowed to receive approval for other clinical uses of the reference biologic if the mechanism of action, target receptors, and clinical setting are sufficiently supported. This principle is especially important in oncology, in which biologics like bevacizumab are applicable in a broad spectrum of malignant tumors, such as sarcomas and metastatic bone disease. Extrapolation will save the duplicative clinical trials, enhancing the speed with which patients access services and scientific and regulatory rigor (Li et al., 2017; Biosimilar Extrapolation of Indications: The Complete Regulatory, IP, and Investment Playbook, 2026).

11.4.2 Regulatory landscape of biosimilars

Regulatory frameworks on biosimilars differ in regions, but they all focus on stringent comparability activities to promote safety, efficacy, and quality. The Biologics Price Competition and Innovation Act (BPCIA) of 2009 in the United States has created a shorter approval route for biosimilars. FDA adhere to a totality-of-evidence principle, which is that it requires extensive information regarding analytical studies, preclinical research, and clinical trials to establish bio similarity. An additional designation of interchangeability exists, which allows pharmacy-level substitution of a biosimilar for its reference product without prescriber intervention. However, only a limited number of oncology biosimilars have achieved this designation so far, reflecting the high regulatory bar for substitution (Kaur et al., 2025). The European Medicines Agency (EMA) was the first regulatory authority to introduce biosimilar guidelines in 2005, making Europe a pioneer in this field. Since then, oncology-related biosimilars such as epoetin, filgrastim, trastuzumab, and bevacizumab have been widely approved and adopted. With more than a decade of clinical use, Europe remains the largest and most experienced market for biosimilars, offering valuable real-world evidence on their safety and effectiveness. India has been at the forefront of biosimilar development among emerging economies, with the approval of its first biosimilar (recombinant human insulin) in 2000. Updated biosimilar guidelines issued in 2016 were designed to align with World Health Organization (WHO) standards and provide a clear regulatory pathway. Today, India has multiple marketed biosimilars for agents such as filgrastim, peg filgrastim, bevacizumab, and rituximab, which are used both domestically and exported to other regions (Car et al., 2025). Biosimilars have caused a reduction of costs by 40%–60%, which has increased the access of patients in low-resource healthcare environments. The role played by the WHO in influencing policies of the biosimilar world, especially in providing guiding principles to regulatory harmonization in low- and middle-income countries (LMICs), is significant. The WHO system focuses on the use of comparability of reference, as opposed to an unjustifiable duplication of clinical trials, which lowers the development cost and allows more people to access the necessary therapies (Avhad et al., 2025).

11.4.3 Clinical applications of biosimilars in bone cancer

Biosimilars can be used in the clinical management of bone cancer and can be broken down into the supportive care biosimilars, which aid the treatment-related toxicities, and the therapeutic monoclonal antibody biosimilars, which target the tumor biology directly (Bachu et al., 2022).

11.4.3.1 Supportive care biosimilars

The most used G-CSF biosimilars include filgrastim, peg filgrastim, and lenograstim, which are used with great frequency in patients with bone cancer who receive intensive chemotherapy. Randomized clinical trials have always demonstrated that such biosimilars are similar in terms of efficacy and safety to their reference products in reducing the risk of febrile neutropenia, which potentially life-threatening complication of chemotherapy (Nahleh et al., 2022). Furthermore, extensive postmarketing surveillance has confirmed their effectiveness in real-world clinical practice, thereby solidifying their role in oncology supportive care. Erythropoietin biosimilars, including epoetin and darbepoetin, are also used to manage chemotherapy-induced anemia in bone cancer patients. While these agents provide symptomatic relief and reduce transfusion requirements, concerns remain regarding their long-term safety, particularly the risk of thromboembolic events and potential adverse effects on cancer outcomes. Nevertheless, the reduced cost of erythropoietin biosimilars has expanded their use in resource-constrained healthcare systems, improving access to supportive care. In addition, thrombopoietin receptor agonists represent an emerging class of supportive biologics for managing chemotherapy-induced thrombocytopenia. Although no biosimilars have yet reached the market, candidates for agents such as romiplostim and eltrombopag are currently in preclinical development. Their successful introduction could fill an important gap in supportive oncology care.

11.4.3.2 Therapeutic monoclonal antibody biosimilars

Beyond supportive agents, biosimilars of therapeutic monoclonal antibodies are beginning to influence the landscape of bone cancer treatment. Bevacizumab biosimilars—including products such as bevacizumab-awwb and bevacizumab-bvzr—have already been approved for use in colorectal, lung, and breast cancers. Although their routine use in bone sarcomas has not yet been established, these biosimilars are under evaluation in clinical studies of sarcomas and bone metastases, where angiogenesis plays a central role in tumor progression. Importantly, the availability of lower-cost bevacizumab biosimilars could expand treatment options for patients who otherwise face financial barriers (Goli & Arun 2022). In contrast, denosumab, a fully human monoclonal antibody against RANKL, currently has no approved biosimilar as of 2025. Nonetheless, several companies are actively pursuing their development. Given the critical role of denosumab in preventing SREs in patients with bone metastases and giant cell tumor of bone, the introduction of a biosimilar could substantially reduce the economic burden of treatment. Finally, immune checkpoint inhibitor biosimilars are also in the pipeline, with candidates for pembrolizumab and nivolumab undergoing clinical development. Once approved, these biosimilars may play a significant role in bone cancer management, particularly since immune checkpoint inhibitors are increasingly being tested in trials for osteosarcoma and Ewing sarcoma, albeit with variable success (Bachu et al., 2022).

11.4.4 Clinical evidence and trials

Biosimilars are generally evaluated through equivalence trials rather than superiority trials, as the primary objective is to demonstrate that they are neither clinically inferior nor meaningfully different from the reference biologics in terms of efficacy, safety, and immunogenicity. Robust clinical evidence exists for supportive care biosimilars such as G-CSF and erythropoietin products, which have been extensively studied not only in pivotal clinical trials but also through long-term postapproval pharmacovigilance programs (Monga et al., 2025). Similarly, therapeutic monoclonal antibody biosimilars such as bevacizumab have demonstrated clinical equivalence in sensitive tumor types like nonsmall cell lung cancer, thereby providing a scientific basis for extrapolating indications to other cancers, including bone malignancies. Furthermore, insights from real-world registries in Europe and India have reinforced the reliability of biosimilars, with data showing high levels of oncologist confidence and safety outcomes comparable to originator biologics (Mascarenhas-Melo et al., 2024).

11.4.5 Economic impact of biosimilars

Biosimilars are generally priced 20%–40% lower than their reference biologics, creating substantial opportunities for cost savings within oncology care. Global projections suggest that the adoption of oncology biosimilars could generate savings of $50–$100 billion over the next decade, thereby alleviating the economic burden on healthcare systems worldwide. In countries like India, the introduction and widespread use of biosimilars for agents such as G-CSFs and bevacizumab have notably expanded patient access to essential therapies that were previously limited due to cost constraints. In addition, biosimilars have a special potential in the LMICs where they will aid in closing disparities in the availability of cancer treatment and provide more people with access to life-saving biologics.

11.4.6 Barriers to biosimilar adoption

Regardless of their clinical and economic benefits, there are a few obstacles that impede the extensive use of biosimilars. The most significant issue is the level of skepticism about the physicians, where the question of their effectiveness, safety, and the general validity of the extrapolation of the indication is raised. Likewise, the perceptions of the patients tend to create the hesitancy factor, with a significant portion of them being confused by the idea of biosimilars as inferior, nonscientifically proven alternatives. The complexity of regulations also makes adoption more difficult, as the process of approval has highly different requirements and pathways across different countries, resulting in unequal availability and adoption of the tools. Moreover, there is still a necessity to have strong pharmacovigilance systems since the long-term safety and effectiveness information is still being produced and followed (Edgar et al., 2021).

11.4.7 Future prospects of biosimilars in bone cancer

The future of biosimilars in bone cancer is very bright, and there are a few trends that influence the evolution of biosimilars and their usage. It is expected that the next generation of therapeutic innovations will be dominated by denosumab and biosimilars of checkpoint inhibitors, as they are essential to bone health and immunotherapy. The increase in the subcutaneous formulations will also enhance patient convenience and compliance, as it will be easy to administer them in comparison to intravenous ones. Furthermore, the increasing integration of real-world data (RWD) and registry studies will play a pivotal role in strengthening physician and patient confidence in the safety and effectiveness of biosimilars. In addition, ongoing advances in biomanufacturing and analytical technologies are likely to streamline production, enhance quality, and further reduce costs, thereby improving accessibility and equity in cancer care worldwide.

11.5 Pharmacovigilance, safety, and economic/ethical considerations

11.5.1 Pharmacovigilance (PV) of biologics and biosimilars

Pharmacovigilance is the science of detecting, assessing, and preventing adverse effects related to drugs, and it plays a particularly critical role in biologics. Since biologics are produced in living systems, they carry unique risks that differ from small-molecule drugs. For bone cancer patients, both biologics and their biosimilars require rigorous pharmacovigilance systems due to long treatment durations, the potential for immunogenicity, and the fact that these agents are used in highly vulnerable populations (Singh et al., 2025). Immunogenicity refers to the ability of a biologic to trigger an immune response, which may lead to the development of neutralizing antibodies or hypersensitivity reactions. Several factors influence immunogenicity. Product-related factors include glycosylation patterns, the presence of aggregates, or impurities in the formulation. Patient-related factors such as genetic background, underlying disease, and concomitant therapies also play a role. In addition, treatment-related aspects like dosing regimens and the route of administration can affect immunogenicity. The consequences are significant, ranging from loss of efficacy as seen with antibody neutralization of epoetin to hypersensitivity reactions, including anaphylaxis with monoclonal antibodies, and in rare cases, the development of autoimmunity (Cordeiro et al., 2024).

While biosimilars undergo extensive preapproval testing, RWD remain crucial for long-term safety evaluation. Postmarketing pharmacovigilance programs include spontaneous reporting of adverse drug reactions (ADRs), structured risk management plans as mandated by regulatory agencies like the EMA and FDA, and large-scale patient registries such as European biosimilar registries for epoetin and G-CSF. Traceability is also emphasized, with biosimilars required to be identified by brand name and batch number to ensure accurate tracking of ADRs (O'Sullivan, 2015). Postmarketing studies of filgrastim biosimilars have revealed no unexpected safety signals, reinforcing their comparability with reference biologics. Long-term follow-up in cancer patients further confirmed sustained safety in neutropenia prophylaxis, highlighting the robustness of biosimilar pharmacovigilance systems (Koenig et al., 2025).

11.5.2 Safety monitoring in bone cancer biologics

Specific biologics used in bone cancer require ongoing safety monitoring due to their unique risk profiles. Denosumab is generally considered safe but has been associated with long-term risks such as ONJ and hypocalcemia. PV programs tracking ONJ incidence in real-world practice have led to improved dental screening guidelines. Bevacizumab is associated with adverse effects, including hypertension, bleeding complications, and impaired wound healing (Park et al., 2022). Postmarketing pharmacovigilance data have also highlighted an increased risk of thromboembolic events, particularly in older patients, which necessitates careful patient selection in bone cancer. Immune checkpoint inhibitors present another safety challenge, as they are frequently linked to irAEs such as colitis, hepatitis, thyroiditis, and pneumonitis. Effective pharmacovigilance in this context requires cross-disciplinary collaboration to ensure early detection and management of irAEs (Coleman et al., 2020; Teissonnière et al., 2025).

11.5.3 Economic impact of biologics and biosimilars

Biologics constitute a major share of oncology costs, accounting for more than 30% of cancer drug spending in many countries. For example, bone cancer therapies like denosumab can cost between $20,000–$30,000 annually, creating significant financial strain and often exceeding patient affordability (Blackstone et al., 2013). Biosimilars have the potential to reduce costs by 20%–40% compared with reference biologics. In Europe, biosimilars generated savings exceeding €12 billion between 2006 and 2020. In India, biosimilars of G-CSF and bevacizumab reduced treatment costs by nearly 50%, thereby expanding access for low-income patients. Market competition among multiple biosimilar manufacturers further drives down prices. However, uptake depends significantly on physician confidence, reimbursement policies, and government procurement strategies. In the United States, adoption has been slower due to patent litigation and regulatory hurdles surrounding interchangeability (Song et al., 2023).

11.5.4 Ethical considerations

Ethical concerns are central to the discussion of biologics and biosimilars. Access and equity represent a pressing issue, as life-saving biologics often remain unaffordable for large patient populations. Biosimilars provide an opportunity to democratize access, particularly in LMICs, and denying patients effective treatment purely on cost grounds contradicts fundamental principles of justice and equity in healthcare. (Mandal et al., 2017)Informed consent is also vital, as patients must be clearly informed whether they are receiving a reference biologic or a biosimilar. This transparency fosters trust and dispels misconceptions about biosimilars being "inferior generics." Finally, physician's responsibility plays an important role. Oncologists must carefully balance cost-effectiveness with clinical benefit, and by prescribing biosimilars responsibly, they can help ensure the sustainability of healthcare systems without compromising patient outcomes (Munshi et al., 2018).

11.5.5 Policy and global perspectives

Biosimilar adoption is shaped by varying policy frameworks across regions. In the United States, adoption remains variable, although Medicare reforms and value-based care models are incentivizing hospitals to switch to biosimilars. The European Union has achieved the highest adoption rates globally, with market penetration exceeding 70% for some oncology biosimilars. Joint procurement and tendering systems have been particularly effective in driving competition and affordability. India is home to one of the largest biosimilar industries worldwide, with more than 60 biosimilars approved for oncology and supportive care. While policies emphasize affordability, pharmacovigilance infrastructure requires strengthening to keep pace with rapid expansion. In Africa and other LMICs, limited regulatory capacity remains a key barrier. However, initiatives such as the WHO prequalification program are expected to accelerate biosimilar adoption and improve access (Subramanian et al., 2024).

11.5.6 Future directions

The future of biologics and biosimilars in bone cancer is poised for significant transformation. Digital pharmacovigilance tools, such as AI-driven adverse event detection and integration with electronic health records (EHRs), will strengthen safety monitoring (Mascarenhas Melo et al., 2024). The development of next-generation biosimilars, including those for checkpoint inhibitors, denosumab, and novel monoclonal antibodies, is expected to redefine therapeutic options in bone cancer. Global collaboration among regulators may further streamline approval processes, reduce duplication, and lower development costs. Importantly, ensuring universal access to biosimilars should be considered an ethical imperative and a global health priority. Collectively, these advances mark the beginning of a new era in bone cancer treatment, one characterized by affordability, innovation, and patient-centered care (Bachu et al., 2022).

11.6 Conclusion and future perspectives

Bone cancers, including osteosarcoma, Ewing sarcoma, chondrosarcoma, and metastatic bone disease, remain among the most difficult malignancies to manage due to their heterogeneity, aggressive progression, and resistance to conventional treatments. Despite advances in multimodal therapy involving surgery, chemotherapy, and radiotherapy, patient survival has plateaued in many cases, and relapse or treatment resistance remains common. The clinical challenges posed by bone cancers underscore the urgent need for innovative therapies that can not only prolong survival but also improve quality of life for patients. Biologic therapies have greatly reshaped the treatment environment in the last 20 years. Contrary to the classic cytotoxic chemotherapy, where all active cells that divide quickly are killed, biologics are meant to target all molecular drivers of tumor growth, angiogenesis, immune evasion, and bone resorption. This specificity provides the possibility of increased efficacy and less systemic toxicity. Even in bone cancers, biologics have already played a significant role, with denosumab emerging as one of the standard treatment options used to prevent SREs in metastatic disease, and bevacizumab and other antiangiogenic therapies proving to be effective in interfering with tumor vascularization. Osteosarcoma Immunomodulators like mifamurtide have given a survival advantage in selective environments. Taken together, these developments indicate the paradigm shift between nonspecific cytotoxicity and targeted and immune-based approaches. Nevertheless, regardless of these innovations, there are several limitations. Many biologics demonstrate heterogeneous activity in bone cancers, in part because the bone tissue tumor microenvironment is highly enriched with growth factors and signaling molecules that also promote tumor survival. Resistance to biologic therapies has been reported, and the high costs of treatment pose major barriers, especially in LMICs. Furthermore, long-term safety concerns, such as immunogenicity, ONJ with denosumab, or irAEs with checkpoint inhibitors, require vigilant monitoring through robust pharmacovigilance systems. This is where biosimilars offer a promising future. As highly similar but more affordable versions of reference biologics, biosimilars have the potential to expand access to advanced therapies across the globe. In supportive care, biosimilars of G-CSF and erythropoietin have already demonstrated safety, efficacy, and significant cost savings, thereby improving treatment continuity for patients undergoing intensive chemotherapy. The next wave of therapeutic biosimilars, particularly those targeting monoclonal antibodies such as bevacizumab and future candidates like denosumab or immune checkpoint inhibitors, could dramatically reshape the economic and clinical landscape of bone cancer treatment. The adoption of biosimilars also carries an ethical imperative, ensuring that life-saving biologic therapies are not limited to high-income populations but are equitably available worldwide.

Looking ahead, several future perspectives emerge. First, the integration of real-world evidence, patient registries, and digital pharmacovigilance systems will enhance confidence in biosimilars by providing long-term safety and effectiveness data.

Second, advances in biomanufacturing technologies, including continuous processing and next-generation analytical tools, will further reduce production costs while maintaining quality. Third, new biologics (bispecific antibodies, CAR-T cell therapies, and next-generation immune checkpoint inhibitors) have the potential to overcome resistance mechanisms and increase precision in bone cancer pathways targeting. Fourth, the use of biologics, biosimilars, and conventional chemotherapy or radiotherapy can be combined to achieve synergistic effects to maximize the therapeutic effect and reduce toxicity. Policymaking-wise, harmonization of biosimilar policies worldwide and coordination of work between regulators, healthcare, and industry stakeholders will be essential. This will not only ensure faster approvals, but it will also reduce the amount of redundant clinical trials, which will reduce the total development costs. Notably, patient education and physician involvement will be needed to eliminate the misconceptions about the use of biosimilars and to make sure that they become widespread in clinical practice. To sum up, bone cancers remain an extremely difficult problem, yet the increasing array of biologic and biosimilar tools can provide new opportunities to achieve better results. The future of treatment of bone cancer is in a patient-centered approach, which would be innovative and accessible, so that the best treatments are efficient and not costly. The coming decade is bound to bring a new era of new therapeutic methods that can not only be scientifically superior but also ethically and economically sustainable for patients with bone cancers by adopting biologics, promoting a greater use of biosimilars, enhancing pharmacovigilance, and engaging in global cooperation.

References

Aden, D., Zaheer, S., Sureka, N., Trisal, M., Chaurasia, J. K., & Zaheer, S. (2025). Exploring immune checkpoint inhibitors: Focus on PD-1/PD-L1 axis and beyond. *Pathology, Research and Practice, 269*, 155864. https://doi.org/10.1016/j.prp.2025.155864.

Adewuyi, E., Chorya, H., Muili, A., Moradeyo, A., Kayode, A., Naik, A., Odedele, T., & Opabode, M. (2025). Chemotherapy, immunotherapy, and targeted therapy for osteosarcoma: Recent advancements. *Critical Reviews in Oncology/Hematology, 206*, 104575. https://doi.org/10.1016/j.critrevonc.2024.104575.

Ambrosio, L., Raucci, M. G, Vadalà, G., Ambrosio, L., Papalia, R., & Denaro, V. (2021). Innovative biomaterials for the treatment of bone cancer. *International Journal of Molecular Sciences, 22*(15), 8214. https://doi.org/10.3390/ijms22158214.

Ando, Kosei, Mori, Kanji, Corradini, Nedege, Redini, Françoise, & Heymann, Dominique (2011a). Mifamurtide for the treatment of nonmetastatic osteosarcoma. *Expert Opinion on Pharmacotherapy, 12*(2), 285–292. doi:10.1517/14656566.2011.543129.

Ando, K., Mori, K., Corradini, N., Redini, F., & Heymann, D. (2011b). Mifamurtide for the treatment of nonmetastatic osteosarcoma. *Expert Opinion on Pharmacotherapy, 12*(2), 285–292. doi:10.1517/14656566.2011.543129.

Apfelbaum, A. A., Wrenn, E. D., & Lawlor, E. R. (2022). The importance of fusion protein activity in Ewing sarcoma and the cell intrinsic and extrinsic factors that regulate it: A review. *Frontiers in Oncology, 12*, 1044707. https://doi.org/10.3389/fonc.2022.1044707.

Avhad, V. R., & Patil, S. B. (2025). A Comparison of US, Europe, Japan and India Biosimilar Regulation. *International Journal of Drug Regulatory Affairs, 13*(1), 34–39. doi:10.22270/ijdra.v13i1.745.

Bachu, R. D., Abou-Dahech, M., Balaji, S., Boddu, S. H. S., Amos, S., Singh, V., ... Tiwari, A. K. (2022). Oncology biosimilars: New developments and future directions. *Cancer Rep (Hoboken), 5*(11), e1720. doi:10.1002/cnr2.1720.

Bădilă, A. E., et al.Dragoş Mihai Rădulescu, Adelina-Gabriela Niculescu, Alexandru Mihai Grumezescu, Marius Rădulescu, and Adrian Radu Rădulescu. (2021). Recent Advances in the Treatment of Bone Metastases and Primary Bone Tumors: An Up-to-Date Review. *Cancers, 13*(16), 4229. doi:10.3390/cancers13164229.

Bahrambeigi, S., Yousefi, B., & Rahimi, M. (2019). Vahid Shafiei-Irannejad, Metformin: An old antidiabetic drug with new potentials in bone disorders. *Biomedicine & Pharmacotherapy, 109*, 1593–1601. https://doi.org/10.1016/j.biopha.2018.11.032.

Biosimilar Extrapolation of Indications: The Complete Regulatory, IP, and Investment Playbook (2026). Drug Patent Watch https://www.drugpatentwatch.com/blog/regulatory-considerations-for-biosimilar-extrapolation-of-indications/.

Blackstone, E. A., & Joseph, P. F. (2013). The economics of biosimilars. *American Health and Drug Benefits, 6*(8), 469–478.

Car, E., Barbier, L., Huys, I., Simoens, S., & Vulto, A. G. (2025). Evolving global regulatory landscape for approval of biosimilars: Current challenges and opportunities for convergence. *Expert Opinion on Biological Therapy, 25*(6), 649–668. https://doi.org/10.1080/14712598.2025.2507832.

Castro, F., Cardoso, A. P., Gonçalves, R. M., Serre, K., & Oliveira, M. J. (2018). Interferon-gamma at the crossroads of tumor immune surveillance or evasion. *Frontiers in Immunology, 9*, 847. https://doi.org/10.3389/fimmu.2018.00847.

Chhipa, A. S, Boscaro, V., Gallicchio, M., & Patel, S. (2024). The curious case of type I interferon signaling in cancer. *Biochimica et Biophysica Acta (BBA) - Reviews on Cancer, 1879*(6), 189204. https://doi.org/10.1016/j.bbcan.2024.189204.

Chintalapati, S., Sang, N., Miyahara, M., Iwata, S., Nishida, K., & Miyako, E. (2025). Living drugs: A wonderful evolution for therapeutic applications. *Cell Biomaterials, 1*, 100193. https://doi.org/10.1016/j.celbio.2025.100193.

Coleman, R. E., Croucher, P. I., Padhani, A. R., Clézardin, P., Chow, E., Fallon, M., Guise, T., Colangeli, S., Capanna, R., & Costa, L. (2020). Bone metastases. *Nature Reviews Disease Primers, 6*(1), 83. https://doi.org/10.1038/s41572-020-00216-3.

Colosia, A., Njue, A., Bajwa, Z., Dragon, E., Robinson, R. L., Sheffield, K. M., Thakkar, S., & Richiemer, S. H. (2022). The burden of metastatic cancer–induced bone pain: A narrative review. *Journal of Pain Research, 15*, 3399–3412. https://doi.org/10.2147/JPR.S371337.

Cordeiro, M. A., Vitorino, C., Sinogas, C., & Sousa, J. J. (2024). A regulatory perspective on biosimilar medicines. *Pharmaceutics, 16*(3), 523. https://doi.org/10.3390/pharmaceutics16030321.

Costa, A. G., Lewiecki, E. M., & Bilezikian, J. P. (2024). Denosumab. *Primer on the Metabolic Bone Diseases and Disorders of Mineral Metabolism*, 553–558. doi:10.1002/9781119266594.ch72.

Devi, S. (2024). Rising costs of cancer medicines. *The Lancet Oncology, 25*(10), 1262. https://doi.org/10.1016/s1470-2045(24)00516-3.

DuBois, S. G., Marina, N., & Glade-Bender, J. (2010). Angiogenesis and vascular targeting in Ewing sarcoma: A review of preclinical and clinical data. *Cancer, 116*(3), 749–757. https://doi.org/10.1002/cncr.24844.

Edgar, B. S., Cheifetz, A. S., Helfgott, S. M., Smith, G. P., Bandekar, T., Hoye-Simek, A., Liu, Y. Q., Singh, R. M., Fajardo, K. I., Carter, J., & Simone, L. (2021). Overcoming barriers to biosimilar adoption: Real-world perspectives from a national payer and provider initiative. *Journal of Managed Care and Specialty Pharmacy, 27*(8), 1129–1135. https://doi.org/10.18553/jmcp.2021.27.8.1129.

Fan, X., Krzyzanski, W., Wong, R. S. M., Liu, D., & Yan, X. (2023). Novel combination of erythropoietin and romiplostim to treat chemotherapy-induced anemia and thrombocytopenia via pharmacodynamic interaction on hematopoietic stem and progenitor cells. *ACS Pharmacology and Translational Science, 6*(12), 1884–1897. https://doi.org/10.1021/acsptsci.3c00194.

Forsyth, R., & Hogendoorn, P. C. W. (2022). Chapter 2 - Epidemiology of primary bone tumors and economical aspects of bone metastases. In *Bone Sarcomas and Bone Metastases - From Bench to Bedside (Third Edition)* (pp. 17–23). Academic Press. doi:10.1016/B978-0-12-821666-8.00038-4.

Garcia, J., Hurwitz, H. I., Sandler, A. B., Miles, D., Coleman, R. L., Deurloo, R., & Chinot, O. L. (2020). Bevacizumab (Avastin®) in cancer treatment: A review of 15 years of clinical experience and future outlook. *Cancer Treatment Reviews, 86*, 102017. https://doi.org/10.1016/j.ctrv.2020.102017.

Gilbert, A., Tudor, M., Montanari, J., Commenchail, K., Savu, D. I, Lesueur, P., & Chevalier, F. ç (2023). Chondrosarcoma resistance to radiation therapy: Origins and potential therapeutic solutions. *Cancers, 15*(7), 1962. https://doi.org/10.3390/cancers15071962.

Goli, V. A. R., & Arun, B. (2022). Biosimilar monoclonal antibodies: Challenges and approaches towards formulation. *Chemico-Biological Interactions, 366*, 110116. doi:10.1016/j.cbi.2022.110116.

Gonçalves, E., Ryan, C. J., & Adams, D. J. (2025). Synthetic lethality in cancer drug discovery: Challenges and opportunities. *Nature Reviews Drug Discovery, 25*, 22–38. https://doi.org/10.1038/s41573-025-01273-7.

Hao, X., Jiang, B., Wu, J., Xiang, D., Xiong, Z., Li, C., Li, Z., He, S., Tu, C., & Li, Z. (2024). Nanomaterials for bone metastasis. *Journal of Controlled Release, 373*, 640–651. https://doi.org/10.1016/j.jconrel.2024.07.067.

Hashimoto, K., Nishimura, S., & Goto, K. (2025). PD-1/PD-L1 immune checkpoint in bone and soft tissue tumors (Review). *Molecular and Clinical Oncology, 22*(4), 33. https://doi.org/10.3892/mco.2025.2826.

Hosseini, H., Heydari, S., Hushmandi, K., Daneshi, S., & Raesi, R. (2025a). Bone tumors: a systematic review of prevalence, risk determinants, and survival patterns. *BMC Cancer, 25*(1), 321. https://doi.org/10.1186/s12885-025-13720-0. doi:10.1186/s12885-025-13720-0.

Jang, G. (2023). *Denosumab: Dosing and drug interaction challenges on the path to approval overcoming obstacles in drug discovery and development: Surmounting the insurmountable-case studies for critical thinking* (pp. 63–74). Elsevier. https://doi.org/10.1016/B978-0-12-817134-9.00006-4.

Jiang, T., Zhou, C., & Ren, S. (2016). Role of IL-2 in cancer immunotherapy. *OncoImmunology, 5*(6), e1163462. https://doi.org/10.1080/2162402x.2016.1163462.

Kaur, S., Yadav, S., Sahu, V., Sharma, N., & Shukla, V. K. (2025). Biosimilar regulations: Current framework and future prospects. *Current Drug Safety, 13*, 15–21. https://doi.org/10.2174/0115748863360017250509063745.

Koenig, L., Juglair, L., Tao, T. P., Fischer, S., Clausen, I., Imhof-Jung, S., Janssen, N., Mader, R., Marbach, D., Niewoehner, J., Winter, A., & Schubert, D. (2025). A microfluidic bone marrow chip for the safety profiling of biologics in pre-clinical drug development. *Communications Biology, 8*(1), 754. https://doi.org/10.1038/s42003-025-08137-1.

Koonrungsesomboon, N., Teeyakasem, P., Dukaew, N., Sirikaew, N., Thongkumkoon, P., Yongpitakwattana, P., Thepbundit, V., Chaiyawat, P., Klangjorhor, J., Sakuludomkan, C., Charoentum, C., Chewaskuyong, B., Sathitsamitphong, L., Kiatisevi, P., Sunpawer-avong, P., Sathitruangsak, C., & Pruksakorn, D. (2025). Efficacy and safety of mycophenolate mofetil in patients with high-grade locally advanced or metastatic osteosarcoma (ES-MMO): A multicenter, phase II clinical trial. *International Journal of Clinical Oncology*. https://doi.org/10.1007/s10147-025-02877-0.

Kruger, T. E., Miller, A. H., Godwin, A. K., & Wang, J. (2014). Bone sialoprotein and osteopontin in bone metastasis of osteotropic cancers. *Critical Reviews in Oncology/Hematology, 89*(2), 330–341. https://doi.org/10.1016/j.critrevonc.2013.08.013.

Kumar, R., Goel, H., Solanki, R., Rawat, L., Tabasum, S., Tanwar, T., Pal, S., & Sabarwal, A. (2024). Recent developments in receptor tyrosine kinase inhibitors: A promising mainstay in targeted cancer therapy. *Medicine in Drug Discovery, 23*, 100195. https://doi.org/10.1016/j.medidd.2024.100195.

Lee, J. A., Lim, J., Jin, H. Y., Park, M., Park, H. J., Park, J. W., ... Won, Y. J. (2021). Osteosarcoma in adolescents and young adults. *Cells, 10*(10), 2684. https://doi.org/10.3390/cells10102684.

Li, E., & Lobaina, E. (2017). Application of the FDA Biosimilar Extrapolation Framework to Make Off-Label Determinations. *Journal of Managed Care & Specialty Pharmacy, 23*(12), 1227–1232. doi:10.18553/jmcp.2017.23.12.1227.

Lu, J., Hu, D., Zhang, Y., Ma, C., Shen, L., & Shuai, B. o (2023). Current comprehensive understanding of denosumab (the RANKL neutralizing antibody) in the treatment of bone metastasis of malignant tumors, including pharmacological mechanism and clinical trials. *Frontiers in Oncology, 13*, 690855. https://doi.org/10.3389/fonc.2023.1133828.

Mandal, J., Ponnambath, D. K., & Parija, S. C. (2017). Bioethics: A brief review. *Tropical Parasitology, 7*(1), 5–7. https://doi.org/10.4103/tp.TP_4_17.

Mascarenhas-Melo, F., Diaz, M., Gonçalves, M. B., Vieira, P., Bell, V., Viana, S., ... Veiga, F. (2024). An Overview of Biosimilars-Development, Quality, Regulatory Issues, and Management in Healthcare. *Pharmaceuticals (Basel), 17*(2), 235. doi:10.3390/ph17020235.

Miao, H., Fang, Y., Pan, C., Yang, H., Wang, Z., Qi, Y. e, Wu, Y., Zhang, Y., Liu, F., Huang, H., Tang, Y. u, Wu, D., & Li, N. (2025). Transforming the landscape of cancer treatment with seven promising novel therapies: Evolution and future perspectives. *Medicine Plus, 2*(2), 100087. https://doi.org/10.1016/j.medp.2025.100087.

Monga, A., Gagan, Jamwal, P., Sharma, S., & Kaur, A. (2025). Biosimilars: A Critical Review of Development, Regulatory Landscape, and Clinical Implications. *AAPS PharmSciTech, 26*(1), 46. doi:10.1208/s12249-025-03038-2.

Munshi, A., & Sharma, V. (2018). Chapter 25 - Safety and Ethics in Biotechnology and Bioengineering: What to Follow and What Not to. In *Omics Technologies and Bio-EngineeringTowards Improving Quality of Life* (pp. 577–590). Academic Press. doi:10.1016/B978-0-12-804659-3.00025-7.

Nahleh, Z., Lyman, G. H., Schilsky, R. L., Peterson, D. E., Tagawa, S. T., Chavez-MacGregor, M., Rumble, R. B., & Gupta, S. (2022). Use of biosimilar medications in oncology. *JCO Oncology Practice, 18*(3), 177–186. https://doi.org/10.1200/OP.21.00771.

Nishida, A., & Andoh, A. (2025). The Role of Inflammation in Cancer: Mechanisms of Tumor Initiation, Progression, and Metastasis. *Cells, 14*(7), 488. doi:10.3390/cells14070488.

Ono, T., Hayashi, M., Sasaki, F., & Nakashima, T. (2020). RANKL biology: Bone metabolism, the immune system, and beyond. *Inflammation and Regeneration, 40*(1), 2. https://doi.org/10.1186/s41232-019-0111-3.

O'Sullivan, G., Carty, F., & Cronin, C. (2015). Imaging of bone metastasis: An update. *World Journal of Radiology, 7*(8), 202–211. doi:10.4329/wjr.v7.i8.202.

Park, J. J., & Wong, C. (2022). Pharmacological Prevention and Management of Skeletal-Related Events and Bone Loss in Individuals with Cancer. *Seminars in Oncology Nursing, 38*(2), 151276. doi:10.1016/j.soncn.2022.151276.

Rahman, T., Das, A., Abir, M. H., Nafiz, I. H., Mahmud, A. R., Sarker, M. R., Emran, T. B., & Hassan, M. M. (2023). Cytokines and their role as immunotherapeutics and vaccine adjuvants: The emerging concepts. *Cytokine, 169*, 156212. https://doi.org/10.1016/j.cyto.2023.156268.

Rajan, A., Kim, C., Heery, C. R., Guha, U., & Gulley, J. L. (2016). Nivolumab, anti-programmed death-1 (PD-1) monoclonal antibody immunotherapy: Role in advanced cancers. *Human Vaccines and Immunotherapeutics, 12*(9), 2219–2231. https://doi.org/10.1080/21645515.2016.1175694.

Rathod, J. S., Jumde, K. S., Hatwar, P. R., Bakal, R. L., & Mangle, A. P. (2025). Targeting bone cancer: Emerging therapies and treatment strategies. *GSC Biological and Pharmaceutical Sciences, 32*(3), 049–058. https://doi.org/10.30574/gscbps.2025.32.3.0308.

Salmon, V., Gondim Teixeira, P. A., & Blum, A. (2025). Bone lesions of the tibia: Multimodal iconographic review and diagnostic algorithms, Part 2: Metaphyseal and epiphyseal lesions. *European Journal of Radiology Open, 14*, 100589. https://doi.org/10.1016/j.ejro.2025.100654.

Sangwan, N., Gangwal, A., Jain, P., Langtso, C., Srivastava, S., Dhawan, U., Baweja, R., & Singh, Y. (2025). Anthrax: Transmission, pathogenesis, prevention and treatment. *Toxins, 17*(2), 56. https://doi.org/10.3390/toxins17020056.

Senapati, S., Mahanta, A. K., Kumar, S., & Maiti, P. (2018). Controlled drug delivery vehicles for cancer treatment and their performance. *Signal Transduction and Targeted Therapy, 3*(1), 7. https://doi.org/10.1038/s41392-017-0004-3.

Singh, A. A., Jadhav, M. M., Singh, T. S., Mishra, S. R., Sange, V. B., & Yadav, K. R. (2025). Pharmacovigilance: Emerging trends, ongoing challenges, AI and future outlook in pharmacovigilance. *World Journal of Biology, Pharmacy and Health Sciences, 23*(3), 060–073. https://doi.org/10.30574/wjbphs.2025.23.3.0785.

Song, M. K., Park, S. I., & Cho, S. W. (2023). Circulating biomarkers for diagnosis and therapeutic monitoring in bone metastasis. *Journal of Bone and Mineral Metabolism, 41*(3), 337–344. https://doi.org/10.1007/s00774-022-01396-6.

Suaste-Vera, H. U., & Villalobos-Molina, R. (2025). Ewing Sarcoma: A Comprehensive Review of Its Classification, Diagnosis, and Emerging Treatments. *Journal of Biosciences and Medicines, 13*(8), 158–187. doi:10.4236/jbm.2025.138014.

Subramanian, A., Muthukumarasamy, G. D, Venkataraman, S., Rangasamy, K., Kanade, A. N, Arulrajan, P. V, Saravanan, K., Shanmuganathan, S., Gopal, G., & Ambi, S. V (2024). Future scope of biosimilars for the treatment of various cancers. In *Springer Nature* (pp. 287–309). https://doi.org/10.1007/978-981-97-3001-8_16.

Taylor, C., & Henry, D. (2014). Pharmacoeconomics of cancer therapies: considerations with the introduction of biosimilars. *Seminars in Oncology, 41*(3), S13–S20. doi:10.1053/j.seminoncol.2014.03.009.

Teissonnière, M., Point, M., Biver, E., Hadji, P., Bonnelye, E., Ebeling, P. R., Kendler, D., de Villiers, T., Holzer, G., Body, J. J., Fuleihan, G. E. H., Brandi, M. L., Rizzoli, R., & Confavreux, C. B. (2025). Bone effects of anti-cancer treatments in 2024. *Calcified Tissue International, 116*(1), 14. https://doi.org/10.1007/s00223-025-01362-0.

Thomson, R.J., Moshirfar, M., Ronquillo, Y. 2023. Tyrosine Kinase Inhibitors. In: StatPearls [Internet]. Treasure Island (FL): StatPearls Publishing; 2026. https://pubmed.ncbi.nlm.nih.gov/33090752/

Tian, H., Cao, J., Li, B., Nice, E. C., Mao, H., Zhang, Y., & Huang, C. (2023). Managing the immune microenvironment of osteosarcoma: The outlook for osteosarcoma treatment. *Bone Research, 11*(1), 32. https://doi.org/10.1038/s41413-023-00246-z.

Ventura, C., Vanesa Núñez, S., Gonçalves, A., Abreu, C., & Costa, L. (2022). Bone health in metastatic cancer. *Seminars in Oncology Nursing, 38*(2), 151278. https://doi.org/10.1016/j.soncn.2022.151278.

Wang, L., Liu, W.-Q., Broussy, S., Han, B., & Fang, H. (2024). Recent advances of anti-angiogenic inhibitors targeting VEGF/VEGFR axis. *Frontiers in Pharmacology, 14*, 1307860. doi:10.3389/fphar.2023.1307860.

Wang, M., Xia, F., & Wei, X. (2020). Molecular mechanisms and clinical management of cancer bone metastasis. *Bone Res, 8*(1), 30. doi:10.1038/s41413-020-00105-1.

Wojtukiewicz, M. Z., Rek, M. M., Karpowicz, K., Górska, M., Polityńska, B., Wojtukiewicz, A. M., Moniuszko, M., Radziwon, P., Tucker, S. C., & Honn, K. V. (2021). Inhibitors of immune checkpoints—PD-1, PD-L1, CTLA-4—new opportunities for cancer patients and a new challenge for internists and general practitioners. *Cancer and Metastasis Reviews, 40*(3), 949–982. https://doi.org/10.1007/s10555-021-09976-0.

Xiao, S., Han, C., Ma, L., Geng, W., Gao, Y., Adeli, M., Xu, L., Ye, L., Qiu, L. i, & Cheng, C. (2025). Enzyme-reaction-system inspired nanobiocatalysts for bioadaptively eradicating primary and metastatic cancers. *Cell Biomaterials, 1*, 100194. https://doi.org/10.1016/j.celbio.2025.100194

CHAPTER

12 Biologics and biosimilars in endometrial cancer

Mamtha Helchel[a], Menaka Reddy Mukkala[b], Vidyasrilekha Sanapalli[c], Bharat Kumar Reddy Sanapalli[d], Prabakaran Annadurai[e], Dilep Kumar Sigalapalli[b] and Ramya Tokala[f]

[a] *Department of Pharmacology, RBVRR Women's College of Pharmacy, Hyderabad, Telangana, India,* [b] *Department of Pharmaceutical Chemistry, Vignan Pharmacy College, Jawaharlal Nehru Technological University, Guntur, Andhra Pradesh, India,* [c] *Department of Pharmaceutical Chemistry, School of Pharmacy and Technology Management, SVKM's Narsee Monjee Institute of Management Studies (NMIMS) (Deemed to be University), Hyderabad, Telangana, India,* [d] *Department of Pharmacology, School of Pharmacy and Technology Management, SVKM's Narsee Monjee Institute of Management Studies (NMIMS) (Deemed to be University), Hyderabad, Telangana, India,* [e] *Department of Pharmaceutics, School of Pharmacy and Technology Management, SVKM's Narsee Monjee Institute of Management Studies (NMIMS) (Deemed to be University), Hyderabad, Telangana, India,* [f] *Athinoula A. Martinos Center for Biomedical Imaging, Department of Radiology, Massachusetts General Hospital, Harvard Medical School, Charlestown, MA, United States*

12.1 Introduction

Endometrial cancer (EC) is a gynecologic malignancy, rising as one of the prevalent cancers in women in terms of incidence and mortality. EC is a tumor originating within the mucous membrane lining the inner side of the uterus (known as endometrium) (Brooks et al., 2019). The incidence is higher in postmenopausal women, most commonly diagnosed at the age of 60–75 years (Mahdy et al., 2025). The endometrium is essential for implantation (embedding of the fertilized egg into the lining of the uterus) and for the maintenance of the pregnancy. In the absence of implantation, hormonal withdrawal triggers endometrial shedding, resulting in menstruation (Thiyagarajan et al., 2025). Estrogen and progesterone are the key hormones that regulate the process of endometrial growth and shedding, respectively. EC is primarily a hormone-driven malignancy, and estrogen signaling via its receptors functions as an oncogenic stimulus (Rodriguez et al., 2019). It usually develops from a precursor lesion, followed by a series of molecular and histological changes leading to malignant transformation. Several risk factors that contribute to the development of EC include overweight, obesity (EC risk factors & biomarkers for natural history & early detection, 2019) endogenous or exogenous unopposed estrogen exposure, (Bassette & Ducie, 2024)

Biosimilars and Biologics in Cancer Treatment: An Era in Therapeutic Strategies.
DOI: https://doi.org/10.1016/B978-0-443-43824-0.00014-6

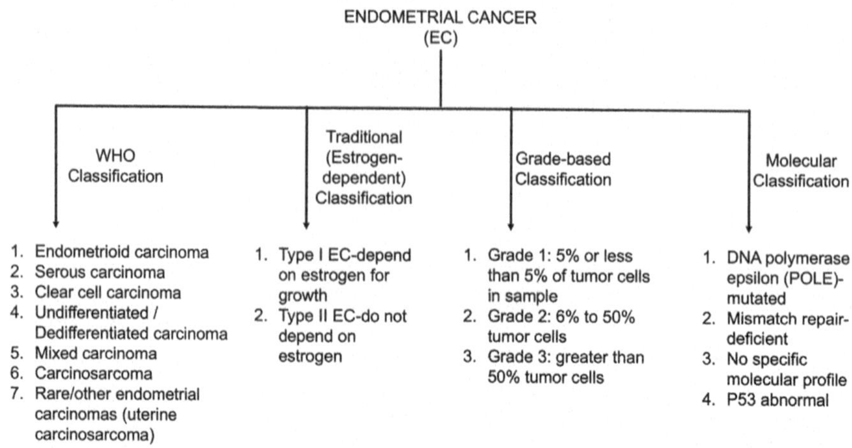

FIGURE 12.1

Classification of endometrial cancers.

genetic predisposition, high blood pressure and other factors like hormone therapy and polycystic ovarian syndrome lower parity, type II diabetes and insulin resistance (Makker et al., 2021). Additional factors, such as menstrual disruption, amenorrhea, and infertility, family history of endometrial, colorectal, ovarian, and other which are characterized through Lynch syndrome and related MMRd (Mutation in one of the *MMR* genes) hereditary diseases, increase the risk of EC (Dörk et al., 2020). The somatic PTEN mutations are common in EC, whereas germline *PTEN* mutations are associated with Cowden syndrome (characterized by increased risk of breast cancer, thyroid cancer, and EC, along with other diseases) (Tan et al., 2012). The most common symptoms of EC include, but are not limited to, abnormal bleeding, pain during urination, pelvic pain, and pain during intercourse (Makker et al., 2021). Clinical assessment involves histopathologic evaluation, imaging, and laboratory tests, which aid in assessment of the preoperative staging and guiding disease management strategies (Török et al., 2020).

12.2 Classification

EC is classified according to various criteria, as shown in Fig. 12.1.

12.2.1 WHO classification

According to the WHO, endometrial carcinomas are classified based on their *histologic types* (i.e., how the cells appear under the microscope). The major histologic subtypes include:

1. Endometrioid carcinoma
2. Serous carcinoma
3. Clear cell carcinoma
4. Undifferentiated carcinoma/Dedifferentiated carcinoma
5. Mixed carcinoma
6. Carcinosarcoma
7. Rare/other endometrial carcinomas

Uterine carcinosarcomas are also categorized as advanced EC as they originate in the endometrium and exhibit features of both endometrial carcinoma and sarcoma.

12.2.2 Traditional classification

EC were traditionally categorized into two types based on their *dependency on estrogen for tumor growth.*

1. *Type 1 ECs:* These are estrogen-dependent, and less aggressive.
2. *Type 2 ECs:* These are nonestrogen dependent, aggressive, and often serous carcinomas.

12.2.3 Grade-based classification

This classification is based on the *degree of tumor differentiation*, determined by the percentage of solid (nondifferentiated) tumor components present in the tissue sample.

1. *Grade 1:* ≤5% of tumor cells (well-differentiated)
2. *Grade 2:* 6%–50% tumor cells (moderately differentiated)
3. *Grade 3:* >50% tumor cells (poorly differentiated and more aggressive)

12.2.4 Molecular classification

These are categorized into subtypes based on the *mutations* observed in cancer cells.

1. DNA polymerase epsilon (POLE)-mutated
2. Mismatch repair-deficient (MMRd)
3. No specific molecular profile
4. P53 abnormal

Based on recent advancements, EC is further classified into four molecular subgroups based on genetic profiles established by The Cancer Genome Atlas (TCGA). The ProMise (proactive molecular risk classifier for EC) proposed by Talhouk et al., is a simple, genomics-based clinical classifier that allows overcoming both clinical and economic difficulties associated with whole genome sequencing (Talhouk et al., 2017; Talhouk & McAlpine, 2016). The four TCGA subgroups are (Loukovaara et al., 2022):

Table 12.1 Staging details of endometrial cancer by FIGO, 2023.

Stage	Details
I	Confined to the ovary and uterine corpus
II	Invasion of cervical stroma without extrauterine extension, OR substantial lymphovascular space involvement, OR aggressive histological types with myometrial invasion
III	Local and/or regional spread of the tumor of any histological subtype. Further divided into substages based on the organ involved
IV	Spread to the bladder mucosa and/or intestinal mucosa and/or distant metastasis

1. DNA polymerase ε (POLE, ultramutated)
2. Microsatellite instability (MSI, hypermutated)
3. Copy number low
4. Copy number high

According to the latest National Comprehensive Cancer Network (NCCN) guidelines, all tumors should be tested for MMR or MSI status to guide therapy decisions (Corr et al., 2022).

12.2.5 EC staging

The International Federation of Gynecology and Obstetrics (FIGO) updated the staging system of EC in 2023 to improve diagnosis and prognosis. The brief 2023 FIGO staging is summarized in, while Berek et al. (2023) provided the comprehensive details on staging and sub-staging. Table 12.1.

12.3 Treatment

The treatment landscape for EC has undergone remarkable changes in recent years, particularly with the integration of biologics and the emergence of molecular classification-driven therapeutic approaches. The available treatment modalities for EC include:

1. Surgery
2. Radiation therapy
3. Chemotherapy
4. Hormone therapy
5. Targeted therapy
6. Immunotherapy

Amongst these, surgery, radiation, and chemotherapy are renowned treatment options in most types of tumors. Hormone therapy, which uses progestins, aromatase inhibitors, antiestrogen drugs, and CDK 4/6 inhibitors, functions by regulating the hormone levels involved in the endometrial carcinogenesis. These agents are fundamentally synthetic molecules (Tuninetti et al., 2024). Targeted and immunotherapy constitute biologics and their biosimilars, forming the main focus of this book chapter. Biologic agents currently approved or in the pipeline for the treatment of EC are discussed in detail. The list of biosimilars is provided as their mechanism of action is comparable to that of their referenced biologics, and their clinical trial information remains limited.

12.3.1 Biologicals

Biological medicines comprise substances produced using living cells or organisms. They are generally used to treat life-threatening diseases such as autoimmune disorders and cancers. Representative examples of biologics include monoclonal antibodies (mAb), antibody-drug conjugates, vaccines, stem cell therapies, transplant tissues, and gene/stem cell therapies. Targeted therapies, like hormonal therapy and immune checkpoint inhibitors (ICIs), are being developed owing to their prime therapeutic efficacy. Biological medicines differ from traditional drugs in terms of structural characteristics, production process, material sources, and compliance standards. Based on the differences in the development of biologics from the conventional drug discovery process, approval of the biological medicines is often costly and time-consuming.

Biologicals used in EC are categorized in this chapter as follows:

1. Monoclonal antibodies
2. Antibody-drug conjugates
3. Vaccines
4. Recombinant proteins

12.3.2 Monoclonal antibodies (mAbs)

mAb are immunoglobulin molecules produced in the laboratory from a single cell lineage (cloned cells). They are more targeted than other forms of treatment. They possess identical (homogenous) structure, and as the name suggests, they exhibit monovalent binding affinity to a specific epitope (part of an antigen). The common route of administration for mAbs is intravenous infusion. The main advantages of using mAbs are their precision over other treatments and their improved therapeutic effectiveness. The mechanisms by which mAbs exhibit their action are as follows:

- *Blocking antigen binding:* mAbs bind to specific antigens on the surface of the cells, thereby blocking the cellular processes, such as cell growth, division, and signaling pathways.
- *Antibody-dependent cellular cytotoxicity (ADCC):* mAbs bind to target cells and recruit immune effector cells such as natural killer (NK) cells, macrophages, and monocytes to destroy the target cells.

- *Antibody-dependent cell-mediated phagocytosis (ADCP)*: A monoclonal antibody has two important regions, the Fab region and the Fc region. The Fab binds to the antigen on the surface of the target cell, while the Fc binds to the respective receptor on the surface of phagocytic cells, such as macrophages, thereby causing phagocytosis (Aboul-Ella et al., 2024).
- *Complement-dependent cytotoxicity (CDC):* Certain mAbs activate the complement system, a series of proteins that can cause lysis of target cells.

mAbs that are approved or under development for EC, along with their mechanistic insights, and clinical trial status, are discussed in detail below.

12.3.2.1 Avelumab

Avelumab (Bavencio, MSB0010718C, Merck) is a fully human monoclonal antibody currently under clinical evaluation in multiple cancers. It is an anti-PD-L1 IgG1 antibody that blocks programmed cell death protein-ligand 1 (PD-L1) protein on cancer cells, allowing the immune system to recognize and attack them. Avelumab binds to PD-L1 and prevents its interaction with T cells-PD-1 receptors. By blocking the inhibitory PD-1/PD-L1 signaling pathway, the body's immune response is restored, and cytotoxic T-cell activity against tumor cells is reactivated. Avelumab retains an active Fc region, which allows it to interact with NK cells and promote direct tumor cell destruction through ADCC, in contrast to other PD-L1 inhibitors such as pembrolizumab and nivolumab (Boyerinas et al., 2015). In a phase II clinical study by Konstantinopoulos et al. (2025), Avelumab was evaluated in patients with mismatch repair deficient (MMRD) and mismatch repair proficient (MMRP) recurrent/persistent EC, with the cohorts defined by immunohistochemical (IHC) expression. Avelumab displayed promising potential in MMRD EC irrespective of PD-L1 status. However, it did not exhibit significant activity in MMRP/non-POLE-mutated ECs, indicating the need for alternative approaches such as combination trials. The study also suggests the routine use of IHC for establishing the MMRD status in EC (Konstantinopoulos et al., 2019). Therefore, Avelumab has been assessed in a separate phase II clinical trial for its effectiveness in patients with MSS (microsatellite stable), MSI-H (microsatellite instability-high), and POLE (polymerase epsilon)-mutated recurrent or persistent EC. It is also being investigated in combination with other medications such as talazoparib (a poly ADP ribose polymerase inhibitor), axitinib (a vascular endothelial growth factor inhibitor), in patients with MSS recurrent or persistent EC (NCT02912572). Fatigue, nausea, diarrhea, decreased appetite, vomiting, and weight loss are the unwanted effects of avelumab in EC patients. Anemia, lymphopenia, and thrombocytopenia are some of the hematologic disorders that are frequently observed. Chills, fever, flushing, hypotension, coughing, or dyspnea are observed with infusion-related symptoms. Further, inflammatory processes overseen by immunity, like pneumonitis, hepatitis, nephritis, colitis, and endocrinopathies (thyroid dysfunction and adrenal insufficiency), have been reported. The other adverse effects include skin rash, pruritus, dry skin, joint or muscle pain, hypertension, and peripheral edema (Disis et al., 2019).

12.3.2.2 Bevacizumab

Bevacizumab (Avastin, Genentech, Roche) is a humanized, monoclonal antibody used off-label for the treatment of advanced or recurrent cases of EC. The mechanism of action involves targeting vascular endothelial growth factor A (VEGF-A), thereby inhibiting angiogenesis and tumor vascularization. Aghajanian et al. (2011) conducted a phase II clinical trial to assess the activity and tolerability of bevacizumab in recurrent or persistent EC. Bevacizumab was injected into patients with persistent disease who had received at least one chemotherapy regimen and had a performance status of ≤ 2. The primary endpoints of the study were overall response rate (ORR) and progression free survival (PFS). Bevacizumab was well tolerated in recurrent or persistent EC after 6 months of treatment (Aghajanian et al., 2011). Rubinstein et al. (2021) reported a retrospective study using bevacizumab involving 101 patients with endometrial cancer. The efficacy of the treatment was evaluated with the endpoints of overall survival (OS), ORR, clinical benefit rate (CBR), progress free survival (PFS), and duration of response (DOS). None of the patients exhibited a complete or partial response, although stable disease was observed in 19 patients. Bevacizumab exhibited a moderate clinical efficacy in heavily pretreated patients with advanced EC. Hypertension was identified as the most common treatment-related adverse effect (Rubinstein et al., 2021). Currently, a phase II clinical trial using the combination of atezolizumab and bevacizumab in patients with recurrent or persistent EC is in progress. This single-arm study injects the drug combination intravenously, with the expected outcomes of anticancer effect and manageable toxicities (NCT03526432).

12.3.2.3 Dostarlimab

Dostarlimab (Jemperli, GlaxoSmithKline) is approved by the FDA under the accelerated approval program for the treatment of adults with EC that has advanced outside the uterus or returned cancer (National Cancer Institute NCI, 2024). Dostarlimab is a humanized IgG4 monoclonal antibody and an immune checkpoint inhibitor targeting the programmed death-1 (PD-1) receptor. It was developed by GSK and represents a major advancement in the immunotherapy of endometrial and other mismatch repair-deficient (dMMR) or microsatellite instability high (MSI-H) solid tumors. Dostarlimab acts by selectively binding to PD-1 receptors on activated T lymphocytes, preventing their interaction with PD-L1 and PD-L2 ligands on tumor and antigen-presenting cells. This blockade reverses T-cell inhibition, reactivating immune responses that lead to tumor cell destruction, resulting in durable clinical responses and improved survival (Oaknin et al., 2023). Structurally, dostarlimab includes a stabilizing S228P mutation in the IgG4 hinge region, ensuring structural integrity and preventing Fab-arm exchange. This mechanism contributes to sustained PD-1 blockade and consistent immune activation, particularly in tumors with dMMR or MSI-H genotypes that harbor high neoantigen loads and increased immunogenicity (Costa & Vale, 2022). The phase I GARNET trial (NCT02715284) established dostarlimab's clinical efficacy and tolerability in patients with advanced or recurrent EC. Patients with dMMR/MSI-H disease achieved objective responses exceeding

40%, with durable clinical benefits observed (Oaknin et al., 2023). The pivotal phase III RUBY trial (ENGOT-EN6/GOG-3031) by Mirza et al. (2023) further confirmed dostarlimab's benefit as first-line therapy when combined with carboplatin and paclitaxel, showing a 24-month progression-free survival (PFS) hazard ratio of 0.28 in dMMR/MSI-H patients and an OS rate of 71.3% compared with 56% in the control group (Mirza et al., 2023). The recommended route and dosing for Dostarlimab is intravenously at 500 mg every 3 weeks for four doses, followed by 1000 mg every 6 weeks until disease progression or unacceptable toxicity. Dostarlimab displays linear pharmacokinetics, with a mean terminal half-life of approximately 25 days and a clearance rate of 0.007 L/h. Drug exposure is dose-proportional and unaffected by demographic factors such as age or body weight. As of 2025, dostarlimab is approved for monotherapy in dMMR/MSI-H recurrent or advanced EC and for dMMR advanced solid tumors progressing after prior therapy. It is also approved in combination with carboplatin and paclitaxel as first-line therapy for advanced or recurrent ECF regardless of mismatch repair status (Oaknin et al., 2023). It is used in individuals with dMMR cancer and got worse during/after platinum-based chemotherapy. These approvals underscore its versatility as a PD-1 inhibitor and its expanding role in precision immunotherapy (Oaknin et al., 2023). Dostarlimab demonstrates a safety profile consistent with PD-1 blockade. The strongest benefit from dostarlimab is observed in dMMR/MSI-H tumors due to their high mutational burden and strong immune infiltration, which create an ideal environment for checkpoint inhibition. In contrast, tumors with proficient mismatch repair (pMMR/MSS) exhibit lower response rates, prompting investigation of combination regimens with PARP inhibitors, antiangiogenic agents, and chemotherapy to extend benefit beyond dMMR populations (Costa & Vale, 2022). The most common treatment-related adverse events (TRAEs) include fatigue, nausea, pruritus, diarrhea, and rash, which are generally grade 1–2 and manageable. Immune-related adverse events such as thyroid dysfunction, colitis, hepatitis, and pneumonitis occur in about one quarter of patients and are typically reversible with corticosteroid therapy and temporary drug interruption (Mirza et al., 2023).

12.3.2.4 Durvalumab

Durvalumab (Imfinzi, MEDI4736, AstraZeneca) is the first FDA-approved biologic for the treatment of adults with dMMR primary advanced or recurrent EC. The approved regimen is durvalumab with carboplatin and paclitaxel, followed by durvalumab monotherapy. Durvalumab is a human IgG1 monoclonal antibody that targets the PDL-1 (CD274) protein. PD-L1 is a transmembrane protein expressed on dendritic cells (DCs), macrophages, and certain cancer cells, while PD-1 is expressed on the activated T cells. The PD-1/PD-L1 interaction inhibits T-cell functions in the tumor microenvironment, thus enabling the cancer cells to evade immunity. Durvalumab inhibits this interaction and restores T-cell-mediated antitumor immunity and enhances immune recognition of cancer cells. The PHAEDRA trial (NCT03015129), a Phase II study initiated in 2017, was the first clinical trial to evaluate durvalumab in the treatment of EC. The study examined the effects of durvalumab administered individually and in combination with tremelimumab in persistent or recurrent endometrial carcinoma and endometrial carcinosarcoma patients. In a clinical trial report by Antill et al. (2021), the activity of durvalumab was assessed in two cohorts of mismatch repair proficient (pMMR) and mismatch repair deficient (dMMR) women with advanced endometrial cancers (AEC). The study demonstrated acceptable safety and promising activity of Durvalumab monotherapy in AEC with dMMR, independent of previous chemotherapy, whereas in the pMMR cohort, the activity remained limited (Antill et al., 2021). Larger follow-up studies like the DUO-E (NCT04269200) Phase III study, which supported durvalumab's FDA approval in 2024, were prompted by an early trial that showed encouraging outcomes for durvalumab monotherapy in advanced dMMR EC. Westin et al. (2024) reported that in patients with advanced or recurrent EC, a regimen of durvalumab and PARP inhibitor (carboplatin/paclitaxel) followed by durvalumab individually or in combination with olaparib for maintenance significantly prolonged PFS (Westin et al., 2024). The recommended dosing of durvalumab is based on the body weight, (a) For adults ≥30 Kg: 1120 mg of durvalumab with carboplatin/paclitaxel every 3 weeks for six cycles, continued by durvalumab monotherapy 1500 mg every 4 weeks. (b) For adults <30 Kg: 15 mg/kg of durvalumab with carboplatin and paclitaxel every 3 weeks for six cycles, continued by durvalumab monotherapy 20 mg/kg every 4 weeks (U.S. Food and Drug Administration 2024a). The most frequently reported adverse effects include fatigue, nausea, anemia, diarrhea, loss of appetite, vomiting, altered taste, renal complications, and flu-like symptoms, with anemia being the most common (U.S. Food and Drug Administration 2024a).

12.3.2.5 Nivolumab

Nivolumab (Opdivo, Bristol Myers Squibb) is a fully humanized monoclonal antibody targeting the PD-1 receptor. There is an increased interest in investigating nivolumab for the treatment of EC. Friedman et al. reported a single-arm, phase II clinical trial using nivolumab for mismatch-repair-deficient (dMMR) or hypermutated gynecologic cancers, including EC. This study investigated the PFS at 24 weeks (PFS24) and objective response rate (ORR), OS, duration of response (DOR), safety, and disease control rate (DCR) in patients with EC receiving nivolumab every 2 or 4 weeks till the unacceptable toxicity or progression of the disease. The biomarker analyses revealed the presence of dysfunctional T cells, and their involvement with PD-L1 cells was linked to longer PFS24, particularly in patients with MEGF8 or SETD1B mutations. Further, the trial demonstrated that both genetic and tumor microenvironment factors may influence response to PD-1 blockade in dMMR cancers (Friedman et al., 2024). Lee et al. (2023) reported a multicenter, single-arm, phase II study (NIVEC) to evaluate nivolumab in patients with surgically resectable dMMR EC. Remarkably, Nivolumab exhibited complete resolution of cancer as evidenced by imaging and biopsy in two patients with grade two and three tumors after six treatment cycles. Further, Lheureux et al. (2022) reported results from a phase II trial using the combination of nivolumab with cabozantinib in patients with advanced, recurrent, or metastatic EC.

The combination therapy resulted in improved outcomes in comparison to nivolumab alone, highlighting a potential synergistic advantage (Lheureux et al., 2022). Additionally, another clinical trial (NCT05112601) is currently recruiting subjects to assess nivolumab's efficacy in dMMR recurrent EC individually and in combination with ipilimumab (National Cancer Institute NCI, 2025a).

12.3.2.6 Pembrolizumab

Pembrolizumab (Keytruda, Merck) in combination with carboplatin and paclitaxel has been approved by the FDA for the treatment of adult patients with primary advanced or recurrent endometrial cancer (U.S. Food and Drug Administration 2024b). It is a fully humanized IgG4 monoclonal antibody directed against the programmed cell death-1 (PD-1) receptor. Pembrolizumab restores antitumor T-cell function by blocking PD-1 interaction with its ligands, PD-L1 and PD-L2, thereby reversing immune tolerance within the tumor microenvironment (Du Rusquec et al., 2019). The PD-1 receptor, expressed on activated T cells, acts as a checkpoint on immune activation when engaged by PD-L1 or PD-L2 on tumor or antigen-presenting cells. This interaction transmits inhibitory signals that suppress cytotoxic T-cell activity and cytokine production, allowing tumor cells to escape immune surveillance. Pembrolizumab binds PD-1 with high affinity, preventing its ligation and re-enabling T-cell effector function and tumor cell killing (Du Rusquec et al., 2019). Pembrolizumab was approved for two major indications in EC in the US. Its clinical development was led through the extensive KEYNOTE clinical trial program, encompassing more than 160 studies across tumor types. Merck Provides Update on Phase 3 KEYNOTE-B21 trial evaluating KEYTRUDA (pembrolizumab) plus chemotherapy, with or without radiotherapy, for patients with newly diagnosed, high-risk EC after surgery with curative intent. Makker et al. (2022) reported that pembrolizumab, in combination with Lenvatinib, resulted in substantially longer disease PFS in advanced EC patients. Further, the OS was also increased with this combination as described in the study 309/KEYNOTE-775 (NCT03517449) (Makker et al., 2022). O'Malley et al. (2022) reported the efficacy of pembrolizumab in microsatellite instability-high (MSI-H) advanced EC patients. In the multicohort, nonrandomized, open-label, phase II KEYNOTE-158 study (NCT02628067), pembrolizumab displayed a significant antitumor potential and a manageable toxicity profile along with a promising survival outcome (O'Malley et al., 2022). In addition, Van Gorp et al. (2024) reported a randomized, double-blind, phase III study (ENGOT-en11/GOG-3053/KEYNOTE-B21) for safety and efficacy of pembrolizumab in newly diagnosed, high-risk EC patients. In this study, pembrolizumab was studied along with adjuvant chemotherapy with or without radiotherapy. Although the safety profile was acceptable, no improvement was observed in disease-free progression with chemotherapy in combination with pembrolizumab (Van Gorp et al., 2024).

Pembrolizumab is administered as an intravenous infusion at a dose of 200 mg every 3 weeks or 400 mg every 6 weeks, continued until disease progression or unacceptable toxicity or up to 24 months (Research, 2024b). It shows linear pharmacokinetics with complete systemic bioavailability and a terminal half-life of

~26 days. The clearance rate (~0.22 L/day) is independent of age, sex, and mild hepatic or renal impairment (Dang et al., 2016). The safety profile of pembrolizumab is consistent with other ICIs. Across more than 4000 patients, any-grade TRAEs occur in approximately 70%–75%, and grade 3–4 events in about 10%. The common side effects include fatigue, pruritus, rash, diarrhea, and arthralgia. Immune-related adverse events (irAEs) are unique to this class and may involve thyroiditis, colitis, hepatitis, pneumonitis, and hypophysitis. Endocrine disorders such as hypothyroidism occur in up to 15% of patients, while severe pneumonitis or myocarditis remains rare (<2%) but potentially fatal (Kim & Prasad, 2023). Early diagnosis and corticosteroid management are essential for toxicity management. Real-world evidence suggests that approximately 2 years of pembrolizumab therapy is generally sufficient to induce durable immune memory, with retreatment possible upon relapse.

12.3.2.7 Trastuzumab

Trastuzumab (Herceptin, Genentech, Roche) is a unique, humanized monoclonal antibody designed to specifically target the Human Epidermal Growth Factor Receptor 2 (HER2) receptor. It has been extensively studied and FDA-approved for the treatment of breast cancer with HER2 expression; however, limited information is available for its use in Uterine serous carcinoma (USC). USC is a rare, high-grade, aggressive subtype of EC, often considered as Type II EC. The HER2 receptor, a transmembrane tyrosine kinase, regulates important cellular functions such as growth and survival and is known to be overexpressed in breast and ECs. As a biologic therapy, Trastuzumab is designed to imitate the body's immune response to cancer, specifically to target and eliminate cells that overexpress HER2. Mechanistically, Trastuzumab binds to the HER2 receptor, specifically at the extracellular subdomain IV, thereby inhibiting the HER2 overexpressing tumor cell signaling pathways and cell proliferation. It also mediates ADCC by attracting immune cells to tumor cells overexpressing HER2 receptors. Furthermore, Trastuzumab destroys HER2 signaling for tumor cells and reduces the levels of proangiogenic VEGF and inactivates the PI3K/Akt pathway, which ultimately reduces tumor cell survival and growth. The use of Trastuzumab in gynecologic oncology emerged after the identification of HER2 overexpression in at least one-third of USC cases, resulting in poor prognosis. Clinical trials combining Trastuzumab with the standard chemotherapy regimen of carboplatin and paclitaxel have demonstrated positive results in patients with advanced recurrent metastatic HER2-positive USC. Fader et al. (2020) in a randomized phase II clinical trial (NCT01367002), focusing on advanced stage III and IV USC that had metastasized beyond the uterus or recurred after prior treatment, following a hysterectomy, studied the advantage of adding trastuzumab to the standard treatment regimen was studied. The primary endpoint was PFS. Women who received trastuzumab along with chemotherapy have higher PFS over who received chemotherapy alone. All participants reported side effects, but none stopped the treatment; women taking the trastuzumab combination experienced high blood pressure. This study demonstrated a substantial impact from adding trastuzumab to the treatment regimen, improving both progression-free survival and OS. This study was the first to establish trastuzumab as a tolerable and effective first-line targeted therapy for USC (Fader et al., 2020; McGrath, 2020). In continuation, Mao et al. (2023) published a case report describing the use of Trastuzumab in combination with chemo and radiation therapy in patients with HER2-positive EC. The multimodal treatment was well-tolerated without any severe adverse effects, highlighting the promising potential of trastuzumab in the management of HER2-positive EC (Mao et al., 2023). Trastuzumab is a safe and well-tolerated option for patients with HER2-positive USC, displaying a safety profile similar to other HER2-targeted therapies for breast and gastric cancers. Clinical studies evaluating the combination of trastuzumab with carboplatin and paclitaxel for the treatment of advanced recurrent or metastatic HER2-positive USC reported manageable tolerability and low toxicity. The most common side effects are Infusion-related fever or chills, fatigue, and minor gastrointestinal symptoms. Although the risk of cardiotoxicity associated with trastuzumab was monitored closely, it remained low. Given the strong clinical evidence of efficacy, manageable adverse effects, and excellent safety profile, trastuzumab has been incorporated as a first-line option for HER2-positive USC in combination with chemotherapy, according to the NCCN 2024 guidelines (Papageorgiou et al., 2024).

12.3.3 Antibody-drug conjugates

ADCs represent an innovative and rapidly evolving therapeutic strategy in oncology. ADCs typically consist of a monoclonal antibody, a cytotoxic drug molecule (payload), and a cleavable linker that connects the two components. These agents are classified as biologics due to the presence of an antibody component, which is derived from living organisms. Consequently, ADCs are regulated by the FDA under the category of biologics and are approved via a Biologics License Application, rather than as chemically synthesized drugs. A number of ADCs are currently under investigation for the treatment of EC. Tu et al. (2025) reported comprehensive details in recent advances in ADC for EC (Tu et al., 2025). The following section summarizes the ADCs evaluated in EC, along with their respective phases of clinical development Table 12.2 .

12.3.3.1 Vaccines

Conventional vaccines have played a vital role in eradicating various infectious diseases by producing immune reactions against antigenic subunits or attenuated, or inactivated pathogens in order to provide protection against the infection. In oncology, anticancer vaccines, specifically DC vaccines, rely on the superefficient ability of DCs as antigen-presenting cells and thereby induce CD8+ T-cell immunity. One of the most common approaches of DC vaccination is the application of antigen peptides or whole tumor cells onto the DCs, followed by reinfusion into the patients (Najafi & Mortezaee, 2023). The vaccines were reported to be safe with a remarkable clinical response in more than 200 clinical trials. DCs can be obtained by ex vivo differentiation from CD34+ hematopoietic stem and progenitor

Table 12.2 ADCs along with targets and clinical status in endometrial cancer.

Target	Antibody-drug conjugates	Clinical status in EC	Study status
HER2 (Human epidermal growth factor receptor 2)	Trastuzumab-DXd	Approved and Phase II	Approved for endometrial cancer and other solid tumors with HER2 3+ expression by IHC. Combination treatment studies are in progress
	Trastuzumab-DM1	Phase II basket trial	Trials are ongoing for multiple HER2-positive solid tumors
	Trastuzumab duocarmazine	Phase I	Trials are ongoing in patients with advanced solid tumors with at least HER2 IHC 1+ expression
	Disitamab vedotin	Phase II basket trial	Trials are ongoing in patients with multiple HER2-positive solid tumors
	DB-1303	Phase I/II	Trials are ongoing in patients with advanced HER2-positive solid tumors
FRα	Mirvetuximab soravtansine (IMGN853)	Phase I	Trials are ongoing for FOLR1-positive solid tumors, and combination treatment studies are also in progress
	IMGN151	Phase I	A trial is ongoing for recurrent gynecologic malignancies and other primary tumors, regardless of the FRα status

(continued on next page)

Table 12.2 ADCs along with targets and clinical status in endometrial cancer—cont'd

Target	Antibody-drug conjugates	Clinical status in EC	Study status
	Farmetuzumab ecteribulin	Phase I/II	A Trial is ongoing in endometrial cancer, ovarian cancer, NSCLC, and triple-negative breast cancer, regardless of FRα expression
	Luveltamab tazevibulin	Phase I	The dose extension cohort trial demonstrated clinical potential in recurrent/progressive FRα-positive endometrial cancer
	PRO1184	Phase I	Trials are ongoing in various tumors, including epithelial, ovarian, endometrial, and breast cancers, NSCLC, and mesothelioma
	AMT-151	Phase I	A trial is ongoing in advanced solid tumors, including gynecologic malignancies and breast cancer
TROP-2	Sacituzumab govitecan	Phase I/II	Therapeutic potential was established independent of Trop-2 expression in advanced/metastatic endometrial cancer. Trials evaluating combination treatments are in progress

(continued on next page)

	Datopotamab-DXd	Phase II	Trials as a monotherapy and in combination for advanced/metastatic solid tumors are in progress
	Sacituzumab tirumotecan	Phase III	Trials comparing the efficacy of Sacituzumab tirumotecan monotherapy with platinum-based chemotherapy and immunotherapy are in progress
TF (Tissue Factor)	Tisotumab vedotin	Phase II/III	Trial showed the safety profile was acceptable
B7-H4	XMT-1660	Phase I	Trials are ongoing in solid tumors, including breast, endometrial, and ovarian cancers
	AZD8205	Phase I/IIa	Trials are ongoing in advanced or metastatic solid tumors, including breast, ovarian, endometrial, and biliary tract cancers
	HS-20089	Phase IIa/b	Efficacy was demonstrated in advanced solid tumors, and a trial is ongoing in recurrent or metastatic ovarian and endometrial cancer

Tu, P., Li, G., Zou, W., Xu, C., & Wang, J. (2025). Advances in antibody–drug conjugates for endometrial cancer. Molecular Cancer Therapeutics, *24(7), 993–1004. https://doi.org/10.1158/1535-7163.mct-24-0763.*

cells (HSPCs) or CD14+ monocyte precursors and pulsing them with cancer-specific antigen from the patient. The resulting antigen-loaded, mature DCs are injected into the patient to induce a targeted immune response against cancer cells (Lee et al., 2023). Koeneman et al. (2024) reported the immunological efficacy of DC vaccination in combination with chemotherapy using carboplatin/paclitaxel in patients with metastatic EC. This exploratory phase I/II trial uses metastatic EC antigens, namely Mucin-1 and Survivin, pulsed onto the blood-derived plasmacytoid dendritic cells to form the DC vaccine. The DC vaccine is injected into the respective patients, and the results demonstrated the synergistic antitumor effect and an acceptable safety profile, supporting the feasibility of the combination treatment with DC vaccination (Koeneman et al., 2024). The **AdHER2DC** vaccine is another DC-based vaccine that targets the HER2 protein, which is elevated in 30% of EC patients (Kilpatrick, 2024). The clinical trials are evaluating the combination of AdHER2DC with two FDA-approved drugs, namely pembrolizumab, Lenvatinib, as well as IL15- super agonist N-803 in advanced/metastatic EC patients (National Cancer Institute NCI, 2025b). Furthermore, Harari et al. (2021) reported the robust immunogenic response induced by peptide neoantigens pulsed autologous DC vaccine in combination with platinum-based chemotherapy. This first peptide DC personalized vaccine induces immunogenicity by T-cell responses, indicating the potential strategy in EC management (Harari et al., 2021).

12.3.3.2 Recombinant proteins

Recombinant proteins are engineered proteins produced by cells that have been genetically engineered to express a specific target protein. These recombinant proteins can either match the same amino acid sequence as their naturally occurring counterparts or may be modified with sequence modifications to enhance desirable characteristics such as solubility, stability, or production efficiency. As they are generated using biological synthesis, recombinant proteins are classified as biologics rather than chemically synthesized compounds.

12.3.3.2.1 Osteopontin (OPN)

OPN plays a significant role in physiological processes such as angiogenesis and cell adhesion. It is an arginine–glycine–aspartic acid glycoprotein, an extracellular matrix molecule, also involved in pathologic processes such as tumor progression and metastasis (Aquino et al., 2025). Li et al. (2015) demonstrated that recombinant human OPN (rhOPN) promotes cell proliferation, migration, and invasion in human endometrial carcinoma through AKT and ERK1/2 signaling. These findings underscore the role of rhOPN in tumor growth and metastasis, opening new avenues for treatment in EC (Li et al., 2015).

12.3.3.2.2 Luteinizing hormone (LH)

LH, secreted by the anterior pituitary gland, plays a significant role in regulating reproductive processes and is important in the production of other hormones, maintains the menstrual cycle, and supports pregnancy. On the other hand, EC, being a hormone-dependent malignancy, is influenced by gonadotropins such as LH and human chorionic gonadotropin (hCG). Dabizzi et al. (2003), demonstrated that LH, along with hCG, increases the invasiveness of cancer cells through activation of protein kinase A signaling pathway, pronouncing their role in the EC cell invasiveness and tumor progression (Dabizzi et al., 2003). The inhibition of LH secretion using GnRH analogues may represent a promising therapeutic strategy in the treatment of EC.

12.3.3.2.3 Human interferon alpha (hIFNα)-2b

hIFNα-2b is a glycoprotein consisting of five alpha helices (helices A to E). It has 166 amino acids, and Phe36, Tyr122, and Tyr129 are required to maintain structural integrity. Functionally, hIFNα-2b regulates gene expression pathways that lead to inhibition of apoptosis, differentiation, and cell cycle arrest, contributing to its antiproliferative activity (Ningrum, 2014). Scambia et al. (1991) reported the effect of recombinant human IFNα-2b on steroid hormone receptors and epidermal growth factor receptors expression in patients with EC. rh IFN-alpha 2b treatment in patients with EC led to substantial increased levels of a substantial increase in progesterone and estrogen levels and a reduction in EGF, suggesting the influence of rh IFN-alpha 2b in EC cell differentiation (Scambia et al., 1991).

12.3.4 Biosimilars

Biosimilars are a class of complex molecules, with a minor difference in structure, and have "*no clinically meaningful differences*" to an existing FDA-approved biologics in terms of safety, purity, and potency, and have the same therapeutic indications, mechanism of action, and route of administration, dosage form (Pfizer, 2022). Although they possess identical amino acid sequences to their reference, they may differ in 3D structure, glycosylation sites, isoform profiles, and protein aggregation. Therefore, pharmacodynamic and pharmacokinetic studies are essential to prove their similarity (Konstantinidou et al., 2020). The first biosimilar was approved in 2006 in Europe, and regulatory guidelines for safety and efficacy were specified. These guidelines were adopted by other countries, facilitating the global development and market entry of biosimilars. Given that the development of biologicals is both time-consuming and expensive, the pharmaceutical industry has been shifting its focus toward developing "*Biosimilars*," thereby providing access to cancer treatments and medications at an affordable price (Joshi et al., 2022). Many mAbs biosimilars are currently under development for therapeutic use. Clinical studies should recruit a homogeneous patient population to ensure reliable comparison with the reference biologic. The efficacy endpoints include ORR and pathological complete response. It is important to note that biologics with unknown quality result in increased risk to patient safety and may fail to demonstrate clinical efficacy, emphasizing the prominence of stringent analytical characterization and clinical validation during biosimilar development (Kos et al., 2018). The details of the biosimilars are presented in Table 12.3.

Table 12.3 Biosimilars in endometrial cancer.

Biologic (reference product)	Target	FDA approved biosimilars	Clinical indications	Endometrial cancer relevance
Trastuzumab (Herceptin)	HER2	Herzuma (Trastuzumab-pkrb) Kanjinti (Trastuzumab-anns) Ogivri (Trastuzumab-dkst) Ontruzant (Trastuzumab-dttb) Trazimera (Trastuzumab-qyyp) Hercessi (Trastuzumab-strf)	Breast, CNS, leptomeningeal metastases, colorectal, esophageal, esophagogastric junction or gastric, hepatobiliary, cholangiocarcinoma or gallbladder, salivary gland, uterine neoplasms—endometrial carcinoma	Approved for use in endometrial cancer. Route of administration: intravenous Initial: 8 mg/kg once (21-d cycle) Maintenance: 6 mg/kg every 3 wk (Trastuzumab Products (Herceptin, Ontruzant, Herzuma, Ogivri, Trazimera, Kanjinti, Hercessi))

(*continued on next page*)

Bevacizumab (Avastin)	VEGF	Alymsys (Bevacizumab-maly) Mvasi (Bevacizumab-awwb) Vegzelma (Bevacizumab-adcd) Zirabev (Bevacizumab-bvzr) Avzivi (Bevacizumab-tnjn)	Cervical, hepatocellular malignant (pleural, peritoneal) mesothelioma, pericardial or tunica vaginalis testis mesothelioma, nonsmall cell lung cancer (NSCLC), soft tissue sarcoma: angiosarcoma, uterine neoplasms: endometrial carcinoma, vaginal cancer, vulvar carcinoma	Approved for use in endometrial cancer. Route of administration: intravenous 15 mg/kg every 3 wk (Bevacizumab products (Avastin; Mvasi; Zirabev; Alymsys; Vegzelma; Avzivi))
Avelumab (Bavencio)	PD-L1 1)	No FDA-approved therapeutic biosimilars	-	-
Dostarlimab (Jemperli)	PD-L1	No FDA-approved therapeutic biosimilars	-	-
Durvalumab (Imfinzi)	PD-L1	No FDA-approved therapeutic biosimilars	-	-
Nivolumab (Opdivo)	PD-L1	No FDA-approved therapeutic biosimilars	-	-
Pembrolizumab (Keytruda)	PD-L1	No FDA-approved therapeutic biosimilars	-	-

12.4 Conclusion

Biologics and biosimilars have advanced the therapeutic landscape of EC, moving treatment from cytotoxic regimens toward targeted and immune-based precision approaches. Biologics offer mechanisms to overcome resistance and improve patient survival, and biosimilars broaden access to these innovations through more affordable alternatives. Ongoing translational research and comprehensive regulatory frameworks are essential to optimize their combination into clinical practice. These agents represent a critical step toward effective and personalized care for patients with EC.

References

Aboul-Ella, H., Gohar, A., Ali, A. A., Ismail, L. M., Mahmoud, A. E. E. R., Elkhatib, W. F., & Aboul-Ella, H. (2024). Monoclonal antibodies: From magic bullet to precision weapon. *Molecular Biomedicine, 5*(1), 12. doi:10.1186/s43556-024-00210-1.

Aghajanian, C., Sill, M. W., Darcy, K. M., Greer, B., McMeekin, D. S., Rose, P. G., Rotmensch, J., Barnes, M. N., Hanjani, P., & Leslie, K. K. (2011). Phase II trial of bevacizumab in recurrent or persistent endometrial cancer: A gynecologic oncology group study. *Journal of Clinical Oncology, 29*(16), 2259–2265. doi:10.1200/JCO.2010.32.6397United States.

Antill, Y., Kok, P. S., Robledo, K., Yip, S., Cummins, M., Smith, D., Spurdle, A., Barnes, E., Lee, Y. C., Friedlander, M., Baron-Hay, S., Shannon, C., Coward, J., Beale, P., Goss, G., Meniawy, T., Lombard, J., Andrews, J., Stockler, M. R., & Mileshkin, L. (2021). Clinical activity of durvalumab for patients with advanced mismatch repair-deficient and repair-proficient endometrial cancer. A nonrandomized phase 2 clinical trial. *Journal for ImmunoTherapy of Cancer, 9*(6), e002255. doi:10.1136/jitc-2020-002255.

Aquino, C. I., Venkatesan, S., Ligori, A., Tinelli, R., Grossini, E., & Surico, D. (2025). Osteopontin expression and its role in endometrial cancer: A systematic review. *Cancers, 17*(13), 2245. doi:10.3390/cancers17132245.

Bassette, E., & Ducie, J. A. (2024). Endometrial cancer in reproductive-aged females: Etiology and pathogenesis. *Biomedicines, 12*(4), 886. doi:10.3390/biomedicines12040886.

Berek, J. S., Matias-Guiu, X., Creutzberg, C., Fotopoulou, C., Gaffney, D., Kehoe, S., Lindemann, K., Mutch, D., Concin, N., & Endometrial Cancer Staging Subcommittee, FIGO Women's Cancer Committee. (2023). FIGO staging of endometrial cancer: 2023. Journal of Gynecologic Oncology, 34(5), e85. https://doi.org/10.3802/jgo.2023.34.e85

Boyerinas, B., Jochems, C., Fantini, M., Heery, C. R., Gulley, J. L., Tsang, K. Y., & Schlom, J. (2015). Antibody-dependent cellular cytotoxicity activity of a novel anti-PD-L1 antibody avelumab (MSB0010718C) on human tumor cells. *Cancer Immunology Research, 3*(10), 1148–1157. doi:10.1158/2326-6066.CIR-15-0059.

Brooks,, R. A., Fleming, G. F., Lastra, R. R., Lee, N. K., Moroney, J. W., Son, C. H., Tatebe, K., & Veneris, J. L. (2019). Current recommendations and recent progress in endometrial cancer. *CA Cancer Journal for Clinicians, 69*(4), 258–279. doi:10.3322/caac.21561.

Corr, B., Cosgrove, C., Spinosa, D., & Guntupalli, S. (2022). Endometrial cancer: Molecular classification and future treatments. *BMJ Medicine, 1*(1), e000152. doi:10.1136/bmjmed-2022-000152.

Costa, B., & Vale, N. (2022). Dostarlimab: A review. *Biomolecules, 12*(8), 1031. doi:10.3390/biom12081031.

Dabizzi, S., Noci, I., Borri, P., Borrani, E., Giachi, M., Balzi, M., Taddei, G. L., Marchionni, M., Scarselli, G. F., & Arcangeli, A. (2003). Luteinizing hormone increases human endometrial cancer cells invasiveness through activation of protein kinase A. *Cancer Research, 63*(14), 4281–4286.

Dang, T. O., Ogunniyi, A., Barbee, M. S., & Drilon, A. (2016). Pembrolizumab for the treatment of PD-L1 positive advanced or metastatic non-small cell lung cancer. *Expert Review of Anticancer Therapy, 16*(1), 13–20. doi:10.1586/14737140.2016.1123626.

Disis, M. L., Taylor, M. H., Kelly, K., Beck, J. T., Gordon, M., Moore, K. M., Patel, M. R., Chaves, J., Park, H., Mita, A. C., Hamilton, E. P., Annunziata, C. M., Grote, H. J., Von Heydebreck, A., Grewal, J., Chand, V., & Gulley, J. L. (2019). Efficacy and safety of avelumab for patients with recurrent or refractory ovarian cancer: Phase 1b results from the JAVELIN solid tumor trial. *JAMA Oncology, 5*(3), 393–401. doi:10.1001/jamaoncol.2018.6258.

Dörk, T., Hillemanns, P., Tempfer, C., Breu, J., & Fleisch, M. C. (2020). Genetic susceptibility to endometrial cancer: Risk factors and clinical management. *Cancers, 12*(9), 2407. doi:10.3390/cancers12092407.

Du Rusquec, P., De Calbiac, O., Robert, M., Campone, M., & Frenel, J. S. (2019). Clinical utility of pembrolizumab in the management of advanced solid tumors: An evidence-based review on the emerging new data. *Cancer Management and Research, 11*, 4297–4312. doi:10.2147/CMAR.S151023.

Fader, A. N., Roque, D. M., Siegel, E., Buza, N., Hui, P., Abdelghany, O., Chambers, S., Secord, A. A., Havrilesky, L., O'Malley, D. M., Backes, F. J., Nevadunsky, N., Edraki, B., Pikaart, D., Lowery, W., ElSahwi, K., Celano, P., Bellone, S., Azodi, M., … Santin, A. D. (2020). Randomized phase II trial of carboplatin–paclitaxel compared with carboplatin–paclitaxel–trastuzumab in advanced (Stage III–IV) or recurrent uterine serous carcinomas that overexpress her2/Neu (NCT01367002): Updated overall survival analysis. *Clinical Cancer Research, 26*(15), 3928–3935. doi:10.1158/1078-0432.CCR-20-0953.

Friedman, C. F., Manning-Geist, B. L., Zhou, Q., Soumerai, T., Holland, A., Cruz Paula, A. D., Green, H., Ozsoy, M. A., Iasonos, A., Hollmann, T., Leitao, M. M., Mueller, J. J., Makker, V., Tew, W. P., O'Cearbhaill, R. E., Liu, Y. L., Rubinstein, M. M., Troso-Sandoval, T., Lichtman, S. M., … Zamarin, D. (2024). Nivolumab for mismatch-repair-deficient or hypermutated gynecologic cancers: A phase 2 trial with biomarker analyses. *Nature Medicine, 30*(5), 1330–1338. doi:10.1038/s41591-024-02942-7.

Harari, A., Sarivalasis, A., de Jonge, K., Thierry, A. C., Huber, F., Boudousquie, C., Rossier, L., Orcurto, A., Imbimbo, M., Baumgaertner, P., Bassani-sternberg, M., & Kandalaft, L. E. (2021). A personalized neoantigen vaccine in combination with platinum-based chemotherapy induces a t-cell response coinciding with a complete response in endometrial carcinoma. *Cancers, 13*(22), 4512. doi:10.3390/cancers13225801.

Joshi, D., Khursheed, R., Gupta, S., Wadhwa, D., Singh, T. G., Sharma, S., Porwal, S., Gauniyal, S., Vishwas, S., Goyal, S., Gupta, G., Eri, R. D., Williams, K. A., Dua, K., & Singh, S. K. (2022). Biosimilars in oncology: Latest trends and regulatory status. *Pharmaceutics, 14*(12), 2512. doi:10.3390/pharmaceutics14122721.

Kilpatrick, C. (2024). *ANKTIVA and AdHER2DC trial tackles endometrial cancer.* https://vaccinenation.org/therapeutic/anktiva-and-adher2dc-trial-tackles-endometrial-cancer/ (accessed: November 7, 2025).

Kim, M. S., & Prasad, V. (2023). Pembrolizumab for all. *Journal of Cancer Research and Clinical Oncology, 149*(3), 1357–1360. https://doi.org/10.1007/s00432-022-04412-4.

Koeneman, B. J., Schreibelt, G., Gorris, M. A. J., Hins-de Bree, S., Westdorp, H., Ottevanger, P. B., & de Vries, I. J. M. (2024). Dendritic cell vaccination combined with carboplatin/paclitaxel for metastatic endometrial cancer patients: Results of a phase I/II trial. *Frontiers in Immunology, 15*, 1368103. https://www.frontiersin.org/journals/immunology#. doi:10.3389/fimmu.2024.1368103.

Konstantinidou, S., Papaspiliou, A., & Kokkotou, E. (2020). Current and future roles of biosimilars in oncology practice. *Oncology Letters, 19*(1), 45–51. https://doi.org/10.3892/ol.2019.11105.

Konstantinopoulos, P. A., Liu, W., Luo, J. F., Gulhan, D. C., Krasner, C., Ibrahim, J. J., … Garcia, K. P. (2019). Phase II study of avelumab in patients with mismatch repair deficient and mismatch repair proficient recurrent/persistent endometrial cancer. *Journal of Clinical Oncology, 37*(30). https://doi.org/10.1200/JCO.19.01021.

Konstantinopoulos, P.A., 2025. Avelumab in Patients With MSS, MSI-H and POLE-mutated Recurrent or Persistent Endometrial Cancer and of Avelumab/Talazoparib and Avelumab/Axitinib in Patients With MSS Recurrent or Persistent Endometrial Cancer. Clinical trial registration NCT02912572. ClinicalTrials.gov. https://clinicaltrials.gov/study/NCT02912572 (accessed: 07 -11-2025).

Kos, I. A., Azevedo, V. F., Neto, D. E., & Kowalski, S. C. (2018). The biosimilars journey: Current status and ongoing challenges. *Drugs in Context, 7*. doi:10.7573/dic.212543.

Lee, K.-W., Yam, J. W. P., & Mao, X. (2023). Dendritic cell vaccines: A shift from conventional approach to new generations. *Cells, 12*(17), 2147. https://doi.org/10.3390/cells12172147.

Lheureux, S., Matei, D. E., Konstantinopoulos, P. A., Wang, B. X., Gadalla, R., Block, M. S., Jewell, A., Gaillard, S. L., McHale, M., McCourt, C., Temkin, S., Girda, E., Backes, F. J., Werner, T. L., Duska, L., Kehoe, S., Colombo, I., Wang, L., … Fleming, G. F. (2022). Translational randomized phase II trial of cabozantinib in combination with nivolumab in advanced, recurrent, or metastatic endometrial cancer. *Journal for ImmunoTherapy of Cancer, 10*(3). doi:10.1136/jitc-2021-004233.

Li, Y., Xie, Y., Cui, D., Ma, Y., Sui, L., Zhu, C., Kong, H., & Kong, Y. (2015). Osteopontin promotes invasion, migration and epithelial-mesenchymal transition of human endometrial carcinoma cell HEC-1A through AKT and ERK1/2 signaling. *Cellular Physiology and Biochemistry, 37*(4), 1503–1512. doi:10.1159/000438518.

Mahdy, H., Vadakekut, E. S., & Crotzer, D. (2025). Endometrial Cancer. StatPearls. FL: StatPearls Publishing: Treasure Island.

Loukovaara, M., Pasanen, A., & Bützow, R. (2022). Molecular classification of endometrial carcinoma: A clinically oriented review. Journal of Clinical Pathology. Advance online publication. https://doi.org/10.1136/jclinpath-2022-208345

Makker, V., Colombo, N., Herráez, A. C., Santin, A. D., Colomba, E., Miller, D. S., Fujiwara, K., Pignata, S., Baron-Hay, S., Ray-Coquard, I., Shapira-Frommer, R., Ushijima, K., Sakata, J., Yonemori, K., Kim, Y. M., Guerra, E. M., Sanli, U. A., McCormack, M. M., Smith, A. D., … Lorusso, D. (2022). Lenvatinib plus pembrolizumab for advanced endometrial cancer. *New England Journal of Medicine, 386*(5), 437–448. doi:10.1056/NEJMoa2108330.

Makker, V., MacKay, H., Ray-Coquard, I., Levine, D. A., Westin, S. N., Aoki, D., & Oaknin, A. (2021). Endometrial cancer. *Nature Reviews Disease Primers, 7*(1), 88. doi:10.1038/s41572-021-00324-8.

Mao, S. P. H., Desravines, N., Zarei, S., Viswanathan, A. N., & Fader, A. N. (2023). Combined trastuzumab and radiation therapy for HER2-positive uterine serous carcinoma: A case report. *Gynecologic Oncology Reports, 49*. doi:10.1016/j.gore.2023.101250.

McGrath, D. (2020). *Trastuzumab may improve survival in women with rare endometrial cancer*. National Cancer Institute https://www.cancer.gov/news-events/cancer-currents-blog/2020/endometrial-cancer-usc-her2-trastuzumab.

Mirza, M. R., Chase, D. M., Slomovitz, B. M., Depont Christensen, R., Novák, Z., Black, D., Gilbert, L., Sharma, S., Valabrega, G., Landrum, L. M., Hanker, L. C., Stuckey, A., Boere, I., Gold, M. A., Auranen, A., Pothuri, B., Cibula, D., McCourt, C., Raspagliesi, F., … Powell, M. A. (2023). Dostarlimab for primary advanced or recurrent endometrial cancer. *New England Journal of Medicine, 388*(23), 2145–2158. doi:10.1056/NEJMoa2216334.

Najafi, S., & Mortezaee, K. (2023). Advances in dendritic cell vaccination therapy of cancer. *Biomedicine & Pharmacotherapy, 164*, 114954. https://doi.org/10.1016/j.biopha.2023.114954.

National Cancer Institute (NCI). (2024). *Dostarlimab-gxly*. https://www.cancer.gov/about-cancer/treatment/drugs/dostarlimab-gxly.

National Cancer Institute (NCI). (2025a). *A Randomized Phase II Trial of Nivolumab and Ipilimumab Compared to Nivolumab Monotherapy in Patients With Deficient Mismatch Repair System Recurrent Endometrial Carcinoma*; Clinical trial registration NCT05112601; clinicaltrials.gov. https://clinicaltrials.gov/study/NCT05112601 (accessed: November 7, 2025).

National Cancer Institute (NCI). (2025b). *Pembrolizumab, Lenvatinib and IL-15 Superagonist N-803 in Combination With HER2 Targeting Autologous Dendritic Cell (AdHER2DC) Vaccine in Participants With Advanced or Metastatic Endometrial Cancer*; Clinical trial registration NCT06253494; clinicaltrials.gov. https://clinicaltrials.gov/study/NCT06253494 (accessed: November 7, 2025).

Ningrum, R. A. (2014). Human interferon alpha-2b: A therapeutic protein for cancer treatment. *Scientifica, 2014*, 1–8. doi:10.1155/2014/970315.

Oaknin, A., Pothuri, B., Gilbert, L., Sabatier, R., Brown, J., Ghamande, S., Mathews, C., O'Malley, D. M., Kristeleit, R., Boni, V., Gravina, A., Banerjee, S., Miller, R., Pikiel, J., Mirza, M. R., Dewal, N., Antony, G., Dong, Y., Zografos, E., … Tinker, A. V. (2023). Safety, efficacy, and biomarker analyses of dostarlimab in patients with endometrial cancer: Interim results of the phase I GARNET study. *Clinical Cancer Research, 29*(22), 4564–4574. doi:10.1158/1078-0432.CCR-22-3915.

O'Malley, D. M., Bariani, G. M., Cassier, P. A., Marabelle, A., Hansen, A. R., De Jesus Acosta, A., Miller, W. H., Safra, T., Italiano, A., Mileshkin, L., Xu, L., Jin, F., Norwood, K., & Maio, M. (2022). Pembrolizumab in patients with microsatellite instability–high advanced endometrial cancer: Results from the KEYNOTE-158 study. *Journal of Clinical Oncology, 40*(7), 752–761. doi:10.1200/JCO.21.01874.

Papageorgiou, D., Liouta, G., Sapantzoglou, I., Zachariou, E., Pliakou, D., Papakonstantinou, K., Floros, T., & Pliakou, E. (2024). HER2-positive serous endometrial cancer treatment: Current clinical practice and future directions. *Medicina, 60*(12), 2012. doi:10.3390/medicina60122012.

Pfizer. (2022). *Biologics vs. biosimilars: Understanding the difference*. https://www.pfizer.com/news/articles/biologics_vs_biosimilars_understanding_the_differences.

Rodriguez, A. C., Blanchard, Z., Maurer, K. A., & Gertz, J. (2019). Estrogen signaling in endometrial cancer: A key oncogenic pathway with several open questions. *Hormones and Cancer, 10*(2-3), 51–63. doi:10.1007/s12672-019-0358-9.

Rubinstein, M. M., Dickinson, S., Narayan, P., Zhou, Q., Iasonos, A., Ma, W., Lakhman, Y., & Makker, V. (2021). Bevacizumab in advanced endometrial cancer. *Gynecologic Oncology, 161*(3), 720–726. doi:10.1016/j.ygyno.2021.04.016.

Scambia, G., Panici, P. B., Battaglia, F., Ferrandina, G., Baiocchi, G., Gallo, A., Sica, G., & Mancuso, S. (1991). Effect of recombinant human interferon-alpha2b on receptors for steroid hormones and epidermal growth factor in patients with endometrial cancer. *European Journal of Cancer and Clinical Oncology, 27*(1), 51–53. doi:10.1016/0277-5379(91)90060-Q.

Talhouk, A., & McAlpine, J. N. (2016). New classification of endometrial cancers: The development and potential applications of genomic-based classification in research and clinical care. *Gynecologic Oncology Research and Practice, 3*(1), 2053–6844. doi:10.1186/s40661-016-0035-4.

Talhouk, A., McConechy, M. K., Leung, S., Yang, W., Lum, A., Senz, J., Boyd, N., Pike, J., Anglesio, M., Kwon, J. S., Karnezis, A. N., Huntsman, D. G., Gilks, C. B., & McAlpine, J. N. (2017). Confirmation of ProMisE: A simple, genomics-based clinical classifier for endometrial cancer. *Cancer, 123*(5), 802–813. doi:10.1002/cncr.30496.

Tan, M. H., Mester, J. L., Ngeow, J., Rybicki, L. A., Orloff, M. S., & Eng, C. (2012). Lifetime cancer risks in individuals with germline PTEN mutations. *Clinical Cancer Research, 18*(2), 400–407. doi:10.1158/1078-0432.CCR-11-2283.

Thiyagarajan, D. K., Basit, H., & Jeanmonod, R. (2025). Physiology, Menstrual Cycle. StatPearls. FL: StatPearls Publishing: Treasure Island.

Török, P., Krasznai, Z., Molnár, S., Lampé, R., & Jakab, A. (2020). Preoperative assessment of endometrial cancer. *Translational Cancer Research, 9*(12), 7746–7758. doi:10.21037/tcr-20-2068.

Tu, P., Li, G., Zou, W., Xu, C., & Wang, J. (2025). Advances in antibody-drug conjugates for endometrial cancer. *Molecular Cancer Therapeutics, 24*(7), 993–1004. https://doi.org/10.1158/1535-7163.MCT-24-0763.

Tuninetti, V., Farolfi, A., Rognone, C., Montanari, D., De Giorgi, U., & Valabrega, G. (2024). Treatment strategies for advanced endometrial cancer according to molecular classification. *International Journal of Molecular Sciences, 25*(21), 11448. doi:10.3390/ijms252111448.

U.S. Food and Drug Administration. (2024a). *FDA approves durvalumab with chemotherapy for mismatch repair deficient primary advanced or recurrent endometrial cancer*. https://www.fda.gov/drugs/resources-information-approved-drugs/fda-approves-durvalumab-chemotherapy-mismatch-repair-deficient-primary-advanced-or-recurrent.

U.S. Food and Drug Administration. (2024b). *FDA approves pembrolizumab with chemotherapy for primary advanced or recurrent endometrial carcinoma*. https://www.fda.gov/drugs/resources-information-approved-drugs/fda-approves-pembrolizumab-chemotherapy-primary-advanced-or-recurrent-endometrial-carcinoma.

Van Gorp, T., Cibula, D., Lv, W., Backes, F., Ortaç, F., Hasegawa, K., Lindemann, K., Savarese, A., Laenen, A., Kim, Y. M., Bodnar, L., Barretina-Ginesta, M.-P., Gilbert, L., Pothuri, B., Chen, X., Flores, M. B., Levy, T., Colombo, N., Papadimitriou, C., … Slomovitz, B. (2024). ENGOT-en11/GOG-3053/KEYNOTE-B21: A randomised, double-blind, phase III study of pembrolizumab or placebo plus adjuvant chemotherapy with or without radiotherapy in patients with newly diagnosed, high-risk endometrial cancer. *Annals of Oncology, 35*(11), 968–980. doi:10.1016/j.annonc.2024.08.2242.

Westin, S. N., Moore, K., Chon, H. S., Lee, J. Y., Pepin, J. T., Sundborg, M., Shai, A., de la Garza, J., Nishio, S., Gold, M. A., Wang, K., McIntyre, K., Tillmanns, T. D., Blank, S. V., Liu, J. H., McCollum, M., Mejia, F. C., Nishikawa, T., Pennington, K., … Van Nieuwenhuysen, E. (2024). Durvalumab plus carboplatin/paclitaxel followed by maintenance durvalumab with or without olaparib as first-line treatment for advanced endometrial cancer: The phase III DUO-E trial. *Journal of Clinical Oncology, 42*(3), 283–299. doi:10.1200/JCO.23.02132.

CHAPTER

Biosimilars and biologics in urothelial carcinoma 13

Esha Sarkar[a], Tariq Mahmood[a], Jyoti Dwivedi[a], Aparna Misra[a], Sujith Kumar Pulukool[b], Adhrit Chandra Pati Tripathi[c] and Pratima Tripathi[d,e]

[a]*Department of Biochemistry, Era's Lucknow Medical College and Hospital, Era University, Lucknow, Uttar Pradesh, India,* [b]*Department of Biosciences, Sri Sathya Sai Institute of Higher Learning, Prasanthi Nilayam, Anantapur, Andhra Pradesh, India,* [c]*Amity School of Communication, Amity University, Lucknow, Uttar Pradesh, India,* [d]*National Institute of Pharmaceutical Education and Research Raebareli (NIPER-R), Lucknow, Uttar Pradesh, India,* [e]*Kaustav's Careerica & Prakdrit Edtech Pvt. Ltd., Lucknow, Uttar Pradesh, India*

13.1 Overview of urothelial carcinoma

Urothelial carcinoma (UC), referred to as transitional cell carcinoma, is the most common malignancy of the urinary tract and arises from the urothelium, which is the epithelial lining of the bladder, ureters, and renal pelvis (Dalghi et al., 2020). UC is the most common histological subtype, and treatment options are guided by whether the disease is muscle-invasive or nonmuscle-invasive, with subsequent risk group stratification (Lobo et al., 2024). This is the predominant type of bladder cancer, exhibiting a wide clinical range from slow-growing, noninvasive lesions to aggressive forms that invade the bladder muscle. It is a major global health issue, with a clear male predominance. Although the incidence of urothelial carcinoma is highest in industrialized countries, a steady increase is also being seen in developing regions, including India. This rise is largely linked to longer life expectancy, lifestyle changes, and continued exposure to preventable risk factors such as tobacco use and occupational contact with carcinogenic substances (Torre et al., 2016). In India, the overall incidence remains lower than in most Western nations, but the numbers are gradually climbing. Current projections suggest that the number of cases will rise sharply over the next few decades. The most common symptom is painless, visible blood in the urine (gross hematuria) (Mishra & Balasubramaniam, 2020). Other possible complaints include an increased urge to urinate, frequent urination, painful urination, or pelvic discomfort, particularly when the disease has progressed. Because early stages often produce mild or vague urinary symptoms, many patients are diagnosed late, making community awareness and timely medical assessment crucial. UC occurs most often in older adults, with the highest rates seen between 60 and 70 years of age, and affects men about three times more often than women. Worldwide, urothelial carcinoma is regarded as one of the more frequently occurring

Biosimilars and Biologics in Cancer Treatment: An Era in Therapeutic Strategies.
DOI: https://doi.org/10.1016/B978-0-443-43824-0.00013-4

cancers. In 2020, it was estimated that over 573,000 people were newly diagnosed, and more than 200,000 lives were lost to the disease. When the figures are adjusted for age, the incidence works out to about 5.6 cases per 100,000 people, with a mortality rate of roughly 1.9 per 100,000 (Sung et al., 2021). By 2040, the annual number of new UC cases could near one million, with deaths expected to rise alongside. This trend is largely attributed to an aging population and the continued presence of avoidable risk factors. In India, records from 2020 show about 22,500 new diagnoses and more than 12,000 deaths, with over 60,000 people living with the disease at that time (Mathur et al., 2020). The incidence, at about 2.9 per 100,000 population, is still lower than in many Western nations, but current trends point to the possibility of cases almost doubling by 2040. Among the established risk factors, tobacco use remains the leading preventable cause, with certain occupational exposures also playing a significant role (Cumberbatch et al., 2016).

13.2 Evolution of treatment paradigms

Over recent decades, the treatment of urothelial carcinoma has undergone substantial change. Today, however, therapeutic strategies have broadened to include immunotherapy, targeted agents, and precision-based approaches. For nonmuscle-invasive disease, the usual approach traditionally involved transurethral resection of the bladder tumor (TURBT), often followed by intravesical administration of chemotherapy or Bacillus Calmette-Guérin (BCG) to reduce the risk of recurrence and disease progression (Babjuk et al., 2022). Historically, metastatic urothelial carcinoma (MUC) was treated mainly with platinum-based chemotherapy, most often using a gemcitabine–cisplatin combination (Von Der Maase et al., 2005). While these treatments offered only modest improvements in survival, their overall benefits were limited, and many patients could not receive them due to kidney problems or other serious health conditions. The advent of immune checkpoint inhibitors targeting the PD-1 and PD-L1 pathways, such as atezolizumab, pembrolizumab, and avelumab, has marked a significant change in therapeutic approaches, providing sustained responses in selected patient groups (Burt et al., 2019). Avelumab, in particular, has been approved for use as a maintenance therapy following completion of first-line chemotherapy (Powles et al., 2020a). Advances in molecular profiling have widened the scope of treatment for UC, supporting the use of targeted agents such as erdafitinib, a fibroblast growth factor receptor (FGFR) inhibitor effective in carefully selected patients (Loriot et al., 2019), and antibody–drug conjugates like enfortumab vedotin, which have demonstrated benefit in cases resistant to both chemotherapy and immunotherapy (Rosenberg et al., 2019). In recent years, the use of combinations such as enfortumab vedotin with pembrolizumab has produced promising outcomes, including in earlier stages of urothelial carcinoma (Powles et al., 2024). These advances mark a shift toward treatment strategies tailored to the biological characteristics of the disease, intending to improve survival, maintain quality of life, and reduce therapy-related complications.

13.3 Significance of biologics and biosimilars in oncology

Biologic therapies have changed the landscape of cancer treatment, offering targeted strategies that focus on destroying malignant cells while preserving most surrounding healthy tissue. This selectivity has improved both treatment effectiveness and tolerability compared with conventional chemotherapy (Scott et al., 2012). In the management of UC, the introduction of immune checkpoint inhibitors and targeted antibodies has significantly expanded therapeutic options. In many parts of the world, the price of biologic cancer drugs is one of the biggest hurdles to patient access (Mellman et al., 2011). To reduce this burden, drug manufacturers have introduced biosimilars-closely comparable versions of original biologic medicines that deliver the same safety, quality, and effectiveness at a far lower cost (Aggarwal et al., 2017). In Oncology, commonly used examples include biosimilar forms of bevacizumab, trastuzumab, and rituximab (Blackstone & Fuhr, 2012). By offering the same clinical benefit at a fraction of the price, these medicines have opened the door for more patients to receive potentially life-saving care (Tabernero, 2018).

13.4 Biosimilar concerns and knowledge deficits

Biosimilars are increasingly shaping the field of oncology, yet their broader use remains limited because of persistent doubts and a lack of awareness. By early 2025, the U.S. Food and Drug Administration had cleared about 60 biosimilars, including key oncology agents such as trastuzumab, rituximab, and bevacizumab. Even so, their acceptance has not been Uniform across healthcare systems (Itoshima et al., 2024). Many oncologists still approach biosimilars with caution. Concerns often focus on immune-related reactions, the extension of approvals to multiple indications without separate trials, and whether these medicines can be substituted directly for their original biologics (McAuliff et al., 2024). Studies and professional surveys indicate that a large proportion of cancer specialists have limited hands-on experience with prescribing biosimilars. In several countries, including parts of Europe, health agencies have not issued clear guidance on substitution or on monitoring for side effects, which leaves oncologists uncertain about the most appropriate way to use biosimilars in routine care (Chaplin et al., 2024). The economic case for biosimilars is, however, undeniable. In the United States alone, oncology biosimilars have generated close to 24 billion dollars in savings since 2015, with annual savings increasing by about 2 billion dollars each year (Cancer Therapy Advisor). Despite this impact, actual uptake has lagged because of the same knowledge gaps, inconsistent regulatory communication, and hesitation among prescribers. Biosimilars in oncology have strong scientific support and clear economic benefits (Joshi et al., 2022).

13.5 Understanding biologics and biosimilars

13.5.1 Definitions and key differences

Biologics: Biologics, or biologicals, refer to a wide variety of biological products that are used as therapeutic agents. They are complex molecules produced by living systems. Unlike chemical therapeutic agents, which are uniform in nature, as they are produced by uniform chemical processes, biologics show a certain degree of variance, but the tolerance limits of these variances are strictly defined and controlled (US, FDA, 2023).

Though new age biologics are marvels of genetic manipulation of living beings, from single-celled bacteria like *Escherichia coli* producing insulin to transgenic mammals like goats producing human proteins, the story of biologics has a much humbler beginning. The first biologics were antisera developed at the turn of the 19th century, when Kitasato Shibasaburō and Emil von Behring in 1890 published their work on the development of tetanus and diphtheria antitoxin (Okui, 2024). They did so by injecting the toxins from these bacteria into guinea pigs, goats, and horses, and then purifying the "anti-toxins" from their serum. These biologics are still used as antivenom serum, antitetanus immunoglobulins, antidiphtheria immunoglobulins, etc.

Biologics include a diverse range of therapeutic products, but the US FDA broadly categorizes them as follows based on their applications (US, FDA, 2024a):

- Monoclonal antibodies for in vivo use.
- Cytokines, growth factors, enzymes, immunomodulators, and thrombolytics.
- Proteins intended for therapeutic use that are extracted from animals or microorganisms, including recombinant versions of these products (except clotting factors).
- Other nonvaccine therapeutic immunotherapies.

Based on their biological nature, these biologics can be classified as follows (Morrow & Felcone, 2004):

- Hormones
- Interferons
- Interleukins
- Growth factors
- Monoclonal antibodies
- Polypeptides
- Proteins
- Vaccines

Biosimilars: These are molecules with structural and functional similarity to the original biologics, which are known as the reference product or reference biologics. Biosimilars are made from the same process as the reference biologics, i.e., from living systems, and have similar variance from a biologic as would be expected between different lots of the reference biologic because of the inherent variability in biological systems.

Therapeutically, biosimilars are expected to have a similar therapeutic profile as the reference biologic, with the acceptable variance being well defined and controlled. They are generally seen as a generic, cheaper alternative to biologics and can be manufactured after the initial market protection on the biologics expires.

As per regulatory regulations in most countries, biosimilars do not need to undergo complete clinical trials like any novel biologic. They need to prove their efficacy and safety as compared to the reference biologics to be allowed to enter the market. This allows for a truncated pathway for the development of biosimilars, further lowering the cost of such drugs.

13.5.2 Mechanism of action in cancer therapy

One of the outstanding success stories in the field of biologics has been the development of cancer immunotherapies. Today, monoclonal antibody-based immune therapies play a central role in the treatment of both solid and liquid tumors. Another fast-growing field of biologics-based cancer therapy, which has shown remarkable success in the treatment of liquid tumors, is CAR-T cell-based therapies. Apart from immunotherapies and CAR-T cell-based therapies, certain other biologics also have a supportive role in cancer therapies, such as interleukins, cytokines, colony-stimulating factors, and cancer vaccines. Cancer immunotherapies have been traditionally divided into active or passive immunotherapies. Immunotherapies that were believed to directly attack the tumor cells and destroy them were called passive immunotherapies. These included monoclonal antibodies that directly caused tumor lysis, CAR-T cell therapies, and Oncolytic viral therapies. Cancer immunotherapies that helped activate the host immune system against the cancer cells were classified as active immunotherapies. These include monoclonal antibodies that acted as checkpoint inhibitors, cancer vaccines, etc. (US, FDA, 2024b). But as studies have shown that almost all immunotherapies recruit, to one extent or another, the immune system of the patient, making the above classification redundant, hence a better classification would be to classify the biologics based on their nature. For example, monoclonal antibody-based therapies, CAR-T-based therapies, dendritic cell-based therapies, oncolytic virus-based therapies, etc.

13.5.2.1 Monoclonal antibodies-based cancer therapies

Monoclonal antibody-based immune therapies are the single largest and oldest group of biologic-based therapeutic agents in cancer immunotherapy. Monoclonal antibodies used in cancer therapies can be divided into two classes (Zahavi & Weiner, 2020):

Cell surface protein binding monoclonal antibodies: These antibodies bind antigens that are either exclusively or predominantly expressed by the cancer cells and cause antibody-dependent cell-mediated cytotoxicity, complement-mediated

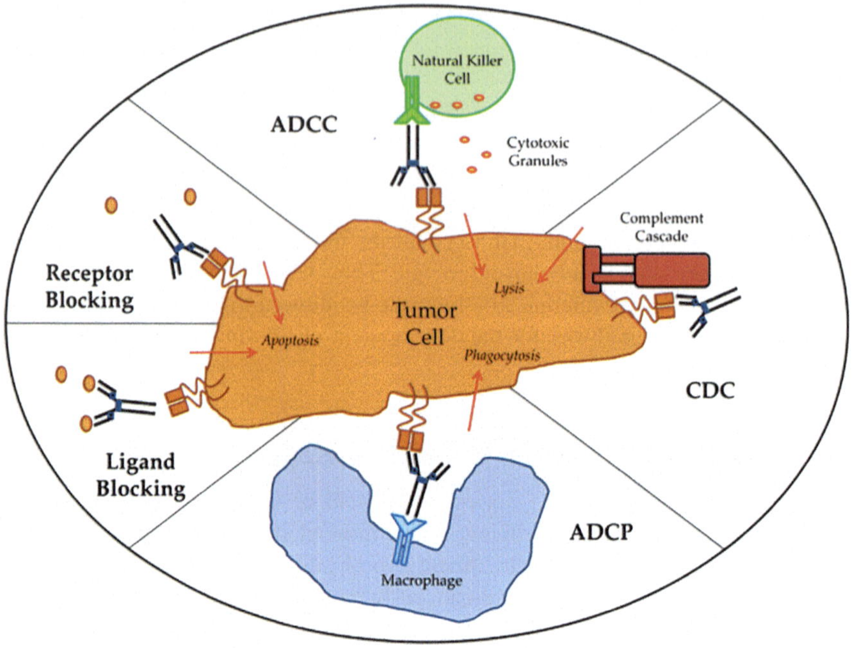

FIGURE 13.1

Antibody effector mechanisms. *ADCC*, antibody-dependent cellular cytotoxicity; *ADCP*: antibody-dependent cellular phagocytosis; *CDC*: complement-dependent cytotoxicity (Zahavi & Weiner, 2020).

cytotoxicity, antibody-dependent phagocytosis, cell death by direct effects of binding of the monoclonal antibodies to the cell surface protein causing disruption of the cells physiological functions or by binding of antibody conjugated toxins or radionuclides to the tumor cells (Fig. 13.1).

Examples of such monoclonal antibody-based immunotherapy are as follows:

1. Rituximab, a monoclonal antibody against the CD20 surface antigen of B cells, is used in chronic lymphocytic leukemia and non-Hodgkin lymphoma. It acts by binding to the CD20 trans membranous protein, which acts as a calcium channel, disrupting its function, and also by immune-mediated mechanisms such as, including antibody-dependent cell-mediated cytotoxicity, complement-mediated cytotoxicity, and antibody-dependent phagocytosis (Hanif & Anwer, 2025).
2. Gemtuzumab ozogamicin, an anti-CD33 calicheamicin conjugate, currently approved for use in acute myeloid leukemia, binds to the tumor cells, directly delivering calicheamicin, a potent chemotherapeutic agent, to the tumor cells (Baron & Wang, 2018).

3. Tigatuzumab, an experimental monoclonal antibody that binds to and activates necrosis factor receptor superfamily, member 10B (TNFRSF10B, also known as TRAILR2 or DR5), causing necrosis of tumor cells (Greenblatt & Khaddour, 2025).
4. Trastuzumab, a monoclonal antibody against human epidermal growth factor receptor 2 (HER2), which acts by facilitating antibody-dependent cellular cytotoxicity along with blocking HER2-mediated signaling pathways (Greenblatt & Khaddour, 2025).
5. Cell surface receptors or trophic factors binding antibodies: The monoclonal antibodies prevent tumor growth by binding to the factors needed for cancer cell survival, such as the cell surface receptors or trophic factors or factors that suppress cell cycle controls.

   Examples of such therapies are as follows:

1. Cetuximab, an epidermal growth factor receptor-specific monoclonal antibody, is used for the treatment of head and neck cancer and colorectal carcinoma (Chidharla et al., 2025).
2. Bevacizumab inhibits vascular endothelial growth factor A (VEGF-A) and prevents angiogenesis. It is used in several cancers such as colorectal carcinoma, nonsmall cell lung cancers, metastatic breast carcinoma, renal cell carcinoma, glioblastoma, ovarian cancer, and cervical cancer. It is also used in noncancerous conditions such as macular degeneration (Gerriets & Kasi, 2025).
3. Pembrolizumab: Binds to programmed death receptor 1 (PD1) on lymphocytes to prevent their binding to programmed death receptor ligand 1 (PDL1) that is overexpressed on cancer cells. This helps immune activation against cancer cells. It is currently used for unresectable or metastatic melanoma, nonsmall cell lung carcinoma, head and neck squamous cell carcinoma, renal cell carcinoma, and urothelial carcinoma (Flynn & Gerriets, 2025).
4. Blinatumomab is a "bi-specific T cell engager" that binds to the cd19 expressing cell of the B cell lineage and CD3 of the cytotoxic T cells, resulting in the triggering of the signaling cascade in the cytotoxic T cells, causing the release of perforin and granzyme, which ultimately cause the lysis of the target cell (Burt et al., 2019).

13.5.2.2 CAR-T cells or chimeric antigen receptor T cells

CAR-T cells are genetically modified autologous T cells. These cells are modified to express a transmembrane signal transducer molecule that, on its extracellular domain, expresses a chimeric antigen receptor and its intracellular domain expresses several signaling domains that activate the cytotoxic pathways through the release of perforin and granzyme (Rabie et al., 2024) (Fig. 13.2).

CART-T cell therapies are personalized therapies where the patient's autologous T cells are modified, cultured, and reintroduced into the patient's body (Fig. 13.3).

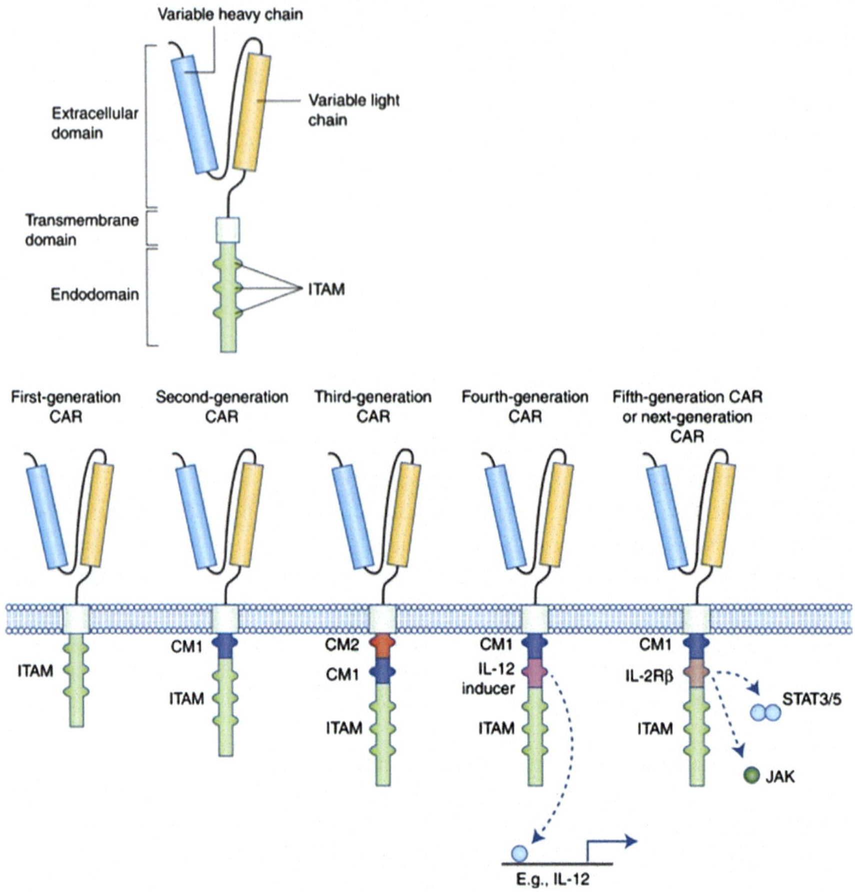

FIGURE 13.2

The above figure shows the basic architecture of chimeric antigen receptor (CAR) along with the various intracellular signal transduction and costimulatory domains of the different generations of CAR (Tokarew et al., 2019). *CM 1*, Costimulatory molecule 1 (CD28 or CD137); *CM 2*, Costimulatory molecule 2 (CD134 or CD137); *ITAM*, Immunoreceptor tyrosine-based activation motifs of the CD 3ζ signaling domain; *IL 12*, Interleukin 12; *IL 2Rβ*, Interleukin 2Rβ; *JAK/STAT*, Janus Kinase and signal transducer and activator of transcription transduction pathway.

CAR-T therapies have shown excellent results against B-cell Acute Lymphocytic Leukemias (B cell ALL), B-cell non-Hodgkin lymphoma (B cell NHL), Mantel cell lymphoma, Follicular Lymphoma, and multiple myeloma. The holy grail of CAR-T cell therapies is the development of efficacious and well-tolerated allogenic CAR-T cell therapies that can be used off the shelf.

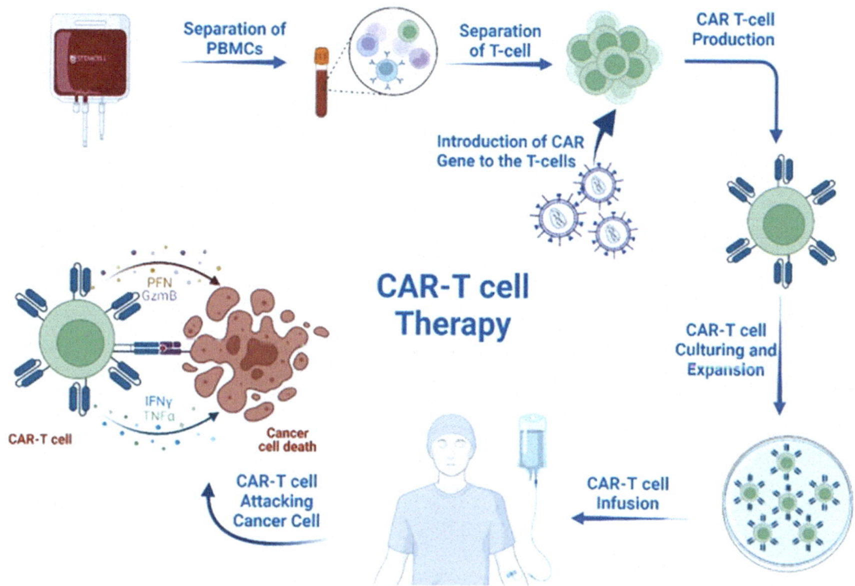

FIGURE 13.3

Schematic representation of the process of CAR T cell-based therapy (Rabie et al., 2024).

13.5.2.3 Dendritic cell-based therapies

Dendritic cell-based therapy used the patient's own antigen-presenting cells (APCs), dendritic cells, to elicit an immune response against the tumor cells. This is done by incubating the patient's own dendritic cells with a fusion protein that contains a tumor antigen and a granulocyte-macrophage colony-stimulating factor, which is an activator of immune cells (Fig. 13.2). After the sensitization of the dendritic cells to present the tumor antigen, the dendritic cells are injected back into the patients. These cells now present the tumor antigen to the patient's own immune system, which then attacks the tumor cells (Shang et al., 2017).

The only current dendritic cell-based therapy commercially available is Sipuleucel-T, an autologous dendritic cell-based therapy approved by the US FDA for treatment of castration resistant prostate carcinoma (Kantoff et al., 2010).

13.5.2.4 Oncolytic virus-based cancer therapies

Scientists had been reporting tumor remissions in patients suffering from viral infections right at the turn of the 19th century. Interestingly, such reports were available in literature even before the discovery of viruses, and knowing that the infection causing the remission was a viral infection (Dock, 1904). Though these remissions were short-lived, these findings suggested the possibility of using viruses in cancer therapy. With the advent of genetic engineering, the idea of using genetically engineered viruses to selectively attack tumor cells gained more ground.

Currently commercially available oncolytic viral therapies, as well as such therapies in development or under trail are designed to function in the following three ways (Lin et al., 2023).

1. *Direct oncolytic action*: These viruses selectively infect tumor cells and then, through a lytic cycle, cause tumor lysis.
2. *Activation of the immune system*: The virus-infected tumor cells can present tumor proteins to the immune system and sensitize the immune system against the tumor cells.
3. *Sensitizing the tumor cells to other therapies*: The oncolytic virus can carry to the tumor cells genes that disrupt the functions of proteins that make the tumor cells immune to chemotherapies, such as blocking the formation of PDL-1 protein.

The following are the viral oncolytic therapies commercially available:

1. Talimogene laherparepvec (TVEC) is a USFDA-approved, HSV type 1 vaccine, approved for treatment of nonresectable melanomas and in cases of recurrence (Ferrucci et al., 2021).
2. Nadofaragene firadenovec-vncg (ADSTILADRIN) is a USFDA-approved intravesical, nonreplicating adeno virus-based gene therapy for the treatment of high-risk Bacillus Calmette-Guérin (BCG)-unresponsive nonmuscle invasive bladder cancer with carcinoma in situ with or without papillary tumors. It acts by inserting cDNA for IFNα2b into the tumor cells, causing their lysis and also by pleotropic effects that sensitize the immune system against the tumor cells (Lee, 2023).
3. G47 (Delytact) is a Herpes simplex virus type 1-based oncolytic therapy for residual or recurrent glioblastoma, marketed by Daiichi Sankyo Company (Todo et al., 2022).
4. Oncorine (H101) is an adeno virus-based oncolytic viral therapy, currently marketed in China by Shanghai Sunway Biotech, for the treatment of head and neck cancers and acts by direct oncolytic effect as well as by activating the immune system against the tumor antigens (Liang, 2018).

13.5.2.5 Anticancer vaccines

Anti-cancer vaccines are another important biologics-based product used both in cancer prevention and cancer treatment. They can be divided into two types:

1. *Cancer preventive vaccines*: The most important among these is the HPV vaccine, which is successful in preventing HPV infections, the single most important cause of cervical carcinoma. Another example would be the Hepatitis B vaccine, which prevents Hep B infection, an important cause of hepatocellular carcinoma.
2. *Immunomodulatory cancer vaccines*: Coley's toxin, the progenitor of all cancer immunotherapies, falls in this group. Coley's toxin was a mixture containing toxins filtered from killed bacteria of the species *Streptococcus pyogenes* and *Serratia marcescens*, which stimulated an immune response and caused remission

in soft tissue sarcomas and bone cancers. The treatment was used from 1893 to as late as 1963. A modern example of an immunomodulatory vaccine would be the use of intravesical Bacillus Calmette-Guérin (BCG) for the treatment of early-stage vesicular carcinoma, especially nonmuscle invasive bladder cancer.

13.5.3 Regulatory environment in India and abroad

As biologics are synthesized using biological processes rather than strictly controlled chemical processes, as in the case of other therapeutic agents, there are issues of variance and contamination. These issues have shown up with devastating effects since the early days of biologics, as early as the dawn of the 20th century. In 1901, in St. Louis, Missouri, contamination of antidiphtheria serum with tetanus toxin resulted in the death of 13 children and was traced to a horse named Jim, who died of tetanus, and before his death, he was used to harvest anti-diphtheria serum (DeHovitz, 2014). The uproar surrounding this incident led to the passage of the Biologics Control Act of 1902 in the USA and, consequently, the formation of the United States Food and Drug Administration (USFDA) in 1906. The Bundaberg tragedy in Australia caused the death of 12 children due to contamination of the anti-diphtheria serum with Staph aureus (Hooker, 2000)· The first mass vaccine-related incident happened in 1955 in the western and midwestern states of the US, with over 200,000 children receiving an ineffectively inactivated Salk polio vaccine. Subsequent investigations showed that this batch of vaccine resulted in 40,000 cases of polio, leaving 200 children with varying degrees of paralysis and killing 10 (Offit, 2005).

Even today, as advanced biologics are bursting onto the scene, directed at a large number of neoplastic and non-neoplastic disorders, the risks of adverse events persist. Several biologics are associated with life-threatening and debilitating adverse effects (Campar & Isenberg, 2011).

In view of the above facts, biologics are strictly controlled by the drug regulatory bodies of the respective countries. In the US, the biologics are controlled by the Center for Biologics Evaluation and Research under the USFDA. In India, biologics are governed by the Central Drugs Standard Control Organization (CDSCO) under the Ministry of Health and Family Welfare and regulated under the Drugs and Cosmetics Act, 1940, and the Drugs and Cosmetics Rules, 1945.

Biologics, during their development and marketing phase, need to undergo the five steps of drug development like any other drug:

1. *Drug development or discovery*: This comprises largely of bench-top research where scientists, based on a hypothesis, develop a biologic using either existing techniques or using novel techniques.
2. *Preclinical testing*: In this phase, the new molecule is tested on cell culture and/or animal models to study their safety profile and their effect on the disorder of interest.
3. *Clinical trials:* Clinical trials are the testing of the drug on human subjects. Clinical trials are conducted in three phases: I, II, and III. These are premarketing studies

and are concerned with studying the pharmacokinetics and pharmacodynamics, as well as the therapeutic potential of the molecule of interest.

4. *The review of data*: Thereafter, the regulatory agencies, such as the USFDA and CDSCO, review the trial data and decide whether the drug can be marketed for the particular ailment or not.
5. *Postmarketing safety monitoring*: If the drug is allowed to be marketed, the regulatory agencies keep a tab on the safety profile of the drug by means of ADR reporting, etc. These are also known as the phase IV postmarketing trial.

In the US, the biologics are provided a period of 12 years of exclusivity from the date of USFDA approval. This is irrespective of the expiry of their patent, which lasts for 20 years. In the European Union, the marketing exclusivity period is 10 years with a possible extension of another year, if a new indication is added to the therapeutic indication of the biologic in the first 8 years. India does not have an exclusivity interval designated for biologics, but allows for a patent for 20 years with no provisions for extension.

Most countries, unlike India, also have a data exclusivity provision during which no company can use the data submitted to the regulator by the original manufacturer for their own generic version. The USA has a data exclusivity period of 4 years, while the EU has a data exclusivity provision of 8 years. This provision is important in the case of biosimilars, as their approval process is expedited based on the data of the reference biologics, with the biosimilars only needing to prove their similarity to the reference biologics rather than prove their safety and efficacy, as was done by the reference biologic, by conducting clinical trials.

Biosimilars are allowed to be marketed postlapse of the patent and marketing exclusivity periods of the reference biologics. These biosimilars have an expedited path to approval by the drug regulatory authority's world over. Drug companies manufacturing biosimilars are allowed to use the safety and efficacy data of the reference biologics after the lapse of the data exclusivity period. They are allowed to use this data to argue the efficacy and the safety of their biosimilar instead of going through the lengthy and expensive process of preclinical and clinical trials. These companies are only expected to show efficacy and safety profiles of the biosimilar to be similar as that of the reference biologics.

13.5.4 Quality control in biologics and biosimilars

As with any biological product, there is an expected variance in the biologics. These variances need to be controlled within a narrow range between the biologics and biosimilars, and also between different lots of the biologics and biosimilars themselves.

The quality and efficacy testing include the following steps:

1. *Identity testing*: Using different techniques such as iso electric focusing, chromatography, HPLC, ELISA, western blotting, etc., to prove molecular equivalence

between reference biologics and biosimilars and among the different lots of biologics and biosimilars themselves.

2. Purity and impurity testing using methods such as chromatographic and spectroscopic methods.
3. *Potency testing*: using ELISA, Bioassays, and ligand binding assays, etc.
4. *Stability testing*: By testing the stability of the product by storing it over time in different storage conditions.
5. Contamination control: by purity testing methods and by implementing quality by design parameters that prevent contamination during the manufacturing process.

The space for biologics in therapy will only increase in the future as new tools of genetic manipulation, like CRISPR/Cas9, make it easier to create transgenic organisms and even create organisms that produce proteins that are not native to any living system. Cancer therapies are becoming more and more driven by biologics, with biologics now becoming first-line cancer treatments. Biologics are being looked at as a hope for the future in a postantibiotic world. In view of this, physicians need to understand the possibilities and limitations of this field of therapeutics.

As the patents of more and more biologics are lapsing, the field of biosimilars is also becoming wide open. Hence, our country needs to be ready to exploit this opening and live up to its reputation of being the "pharmacy of the world."

13.6 Biologics in urothelial carcinoma: Current landscape

Immunotherapies used to play a crucial role in targeting urothelial carcinoma, which function by directing the immune system to recognize and expel the cancerous cells. Immune checkpoint inhibitors, intravesical therapy, and cytokine therapies are several types approved as the most effective strategies. Other biological techniques involve antibody-drug conjugates and oncotic viruses as well.

13.6.1 Immune checkpoint inhibitors

Multiple immune checkpoint inhibitors are a type of immunotherapy that works by blocking different receptor proteins (like PD-1, PD-L1, and CTLA-4, etc.), which are either small molecules or monoclonal antibodies, authorized for treating locally advanced and metastatic urothelial carcinoma, especially recurrence after platinum-drug induced chemotherapy or neoadjuvant cisplatin-based multi-chemotherapies (Roviello et al., 2021). Alongside the chemotherapy, numerous antiangiogenic drugs (atezolizumab, bevacizumab, avelumab, sorafenib, and sunitinib), growth factor-targeted inhibitors (anti-EGFR: gefitinib; anti-HER-1/2: lapatinib) were also implemented as the first-line systemic therapy (Rizzo et al., 2021).

According to various clinical reports, patients undergoing platinum-induced chemotherapy frequently experience an adverse progression of bladder cancer in 70% of cases. Where fibroblast growth factor receptor inhibitor (anti-FGFR: erdafitinib)

(Necchi et al., 2019), and vascular endothelial growth factor receptor-2 inhibitor (anti-VEGFR-2: ramucirumab) (Petrylak et al., 2020) are considered for second-line targeted therapy for advanced metastasis.

Mechanism of action: Lymphocytes and WBCs are the major components of the immune system, which mature in bone marrow from lymphoblasts and fight against infections. B-lymphocytes, also known as B-cells, and T-lymphocytes, sometimes known as T-cells, are the two primary forms of lymphocytes (Cano & Lopera, 2013). Among several types of T-cells, helper T-cells, that recognizes the nonself/foreign antigenic particles and activate B-cells and T-cells for attack. The killer T-cells specifically target cancer cells or infected cells by introducing proteins and accelerate necrosis. After an immune reaction, regulatory T-cells help the immune system to slow down, and lastly, memory T-cells store the information of the antigen for years (Alberts et al., 2002).

However, by slowing down the active immunity via regulatory T-cells, checkpoints prevent the immune system from attacking healthy cells. Once the immune response is completed, a sequence of impulses between appropriate proteins and surface proteins regulates the active T-cells to reduce hypersensitivity. Under normal conditions, immune checkpoints are essential for maintaining self-tolerance of the body and preventing autoimmunity. Although, checkpoints help shield tissues from harm when the immune system is reacting to a pathogenic infection, the dysregulated expression of checkpoint receptors (CTLA4: Cytotoxic T-lymphocyte associated protein-4; PD-1: Programmed cell death protein-1; PD-L1: Programmed cell death-ligand 1; LAG3: Lymphocyte activation gene-3; TIM3: T-cell membrane protein-3, etc.) in tumors also end up with immunological resistance strategy (Pardoll, 2012). Hence, the immune checkpoint blockers unlock the potential of anti-tumor immune response and revolutionize cancer therapeutics (Table 13.1).

Besides small molecular agents, monoclonal antibodies are also designed to act as immune checkpoint inhibitors, used mainly for restoring immune surveillance in cancers (Zamani & Šácha, 2025). Among them, pembrolizumab, atezolizumab, durvalumab, avelumab, ipilimumab, and nivolumab are the recent FDA-approved monoclonal antibodies that function as ICI by targeting various immune checkpoint receptor proteins (Ribas & Wolchok, 2018).

Common side effects: Immune therapies come with several mild to adverse consequences, like other cancer treatments. As these therapies over-stimulate the immune system, mild symptoms include high fever, swelling, redness, inflammation, shortness of breath, cough, constipation, headache, nausea, diarrhea, and mild to severe abdominal pain. The most severe adverse effects of immune checkpoint inhibitors are immune-related disorders, such as cardiomyositis, pneumonitis, gastrointestinal abnormalities, hepatitis, encephalitis, endocrinopathies, and dermatitis, etc.

13.6.2 Monoclonal antibodies and antibody-drug conjugates

While UC remains an emerging and fourth common global burden, the therapeutic landscape has significantly upgraded in recent years. Other than the immune checkpoint inhibitors, the FDA also approved monoclonal antibodies and novel

Table 13.1 Common immune checkpoint inhibitor (ICI) small molecules used in recent clinical trials.

ICI Drugs	Molecular Structure	Target	Clinical Trial	Cancers to Treat
CA-170		PD-L1, PD-L2 and VISTA	*NCT02812875* (Phase-I)	Solid tumor and lymphoma (Lee et al., 2017)
IMMH-010 (YPD-30)		PD-L1	NCT04343859 (Phase-I)	Advanced malignant solid tumors (Tianjin Chasesun Pharmaceutical Co , LTD, 2023)
MAX-10181		PD-L1	NCT04122339 (Phase-I)	Advanced malignant solid tumors (Maxinovel Pty., Ltd., 2022)
GS-4224		FD-L1	NCT04049617 (Phase-I)	Advanced malignant solid tumors (Odegard et al., 2024)
BMS-202		PD-L1	Preclinical	Human glioblastoma (Yang et al., 2024)
BMS-986189		PD-L1	NCT02739373 (Phase-I)	Healthy volunteers (Gaynor et al., 2022)

antibody-drug conjugates for treating various metastatic cancers (Lattanzi & Rosenberg, 2020). Antibodies are initially defined for neutralizing external substances found in blood. However, the laboratory-made mAbs are designed from unique B-cells to function as the substitute of antibodies, restoring, enhancing, and modifying the immune activation against cancer cells by several mechanisms (Zahavi & Weiner, 2020).

Structure and Mechanism of mAbs: Antibodies are the glycoproteins from the immunoglobulin superfamily (Ig), synthesized in the body to recognize antigens, stimulate immune response, and neutralize them. Structurally, it contains two light and two heavy chains, along with the fragment antigen-binding site (Fab), and a fragment for crystallization (Fc) region for interacting with other immune cells. Based on the various structures of heavy chains, IgGs can be subclassified into IgA, IgG, IgM, IgD, and IgE. However, IgGs are most often used for antibody therapies, as they can be associated with multiple antigen-presenting cells (APCs), like lymphocytes, NK cells, dendritic cells, and complement-dependent cytotoxic cells (CDCs). Depending on the Fc region structures, IgGs can be further classified into IgG1, IgG2, IgG3, and IgG4 (Zahavi & Weiner, 2020).

A clonal form of a particular antibody isotype that is directed against a distinct antigen epitope is known as a monoclonal antibody. These specific mAbs are targeted to block the growth factor receptor-mediated signaling mechanism. While binding with the growth factor receptor, mAbs alter their activation state or prevent ligand binding, which disrupts pro-tumor survival and proliferation signals, leading to tumor cell death (Fig. 13.4). The Fab arm is crucial to the effectiveness of mAb therapy. Furthermore, the Fc tail, which controls an antibody's downstream effector actions, is also crucial. Consequently, the final result of an antibody's binding to its target is likewise impacted by the isotype that is selected (Vukovic et al., 2021).

For example, cetuximab (anti-EGFR mAb) induces apoptosis in tumor cells by targeting receptor dimerization and ligand binding of EGFR growth factor (Patel et al., 2009), where trastuzumab (anti-HER2 mAb) is applied for treating $HER2^+$ early breast cancer (Sanglier et al., 2022) (Table 13.2).

Antibody-drug conjugates (ADCs): Usually, the monoclonal antibodies (mAbs) are covalently attached to some cytotoxic chemical agent, which facilitates the elimination of cancer cells more accurately and efficiently (Fu et al., 2022). The very first FDA-approved ADC, used in acute myeloid leukemia (AML), is Gemtuzumab ozogamicin (Mylotarg) in 2000 (Norsworthy et al., 2018). Among all the subtypes of IgG antibodies, IgG1 is the most common to be used for ADC-linked immunotherapy and IgG3 is rarely employed due to its rapid clearance rate (half-life) in serum (Stapleton et al., 2011).

It plays a synergistic role to target and kill the cancer cells in a specific way, called a precision-guided "biological missile," with minimal off-target side effects. Once the ADC-bound mAb recognizes the target receptor/antigen, expressed on cancerous cells, these conjugates get released and endocytosed by the lysosome, forming endosomes. Targeting DNA or microtubules, the cytotoxic payloads are ultimately released into the lysosomes via chemical or enzyme means, causing cell death or apoptosis (Fu et al., 2022) (Table 13.3).

Table 13.2 List of FDA-approved monoclonal antibodies (mAbs) and their target receptor proteins in different cancers.

Generic Name	Trade Name	Unconjugated Antibody	Target Antigen	Cancer to Treat
Atezolizumab (mAb acts as both ICI)	Tecentriq	Humanized IgG1	PD-L1	Urothelial carcinoma (bladder cancer), small cell, nonsmall cell lung carcinoma, and liver carcinoma
Avelumab (mAb acts as both ICI)	Bavencio	Humanized IgG1	PD-L1	Markel cell carcinoma (skin cancer) and urothelial carcinoma
Bevacizumab	Avastin	Humanized IgG1	VEGF	Colorectal cancer, nonsmall cell lung cancer, renal carcinoma, Glioblastoma, and cervical carcinoma
Cemiplimab	Libtayo	Humanized IgG4	PD-1	Cutaneous squamous-cell carcinoma, nonsmall cell lung cancer
Cetuximab	Erbitux	Humanized IgG1	EGFR	Colorectal cancer, head and neck squamous carcinoma
Daratumumab	Darzalex	Human IgG1	CD38	Multiple myeloma
Dinutuximab	Unituxin	Chimeric IgG1	GD2	Neuroblastoma
Durvalumab (mAb acts as both ICI)	Imfinzi	Human IgG1	PD-L1	Urothelial carcinoma, small cell, nonsmall cell carcinoma, and liver carcinoma
Elotuzumab	Empliciti	Humanized IgG1	SLAMF7	Multiple myeloma
Ipilimumab (act as both ICI and mAbs)	Yervoy	Human IgG1	CTLA-4	Melanoma, renal cell carcinoma, colorectal cancer, hepatocellular (liver) carcinoma, oesophageal carcinoma, nonsmall cell lung carcinoma, pleural mesotheliomas

(continued on next page)

Table 13.2 List of FDA-approved monoclonal antibodies (mAbs) and their target receptor proteins in different cancers—cont'd

Generic Name	Trade Name	Unconjugated Antibody	Target Antigen	Cancer to Treat
Isatuximab	Sarclisa (Sanofi-Aventis)	Chimeric IgG1	CD38	Multiple Myeloma
Mogamulizumab	Poteligeo (mogamulizuman-kpkc)	Humanized IgG1	CCR4	Cutaneous T-cell lymphoma
Necitumumab (mAb acts as both ICI)	Portazza	Human IgG1	EGFR	Nonsmall cell lung cancer
Nivolumab (mAb acts as both ICI)	Opdivo	Human IgG4	PD-1	Melanoma, nonsmall cell lung carcinoma, renal cell carcinoma, classical Hodgkin lymphoma, colorectal cancer
Obinutuzumab (mAb acts as both ICI)	Gazyva	Humanized IgG2	CD20	Chronic lymphocytic leukemia
Ofatumumab (mAb acts as both ICI)	Arzerra	Human IgG1	CD20	Chronic lymphocytic leukemia
	Kesimpta			Multiple sclerosis
Olaratumab	Lartruvo	Human IgG1	PDGFRα	Sarcoma
Panitumumab	Vectibix	Human IgG2	EGFR	Colorectal Cancer
Pembrolizumab (mAb acts as both ICI)	Keytruda	Humanized IgG4	PD-1	Melanoma
Pertuzumab	Perjeta	Humanized IgG1	HER2	Breast cancer
Ramucirumab	Cyramza	Human IgG1	VEGFR2	Gastric cancer
Rituximab	Rituxan	Chimeric IgG1	CD20	B-Cell Lymphoma
Trastuzumab	Herceptin	Humanized IgG1	HER2	HER2+ breast cancer and gastric cancer

Table 13.3 FDA, EMA, PDMA, and NMPA approved antibody-drug conjugates (ADCs) (Fu et al., 2022).

Drugs	Trade Name	Unconjugated Antibody	Target Antigen	Linkers	Payloads	Approved to Treat Cancer (Approval Year)
Gemtuzumab ozogamici	Mylotarg,	Humanized ADC	CD33	Hydrazone	N-acetyl-γ-calicheamicin	$CD33^+$ AML
Brentuximab vedotin	Adcetris	Chimeric ADC	CD30	mc-VC-PAEC	MMAE	$CD30^+$ Hodgkin's lymphoma and Anaplastic large-cell lymphoma
Ado-Trastuzumab emtansine	Kadcyla	Humanized ADC	HER2	SMCC	DM1	$HER2^+$ Breast cancer
Inctuzumab ozogamicin	Besponsa	Humanized ADC	CD22	Hydrazone	N-acetyl-γ-calicheamicin	Acute lymphoblastic leukemia
Moxetumomab pasudotox	Lumoxiti	Mouse ADC	CD22	mc-VC-PABC	PE38	Hairy-cell leukemia
Fam-trastuzumab deruxtecan	Enhertu	Humanized ADC	HER2	Tetrapeptide	DXd	$HER2^+$ Breast cancer
Polatuzumab vedotin	Polivy	Humanized ADC	CD79B	mc-VC-PABC	MMAE	B-Cell Lymphoma

(continued on next page)

Enfortumab vedotin	Padcev	Human ADC	Nectin-4	mc-VC-PABC	MMAE	Urothelial (bladder) cancer
Sacituzumab govitecan	Trodelvy	Humanized ADC	Trop-2	CL2A	SN38	Triple negative breast cancer
Belantamab mafodotin	Blenrep	Humanized ADC	BCMA	mc	MMAF	Multiple myeloma
Cetuximab sarotalocan	Akalux	Chimeric fluorescent probe conjugate	EGFR	NA	IRDye700DX	Locally advanced or recurrent Head and neck cancer
Loncastuximab tesirine	Zynlota	Humanized ADC	CD19	Dipeptide	PBD dimer (SG3199)	B-cell lymphoma
Disitamab vedotin	Aidixi	Humanized ADC	HER2	mc-VC-PABC	MMAE	Metastatic gastric cancer
Tisotumab vedotin	Tivdak	Humanized ADC	TF	mc-VC-PABC	MMAE	Metastatic cervical cancer

CL2A, *cleavable complicated;* PEG8, *and triazole-containing;* PABC, *peptide-mc linker;* DM1, *derivative of maytansine1;* DXd, *exatecan derivative for AD;* EMA, *European Medicine agency;* FDA, *US Food and Drug administration;* IRDye700DX, *near infrared photosensitizing dye;* mc, *maleimidocaproyl;* mc-VC-PABC, *Maleimidocaproyl-valine-citrulline-p-amino benzoyl oxycarbonyl;* MMAE, *monomethyl auristatin E;* MMAF, *Monomethyl auristatin-F;* NMPA, *National Medical Products administration of China;* PBD, *pyrrolobenzodiazepine;* PDMA, *pharmaceuticals and medical devices agency ofJapan;* PE38, *38kD fragment of pseudomonas exotoxin A;* SMCC, *succinimidyl-4-(N-maleimidomethyl) cyclohexane-1-carboxylate;* SN38, *active metabolite of irinotecan.*

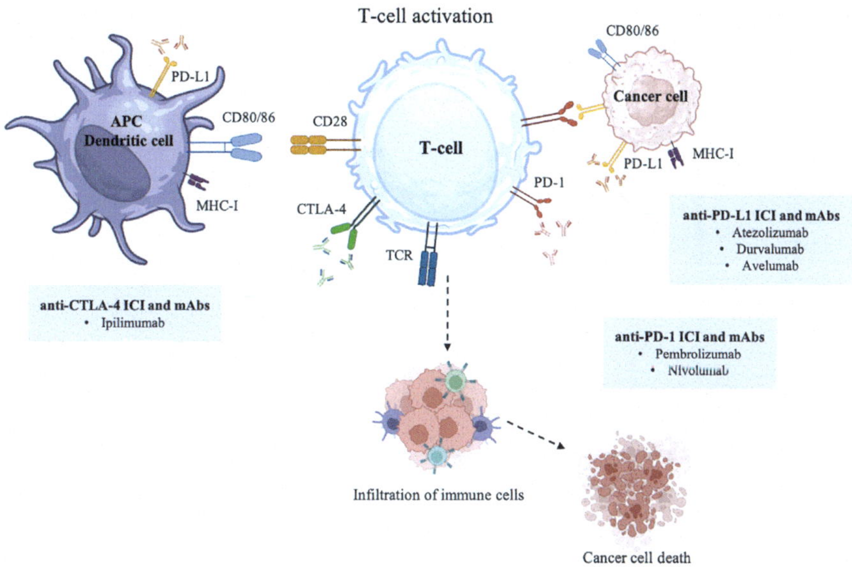

FIGURE 13.4

Mechanism of T-cell activation via immune checkpoint inhibitors (ICIs) and monoclonal antibodies (mAbs) against PD-1, PD-L1, and CTLA-4 receptor proteins (Created using Biorender.com).

13.6.3 Clinical trials supporting urothelial carcinoma therapies

(Table 13.4)

13.7 Biosimilars in oncology: Emergence and expansion

13.7.1 Concept and development of biosimilars

Biologics are medicinal products derived from living organisms, typically produced through recombinant DNA (rDNA) technology, cell culture, or other advanced biotechnological processes. Unlike small-molecule drugs, which are chemically synthesized and structurally simple, biologics are large, complex macromolecules with high molecular weights, often requiring intricate folding and posttranslational modifications for functional activity. These attributes make biologics inherently more variable and less stable than their small-molecule counterparts (Walsh, 2018). The advent of biologics has revolutionized modern medicine. They have become an integral part of the standard of care for a wide range of diseases, including oncological conditions, autoimmune disorders, cardiovascular diseases, and hematological disorders such as hemophilia (Lyman et al., 2018). Despite their remarkable therapeutic impact, the

Table 13.4 Immunotherapy clinical trials using monoclonal antibody (mAb) and an immune checkpoint inhibitor (ICI) in urothelial carcinoma.

Trials	Line of Treatment	Phase	FDA Approval	CTX	ICI agent	ICI Target	No. of Patients (N)	ORR (%)	PFS (mo)	OS (mo)
IMvigor211 NCT02302807 (Powles et al., 2018)	Second line	III	2016	Yes	Atezolizumab (Tecentriq)	PD-L1 inhibitor	N=931 Atezolizumab (467)	13.4%	2.1 mo	11.1 mo
							Chemotherapy (464)	13.4%	4.0 mo	10.6 mo
IMvigor210 NCT02108652 (Balar et al., 2017)	First line + PD-L1 + Platinum ineligible patients	II	2017	No	Atezolizumab (Tecentriq)	PD-L1 inhibitor	N = 123	23%	2.7 mo	15.9 mo
IMvigor130 NCT02807636 (Galsky et al., 2020)	First line	III	2020	No	Atezolizumab (Tecentriq)	PD-L1 inhibitor	N = 1213 Atezolizumab+ CDDP/BCDCA+ Gemcitabine (451)	47%	8.2 mo	16 mo
							Atezolizumab (362)	23%	6.3 mo	15.7 mo
							CDDP/CBDCA+ Gemcitabine + Placebo (400)	44%	6.2 mo	13.4 mo
JAVELINSolid Tumor NCT01772004 (Apolo et al., 2017)	Second line	I	2017	Yes	Avelumab (Bavencio)	PD-1/PD-L1 inhibitor	N = 44	18.2%	11.6 wk	13.7 mo
JAVELIN Bladder 100 NCT02603432 (Powles et al., 2020a)	Maintenance therapy	III	2020	Yes	Avelumab (Bavencio)	PD-1/PD-L1 inhibitor	N = 700	-	3.7 mo	21.4 mo

(continued on next page)

CheckMate-275 NCT02387996 (Sharma et al., 2017)	Second line	II	2017	Yes	Nivolumab (Opdivo)	PD-1 inhibitor	N = 265	19.6%	2.0 mo	8.7 mo
KEYNOTE-052 NCT02335424 (Vuky et al., 2020)	Fist line + PD-L1 + Platinum ineligible patients	II	2017	No	Pembrolizumab (Keytruda)	PD-1/PD-L1 inhibitor	N = 370	29%	2.2 mo	11.3 mo
KEYNOTE-045 NCT02256436 (Bellmunt et al., 2017)	Second line	III	2017	Yes	Pembrolizumab (Keytruda)	PD-1/PD-L1 inhibitor	N = 542	21%	2.1 mo	10.3 mo
KEYNOTE-361 NCT02853305 (Alva et al., 2020)	First line + Platinum + PD-L1 combined therapy	III	2020	No	Pembrolizumab (Keytruda)	PD-1/PD-L1 inhibitor	N = 1010 Pembrolizumab (307)	30.3%	3.9 mo	15.6 mo
							Pembrolizumab + CTX (351)	54.7%	8.3 mo	17.0 mo
							Platinum-based CTX (352)	44.9%	7.1 mo	14.3 mo
MEDI4736 NCT01693562 (Powles et al., 2017)	Second line	I/II	2017	Yes	Durvalumab (Infinzi)	PD-L1 inhibitor	N = 191	18%	1.5 mo	18.2 mo
DANUBE NCT02516241 (Powles et al., 2020b)	First line	III	2020		Durvalumab (Imfinzi)	PD-L1 inhibitor	N = 1032 Durvalumab (346)	25.7%	9.3 mo	14.4 mo
							Durvalumab + Tremelimumab (342)	36.3%	11.1 mo	15.1 mo
							Chemotherapy (344)	49.1%	5.7 mo	12.1 mo

CBDCA, *carboplatin;* CDDP, *cisplatin;* CTX, *platinum-based chemotherapy;* FDA, *US Food and Drug Administration;* mo, *months;* ORR, *objective response rate;* OS, *overall survival;* PFS, *progression free survival;* wk, *Weeks.*

high cost of biologics has imposed significant financial burdens on healthcare systems and restricted patient access, particularly in low- and middle-income countries. To address these challenges, biosimilars were developed as cost-effective alternatives. A biosimilar is defined as a biologic product that is highly similar to an already approved reference biologic, with no clinically meaningful differences in terms of quality, safety, and efficacy (U.S. Department of Health and Human Services Food and Drug Administration Center for Drug Evaluation and Research [CDER], Center for Biologics Evaluation and Research [CBER], 2015). Unlike generics of small molecules, biosimilars are not identical copies because biologics are produced in living systems such as bacteria, yeast, or mammalian cells, which introduce inherent variability in the final product. Instead, biosimilars are designed to demonstrate a high degree of similarity through a rigorous comparability exercise. The first biosimilar approved was somatropin (Omnitrope) by the European Medicines Agency (EMA) in 2006, marking a historic milestone in global drug development (European Medicines Agency (EMA), 2018). Since then, biosimilars have gained regulatory approval across multiple therapeutic classes, including monoclonal antibodies, growth factors, and cytokine inhibitors. Their introduction has fostered increased competition, reduced treatment costs, and improved patient access to advanced therapies (Blackstone & Joseph, 2013).

The manufacturing of biosimilars is particularly complex, often described as a process of reverse engineering. Unlike generics, where the chemical structure can be reproduced exactly, biosimilar development begins with the final therapeutic protein of interest and then reconstructs the manufacturing pathway through detailed structural, functional, and process characterization. The variability introduced by expression systems such as yeast, bacteria, or mammalian cell lines means that an identical copy of a reference biologic is scientifically unattainable. Nevertheless, through stringent analytical characterization, functional assays, and confirmatory clinical studies, biosimilars can be demonstrated to be clinically equivalent to their reference products (Schellekens, 2009). The regulatory framework for biosimilars has evolved significantly. The European Medicines Agency (EMA) was the first to establish biosimilar approval pathways in 2005, followed by the U.S. Food and Drug Administration (FDA) under the Biologics Price Competition and Innovation Act (BPCIA) in 2009. India has also emerged as a global leader, with its Guidelines on Similar Biologics first issued in 2012 and later revised in 2016 and 2019, contributing to a rapidly growing biosimilar industry (Meher et al., 2019). Today, India is among the largest suppliers of biosimilars globally, enhancing access to affordable biologics in both domestic and international markets. In summary, the concept and development of biosimilars represent both a scientific and economic breakthrough. Scientifically, it has required the advancement of high-resolution analytical technologies and robust comparability studies. Economically, biosimilars have contributed to cost containment, improved access, and sustainability of healthcare systems, particularly in oncology and chronic diseases. Their continued development is expected to play a pivotal role in global healthcare delivery.

13.7.2 Approved oncology biosimilars: Global and Indian market

India has emerged as a global leader in biosimilar development, locally referred to as "similar biologics." These products are developed to demonstrate high similarity to reference biologics approved by stringent regulatory agencies such as the U.S. Food and Drug Administration (FDA) and the European Medicines Agency (EMA). While biosimilars cannot be identical to the original biologics due to the complexity of living cell–based production and inherent variability in posttranslational modifications (PTMs), they are required to establish comparability in terms of safety, efficacy, potency, and immunogenicity (.; "U.S. Department of Health and Human Services Food and Drug Administration Center for Drug Evaluation and Research [CDER], Center for Biologics Evaluation and Research [CBER], 2015). Globally, the approval of Omnitrope (somatropin) by the EMA in 2006 laid the foundation for the regulatory and commercial acceptance of biosimilars. In oncology, biosimilars have been transformative in both therapeutic and supportive care settings. They are extensively evaluated through analytical characterization, preclinical studies, pharmacokinetic or pharmacodynamic assessments, and comparative clinical trials before approval (Lyman et al., 2018).

13.7.2.1 Global oncology biosimilars

In the oncology space, biosimilars are approved for both direct anticancer activity (e.g., monoclonal antibodies such as rituximab, trastuzumab, and bevacizumab) and for supportive care (e.g., myeloid growth factors).

- Myeloid growth factors such as filgrastim and pegfilgrastim biosimilars were among the first oncology-related biosimilars approved in the U.S. and Europe. These agents are widely used for the prevention of chemotherapy-induced neutropenia, significantly lowering the costs of treatment and improving patient access (Zelenetz et al., 2011).
- Rituximab biosimilars (used in non-Hodgkin lymphoma and chronic lymphocytic leukemia) have been approved in both the EMA and FDA markets (e.g., Truxima, Rixathon, and Riabni).
- Trastuzumab biosimilars (for HER2-positive breast and gastric cancers) have had a major global impact, with multiple approvals (e.g., Ogivri, Herzuma, Kanjinti).
- Bevacizumab biosimilars (anti-VEGF, for colorectal, lung, and other cancers) have been approved by EMA and FDA (e.g., Mvasi, Zirabev).

These biosimilars have been integrated into international oncology guidelines, including those by the European Society for Medical Oncology (ESMO) and the National Comprehensive Cancer Network (NCCN), underscoring their equivalence to reference biologics.

Indian oncology biosimilars: India's journey began with the approval of its first biosimilar for hepatitis B vaccine in 2000, after which the biosimilar market has expanded rapidly (Meher et al., 2019). The Indian biosimilar landscape in oncology is now robust, with approvals across major therapeutic classes:

- Rituximab biosimilars (e.g., Reditux by Dr Reddy's Laboratories, 2007) were among the earliest monoclonal antibody biosimilars worldwide, approved before the establishment of formal biosimilar guidelines.
- Trastuzumab biosimilars (e.g., Herceptin biosimilars by Biocon/Mylan, 2014) have provided cost-effective options for HER2-positive cancers, making targeted therapy accessible to a larger patient population in India.
- Bevacizumab biosimilars (launched by multiple Indian manufacturers) are now available for colorectal and lung cancers.
- Filgrastim and Pegfilgrastim biosimilars are widely used in Indian oncology practice, reducing the cost of neutropenia prophylaxis.

India currently has over 100 approved biosimilars, with a significant portion dedicated to oncology and supportive care (McCamish & Woollett, 2012). Its regulatory framework, guided by the "Guidelines on Similar Biologics" (2012, revised in 2016 and 2019), has enabled local manufacturers to enter both domestic and international markets, positioning India as a hub for affordable biosimilar production.

The approval and adoption of oncology biosimilars have had profound economic implications, reducing treatment costs by 30%–70% compared to originator biologics (Blackstone & Joseph, 2013). This is particularly critical in oncology, where biologic therapies constitute some of the highest-cost treatments. Biosimilars thus play a dual role:

- Clinically, they expand patient access to life-saving therapies.
- Economically, they reduce the burden on healthcare systems and foster sustainable oncology care.

Challenges and future directions: Despite their promise, biosimilars in oncology face challenges, including:

- Physician and patient confidence regarding interchangeability.
- Pharmacovigilance requirements to monitor long-term safety.
- Regulatory harmonization across countries.
- Market penetration barriers due to brand loyalty and aggressive pricing strategies of innovator companies.

Nevertheless, ongoing real-world evidence and increasing endorsement by global oncology societies are expected to strengthen the role of biosimilars in cancer care worldwide.

13.7.3 Cost effectiveness and health care economics

UC imposes a large and sustained economic burden due to high recurrence rates, long-term surveillance, repeated procedures, and costly systemic therapies. In the United States, national expenditures for bladder cancer have been estimated in the billions annually, and per-patient lifetime costs frequently exceed six figures. These high costs are driven not only by drug acquisition but also by hospital admissions, surgical procedures, and surveillance (Clark et al., 2024).

Cost drivers in contemporary UC care:

1. *Systemic therapies.* The advent of immune-checkpoint inhibitors (ICIs) and antibody–drug conjugates (ADCs) has improved outcomes for many patients, but at high prices. Annual costs for ICI therapy and ADC regimens can reach tens to hundreds of thousands of dollars per patient, pushing many new agents beyond traditional willingness-to-pay thresholds in some health-economics analyses (Walia et al., 2021)
2. *Procedures and surveillance.* Frequent cystoscopies, intravesical treatments, and surgeries (including cystectomy) add substantial direct medical costs, particularly for nonmuscle-invasive disease with repeated recurrences (Bree et al., 2022).

Cost-effectiveness of new agents: Pembrolizumab (PD-1 inhibitor): cost-effectiveness analyses show mixed results across settings and jurisdictions. Some studies conclude that pembrolizumab may be cost-effective relative to chemotherapy under higher willingness-to-pay thresholds, but ICERs are often substantial (Chen & McQueen, 2023).

Enfortumab vedotin and sacituzumab govitecan (ADCs): clinical benefit is clear in selected settings, but several analyses indicate they may not be cost-effective at current prices in multiple countries; pricing and durations of therapy heavily influence ICER estimates (Op 't Hoog et al., 2025).

Role of biosimilars in cost containment and access

Biosimilars can substantially reduce drug acquisition costs and create fiscal space within health systems. Real-world and modelling studies show that biosimilar entry (especially for supportive-care biologics such as filgrastim and for oncology mAbs) produces meaningful savings, higher utilization of guideline therapies, and the potential to expand patient access to effective biologics. These savings can translate into more patients treated, or funds reallocated to diagnostics, surgical capacity, or supportive services.

Health-system and policy factors that affect savings

- Procurement and reimbursement policies (e.g., tendering, reference pricing, formulary placement) determine the degree of biosimilar price erosion and uptake (Cross, 2022).
- Institutional incentives may blunt biosimilar adoption in some settings (for example, discount programs that favour originator products), reducing expected savings (Reed, 2023).
- Physician and patient confidence: educational initiatives and robust real-world evidence help increase prescribing of biosimilars.

Implications for UC care and research priorities

- *Budget impact*: Modeling for UC should include both direct drug savings from biosimilar adoption and secondary effects (e.g., earlier therapy initiation, fewer hospitalizations) (Kvien et al., 2022).
- *Value-based pricing*: aligning price to clinical benefit (e.g., outcomes-based contracts) could improve cost-effectiveness for high-cost ADCs and ICIs (Kaye, 2023).
- *Access in LMICs*: domestic biosimilar manufacture (where feasible) can markedly improve access in middle- and low-income countries and reduce dependence on expensive imports.

13.7.4 Case studies in other cancers (like breast cancer, lymphoma)

A. Breast Cancer: Trastuzumab Biosimilars

1. Budget-Impact Analyses:
    - Croatia (2015): Introducing trastuzumab biosimilars (at 15%–35% discounts) projected drug cost savings ranging from €0.26 million to €0.69 million, enabling treatment of 14–47 additional patients (Cesarec & Likić, 2017).
    - China (2022): Over 5 years, biosimilar adoption could yield net savings of US $10 million to $60 million annually, totalling approximately US $46.7 million, and potentially treat an additional 654 to 13,200 patients, depending on pricing and procurement policy.
2. *Broader European modeling*: A review across 28 European countries estimated that a 30 % discount on trastuzumab biosimilars, with moderate uptake, could save €58–68 million in year one and €0.91–2.27 billion over 5 years. Savings might allow treatment of 55,000–116,000 additional patients (Triantafyllidi & Triantafillidis, 2022).
3. *Real-world cost reductions in the U.S.*: Between 2019 and 2022, the price of trastuzumab's originator dropped by 29 %, while the fifth biosimilar entrant was priced 58 % lower than the 2019 reference. Average treatment costs dropped by about US $45,659, with biosimilars capturing up to 32 % market share by the second year.
4. *Administration setting – IV versus SC formulations*: A Belgian case study showed that IV biosimilar trastuzumab cost €10,244–17,858 annually, depending on patient weight, whereas the subcutaneous (SC) reference cost €15,228. For patients up to 75 kg, the IV biosimilar was less costly; when including administration costs, total savings reached €807 per patient per year. More discounts and vial-sharing could yield €400,000+ savings over 100 patients (Simoens et al., 2021).

B. Lymphoma: Rituximab biosimilars

Cost-effectiveness of R-CHOP versus CHOP (Netherlands, prebiosimilar era): Adding rituximab to CHOP chemotherapy yielded 0.88 additional QALYs and an

ICER of €13,983–17,933 per QALY—considered acceptable in view of disease severity (Groot et al., 2005).

13.8 Scope of biosimilars in urothelial carcinoma

13.8.1 Potential targets and pipeline agents

UC, commonly referred to as bladder cancer, represents the tenth most frequently diagnosed malignancy worldwide, accounting for significant morbidity and mortality (Sung et al., 2021). Traditionally, the standard first-line treatment has been platinum-based chemotherapy, which provides meaningful responses but is often limited by cumulative toxicity and eventual resistance. In recent years, immunotherapy, particularly immune checkpoint inhibitors (ICIs) targeting the PD-1/PD-L1 axis-has offered an alternative therapeutic pathway, especially for patients with high PD-L1 expression (Powles et al., 2018). However, neither conventional chemotherapy nor ICIs alone have consistently demonstrated sufficient long-term survival benefits across the broader UC patient population.

This has accelerated research into novel therapeutic targets and biosimilar development for UC. Several promising molecular targets and drug classes are under active investigation:

13.8.1.1 A. Fibroblast growth factor receptor (FGFR) pathway

FGFR alterations, including mutations and fusions, are observed in approximately 15%–20% of UC patients (Loriot et al., 2019). Targeting FGFR signaling has emerged as a viable therapeutic option, with FGFR inhibitors like erdafitinib receiving FDA approval. Biosimilar development for FGFR-directed therapies is anticipated to expand access, given the high costs of original biologics.

13.8.1.2 B. Nectin-4 pathway

Nectin-4, a cell adhesion molecule, is overexpressed in UC and serves as the target for enfortumab vedotin, an antibody–drug conjugate (ADC) that has shown robust efficacy in patients who progressed after chemotherapy and immunotherapy (Rosenberg et al., 2019). Given the global success of enfortumab vedotin, biosimilar ADC development directed at Nectin-4 could significantly reduce treatment costs and improve accessibility.

13.8.1.3 C. Trophoblast cell-surface antigen 2 (Trop-2)

Trop-2 is another promising target in UC. ADCs such as sacituzumab govitecan, which link anti–Trop-2 antibodies with cytotoxic payloads, have demonstrated encouraging outcomes in refractory UC populations (Tagawa et al., 2021). The complexity of ADC production poses challenges for biosimilar development, but pipeline efforts in this area are expanding.

13.8.1.4 D. Vascular endothelial growth factor receptor-2 (VEGFR-2)

VEGFR-2 plays a critical role in angiogenesis, a key driver of tumor growth and metastasis. Ramucirumab, a monoclonal antibody against VEGFR-2, has been studied in UC in combination with chemotherapy, showing improved progression-free survival (Clarke & Hurwitz, 2013). With angiogenesis being a validated therapeutic target, biosimilar VEGFR inhibitors are under exploration in oncology, including UC.

13.8.1.5 E. HER2-directed antibody–drug conjugates

HER2 is overexpressed in a subset of UC patients, opening the door for HER2-targeted therapies. ADCs like trastuzumab deruxtecan have shown significant antitumor activity in HER2-positive UC (Yu et al., 2021). Clinical trials of HER2-directed ADCs are ongoing, and the biosimilar landscape for HER2 monoclonal antibodies (such as trastuzumab biosimilars) provides a precedent for future UC-directed biosimilar ADCs.

The evolution of ADCs has reshaped the treatment paradigm in UC, particularly for patients progressing after platinum-based regimens and immunotherapy. Biosimilars of established biologics and next-generation ADCs hold the potential to reduce costs, expand access, and integrate into earlier lines of therapy. However, manufacturing challenges remain due to the complexity of linking monoclonal antibodies with cytotoxic payloads

13.8.2 Extrapolation of indications

One of the most important regulatory and clinical concepts in the approval of biosimilars is extrapolation of indications. This refers to the regulatory decision to grant approval of a biosimilar for use in one or more clinical indications held by the reference biologic, even if the biosimilar has not been directly studied in those specific indications. Instead, approval is based on the totality of evidence—including analytical comparability, pharmacokinetics or pharmacodynamics (PK/PD), immunogenicity, and at least one comparative clinical efficacy study.

In oncology, extrapolation has played a pivotal role in accelerating patient access to affordable therapies. For example, biosimilars of rituximab, trastuzumab, and bevacizumab were initially studied in one cancer type (such as non-Hodgkin lymphoma or breast cancer) but were later approved across all licensed indications of the reference product (Joshi et al., 2022). This approach avoids duplicating large-scale, expensive, and ethically challenging trials across every cancer subtype.

For UC, extrapolation becomes particularly relevant as biologics and monoclonal antibodies developed in other cancers are being repurposed or tested in UC. For instance:

- *Immune checkpoint inhibitors (PD-1/PD-L1 inhibitors)*: Initially approved for melanoma or lung cancer, they were later approved for UC after emerging data demonstrated comparable mechanisms of action and efficacy across tumor types. Biosimilars of these agents may follow a similar extrapolation pathway.

- *Bevacizumab (anti-VEGF antibody)*: Approved in colorectal and lung cancers, it has also been studied in UC in combination regimens. Bevacizumab biosimilars could thus be extrapolated into UC indications based on mechanism and clinical evidence from related cancers.
- *HER2-directed therapies*: Trastuzumab biosimilars, extensively validated in breast and gastric cancers, provide a precedent for extrapolation into HER2-positive UC as clinical trial data accumulate. Some of the advantages are:
- Faster access to cost-effective therapies in oncology settings.
- Reduced trial burden on patients with rare cancer subtypes such as UC.
- Economic benefits, especially in resource-limited settings like India, where biosimilars significantly improve affordability.

Despite its advantages, extrapolation is not without controversy. Concerns arise regarding:

- Potential differences in tumor biology across cancer types.
- Variations in immune microenvironments and biomarker expression.
- Physician and patient hesitancy to accept efficacy in extrapolated indications without direct clinical data.

Regulatory agencies such as the EMA, FDA, and DCGI (India) emphasize that extrapolation is not automatic but must be scientifically justified, ensuring that the mechanism of action, safety, and efficacy are sufficiently similar across indications (EMA, 2017).

13.8.3 Comparative clinical outcomes and safety concerns

Biosimilar development programs include comparative clinical studies designed to establish clinical equivalence with the reference product, demonstrating similar efficacy within predefined margins (Stebbing et al., 2020).

For biosimilars, the benefit–risk profile is established through the totality of evidence, with highly sensitive analytical and functional assays forming the foundation. These methods are generally more reliable than clinical studies in detecting subtle differences, while comparative efficacy trials serve only as confirmatory assessments. Importantly, significant deviations in quality attributes cannot be justified by clinical data alone.

Comparative clinical studies of trastuzumab biosimilars in breast cancer

The development of trastuzumab biosimilars illustrates the current approach to biosimilar evaluation. By December 2019, five trastuzumab biosimilars had been approved in the EU and the United States for intravenous use. Their programs included:

- Single-dose PK similarity studies in healthy volunteers.
- Comparative efficacy and safety studies in patients with HER2-positive breast cancer.

Study designs differed mainly in patient populations:

- ABP 980, CT-P6, and SB3 were evaluated in the neoadjuvant/adjuvant setting of early breast cancer.
- MYL-1401O and PF-05280014 were studied in first-line metastatic breast cancer.
- PF-05280014 was also tested in a comparative PK noninferiority trial in early breast cancer.

Despite variations, all programs confirmed biosimilarity to reference trastuzumab when combined with taxane-based chemotherapy.

The necessity of large comparative efficacy trials in biosimilar development is increasingly debated. Some experts argue they may be unnecessary, as no biosimilar deemed highly similar through analytical and PK data has ever failed clinical equivalence testing. Regulatory guidance reflects this evolving perspective:

- FDA: A comparative clinical study is required only if residual uncertainty remains after analytical, PK/PD, and immunogenicity data are considered.
- EU: Requirements have shifted, with comparative efficacy trials being waived in some product classes (e.g., granulocyte colony-stimulating factors), where structure, function, and PD markers are well characterized.

For monoclonal antibodies, however, comparative trials are still expected, especially in oncology, where curative treatment intent and the absence of robust PD markers make clinical confirmation critical.

13.8.4 Integration in existing treatment algorithms

The integration of biosimilars into standard treatment algorithms for oncology, including urothelial carcinoma (UC), represents an important step toward broadening patient access, reducing healthcare costs, and sustaining long-term cancer care delivery. As the armamentarium for UC expands with platinum-based chemotherapy, immune checkpoint inhibitors, ADCs, and targeted therapies, biosimilars are increasingly positioned to complement and strengthen these treatment paradigms.

In UC, platinum-based chemotherapy remains the first-line standard for fit patients, with immunotherapy reserved for those with PD-L1 expression or as maintenance therapy. Biosimilars have already been successfully integrated into the supportive care domain, particularly through granulocyte colony-stimulating factor (G-CSF) biosimilars (filgrastim, pegfilgrastim). These agents reduce chemotherapy-induced neutropenia, enabling optimal dose intensity without escalating costs (Cohen et al., 2018).

Similarly, erythropoiesis-stimulating agents (ESA) biosimilars help manage chemotherapy-induced anemia, further embedding biosimilars into the supportive care algorithm of UC patients.

Checkpoint inhibitors such as pembrolizumab and atezolizumab, as well as ADCs like enfortumab vedotin, are being tested in UC treatment algorithms. While direct biosimilar versions of these agents are still under development, experience with biosimilars of monoclonal antibodies (rituximab, trastuzumab, bevacizumab) in other cancers provides a regulatory and clinical precedent.

- *Checkpoint inhibitors*: Once PD-1/PD-L1 inhibitor biosimilars enter the market, they can be incorporated into UC immunotherapy regimens, particularly for maintenance and second-line therapy (Rhea & Aragon-Ching, 2021).
- *HER2-directed therapies*: With HER2-positive UC emerging as a defined subset, trastuzumab biosimilars could be integrated into clinical pathways alongside chemotherapy and ADCs (Cuellar, 2020).

Cost-effectiveness is a key driver of biosimilar adoption. Countries like India have already demonstrated how biosimilars can be rapidly integrated into national cancer programs to extend therapy to underserved populations. In Europe, treatment guidelines by ESMO explicitly encourage the use of oncology biosimilars where available, with safety and efficacy established through rigorous comparability studies.

Inclusion of biosimilars in institutional formularies and clinical guidelines for UC would ensure seamless integration, guided by:

- Clinical interchangeability policies, allowing substitution where evidence supports equivalence.
- Education of oncologists and patients, addressing concerns about efficacy and safety.
- Real-world evidence (RWE) generation to monitor outcomes post-adoption.

Despite regulatory approvals, barriers remain in their integration into treatment algorithms:

- Physician hesitancy due to lack of direct UC trial data.
- Variable regulatory and reimbursement policies across regions.
- Concerns regarding immunogenicity in combination regimens.

13.9 Access, affordability, and adoption barriers

13.9.1 Global access disparities in biologic therapies

Biologic therapies, particularly monoclonal antibodies and targeted treatments, have revolutionized the management of complex diseases, including cancer, autoimmune conditions, and metabolic disorders. However, access to these therapies is highly uneven across the globe. High-income countries such as the United States, Japan, and those within the European Union have significantly higher uptake of biologics due to advanced healthcare infrastructure, greater affordability, and established regulatory frameworks. Conversely, in low- and middle-income countries (LMICs), biologics remain inaccessible to the majority of patients due to high costs, inadequate distribution systems, and limited health insurance coverage.

For instance, trastuzumab, a monoclonal antibody against HER2-positive breast cancer, has shown a drastic improvement in survival outcomes, yet its high cost keeps it out of reach for thousands of women in LMICs (Lammers et al., 2014). Furthermore, logistical barriers such as the lack of cold-chain facilities and skilled healthcare providers exacerbate these disparities. The World Health Organization (WHO) continues to emphasize equitable access, but financial and infrastructural gaps remain formidable obstacles.

13.9.2 Role of biosimilars in reducing treatment gaps

Biosimilars—biological products that are highly similar to an already approved reference biologic—have emerged as a cost-effective strategy to address access disparities. With regulatory approvals from agencies such as the U.S. Food and Drug Administration (FDA), European Medicines Agency (EMA), and Central Drugs Standard Control Organization (CDSCO) in India, biosimilars are gaining acceptance as reliable alternatives.

Clinical trials and postmarketing surveillance have demonstrated comparable efficacy and safety of biosimilars to originator biologics, which has encouraged their integration into national formularies (Rugo et al., 2017). In oncology, biosimilars of trastuzumab and bevacizumab have already reduced treatment costs by up to 30%–40%, enabling greater access in countries like India and Brazil. By lowering the financial burden on healthcare systems and patients, biosimilars play a critical role in bridging the treatment divide between high- and low-resource settings.

Moreover, biosimilars have spurred healthy competition in the pharmaceutical industry, compelling originator companies to reduce prices. This dual effect—direct cost reduction through biosimilars and indirect reduction via market competition—further narrows treatment inequities.

13.9.3 Policy initiatives: NEP 2020, Ayushman Bharat, and CDSCO

Policy frameworks play a pivotal role in shaping the accessibility and affordability of biologics and biosimilars in India. The National Education Policy (NEP) 2020 indirectly contributes to healthcare access by strengthening research, innovation, and skill development in biotechnology and pharmaceutical sciences. By promoting interdisciplinary learning and industry-academic collaborations, NEP 2020 creates a robust pipeline of researchers and clinicians capable of advancing biologic and biosimilar technologies.

The Ayushman Bharat-Pradhan Mantri Jan Arogya Yojana (PM-JAY) is one of the world's largest health insurance schemes, providing financial coverage to over 500 million Indians. By including cancer care and chronic disease management under its coverage, Ayushman Bharat has the potential to improve access to biologics and biosimilars for economically vulnerable populations (Garg et al., 2024). However, challenges such as ensuring the availability of advanced therapies at empanelled hospitals and negotiating sustainable pricing remain.

The Central Drugs Standard Control Organization (CDSCO) has issued clear guidelines for the approval of biosimilars, aligning with global standards. The 2019 revised guidelines emphasized stringent preclinical and clinical comparability exercises, pharmacovigilance, and postmarketing commitments to ensure safety and efficacy. These policies enhance trust in biosimilars and promote their wider adoption in India's healthcare ecosystem.

13.9.4 Stakeholder perspectives: Oncologists, patients, policymakers

The adoption and acceptance of biologics and biosimilars depend significantly on stakeholder perspectives:

- *Oncologists and clinicians:* Physicians are increasingly supportive of biosimilars, given the growing body of evidence affirming their safety and efficacy. However, some remain cautious, emphasizing the need for transparent data sharing and continuing education on biosimilar science. Acceptance improves when healthcare providers receive strong regulatory backing and consistent pharmacovigilance reports.
- *Patients*: From the patient perspective, affordability remains the most pressing concern. Biosimilars provide hope by drastically lowering therapy costs, but awareness is still limited in rural and semi-urban populations. Patients often rely on clinicians' guidance, which makes physician advocacy critical for biosimilar adoption.
- *Policymakers*: Governments and regulatory bodies perceive biosimilars as strategic tools to contain healthcare costs and expand universal access. However, ensuring ethical pricing, addressing intellectual property challenges, and maintaining quality control are ongoing priorities. Programs like Ayushman Bharat signify policymakers' intent to integrate affordable biologics into public healthcare systems.

Global disparities in access to biologics continue to pose challenges to equitable healthcare delivery. However, biosimilars represent a transformative solution, significantly reducing cost-related barriers while maintaining therapeutic efficacy. Policy frameworks such as NEP 2020, Ayushman Bharat, and CDSCO's regulatory guidance provide strong foundations for wider adoption in India. Success, however, lies in fostering collaboration between oncologists, patients, policymakers, and industry stakeholders to ensure affordability, trust, and availability. In this context, biosimilars are not merely substitutes but strategic enablers in democratizing advanced healthcare access worldwide.

13.10 Ethical, legal, and regulatory considerations

As biosimilars gain prominence in the treatment of UC, especially in low-resource settings, navigating the ethical, legal, and regulatory frameworks becomes essential for safe, equitable, and effective use. These considerations shape clinical decision-making, patient-physician relationships, innovation strategies, and public health policies. Ethical principles such as autonomy, beneficence, nonmaleficence, and justice intersect with regulatory requirements to create a matrix of rights and obligations.

13.10.1 Interchangeability and physician autonomy

Interchangeability is a unique regulatory designation allowing a biosimilar to be substituted for the originator biologic without prescriber intervention. In the U.S., the FDA requires additional switching studies to demonstrate equivalent safety and efficacy in alternating use. While this designation improves affordability and access, it raises ethical concerns regarding physician autonomy.

Automatic substitution at the pharmacy level, without prescriber knowledge, risks undermining clinician judgment and the therapeutic alliance with patients. Given the immunogenic complexity of biologics, oncologists emphasize the need for oversight in switching decisions, particularly in fragile UC patients. Policies must therefore safeguard physician discretion through shared decision-making, substitution documentation, and override provisions, ensuring cost savings do not compromise individualized care (Montori et al., 2023).

13.10.2 Pharmacovigilance and postmarketing surveillance

Due to the structural intricacies of biologics, postmarketing surveillance (PMS) is critical to detect rare or long-term adverse events. Pre-approval trials are often too small to capture these risks comprehensively. Regulatory frameworks, including the EU's Eudra Vigilance and India's Pharmacovigilance Programme, emphasize real-world evidence gathering through adverse event reporting, registries, and electronic health records (Prakash et al., 2021).

For UC patients, robust pharmacovigilance must track immunogenicity, disease progression, and quality-of-life outcomes. Failures in reporting due to clinician workload or fragmented systems can obscure safety signals. Ethical responsibilities extend to manufacturers, who must transparently disclose adverse data, and regulators, who should enforce compliance via audits and penalties. Strengthened collaboration among clinicians, regulators, patients, and industry—including mobile health tools and patient-reported outcomes—can improve pharmacovigilance capacity and responsiveness (Tabernero et al., 2016).

13.10.3 Intellectual property and market competition

Legal frameworks governing biologics are dominated by intellectual property (IP) protections. Patents, typically lasting 20 years, often extend beyond expiry through "evergreening" practices, where manufacturers file secondary patents on minor modifications (Gurgula, 2020). Such tactics delay biosimilar entry, inflate costs, and raise ethical concerns about prioritizing profits over patient welfare.

For UC patients, delayed access to cost-effective biosimilars translates into inequities in survival outcomes. TRIPS flexibilities and compulsory licensing, endorsed by the WHO and WTO, remain vital instruments to balance IP rights with public health imperatives (McGivern, 2023). Encouraging local biosimilar production via tax incentives, grants, and technology transfers can mitigate reliance on imports and enhance affordability in LMICs. A transparent and competitive legal environment ultimately aligns innovation incentives with societal health needs.

13.10.4 Informed consent and patient education

Informed consent is a cornerstone of ethical oncology care. With biosimilars, informed consent processes must address misconceptions—patients often assume biosimilars are identical to generics, overlooking their structural and clinical complexities. In oncology, where patients face significant psychological stress, insufficient explanations can lead to mistrust and poor adherence (Jacobs et al., 2016).

Healthcare providers should explain biosimilar development, regulatory rigor, equivalence to reference products, and rationale for selection. Consent must be viewed as a continuous dialogue, particularly during therapy switches or adverse events. Educational resources—including decision aids, multilingual materials, and culturally competent communication—are critical for equitable engagement. Patient advocacy groups and caregiver inclusion can further support autonomy in decision making.

13.11 Future directions in urothelial carcinoma treatment

The therapeutic landscape for UC is rapidly evolving, with biologics, immunotherapies, and digital innovations leading the way.

13.11.1 Advances in biologic platforms (Bispecifics, CAR-T)

Bispecific antibodies and CAR-T cell therapies represent transformative biologic platforms. Bispecific antibodies simultaneously engage tumor antigens and immune effectors (e.g., CD3), thereby enhancing T-cell–mediated cytotoxicity. In UC, preclinical models show promise for bispecifics targeting FGFR3, EpCAM, and HER2 (Lum & Thakur, 2011).

CAR-T therapy, successful in hematological malignancies, faces hurdles in UC due to immune suppression in the tumor microenvironment. Innovative strategies, such as safety switches, tandem CARs, and intravesical delivery, aim to improve efficacy and safety (Liu et al., 2020). Development of off-the-shelf allogeneic CAR-T products could expand accessibility, particularly in LMICs.

13.11.2 Personalized medicine and biomarker-driven therapy

Biomarker-guided approaches are central to precision oncology. FGFR3 mutations, PD-L1 expression, and tumor mutational burden (TMB) are already shaping therapeutic decisions in UC. Erdafitinib, an FGFR3 inhibitor, exemplifies success in targeted therapy (Loriot et al., 2019).

Technologies such as next-generation sequencing (NGS) and liquid biopsies enable real-time tumor profiling, resistance monitoring, and adaptive therapy. Multiomic integration, including transcriptomic and proteomic data, promises to identify novel therapeutic vulnerabilities. Innovative trial designs (basket and umbrella trials) are further accelerating biomarker validation in UC.

13.11.3 Digital health and artificial intelligence integration in oncology

AI and digital platforms are revolutionizing UC care. Machine learning models analysing pathology and imaging data aid in early detection and prognostication (Esteva et al., 2019). Clinical decision support systems integrated into electronic health records provide oncologists with evidence-based, personalized therapy recommendations.

Wearable devices and mobile health apps support remote monitoring of symptoms, adherence, and survivorship outcomes, reducing hospitalizations and enhancing quality of life. AI-driven natural language processing also enables mining of unstructured clinical notes for real-world evidence, supporting both clinical care and policy.

13.11.4 Research gaps and innovation opportunities

Despite advancements, research gaps persist. Current biosimilar evidence in UC is largely extrapolated from other cancers, underscoring the need for tumor-specific trials. Tumor microenvironment heterogeneity and barriers to biologic penetration remain unresolved challenges. Investment in innovative delivery platforms (nanoparticles, intravesical biologic delivery) and global registries collecting diverse real-world evidence can accelerate progress. Harmonizing biosimilar regulatory pathways globally will facilitate faster, safer adoption. Crucially, public-private partnerships, local manufacturing incentives, and patient education initiatives can reduce costs and build confidence in biosimilars. Interdisciplinary collaboration across oncology, immunology, bioinformatics, and digital health will define the next era of UC management.

References

Reed, T. Study: Hospital drug discount program hurts biosimilars market, 2023 Retrieved August 31, 2025, from https://www.axios.com/2023/05/02/340b-drug-discount-program-hospitals?utm_source=chatgpt.com.

Aggarwal, A., Fojo, T., Chamberlain, C., Davis, C., & Sullivan, R. (2017). Do patient access schemes for high-cost cancer drugs deliver value to society?-Lessons from the NHS cancer drugs fund. *Annalsof Oncology, 28*, 1738–1750. https://doi.org/10.1093/annonc/mdx110.

Alberts, B., Johnson, A., Lewis, J., Raff, M., Roberts, K., & Walter, P. (2002). Helper T cells and lymphocyte activation. *Molecular biology of the cell* (4th ed.)). Garland Science.

Alva, A., Csőszi, T., Ozguroglu, M., Matsubara, N., Geczi, L., Fradet, Y., Oudard, S., Vulsteke, C., Morales Barrera, R., Fléchon, A., Gunduz, S., Loriot, Y., Rodriguez-Vida, A., Mamtani, R., Yu, E. Y., Nam, K., Imai, K., Moreno, B. H., & Powles, T. B. (2020). LBA23 Pembrolizumab (P) combined with chemotherapy (C) vs C alone as first-line (1L) therapy for advanced urothelial carcinoma (UC): KEYNOTE-361. *Annals of Oncology, 31*, S1155. https://doi.org/10.1016/j.annonc.2020.08.2252.

Apolo, A. B., Infante, J. R., Balmanoukian, A., Patel, M. R., Wang, D., Kelly, K., Mega, A. E., Britten, C. D., Ravaud, A., Mita, A. C., Safran, H., Stinchcombe, T. E., Srdanov, M., Gelb, A. B., Schlichting, M., Chin, K., & Gulley, J. L. (2017). Avelumab, an anti–programmed death-ligand 1 antibody, in patients with refractory metastatic urothelial carcinoma: Results from a multicenter, phase Ib study. *Journal of Clinical Oncology, 35*, 2117–2124. https://doi.org/10.1200/JCO.2016.71.6795.

Babjuk, M., Burger, M., Capoun, O., Cohen, D., Compérat, E. M., Dominguez Escrig, J. L., Gontero, P., Liedberg, F., Masson-Lecomte, A., Mostafid, A. H., Palou, J., Van Rhijn, B. W. G., Rouprêt, M., Shariat, S. F., Seisen, T., Soukup, V., & Sylvester, R. J. (2022). European Association of Urology guidelines on non–muscle-invasive bladder cancer (Ta, T1, and Carcinoma in Situ). *European Urology, 81*, 75–94. https://doi.org/10.1016/j.eururo.2021.08.010.

Balar, A. V., Galsky, M. D., Rosenberg, J. E., Powles, T., Petrylak, D. P., Bellmunt, J., Loriot, Y., Necchi, A., Hoffman-Censits, J., Perez-Gracia, J. L., Dawson, N. A., Van Der Heijden, M. S., Dreicer, R., Srinivas, S., Retz, M. M., Joseph, R. W., Drakaki, A., Vaishampayan, U. N., Sridhar, S. S., … Bajorin, D. F. (2017). Atezolizumab as first-line treatment in cisplatin-ineligible patients with locally advanced and metastatic urothelial carcinoma: A single-arm, multicentre, phase 2 trial. *The Lancet, 389*, 67–76. https://doi.org/10.1016/S0140-6736(16)32455-2.

Baron, J., & Wang, E. S. (2018). Gemtuzumab ozogamicin for the treatment of acute myeloid leukemia. *Expert Review Clinical Pharmacology., 11*, 549–559. https://doi.org/10.1080/17512433.2018.1478725.

Bellmunt, J., De Wit, R., Vaughn, D. J., Fradet, Y., Lee, J.-L., Fong, L., Vogelzang, N. J., Climent, M. A., Petrylak, D. P., Choueiri, T. K., Necchi, A., Gerritsen, W., Gurney, H., Quinn, D. I., Culine, S., Sternberg, C. N., Mai, Y., Poehlein, C. H., Perini, R. F., & Bajorin, D. F. (2017). Pembrolizumab as second-line therapy for advanced urothelial carcinoma. *New England Journal of Medicine, 376*, 1015–1026. https://doi.org/10.1056/NEJMoa1613683.

Blackstone, E. A., & Fuhr, J. P. (2012). Innovation and Competition: Will Biosimilars Succeed?: The creation of an FDA approval pathway for biosimilars is complex and fraught with hazard. Yes, innovation and market competition are at stake. But so are efficacy and patient safety. *Biotechnology Healthcare., 9*, 24–27.

Blackstone, E. A., & Joseph, P. F. (2013). The economics of biosimilars. *American Health and Drug Benefits, 6*, 469–478.

Bree, K. K., Shan, Y., Hensley, P. J., Lobo, N., Hu, C., Tyler, D. S., Chamie, K., Kamat, A. M., & Williams, S. B. (2022). Management, surveillance patterns, and costs associated with low-grade papillary stage T: A non-muscle-invasive bladder cancer among older adults, 2004-2013. *JAMA Network Open, 5*, e223050. https://doi.org/10.1001/jamanetworkopen.2022.3050.

Burt, R., Warcel, D., & Fielding, A. K. (2019). Blinatumomab, a bispecific B-cell and T-cell engaging antibody, in the treatment of B-cell malignancies. *Human Vaccines Immunotherapy, 15*, 594–602. https://doi.org/10.1080/21645515.2018.1540828.

Campar, A., & Isenberg, D. A. (2011). Life-threatening complications of biological therapies. In M. A. Khamashta, & M. Ramos-Casals (Eds.), *Autoimmune diseases* (pp. 375–403). Springer. https://doi.org/10.1007/978-0-85729-358-9_23.

Cano, R. L. E., & Lopera, H. D. E. (2013). Introduction to T and B lymphocytes. *Autoimmunity: From bench to bedside*. El Rosario University Press.

Cesarec, A., & Likić, R. (2017). Budget impact analysis of biosimilar trastuzumab for the treatment of breast cancer in Croatia. Applied Health Economics and Health Polic. *y, 15*, 277–286. https://doi.org/10.1007/s40258-016-0285-7.

Chaplin, S., Van Stiphout, J., Chen, A., & Li, E. (2024). Budget impact analysis of including biosimilar adalimumab on formulary: A United States payer perspective. *J, 30*, 1226–1238. ournal of Managed Care & Specialty Pharmacy https://doi.org/10.18553/jmcp.2024.24036.

Chen, N. C., & McQueen, R. (2023). EE369 Cost-effectiveness analysis of pembrolizumab as a second-line therapy for advanced urothelial carcinoma in the United States. *Value in Health: The Journal of the International Society for Pharmacoeconomics and Outcomes Research, 26*, S122. https://doi.org/10.1016/j.jval.2023.09.635.

Chidharla, A., Parsi, M., & Kasi, A. (2025). Cetuximab. *StatPearls*. StatPearls Publishing.

Clark, O., Sarmento, T., Eccleston, A., Brinkmann, J., Picoli, R., Daliparthi, V., Voss, J., Chandrasekar, S., Thompson, A., & Chang, J. (2024). Economic impact of bladder cancer in the USA. *PharmacoEconomics - Open, 8*, 837–845. https://doi.org/10.1007/s41669-024-00512-8.

Clarke, J. M., & Hurwitz, H. I. (2013). Targeted inhibition of VEGF receptor 2: An update on ramucirumab. *Expert Opinion on Biological Therapy, 13*, 1187–1196. https://doi.org/10.1517/14712598.2013.810717.

Cohen, H. P., Blauvelt, A., Rifkin, R. M., Danese, S., Gokhale, S. B., & Woollett, G. (2018). Switching reference medicines to biosimilars: A systematic literature review of clinical outcomes. *Drugs, 78*, 463–478. https://doi.org/10.1007/s40265-018-0881-y.

Cross, R. K. (2022). Implementation strategies of biosimilars in healthcare systems: The path forward. *American Health & Drug Benefits, 15*, 45.

Cuellar, S. (2020). Integrating trastuzumab biosimilars and HER2-directed therapies into HER2-positive breast cancer management. *American Journal of Managed Care, 26*, S32–S40. https://doi.org/10.37765/ajmc.2020.42900.

Cumberbatch, M. G., Rota, M., Catto, J. W. F., & La Vecchia, C. (2016). The role of tobacco smoke in bladder and kidney carcinogenesis: A comparison of exposures and meta-analysis of incidence and mortality risks. *European Urology, 70*, 458–466. https://doi.org/10.1016/j.eururo.2015.06.042.

Dalghi, M. G., Montalbetti, N., Carattino, M. D., & Apodaca, G. (2020). The urothelium: Life in a liquid environment. *Physiological Reviews, 100*, 1621–1705. https://doi.org/10.1152/physrev.00041.2019.

DeHovitz, R. E. (2014). The 1901 St Louis incident: The first modern medical disaster. *Pediatrics, 133*, 964–965. https://doi.org/10.1542/peds.2013-2817.

G. Dock, 1904. The influence of complicating diseases upon leukæmia. Elsevier BV.

Esteva, A., Robicquet, A., Ramsundar, B., Kuleshov, V., DePristo, M., Chou, K., Cui, C., Corrado, G., Thrun, S., & Dean, J. (2019). A guide to deep learning in healthcare. *Nature Medicine, 25*, 24–29. https://doi.org/10.1038/s41591-018-0316-z.

Omnitrope. European Medicines Agency (EMA) [WWW Document], 2018. Retrieved August 31, 2025, from https://www.ema.europa.eu/en/medicines/human/EPAR/omnitrope.

Ferrucci, P. F., Pala, L., Conforti, F., & Cocorocchio, E. (2021). Talimogene Laherparepvec (T-VEC): An intralesional cancer immunotherapy for advanced melanoma. *Cancers, 13*, 1383. https://doi.org/10.3390/cancers13061383.

Flynn, J. P., & Gerriets, V. (2025). Pembrolizumab. *StatPearls*. StatPearls Publishing.

Fu, Z., Li, S., Han, S., Shi, C., & Zhang, Y. (2022). Antibody drug conjugate: the "biological missile" for targeted cancer therapy. *Signal Transduction and Targeted Therapy, 7*, 93. https://doi.org/10.1038/s41392-022-00947-7.

Galsky, M. D., Arija, J. Á. A., Bamias, A., Davis, I. D., De Santis, M., Kikuchi, E., Garcia-del-Muro, X., De Giorgi, U., Mencinger, M., Izumi, K., Panni, S., Gumus, M., Özgüroğlu, M., Kalebasty, A. R., Park, S. H., Alekseev, B., Schutz, F. A., Li, J.-R., Ye, D., … Grande, E. (2020). Atezolizumab with or without chemotherapy in metastatic urothelial cancer (IMvigor130): A multicentre, randomised, placebo-controlled phase 3 trial. *The Lancet, 395*, 1547–1557. https://doi.org/10.1016/S0140-6736(20)30230-0.

Garg, S., Bebarta, K. K., & Tripathi, N. (2024). The Ayushman Bharat Pradhan Mantri Jan Arogya Yojana (AB-PMJAY) after four years of implementation - is it making an impact on quality of inpatient care and financial protection in India? BMC Health Serv. *Res., 24*, 919. https://doi.org/10.1186/s12913-024-11393-2.

Gaynor, N., Crown, J., & Collins, D. M. (2022). Immune checkpoint inhibitors: Key trials and an emerging role in breast cancer. *Seminars in Cancer Biology, 79*, 44–57. https://doi.org/10.1016/j.semcancer.2020.06.016.

Gerriets, V., & Kasi, A. (2025). Bevacizumab. *StatPearls*. StatPearls Publishing.

Greenblatt, K., & Khaddour, K. (2025). Trastuzumab. *StatPearls*. StatPearls Publishing.

Groot, M. T., Lugtenburg, P. J., Hornberger, J., Huijgens, P. C., & Uyl-de Groot, C. A. (2005). Cost-effectiveness of rituximab (MabThera) in diffuse large B-cell lymphoma in The Netherlands. *European Journal of Haematology, 74*, 194–202. https://doi.org/10.1111/j.1600-0609.2004.00368.x.

Gurgula, O. (2020). Strategic patenting by pharmaceutical companies - should competition law intervene?. International Review of Industrial Property and Copyright La *w, 51*, 1062–1085. https://doi.org/10.1007/s40319-020-00985-0.

Hanif, N., & Anwer, F. (2025). Rituximab. *StatPearls*. StatPearls Publishing.

Hooker, C. (2000). Diphtheria, immunisation and the Bundaberg tragedy: A study of public health in Australia. *Health and History, 2*, 52. https://doi.org/10.2307/40111374.

Itoshima, H., Takada, D., Goto, E., Sasaki, N., Kunisawa, S., & Imanaka, Y. (2024). The impact of financial incentives promoting biosimilar products in oncology: A quasi-experimental study using administrative data. *PLoS ONE, 19*, e0312577. https://doi.org/10.1371/journal.pone.0312577.

Jacobs, I., Singh, E., Sewell, K. L., Al-Sabbagh, A., & Shane, L. G. (2016). Patient attitudes and understanding about biosimilars: an international cross-sectional survey. *Patient Preference Adherence, 10*, 937–948. https://doi.org/10.2147/PPA.S104891.

Joshi, D., Khursheed, R., Gupta, S., Wadhwa, D., Singh, T. G., Sharma, S., Porwal, S., Gauniyal, S., Vishwas, S., Goyal, S., Gupta, G., Eri, R. D., Williams, K. A., Dua, K., & Singh, S. K. (2022). Biosimilars in Oncology: Latest Trends and Regulatory Status. *Pharmaceutics, 14*, 2721. https://doi.org/10.3390/pharmaceutics14122721.

Kantoff, P. W., Higano, C. S., Shore, N. D., Berger, E. R., Small, E. J., Penson, D. F., Redfern, C. H., Ferrari, A. C., Dreicer, R., Sims, R. B., Xu, Y., Frohlich, M. W., &

Schellhammer, P. F.IMPACT Study Investigators. (2010). Sipuleucel-T immunotherapy for castration-resistant prostate cancer. *New England Journal of Medicine, 363*, 411–422. https://doi.org/10.1056/NEJMoa1001294.

Kaye, D. R. (2023). Drug costs: acquisition costs are critical but not the entire story of financial toxicity. *Bju International, 132*, 115–116. https://doi.org/10.1111/bju.16092.

Kvien, T. K., Patel, K., & Strand, V. (2022). The cost savings of biosimilars can help increase patient access and lift the financial burden of health care systems. *Seminars in Arthritis and Rheumatism, 52*, 151939. https://doi.org/10.1016/j.semarthrit.2021.11.009.

Lammers, P., Criscitiello, C., Curigliano, G., & Jacobs, I. (2014). Barriers to the Use of Trastuzumab for HER2+ Breast Cancer and the Potential Impact of Biosimilars: A Physician Survey in the United States and Emerging Markets. *Pharmaceuticals., 7*, 943–953. https://doi.org/10.3390/ph7090943.

Lattanzi, M., & Rosenberg, J. E. (2020). The emerging role of antibody-drug conjugates in urothelial carcinoma. *Expert Review of Anticancer Therapy, 20*, 551–561. https://doi.org/10.1080/14737140.2020.1782201.

Lee, A. (2023). Nadofaragene firadenovec: First approval. *Drugs, 83*, 353–357. https://doi.org/10.1007/s40265-023-01846-z.

Lee, J. J., Powderly, J. D., Patel, M. R., Brody, J., Hamilton, E. P., Infante, J. R., Falchook, G. S., Wang, H., Adams, L., Gong, L., Ma, A. W., Wyant, T., Lazorchak, A., Agarwal, S., Tuck, D. P., & Daud, A. (2017). Phase 1 trial of CA-170, a novel oral small molecule dual inhibitor of immune checkpoints PD-1 and VISTA, in patients (pts) with advanced solid tumor or lymphomas. *Journal of Clinical Oncology, 35*, TPS3099. https://doi.org/10.1200/JCO.2017.35.15_suppl.TPS3099.

Liang, M. (2018). Oncorine, the world first oncolytic virus medicine and its update in China. *Current Cancer Drug Targets, 18*, 171–176. https://doi.org/10.2174/1568009618666171129221503.

Lin, D., Shen, Y., & Liang, T. (2023). Oncolytic virotherapy: Basic principles, recent advances and future directions. *Signal Transduction Targeted Ther., 8*, 156. https://doi.org/10.1038/s41392-023-01407-6.

Liu, E., Marin, D., Banerjee, P., Macapinlac, H. A., Thompson, P., Basar, R., Nassif Kerbauy, L., Overman, B., Thall, P., Kaplan, M., Nandivada, V., Kaur, I., Nunez Cortes, A., Cao, K., Daher, M., Hosing, C., Cohen, E. N., Kebriaei, P., Mehta, R., ... Rezvani, K. (2020). Use of CAR-transduced natural killer cells in CD19-positive lymphoid tumors. *New England Journal of Medicine, 382*, 545–553. https://doi.org/10.1056/NEJMoa1910607.

Lobo, A., Collins, K., Kaushal, S., Acosta, A. M., Akgul, M., Adhya, A. K., Al-Ahmadie, H. A., Al-Obaidy, K. I., Amin, A., Amin, M. B., Aron, M., Balzer, B. L., Biswal, R., Mohanty, S., Browning, L., Chakrabarti, I., Cima, L., Cimadamore, A., Desai, S., ... Mohanty, S. K. (2024). Advances, recognition, and interpretation of molecular heterogeneity among conventional and subtype histology of urothelial carcinoma (UC): A survey among urologic pathologists and comprehensive review of the literature. *Histopathology, 85*, 748–759. https://doi.org/10.1111/his.15287.

Loriot, Y., Necchi, A., Park, S. H., Garcia-Donas, J., Huddart, R., Burgess, E., Fleming, M., Rezazadeh, A., Mellado, B., Varlamov, S., Joshi, M., Duran, I., Tagawa, S. T., Zakharia, Y., Zhong, B., Stuyckens, K., Santiago-Walker, A., De Porre, P., O'Hagan, A., ... Siefker-Radtke, A. O. (2019). Erdafitinib in locally advanced or metastatic urothelial carcinoma. *New England Journal of Medicine, 381*, 338–348. https://doi.org/10.1056/NEJMoa1817323.

Lum, L. G., & Thakur, A. (2011). Targeting T cells with bispecific antibodies for cancer therapy. , Biopharmaceuticals and Gene Therapy. ., *25*, 365–379. https://doi.org/10.2165/11595950-000000000-00000.

Lyman, G. H., Balaban, E., Diaz, M., Ferris, A., Tsao, A., Voest, E., Zon, R., Francisco, M., Green, S., Sherwood, S., Harvey, R. D., & Schilsky, R. L. (2018). American Society of Clinical Oncology Statement: Biosimilars in oncology. *Journal of Clinical Oncology, 36*, 1260–1265. https://doi.org/10.1200/JCO.2017.77.4893.

Mathur, P., Sathishkumar, K., Chaturvedi, M., Das, P., Sudarshan, K. L., Santhappan, S., Nallasamy, V., John, A., Narasimhan, S., & Roselind, F. S. (2020). Cancer Statistics, 2020: Report From National Cancer Registry Programme, India. In *JCO Global Oncology* (pp. 1063–1075). https://doi.org/10.1200/GO.20.00122.

Maxinovel Pty., Ltd., 2022. A phase I study of MAX-10181 given orally to patients with advanced solid tumor. (Clinical trial registration No. NCT04122339). clinicaltrials.gov.

McAuliff, K., Shah, D., Rutter, C., Barnett, T., Avalos-Reyes, E., Grover, R., Feczko, L., Cavers, W., Verbrugge, D. J., & Johnson, K. A. (2024). Real-world evaluation of oncolytic biosimilar adoption on medication adherence and out-of-pocket costs among members of a national payor. *Journal of Clinical Oncology, 42*, e23092. https://doi.org/10.1200/JCO.2024.42.16_suppl.e23092.

McCamish, M., & Woollett, G. (2012). The state of the art in the development of biosimilars. *Clinical Pharmacology & Therapeutics, 91*, 405–417. https://doi.org/10.1038/clpt.2011.343.

McGivern, L. (2023). Trade-related aspects of intellectual property rights flexibilities and public health: implementation of compulsory licensing provisions into national patent legislation. *Milbank Quarterly, 101*, 1280–1303. https://doi.org/10.1111/1468-0009.12669.

Meher, B. R., Balan, S., Mohanty, R. R., Jena, M., & Das, S. (2019). Biosimilars in India; Current status and future perspectives. *J, 11*, 12–15. ournal of Pharmacy andBioallied Sciences https://doi.org/10.4103/jpbs.JPBS_167_18 .

Mellman, I., Coukos, G., & Dranoff, G. (2011). Cancer immunotherapy comes of age. *Nature, 480*, 480–489. https://doi.org/10.1038/nature10673.

Mishra, V., & Balasubramaniam, G. (2020). Urinary bladder cancer and its associated factors – An epidemiological overview. *Indian Journal of Medical Sciences, 73*, 239–248. https://doi.org/10.25259/IJMS_159_2020.

Montori, V. M., Ruissen, M. M., Hargraves, I. G., Brito, J. P., & Kunneman, M. (2023). Shared decision-making as a method of care. *BMJ Evidence-Based Medicine, 28*, 213–217. https://doi.org/10.1136/bmjebm-2022-112068.

Morrow, T., & Felcone, L. H. (2004). Defining the difference: What makes biologics unique. *Biotechnolgy Healthcare., 1*, 24–29.

Necchi, A., Lo Vullo, S., Raggi, D., Gloghini, A., Giannatempo, P., Colecchia, M., & Mariani, L. (2019). Prognostic effect of FGFR Mutations or gene fusions in patients with metastatic urothelial carcinoma receiving first-line platinum-based chemotherapy: Results from a large, single-institution cohort. *European Urology Focus, 5*, 853–856. https://doi.org/10.1016/j.euf.2018.02.013.

Norsworthy, K. J., Ko, C.-W., Lee, J. E., Liu, J., John, C. S., Przepiorka, D., Farrell, A. T., & Pazdur, R. (2018). FDA approval summary: Mylotarg for treatment of patients with relapsed or refractory CD33-positive acute myeloid leukemia. *The Oncologist, 23*, 1103–1108. https://doi.org/10.1634/theoncologist.2017-0604.

Odegard, J. M., Othman, A. A., Lin, K.-W., Wang, A. Y., Nazareno, J., Yoon, O. K., Ling, J., Lad, L., Dunbar, P. R., Thai, D., Ang, E., Waldron, N., & Deva, S. (2024). Oral PD-L1 inhibitor GS-4224 selectively engages PD-L1 high cells and elicits pharmacodynamic responses in patients with advanced solid tumors. *Journal for ImmunoTherapy of Cancer, 12*, e008547. https://doi.org/10.1136/jitc-2023-008547.

Offit, P. A. (2005). *The cutter incident: How America's first polio vaccine led to the growing vaccine crisis*. Yale University Press.

Okui, N. (2024). *Shibasaburo Kitasato (1853-1931): Pioneer of Japanese medicine and global immunology innovator*. Cureus https://doi.org/10.7759/cureus.68276.

Op 't Hoog, C. J. P., Rieborn, A., Moes, D. J. A., Hendrikx, J. J., van der Heijden, M. S., Franken, M. D., van der Hulle, T., van Kruchten, M., Willemsen, A. E., Koolen, S. L., Boerrigter, E., & Ter Heine, R. (2025). Model-informed development of a cost-saving dosing regimen for enfortumab vedotin. *Cancer Chemotherapy and Pharmacology, 95*, 36. https://doi.org/10.1007/s00280-025-04764-x.

Pardoll, D. M. (2012). The blockade of immune checkpoints in cancer immunotherapy. *Nature Reviews Cancer, 12*, 252–264. https://doi.org/10.1038/nrc3239.

Patel, D., Bassi, R., Hooper, A., Prewett, M., Hicklin, D. J., & Kang, X. (2009). Anti-epidermal growth factor receptor monoclonal antibody cetuximab inhibits EGFR/HER-2 heterodimerization and activation. *International Journal of Oncology, 34*, 25–32.

Petrylak, D. P., De Wit, R., Chi, K. N., Drakaki, A., Sternberg, C. N., Nishiyama, H., Castellano, D., Hussain, S. A., Fléchon, A., Bamias, A., Yu, E. Y., Van Der Heijden, M. S., Matsubara, N., Alekseev, B., Necchi, A., Géczi, L., Ou, Y.-C., Coskun, H. S., Su, W.-P., … Lowe, T. E. (2020). Ramucirumab plus docetaxel versus placebo plus docetaxel in patients with locally advanced or metastatic urothelial carcinoma after platinum-based therapy (RANGE): Overall survival and updated results of a randomised, double-blind, phase 3 trial. *The Lancet Oncology, 21*, 105–120. https://doi.org/10.1016/S1470-2045(19)30668-0.

Powles, T., Durán, I., Van Der Heijden, M. S., Loriot, Y., Vogelzang, N. J., De Giorgi, U., Oudard, S., Retz, M. M., Castellano, D., Bamias, A., Fléchon, A., Gravis, G., Hussain, S., Takano, T., Leng, N., Kadel, E. E., Banchereau, R., Hegde, P. S., Mariathasan, S., … Ravaud, A. (2018). Atezolizumab versus chemotherapy in patients with platinum-treated locally advanced or metastatic urothelial carcinoma (IMvigor211): A multicentre, open-label, phase 3 randomised controlled trial. *The Lancet, 391*, 748–757. https://doi.org/10.1016/S0140-6736(17)33297-X.

Powles, T., O'Donnell, P. H., Massard, C., Arkenau, H.-T., Friedlander, T. W., Hoimes, C. J., Lee, J. L., Ong, M., Sridhar, S. S., Vogelzang, N. J., Fishman, M. N., Zhang, J., Srinivas, S., Parikh, J., Antal, J., Jin, X., Gupta, A. K., Ben, Y., & Hahn, N. M. (2017). Efficacy and safety of durvalumab in locally advanced or metastatic urothelial carcinoma: Updated results from a phase 1/2 open-label study. *JAMA Oncology, 3*, e172411. https://doi.org/10.1001/jamaoncol.2017.2411.

Powles, T., Park, S. H., Voog, E., Caserta, C., Valderrama, B. P., Gurney, H., Kalofonos, H., Radulović, S., Demey, W., Ullén, A., Loriot, Y., Sridhar, S. S., Tsuchiya, N., Kopyltsov, E., Sternberg, C. N., Bellmunt, J., Aragon-Ching, J. B., Petrylak, D. P., Laliberte, R., … Grivas, P. (2020a). Avelumab maintenance therapy for advanced or metastatic urothelial carcinoma. *New England Journal of Medicine, 383*, 1218–1230. https://doi.org/10.1056/NEJMoa2002788.

Powles, T., Valderrama, B. P., Gupta, S., Bedke, J., Kikuchi, E., Hoffman-Censits, J., Iyer, G., Vulsteke, C., Park, S. H., Shin, S. J., Castellano, D., Fornarini, G., Li, J.-R., Gümüş, M., Mar, N., Loriot, Y., Fléchon, A., Duran, I., Drakaki, A., … Van Der Heijden, M. S. (2024).

Enfortumab vedotin and pembrolizumab in untreated advanced urothelial cancer. *New England Journal of Medicine, 390*, 875–888. https://doi.org/10.1056/NEJMoa2312117.

Powles, T., Van Der Heijden, M. S., Castellano, D., Galsky, M. D., Loriot, Y., Petrylak, D. P., Ogawa, O., Park, S. H., Lee, J.-L., De Giorgi, U., Bögemann, M., Bamias, A., Eigl, B. J., Gurney, H., Mukherjee, S. D., Fradet, Y., Skoneczna, I., Tsiatas, M., Novikov, A., … Lesniewski-Kmak, K. (2020b). Durvalumab alone and durvalumab plus tremelimumab versus chemotherapy in previously untreated patients with unresectable, locally advanced or metastatic urothelial carcinoma (DANUBE): A randomised, open-label, multicentre, phase 3 trial. *The Lancet Oncology, 21*, 1574–1588. https://doi.org/10.1016/S1470-2045(20)30541-6.

Prakash, J., Sachdeva, R., Shrivastava, T. P., Jayachandran, C. V., & Sahu, A. (2021). Adverse event reporting tools and regulatory measures in India through outcome of Pharmacovigilance Programme of India. *Indian Journal of Pharmacology, 53*, 143–152. https://doi.org/10.4103/ijp.ijp_901_20.

Rabie, L. E., Mohran, A. A., Gaber, K. A., Ali, N. M., Abd El Naby, A. M., Ghoniem, E. A., Abd Elmaksod, B. A., & Abdallah, A. N. (2024). Beyond conventional treatments: Exploring CAR-T cell therapy for cancer stem cell eradication. *Stem Cell Reviews and Reports., 20*, 2001–2015. https://doi.org/10.1007/s12015-024-10786-4.

US, FDA. (2023). *What are "biologics" questions and answers*. FDA.

Rhea, L. P., & Aragon-Ching, J. B. (2021). Advances and controversies with checkpoint inhibitors in bladder cancer. *Clinical Medicine Insights Oncology., 15*, 11795549211044963. https://doi.org/10.1177/11795549211044963.

Ribas, A., & Wolchok, J. D. (2018). Cancer immunotherapy using checkpoint blockade. *Science, 359*, 1350–1355. https://doi.org/10.1126/science.aar4060.

Rizzo, M., Giannatempo, P., & Porta, C. (2021). Biological therapeutic advances for the treatment of advanced urothelial cancers. *Biologics: Targets Therapy., 15*, 441–450. https://doi.org/10.2147/BTT.S290311.

Rosenberg, J. E., O'Donnell, P. H., Balar, A. V., McGregor, B. A., Heath, E. I., Yu, E. Y., Galsky, M. D., Hahn, N. M., Gartner, E. M., Pinelli, J. M., Liang, S.-Y., Melhem-Bertrandt, A., & Petrylak, D. P. (2019). Pivotal trial of enfortumab vedotin in urothelial carcinoma after platinum and anti-programmed death 1/programmed death ligand 1 therapy. *Journal of Clinical Oncology, 37*, 2592–2600. https://doi.org/10.1200/JCO.19.01140.

Roviello, G., Catalano, M., Santi, R., Palmieri, V. E., Vannini, G., Galli, I. C., Buttitta, E., Villari, D., Rossi, V., & Nesi, G. (2021). Immune checkpoint inhibitors in urothelial bladder cancer: State of the art and future perspectives. *Cancers, 13*, 4411. https://doi.org/10.3390/cancers13174411.

Rugo, H. S., Barve, A., Waller, C. F., Hernandez-Bronchud, M., Herson, J., Yuan, J., Sharma, R., Baczkowski, M., Kothekar, M., Loganathan, S., Manikhas, A., Bondarenko, I., Mukhametshina, G., Nemsadze, G., Parra, J. D., Abesamis-Tiambeng, M. L. T., Baramidze, K., Akewanlop, C., … Vynnychenko, I.Heritage Study Investigators. (2017). Effect of a proposed trastuzumab biosimilar compared with trastuzumab on overall response rate in patients with ERBB2 (HER2)-positive metastatic breast cancer: A randomized clinical trial. *JAMA, 317*, 37–47. https://doi.org/10.1001/jama.2016.18305.

Sanglier, T., Ross, R., Shi, T., Mouta, J., Swain, S., & Cardoso, F. (2022). Trastuzumab-based regimens beyond progression: A crucial treatment option for HER2+ advanced/metastatic breast cancer. *The Breast, 66*, 262–271. https://doi.org/10.1016/j.breast.2022.10.008.

Schellekens, H. (2009). Biosimilar therapeutics-what do we need to consider? *NDT Plus, 2*, i27–i36. https://doi.org/10.1093/ndtplus/sfn177.

Scott, A. M., Wolchok, J. D., & Old, L. J. (2012). Antibody therapy of cancer. *Nature Reviews Cancer, 12*, 278–287. https://doi.org/10.1038/nrc3236.

Shang, N., Figini, M., Shangguan, J., Wang, B., Sun, C., Pan, L., Ma, Q., & Zhang, Z. (2017). Dendritic cells based immunotherapy. *A, 7*, 2091–2102 merican Journal of Cancer Research.

Sharma, P., Retz, M., Siefker-Radtke, A., Baron, A., Necchi, A., Bedke, J., Plimack, E. R., Vaena, D., Grimm, M.-O., Bracarda, S., Arranz, J. Á., Pal, S., Ohyama, C., Saci, A., Qu, X., Lambert, A., Krishnan, S., Azrilevich, A., & Galsky, M. D. (2017). Nivolumab in metastatic urothelial carcinoma after platinum therapy (CheckMate 275): A multicentre, single-arm, phase 2 trial. *The Lancet Oncology, 18*, 312–322. https://doi.org/10.1016/S1470-2045(17)30065-7.

Simoens, S., Vulto, A. G., & Dylst, P. (2021). Simulating costs of intravenous biosimilar trastuzumab vs. subcutaneous reference trastuzumab in adjuvant HER2-positive breast cancer: A Belgian case study. *Pharmaceuticals, 14*, 450. https://doi.org/10.3390/ph14050450.

Stapleton, N. M., Andersen, J. T., Stemerding, A. M., Bjarnarson, S. P., Verheul, R. C., Gerritsen, J., Zhao, Y., Kleijer, M., Sandlie, I., De Haas, M., Jonsdottir, I., Van Der Schoot, C. E., & Vidarsson, G. (2011). Competition for FcRn-mediated transport gives rise to short half-life of human IgG3 and offers therapeutic potential. *Nature Communications, 2*, 599. https://doi.org/10.1038/ncomms1608.

Stebbing, J., Mainwaring, P. N., Curigliano, G., Pegram, M., Latymer, M., Bair, A. H., & Rugo, H. S. (2020). Understanding the role of comparative clinical studies in the development of oncology biosimilars. *Journal of Clinical Oncology, 38*, 1070–1080. https://doi.org/10.1200/JCO.19.02953.

Sung, H., Ferlay, J., Siegel, R. L., Laversanne, M., Soerjomataram, I., Jemal, A., & Bray, F. (2021). Global cancer statistics 2020: GLOBOCAN estimates of incidence and mortality worldwide for 36 cancers in 185 countries. *CA. Cancer Journal of Clinicians., 71*, 209–249. https://doi.org/10.3322/caac.21660.

Tabernero, J. (2018). Biosimilars in oncology. *ESMO Open, 3*, e000456. https://doi.org/10.1136/esmoopen-2018-000456.

Tabernero, J., Vyas, M., Giuliani, R., Arnold, D., Cardoso, F., Casali, P. G., Cervantes, A., Eggermont, A. M. M., Eniu, A., Jassem, J., Pentheroudakis, G., Peters, S., Rauh, S., Zielinski, C. C., Stahel, R. A., Voest, E., Douillard, J.-Y., McGregor, K., & Ciardiello, F. (2016). Biosimilars: A position paper of the European Society for Medical Oncology, with particular reference to oncology prescribers. *ESMO Open, 1*, e000142. https://doi.org/10.1136/esmoopen-2016-000142.

Tagawa, S. T., Balar, A. V., Petrylak, D. P., Kalebasty, A. R., Loriot, Y., Fléchon, A., Jain, R. K., Agarwal, N., Bupathi, M., Barthelemy, P., Beuzeboc, P., Palmbos, P., Kyriakopoulos, C. E., Pouessel, D., Sternberg, C. N., Hong, Q., Goswami, T., Itri, L. M., & Grivas, P. (2021). TROPHY-U-01: A phase II open-label study of sacituzumab govitecan in patients with metastatic urothelial carcinoma progressing after platinum-based chemotherapy and checkpoint inhibitors. *Annals of Oncology: Official Journal of the European Society for Medical Oncology/ESMO, 39*, 2474–2485. https://doi.org/10.1200/JCO.20.03489.

Tianjin Chasesun Pharmaceutical Co., LTD, 2023. A phase I clinical trial of IMMH-010 in patients with advanced malignant solid tumors (Clinical trial registration No. NCT04343859). clinicaltrials.gov.

Todo, T., Ito, H., Ino, Y., Ohtsu, H., Ota, Y., Shibahara, J., & Tanaka, M. (2022). Intratumoral oncolytic herpes virus G47Δ for residual or recurrent glioblastoma: A phase 2 trial. *Nature Medicine, 28*, 1630–1639. https://doi.org/10.1038/s41591-022-01897-x.

Tokarew, N., Ogonek, J., Endres, S., von Bergwelt-Baildon, M., & Kobold, S. (2019). Teaching an old dog new tricks: Next-generation CAR T cells. *British Journal of Cancer, 120*, 26–37. https://doi.org/10.1038/s41416-018-0325-1.

Torre, L. A., Siegel, R. L., Ward, E. M., & Jemal, A. (2016). Global cancer incidence and mortality rates and trends—An Update. *Cancer Epidemiology and Prevention Biomarkers, 25*, 16–27. https://doi.org/10.1158/1055-9965.EPI-15-0578.

Triantafyllidi, E., & Triantafillidis, J. K. (2022). Systematic review on the use of biosimilars of trastuzumab in HER2+ breast cancer. *Biomedicines, 10*, 2045. https://doi.org/10.3390/biomedicines10082045.

U.S. Department of Health and Human Services Food and Drug Administration Center for Drug Evaluation and Research (CDER) Center for Biologics Evaluation and Research (CBER) Scientific considerations in demonstrating biosimilarity to a reference product; guidance for industry, 2015. Retrieved August 31, 2025, from, https://www.federalregister.gov/documents/2015/04/30/2015-10062/scientific-considerations-in-demonstrating-biosimilarity-to-a-reference-product-guidance-for.

US, FDA 2024a. Therapeutic biologics applications (BLA). FDA. Retrieved August 18, 2025, from, https://www.fda.gov/drugs/types-applications/therapeutic-biologics-applications-bla.

US, FDA 2024b. Overview for health care professionals. FDA.

Von Der Maase, H., Sengelov, L., Roberts, J. T., Ricci, S., Dogliotti, L., Oliver, T., Moore, M. J., Zimmermann, A., & Arning, M. (2005). Long-term survival results of a randomized trial comparing gemcitabine plus cisplatin, with methotrexate, vinblastine, doxorubicin, plus cisplatin in patients with bladder cancer. *Journal of Clinical Oncology, 23*, 4602–4608. https://doi.org/10.1200/JCO.2005.07.757.

Vukovic, N., van Elsas, A., Verbeek, J. S., & Zaiss, D. M. W. (2021). Isotype selection for antibody-based cancer therapy. *Clinical and Experimental Immunology, 203*, 351–365. https://doi.org/10.1111/cei.13545.

Vuky, J., Balar, A. V., Castellano, D., O'Donnell, P. H., Grivas, P., Bellmunt, J., Powles, T., Bajorin, D., Hahn, N. M., Savage, M. J., Fang, X., Godwin, J. L., Frenkl, T. L., Homet Moreno, B., De Wit, R., & Plimack, E. R. (2020). Long-term outcomes in KEYNOTE-052: Phase II study investigating first-line pembrolizumab in cisplatin-ineligible patients with locally advanced or metastatic urothelial cancer. *Journal of Clinical Oncology, 38*, 2658–2666. https://doi.org/10.1200/JCO.19.01213.

Walia, A. S., Sweis, R. F., Agarwal, P. K., Kader, A. K., & Modi, P. K. (2021). Cost-effectiveness of immune checkpoint inhibitors in urothelial carcinoma: A review. *Cancers, 14*, 73. https://doi.org/10.3390/cancers14010073.

Walsh, G. (2018). Biopharmaceutical benchmarks 2018. *Nature Biotechnology, 36*, 1136–1145. https://doi.org/10.1038/nbt.4305.

Yang, X., Wang, W., & Ji, T. (2024). Metabolic remodeling by the PD-L1 inhibitor BMS-202 significantly inhibits cell malignancy in human glioblastoma. *Cell Death & Disease, 15*, 186. https://doi.org/10.1038/s41419-024-06553-5.

Yu, E. Y., Petrylak, D. P., O'Donnell, P. H., Lee, J.-L., van der Heijden, M. S., Loriot, Y., Stein, M. N., Necchi, A., Kojima, T., Harrison, M. R., Hoon Park, S., Quinn, D. I., Heath, E. I., Rosenberg, J. E., Steinberg, J., Liang, S.-Y., Trowbridge, J., Campbell, M., McGregor, B., & Balar, A. V. (2021). Enfortumab vedotin after PD-1 or PD-L1 inhibitors in cisplatin-ineligible patients with advanced urothelial carcinoma (EV-201): A multicentre, single-arm, phase 2 trial. *The Lancet Oncology, 22*, 872–882. https://doi.org/10.1016/S1470-2045(21)00094-2.

Zahavi, D., & Weiner, L. (2020). Monoclonal antibodies in cancer therapy. *Antibodies (Baseline Switzerland), 9*, 34. https://doi.org/10.3390/antib9030034.

Zamani, M. R., & Šácha, P. (2025). Immune checkpoint inhibitors in cancer therapy: What lies beyond monoclonal antibodies? *Medical Oncology., 42*, 273. https://doi.org/10.1007/s12032-025-02822-1.

Zelenetz, A. D., Ahmed, I., Braud, E. L., Cross, J. D., Davenport-Ennis, N., Dickinson, B. D., Goldberg, S. E., Gottlieb, S., Johnson, P. E., Lyman, G. H., Markus, R., Matulonis, U. A., Reinke, D., Li, E. C., DeMartino, J., Larsen, J. K., & Hoffman, J. M. (2011). NCCN biosimilars white paper: Regulatory, scientific, and patient safety perspectives. *J, 9*(Suppl 4), S1–22. ournal of the National Comprehensive Cancer Network: JNCCN https://doi.org/10.6004/jnccn.2011.0136.

CHAPTER

Biosimilars and biologics in melanoma 14

Patakamuri Niharika[a], Shivani Suryavanshi[a], Jalapa Pradhan[b] and Sujit Kumar Sah[a]

[a] *Department of Pharmaceutical Sciences, School of Health Sciences and Technology, Dr. Vishwanath Karad MIT World Peace University, Pune, Maharashtra, India,* [b] *Department of Population Health and Leadership, School of Health Sciences, University of New Haven, West Haven, CT, United States*

14.1 Introduction

Melanoma is an aggressive, lethal, and heterogenous nature of cutaneous neoplasm also known as melanoma skin cancer (MSC) (Switzer et al., 2022). It arises from dysfunctional and abnormally proliferating melanocytes, which are making cells typically found in the basal surface of skin (Zaidi et al., 2019). Melanocytes are developed from highly migratory neural crest that express many signaling molecules, which enhance the migration of oncolytic cells (Paluncic et al., 2016). Melanomas contribute approximately 1%–4% of skin carcinoma and account for 80% of 80% death (Saginala et al., 2021). Early-stage melanomas have a higher survival rate of approximately 94%, but mortality significantly increased with malignant melanomas (Siegel et al., 2023). In 2020, based on the Global Cancer Observatory, malignant melanoma accounted for 3,25,000 new cases and around 57,000 deaths globally. The highest incidence of melanoma was detected among Australian and North American populations, while the lowest rate was observed among Aisa and Southern America population (De Pinto et al., 2024). Incidence of melanomas differs with the color of skin, age, and sex of populations. White populations have the highest rate of getting melanoma than the American Indian, Hispanic, Asian, and Black populations (Gupta et al., 2016). However, black skin populations have two to three times higher risk of mortality than the white population due to advanced-stage disease and late diagnosis (Park et al., 2012).

In individuals under 50 years, the rate has remained steady in women but has dropped by 1% annually in men since the early 2000s. For those $\geq$ 50 years, new cases in women have increased by 3% per year while rates in men have remained stable. The melanoma-related mortality has significantly declined from 2013 to 2022 (American Cancer Society, 2019). Increased melanin level and broader distribution of melanosomes absorb and deflect ultraviolet (UV) light. Further, excessive UV radiation exposure, specifically UV-A (315–400 nm)/UV-B (280–215 nm), could damage

Biosimilars and Biologics in Cancer Treatment: An Era in Therapeutic Strategies.
DOI: https://doi.org/10.1016/B978-0-443-43824-0.00003-1

skin cells through oxidative stress induced upon DNA and RNA damages, genetic mutations, and p53 protein inactivation, which result in tumorigenesis and cell death (Sample & He, 2018). Melanoma occurrence is often linked with familial atypical mole melanoma syndrome, mutation in CDKN2A genes and, less often, CDK4 genes, affecting tumor suppressor $p14^{ARF}$ and $p16^{INK4A}$. These genetic mutations cause impairment of the normal cell cycle, leading to abnormal skin cell growth (Roky et al., 2024). The four main forms of melanoma include lentigo maligna, acral lentiginous, nodular, and superficial spreading. Less than 10% of cases have lentigo maligna and acral lentiginous types of melanomas, while more than 70% of cases are superficial spreading, which is followed by nodular form (15%–30%). Other less common types of melanomas can develop in the lining of mucosal membranes of the nose, mouth, and other internal passages (American Cancer Society, 2019; Liu & Sheikh, 2014).

Melanoma is less prevalent but highly aggressive and deadly, posing a significant health risk and challenges in its management. Over the past decades, melanoma treatment has dramatically improved through various approaches, including surgery interventions, radiation therapy, advancement of oncologic drugs, targeted small molecule agents, biological therapies, and biosimilars (Boutros et al., 2024; Dixon et al., 2024; Liu et al., 2024). Further, early detection of melanoma has significantly improved patient survival and clinical outcomes. In recent years, advancements in biologics and biosimilars have added new hope in the management of malignant melanomas to enhance long-term efficacy, remission, and survival (Boutros et al., 2024; Sabbatino et al., 2022). Biologics are synthesized from living organisms, while biosimilars are similarly derived from living sources or contain biological components, and are designed to be structurally and functionally similar to existing approved biologics (Malakar et al., 2024; Papież & Krzyściak, 2021).

14.1.1 Biologics in melanoma

Biologics are specially engineered complex biological components synthesized and purified from bacteria, yeast, animals, or plant cells (Pham et al., 2019). They include protein, cells, or genetically modified substances such as monoclonal antibodies, growth factors, nucleic acids, cell-based therapies, vaccines, immune modulators, and drug components derived from human blood and plasma (Jozala et al., 2016). These substances are unique in nature and are synthesized, tested, regulated, and managed differently from conventional pharmaceutical products to ensure enhanced quality and efficacy (Süntar et al., 2021). The biological therapies pose a significant impact in cancer management, either by assisting immune response in targeting cancer cells or by directly attacking the cancer itself (Papież & Krzyściak, 2021; Süntar et al., 2021). Biological therapies play an essential role in melanoma management, either by fighting the cancer directly or by helping the immune system target cancer cells (Vinay et al., 2015).

Advances in biological agents, particularly immunotherapies, which have transformed the management of melanoma, significantly enhance patient outcomes and offer new hope for long-term remission (Knight et al., 2023). In recent years, regulatory bodies have approved an increasing number of biological agents as alternatives to conventional drug regimens. These agents often replace long-term, complex medication requirements, simplifying treatment protocols and improving patient compliance to therapy (Puccetti et al., 2024). In melanoma, immunotherapies have shown a significant impact in enhancing patient outcomes. These are available in different types and natures which used for melanoma treatment (Mooradian & Sullivan, 2023).

14.1.2 Immune checkpoint inhibitors

Checkpoints for immunity proteins are crucial for immune response regulation because they preserve immunological homeostasis (Lim et al., 2017). They promote immune self-tolerance and help to prevent the overactivation of immune response, which leads to the eventual destruction of invasive pathogens, modified cells, overexpressed antibodies, and excessive inflammatory process (Marin-Acevedo et al., 2018). Melanocytes may use immune checkpoint proteins as a mechanism to regulate immune response to prevent excessive downregulation and minimize collateral damage (Pardoll, 2012). However, any disruption in this regulatory process can lead to immune dysregulation, and dysfunction of melanocytes may create a conducive environment for the development of melanocytes (Upadhyay et al., 2021). In view of this, ICIs, also known as monoclonal antibodies (mAbs), are introduced as a therapeutic strategy (Kumar et al., 2020). These targeted immune checkpoint molecules revolutionized the treatment of cancer, including melanoma, and have proven significant improvement in patient health outcomes (Wei et al., 2024). Cytotoxic T-lymphocytes antigen 4 (CTLA-4) and anti-PD-1 specific mAbs approved ICIs for malignant melanoma treatment and have been proven significantly linked with better patient response rate, overall survival (OS), decreased recurrence, and even long term remission (Wojtukiewicz et al., 2021).

14.1.2.1 CTLA-4 specific antibodies

CTLA-4 mAb inhibitors disrupt the binding of B7 ligands to antigen-presenting cells (APCs), and the CTLA-4 receptor in T cells results in inhibition of the downregulatory signal during T-cell response (Sabbatino et al., 2022; Walker & Sansom, 2015). This enhances the CD28-mediated costimulatory signaling (via CD80/CD86 ligands) and promotes T-lymphocyte proliferation to enhance adaptive immune responses (Sharma et al., 2023). Additionally, CTLA-4 blockage may inhibit Treg cells that suppress further amplification of antitumor immunity (Walker, 2013). The combined pooled data analysis of the phase II/III study revealed that CTLA-4 mAb significantly enhances overall survival rates (OSRs) and long-term clinical outcomes among metastatic melanoma (Savoia et al., 2016). At the current time, Ipilimumab has only been approved CTLA-4 specific mAb used to treat advanced malignant melanoma (Lipson & Drake, 2011).

14.1.2.2 Ipilimumab

This human IgG1 mAb was administered in four dosages at 3 mg/kg, repeated every 3 weeks over four treatment cycles. This was a significant breakthrough in clinical treatment of melanoma, which is the first ICI to get FDA approval (Ascierto et al., 2017). In 2015, it was authorized as adjuvant therapy for melanoma with third-stage metastasis at a high dosage of 10 mg/kg (Tarhini et al., 2020). However, recent evidence supports the use of 3mg/kg due to a better safety and efficacy profile. Despite therapeutic benefits, Ipilimumab blockade of the CTLA4-dependent immunological checkpoint can lead to serious adverse effects, including autoimmune and inflammatory reactions, with 10%–15% of those cases being of grade$^{3/4}$ severity (Ascierto et al., 2017; Martins et al., 2019). These risks highlight the necessity for careful assessment of patient needs and vigilant monitoring for potential adverse effects following administration (Tarhini et al., 2018; Yan et al., 2024a).

14.1.2.3 Anti-PD-1 monoclonal antibodies

PD-1 (PDCD or CD279) is a surface receptor predominantly presented in T lymphocytes (Lin et al., 2024). The PD-L1 ligand is present in numerous immune cells, tissues, and melanocytes (Massi et al., 2014). PD-1/PD-L1 inhibition demonstrated an essential role in immune regulation through suppressing T-cell response, cytokine production, and tumor cell clearance, making it a significant therapeutic target in oncology (Han et al., 2020). Clinical studies have demonstrated that blocking this pathway enhances anti-tumor immunity specifically in melanoma (Zhang et al., 2023). Based on these efficacy data, the FDA authorized two PD-1 agonist medications, named pembrolizumab and nivolumab, in 2014 to treat metastatic melanoma (Burns et al., 2016).

14.1.2.4 Anti-PD-1 monotherapies

14.1.2.4.1 Pembrolizumab

Pembrolizumab (also named MK-3475 or lambrolizumab) is a more effective IgG4 kappa mAb that targets PD-1 and stops PD-L1/PD-L2 ligand binding and inhibits reactive T lymphocytes, which helps in apoptosis of melanoma cells. Dosage of 2 mg/kg administered Q3W, 10 mg/kg delivered for Q3W, and 10 mg/kg administered Q2W were evaluated in clinical trials (Huddleston, 2015; Sabbatino et al., 2022). The PFS duration reached 0.7 years, OSRs of 81% in 1 year and 71% in 1.6 years (Lin et al., 2018). This study also revealed OSRs of 52% for 10 mg/kg Q3W, 27%, and 25% at 10 mg/kg Q3W and 2 mg/kg Q3W, respectively. However, the FDA ultimately approved a 2 mg/kg Q2W regimen as the standard regimen for treating metastatic melanoma based on safety and effectiveness profiles (Chatterjee et al., 2016). The long-term further enhances the long-term remission and improved survival of patients (Shalata et al., 2023). However, pembrolizumab is associated with serious immune-mediated adverse effects like pneumonitis, hepatitis, colitis, and endocrine dysfunctions (Husebye et al., 2022).

14.1.2.4.2 Nivolumab

It is also a human IgG4 PD-1 mAb antagonist that specifically stops PD-1 binding with PD-L1/L2 legends (Rajan et al., 2016). In clinical trial phase I, nivolumab was given to 94 melanoma patients at doses of 0.1–10 mg/kg for Q2W (Johnson et al., 2015). At 3 mg/kg Q2W, 41% was the highest observed response rate (Jiang et al., 2022). Long-term data revealed overall survival rates of 62%–42% with side effects like fatigue, diarrhea, itching, nausea, and reduced appetite (Topalian et al., 2014). The majority of adverse events were mild in severity, and the drug is well-tolerated (Jiang et al., 2022).

14.1.3 Combination therapies

14.1.3.1 Relatlimab and nivolumab

Relatlimab is an IgG4 mAb that targets the LAG-3 receptor, which suppresses T cell activation (Tawbi et al., 2022). Combined with nivolumab increases the immunity against tumor through LAG-3 and PD-1 double blockage pathways (Su et al., 2023). PFS was 10.1 months for metastatic melanoma administered in combination of relatlimab (160 mg) and nivolumab (480 mg) Q4W in a phase III trial and compared to patient received nivolumab monotherapy. PFS was noted at 4.6 months with the combination regimen (Ascierto et al., 2023). The 43.7% response rate was noted in combination regimen versus 33.7% response rate in monotherapy with nivolumab alone, median OS at 51 months for the combination regimen versus 34.1 months with nivolumab (Long et al., 2023). Due to the superior effectiveness of this combination, the FDA has approved it. These were also linked to side effects such as diarrhea, pruritus, skin rashes, exhaustion, and increased liver enzymes of grade >/3 toxicities (Lipson et al., 2025).

14.1.3.2 Ipilimumab and nivolumab

Ipilimumab increases T cell response through inhibition of CTLA-4-mediated inhibitory signaling (Camacho, 2015). Nivolumab, as mentioned previously, acts by suppressing the PD-1 pathway. Through complementary mechanisms, the combination of these two medications aims to improve T cells' T-mediated antitumor responses (Buchbinder & Desai, 2016). In the Phase III clinical study, metastatic melanoma were administered a therapeutic regimen consisting of nivolumab 1 mg/kg combined with ipilimumab 3 mg/kg every Q3W for four treatment cycles, followed by nivolumab 3 mg/kg every Q2W (Kooshkaki et al., 2020). During this study, OSRs noted 71.9 months recorded for the combination regimen, 36.9 months for nivolumab, and 19.9 months with ipilimumab monotherapy, which was determined till 10 years of follow-up (Vázquez-Montero et al., 2023). Ipilimumab and Nivolumab combination therapies have been authorized as first-choice treatments in advanced melanoma, particularly for unresected stage III or IV, because of their superior efficacy and safety (Wolchok et al., 2013). The common side effects include colitis, deformities, hepatotoxicity, and endocrine disorders found among 59% patients in the combination therapy group. Grade 3 toxicities were experienced, while 24% patients were administered nivolumab, and 28% patients were administered ipilimumab (Albrecht et al., 2023; Riveiro-Barciela et al., 2024).

14.1.4 Oncolytic virus therapy

Oncolytic virus therapy (OVT) is an originally occurring or genetically engineered virus that specifically infects, multiplies, and destroys cancer cells, leaving healthy cells unharmed (Yan et al., 2024b). Through reprogramming of immunosuppressive tumor microenvironment (TME) into an immunostimulatory state, which potentiates antitumor immunity (Wang et al., 2022). OVs multiply inside cancer cells, causing cell apoptosis and generating tumor antigens (Lemos de Matos et al., 2020). These components stimulate T-cell-mediated immune activities against the tumor and activate APCs (Jin et al., 2021). Talimogene laherparepvec (T-VEC), the first OVT expressly authorized for advanced melanoma management of stage IIIB, IIIC, or IV. It was first authorized by the US FDA in 2015 (Ferrucci et al., 2021; Pol et al., 2016).

14.1.4.1 Talimogene laherparepvec

The genetically engineered oncolytic herpes simplex virus type-1 (HSV-1) is programmed to specifically infect, multiply, and cause death in melanoma cells (Bommareddy et al., 2017). The modification in genes ICP34.5 and ICP47 limits the number of viral replications in tumor cells and lowers neurovirulence (Kaufman et al., 2022). Additionally, it expresses granulocyte macrophage colony-stimulating factor (GM-CSF) that stimulates the formation of antigens which cause systemic antitumor apoptosis (Bommareddy et al., 2017; Kaufman et al., 2022). T-VEC is directly administered under ultrasound guidance in epidermis or melanoma tissue. Initially, T-VEC 4 mL of PFU/mL is used to administer, followed by 4 mL Q2W until tumor remission or intolerable toxicity is noted. Depending upon the patient's response, T-VEC mostly lasts for 6 months or more (Ferrucci et al., 2021; Johnson et al., 2015). Clinical trials evaluating T-VEC against GM-CSF in unresectable stage III/IV melanoma patients demonstrated better outcomes. Polled trial data reported 23.3 months of median OS, 16% response rate, and objective response (ORR) of 26.4% for T-VEC-treated cohorts (Andtbacka et al., 2015). These outcomes represented the most clinically significant benefits for advanced-stage diseases (IIIB, IIIC, IV, and M1a), establishing T-VEC as a therapeutic breakthrough (Andtbacka et al., 2019; O'Donoghue et al., 2016). Regulatory agencies, the FDA granted approval of T-VEC in October 2015, and by the European Medicines Agency (EMA) in December 2015 to use for unresectable stage III/IV melanoma management (Perez et al., 2019). The common adverse events (AEs) included transient symptoms, such as fatigue, chills, fever, nausea, and localized inflammation at the injection site. Notably, 15%–26% of participants experienced grade ¾ toxicities, usually manifested as cellulitis and immune-related reaction (Rehman et al., 2016). Due to a manageable safety profile,

T-VEC may be utilized either monotherapy or synergistically with ICIs for melanoma management (Hong et al., 2024).

14.1.5 Adoptive cell therapy

Adoptive cell therapy (ACT) is a personalized immunotherapy that boosts the immune system to fight against melanoma through tumor-infiltrating lymphocytes (TIL) activation (König et al., 2024). Cancer-reactive T cells are isolated, multiplied, and reinfused to patients who are unresponsive to ICIs therapy (Warner et al., 2023). Clinical trials show objective response rates of 49%–72% in metastatic melanoma with desirable responses lasting years in many patients (Dafni et al., 2019). The success includes infused cell persistence, telomere length, and CD8+ CD27+ cell counts (Olivera et al., 2024). Recent trials reported 28% complete response rate with most patients maintaining remission for 18–75 months (Fang et al., 2022). However, some patients develop resistance or remain refractory, promoting exploration of combination therapies to increase efficacy (Fang et al., 2022; Feldman et al., 2015).

14.1.6 Tumor-infiltrating lymphocytes

TIL therapy relies on isolating naturally occurring tumor-reactive lymphocytes from melanoma tissues. These tumor-reactive lymphocytes were cultured in-vitro using IL-2 to achieve significant numerical expansion. Before infusion, patients undergo lymphodepleting chemotherapy to optimize the therapeutic environment, after which the activated lymphocytes are reintroduced systemically (Warner et al., 2023; Zhao et al., 2022). This process enhances the endurance and efficiency of TILs in killing melanoma. Transgenic T-cell receptors (TCRs) and chimeric antigen receptors (CARs) are examples of genetically modified TIL that have been studied to increase specificity and effectiveness (Hulen et al., 2021; Warner et al., 2023). The treatment usually consists of a lymphodepletion chemotherapy regimen include fludarabine (25 mg/m^2) for 5 days and cyclophosphamide (60 mg/kg) for 2 days, followed by an injection of TILs (109–1011 cells) (Lickefett et al., 2023). Every 8 h after TIL administration, an IL-2 of 720,000 IU/kg dose is provided to encourage T-cell activation and proliferation against the tumor (Owen et al., 2023). Objective response rates range from 40% to 72%, depending on the extent of lymphodepletion, with overall rates showing a disease-free period of 3–5 years (Hirayama et al., 2019). Although ACT is highly effective, toxicities arising from the chemotherapy and IL-2 rather than TIL infusion are common (Rohaan et al., 2018). Adverse effects include thrombocytopenia, anemia, neutropenia, hypotension, and capillary leak syndrome over 50% of patients experience grade ¾ adverse events (Chesney et al., 2022). Lifileucel, an autologous TIL authorized by the FDA in February 2024 to treat metastatic or incurable melanoma (Parums, 2024).

14.1.7 Cytokine therapy

Several clinical studies have tested the potency of cytokines in melanoma management, emphasizing their crucial function in controlling innate as well as adaptive immunity (Safaei et al., 2025). Cytokines affect immune responses by causing APCs, NK cells, and cytotoxic T cells that target and eliminate malignant cells through apoptosis and stopping proliferating (Abel et al., 2018). The way cytokines are used in melanoma treatment has evolved (Paul & Lal, 2017). FDA authorized cytokines for melanoma, especially IL-2 and interferon-alpha (IFN-α), which are mainly used for adjuvant therapy and metastatic disease (Floros & Tarhini, 2015). However, these treatments necessitate high and frequent dosages, which frequently lead to toxicities that limit dosage, including capillary leak syndrome, flu-like symptoms, and organ malfunction (Chen et al., 2022). Despite these challenges, cytokine therapy remains a crucial part of melanoma immunotherapy, offering significant benefits in modulating immune responses against cancer (Nicholas & Lesinski, 2011).

14.1.8 Interleukin-2

Interleukin-2 (IL-2) is a 15 kDa glycoprotein that stimulates expansion and regulates functional activation of NK cells and CD8+ cytotoxic T lymphocytes, enhancing their antitumor capabilities (Liao et al., 2013). This cytokine binds to IL-2 receptor complex (IL-R2R$\alpha\beta\gamma$), triggering downstream signaling cascades (PI3K, MAPK, and JAK-STAT) pathways. These collectively amplify immune-mediated destruction of melanoma cells (Ross & Cantrell, 2018). The higher dosage of IL-2 (HDIL-2) treatment consists of two 5-day cycles spaced 8–12 weeks apart with 600,000 or 720,000 IU/kg injected every 8 h for a 15-min intravenous infusion (Buchbinder et al., 2019). Clinical investigations indicate the complete response rate of 5%–7% and an objective response rate (ORR) of 15%–20% (Ko et al., 2023). However, IL-2 medication is linked to serious adverse events like hepatotoxicity, pulmonary edema, pancytopenia, renal failure, febrile neutropenia, capillary leak syndrome (CLS), and systemic inflammatory symptoms. These side effects are instances of grade ¾ toxicities (Dutcher et al., 2014). Despite these difficulties, the FDA authorized IL-2 in 1998 for metastatic melanoma management because of its long-lasting effects (Rohaan et al., 2018).

14.1.9 Interferon-alpha

It is a highly potent cytokine used for melanoma. It increases tumor immunogenicity by preventing tumor cell growth, triggering apoptosis, and upregulating major histocompatibility complex (MHC) class I molecules (Vidal, 2020). Additionally, it boosts the cytotoxicity of NK cells and dendritic DC activation, that strengthen adaptive immune responses against melanoma (Lee et al., 2021). The higher dosage regimen of IFN-α includes an initial dosage of 20 $MU/m^2/d/IV$ for 5 days/wk for 1 month, followed by a maintenance dose of 10 $MU/m^2/d/SC$ three times a week for 1 year (Xiong et al., 2022). An alternative PEGylated IFN-α (PEG-IFN) has also been approved for stage III melanoma due to its better compliance and tolerability (Ascierto et al., 2010). It is administered once weekly for eight two months of 6 μg/kg initial dose and then weekly for a maintenance dose of 3 μg/kg (Yamazaki et al., 2016). Although the median OS and relapse-free survival (RFS) have not increased significantly, ORR ranges from 5% to 30% depending on the regimen and doses (Dummer & Mangana, 2009). Flu-like symptoms, exhaustion, nausea, hepatoxicity, leukopenia, and thrombocytopenia are the common adverse events. Severe toxicities include depression, mania, confusion, and infrequent suicide instances (Sato, 2014). In 1996, the FDA approved IFN-α to provide for high-risk patients having surgical removal of stage IIB/III melanoma. FDA also authorized PEG-IFN in stage III melanoma in 2011, which provides weekly once to increase patient adherence (Boutros et al., 2024; Davar et al., 2012).

14.1.10 Cancer vaccines

It is investigated to enhance the immune system's ability to detect and eliminate cancer for better therapeutic outcomes (Liu et al., 2023). Traditional vaccines that focused on prevention from infectious disease, but modified cancer vaccines have been developed to activate oncolytic-specific antigens against existing tumors. This approach activates T lymphocytes to eliminate cancerous cells due to cytotoxic effects (Kaczmarek et al., 2023). These vaccines deliver tumor-associated antigens via lipid nanoparticles (LNPs), triggering robust immune responses (Ramadan et al., 2024). Recently, mRNA vaccines have shown promising potential to enhance survival outcomes for cancer-like melanomas, particularly when provided in combination with ICIs (Mehta et al., 2025).

14.1.10.1 mRNA-based cancer vaccines

It deliver tumor associated antigens into cells using lipid nanoparticles (LNPs) or a similar delivery system (Senti et al., 2024). After administration, mRNA is translated into tumor antigens, which are presented in MHC molecules that activate T lymphocytes, resulting in cytotoxicity of tumor cells (Gao et al., 2024). This triggers potent cell-mediated immune responses and effectively destroys melanoma cells (Bidram et al., 2021). These vaccines can be administered intratumorally (IT), subcutaneously (SC), or intramuscularly (IM) with doses varying by formulation and type (Zeng et al., 2022). For instance, clinical studies on MART1 mRNA lipopolyplexes in melanoma suggest a dose of 12.5 mg/injection for tumor protection. Objective response rates differ across vaccine types, but combining mRNA vaccines in conjunction with checkpoint inhibitors has demonstrated better survival outcomes (Bidram et al., 2021; Miao et al., 2021; Ramadan et al., 2024). Cancer vaccines also reported with adverse events include mild to moderate symptoms such as tiredness, fever, and angioedema, with rare severe AEs including autoimmune-like reactions and cytokine release syndrome (CRS). Several candidates, such as BioNTech and Moderna, are presently undergoing phase II and III trials; however, no mRNA-related cancer vaccines have achieved full FDA approval (Géraud et al., 2024). Early phase clinical trials of BioNTech's mRNA vaccine and BNT111 use in the advanced stage of melanoma demonstrated durable responses and manageable safety profiles (Xu & Fisher, 2023). These advancements highlight how mRNA vaccines have the potential to revolutionize immune response against cancer cells (Yaremenko et al., 2025).

14.2 Biosimilars in melanoma

Biosimilars are biological products that are relatively similar to the original biological molecule (innovator or originator) in terms of strength, purity, and safety, providing a more cost-effective choice for supportive care and cancer treatment (Ditani et al., 2021). Biosimilars are complex proteins or monoclonal antibodies derived from living cell lines, making them difficult to reproduce as exact generics (Joshi et al., 2022). They undergo rigorous analytical testing to compare molecular structure, biological activity, and pharmacokinetics with reference molecules, followed by clinical studies to prove similar efficacy and safety (Sheridan et al., 2024). The concept and regulatory guidelines were established in 2003 in Europe, and the first biosimilar was authorized in 2006. This approach and framework were later adopted by the US, Canada, and Japan (Wang & Chow, 2012; Yamaguchi & Arato, 2011). The first approved biosimilar was approved in 2015, and since then, 17 cancer-related biosimilars have been authorized. These biosimilar such as trastuzumab and rituximab, are synthesized from living organisms and share the amino acid sequence and immunogenicity profiles as their reference biologics (Joshi et al., 2022; Rugo et al., 2016). According to clinical research, biosimilars perform similarly to reference molecules in malignancy response and progression-free survival (Tkaczuk & Jacobs, 2014). However, the challenges such as immunogenicity, extrapolation of indications, and interchangeability remain concerns. Despite these issues, biosimilars are the cheapest and provide a suitable option to improve cancer patient outcomes (Chopra & Lopes, 2017). It is anticipated that the incorporation of cutting-edge technology like personalized medicine and next-generation sequencing would further increase their therapeutic potential. Numerous biosimilars are currently authorized for malignancies of the blood, lung, bladder, and melanoma (Satam et al., 2023). Since EMA authorized 5, 7, and 8 biosimilars for different malignancy management, respectively, and anti-CD20 rituximab, anti-HER2 trastuzumab, and anti-VEGF bevacizumab are widely accessible in Europe. There aren't any approved biosimilars for melanoma treatment (Broer et al., 2024). There are numerous biosimilars for monoclonal antibodies (mAbs), which are utilized as immunotherapy to treat various cancers, including melanoma. These agents are still under clinical development and proving their ability (Shepard et al., 2017). There are various types of biosimilars under clinical development phases but not yet approved (Mascarenhas-Melo et al., 2024).

14.2.1 Monoclonal antibodies biosimilar

ICIs, anti-CTLA-4, and anti-PD-1 mAbs are frequently used to treat melanoma. However, biosimilars of these particular medications are not yet generally authorized and accessible (Johnson et al., 2015; Knight et al., 2023; Timis et al., 2025). These biosimilar claims are cost-effective and could be an alternative to expensive biologics after their patent expires (Konstantinidou et al., 2020). These biosimilars undergo rigorous analytical, preclinical, and clinical assessments to ensure similarity to originator drugs (Khraishi et al., 2016). The way mAbs have demonstrated notable effectiveness in melanoma, the same excepting from mAbs biosimilar (Au et al., 2017).

14.2.2 Anti-hCTLA4-hIgG1

It is a biosimilar of the clinically approved monoclonal antibody ipilimumab designed to replicate its structure and function. It combines the various regions of ipilimumab targeting CTLA-4 (CD152) (He et al., 2017). Additionally, ipilimumab also has therapeutic effects via stimulating TNF-α production and inducing antibody-dependent cell-mediated cytotoxicity (Rosskopf et al., 2019). Shanghai Henlius Biotech Inc. declared in April 2020 that the China National Medicine Product Administration (NMPA) had accepted its investigational new drug (IND) application for HLX13, which is a biosimilar of ipilimumab (Rosskopf et al., 2019; Shanghai Henlius Biotech, 2023). A completely human monoclonal antibody called HLX13 is a genetically changed anti-CTLA-4 antibody that may treat metastatic melanoma. HLX13 has demonstrated high similarity to originator ipilimumab in preclinical studies, including pharmacology, pharmacokinetics, toxicokinetics, immunogenicity, and toxicity profiles. It acts as a promising candidate in combination with anti-PD-1 inhibitors, which may reduce the burden among advanced melanoma patients (Cheng et al., 2024; Shanghai Henlius Biotech, 2023).

14.2.3 Pembrolizumab biosimilars

Pembrolizumab biosimilars such as FYB-206 and HLX17 are in development for cancers like melanoma by targeting the PD-1 immune checkpoint inhibitor (Dang et al., 2016; Formycon, 2024). HLX17, developed by Shanghai Henlius Biotech Inc., has shown pharmacological similarity to pembrolizumab in preclinical studies. HLX17 enhances antitumor activity by blocking the PD-1 pathway (Shanghai Henlius Biotech, Inc., 2024). With a 39% PTSR for melanoma Phase III drugs, pembrolizumab biosimilars are competitive and promising for antitumor activities. Henlius is developing a variety of pipelines that target CTLA-4 and PD-1/L1 (Shanghai Henlius Biotech, Inc., 2024; Spain et al., 2015). These developments aim to improve treatment accessibility and foster innovation in cancer therapy. The growing focus on biosimilars is based on their potential in tumor destruction and low cost (Malakar et al., 2024; Rugo et al., 2016).

14.2.4 Nivolumab biosimilar

Amgen is working on a biosimilar to nivolumab. For metastatic melanoma, it is presently in phase III with a 43% phase transition success rate (PTSR) requirement to move on to preregistration (Amgen, 2025). Global data analysis, based on 18 years of historical development, to evaluate the biosimilars PTSR and their likelihood of approval against industry benchmarks (Mascarenhas-Melo et al., 2024). The biosimilar is an injectable human IgG4 mAb that targets PD-1 to boost immune responses against tumors (Liu & Yi, 2020).

14.3 Recommendations for biologics and biosimilar use in melanoma management

- Anti-PD-1 monotherapy (pembrolizumab or nivolumab) is recommended as first-line treatment for advanced or metastatic melanoma due to its high efficacy, durable responses, and manageable toxicity profile.
- Combination therapy with nivolumab and ipilimumab should be considered for unresectable stage III/IV melanoma. This dual checkpoint blockade improves survival but necessitates vigilant monitoring for immune-related adverse events such as colitis or hepatotoxicity.
- Nivolumab combined with relatlimab (LAG-3 inhibitors) offers improved progression-free survival by targeting complementary immune pathways. This regimen is particularly beneficial for patients with limited response to prior therapies.
- Talimogene lagerparepvec (T-VEC) is an oncolytic virus therapy recommended for unresectable stage IIIB-IV melanoma with injectable lesions. Administered intratumorally induces local tumor lysis and systemic immune activation.
- Adoptive cell therapy (ACT) using tumor-infiltrating lymphocytes (TILs) such as lifileucel is reserved for metastatic or refractory lymphodepletion and IL-2 support to enhance TIL efficacy.
- High-dose IL-2 may be considered for selective patients with potential for durable remission despite risks of severe toxicity like capillary leak syndrome. PEGylated IFN-α serves as an adjuvant option for high-risk stage III melanoma to delay recurrence.
- Emerging strategies like mRNA vaccines combined with ICIs show promising effects in clinical trials, leveraging neoantigen boosting long-term immunity.
- Biosimilars of anti-CTLA-4/PD-1 agents (HLX13, HLX13, FYB-206) are under development to improve cost-effectiveness post their patient expires.
- Personalized treatment guided by biomarkers like PD-L1 expression or tumor mutational burden optimizes therapeutic outcomes of patients.
- Participating patients in clinical trial for development of novel therapies, ensuring alignment with disease stage, patient tolerance, and molecular profiles.
- Supportive care must include multidisciplinary monitoring for adverse events with prompt interventions. Early detection through routine and specific examination remains critical for the enhancement of therapeutic outcomes.

14.4 Conclusion

Biologics such as ICIs, which regulate CTLA-4 and PD-1/PD-L1 pathways to improve antitumor immunity and increase survival in advanced-stage melanoma, have revolutionized melanoma management. Double checkpoint blocking mAbs CTLA4/PD-1 or PD-1/LAG-3 greatly increase efficacy but necessitate careful toxicity monitoring because of serious side effects linked to the immune system. Oncolytics viruses and adaptive cell therapies provide novel mechanisms by directly lysing tumors or utilizing immune cells to achieve durable responses in refractory cases. Cytokine therapies remain niche due to severe toxicity but retain utility in specific clinical scenarios. Emerging strategies like mRNA vaccines show promising effects in early trials, specifically in combination with ICIs, although clinical adoptions require further validation. Further, biosimilars sharing similarity with original biological products are meant to improve accessibility to biologics due to their cost-effectiveness, but face challenges in regulatory approval and maintaining therapeutic equivalence. Furthermore, advancements will focus on personalized treatment, optimizing combination regimens, and addressing resistance mechanisms and toxicity to enhance long-term remission and improve patient outcomes. However, early detection and a multidisciplinary approach to eradicating cancer cells remain primary priorities for improving patient survival.

References

Abel, A. M., et al. (2018). Natural killer cells: Development, maturation, and clinical utilization. *Frontiers in Immunology, 9*, e1869. https://doi.org/10.3389/fimmu.2018.01869.

Albrecht, L. J., et al. (2023). The latest option: Nivolumab and relatlimab in advanced melanoma. *Current Oncology Reports, 25*(6), 647–657. https://doi.org/10.1007/s11912-023-01406-4.

American Cancer Society (2019) *'What is melanoma skin cancer', American Cancer Society*. Retrieved February 18, 2025, from https://www.cancer.org/cancer/acs-medical-content-and-news-staff.html.

Amgen. Nivolumab biosimilar by Amgen for metastatic melanoma: Likelihood of approval. Pharmaceutical Technology. Pharmaceutical Technology article. https://www.pharmaceutical-technology.com/data-insights/nivolumab-biosimilar-amgen-metastatic-melanoma-likelihood-of-approval.

Andtbacka, R. H. I., et al. (2015). Talimogene laherparepvec improves durable response rate in patients with advanced melanoma. *Journal of Clinical Oncology, 33*(25), 2780–2788. https://doi.org/10.1200/JCO.2014.58.3377.

Andtbacka, R. H. I., et al. (2019). Final analyses of OPTiM: A randomized phase III trial of talimogene laherparepvec versus granulocyte-macrophage colony-stimulating factor in unresectable stage III-IV melanoma. *Journal for Immunotherapy of Cancer, 7*(1), 145. https://doi.org/10.1186/s40425-019-0623-z.

Ascierto, P. A., et al. (2010). Regulatory T cell frequency in patients with melanoma with different disease stage and course, and modulating effects of high-dose interferon-alpha 2b treatment. *Journal of Translational Medicine, 8*, 76. https://doi.org/10.1186/1479-5876-8-76.

Ascierto, P. A., et al. (2017). Ipilimumab 10 mg/kg versus ipilimumab 3 mg/kg in patients with unresectable or metastatic melanoma: A randomised, double-blind, multicentre, phase 3 trial. *The Lancet Oncology, 18*(5), 611–622. https://doi.org/10.1016/S1470-2045(17)30231-0.

Ascierto, P. A., et al. (2023). Nivolumab and relatlimab in patients with advanced melanoma that had progressed on anti-programmed death-1/programmed death ligand 1 therapy: Results from the phase I/IIa relativity-020 trial. *Journal of Clinical Oncology : Official Journal of the American Society of Clinical Oncology, 41*(15), 2724–2735. https://doi.org/10.1200/JCO.22.02072.

Au, L., O'Reilly, A., & Larkin, J. (2017). *Anti-PD-1 monotherapy versus anti-PD1 plus anti-CTLA4 in advanced melanoma: How do we decide?* (pp. 151–155). *Melanoma management*. https://doi.org/10.2217/mmt-2017-0016.

Bidram, M., et al. (2021). mRNA-based cancer vaccines: A therapeutic strategy for the treatment of melanoma patients. *Vaccines, 9*(10), e1060. https://doi.org/10.3390/vaccines9101060.

Bommareddy, P. K., et al. (2017). Talimogene laherparepvec (T-VEC) and other oncolytic viruses for the treatment of melanoma. *American Journal of Clinical Dermatology, 18*(1), 1–15. https://doi.org/10.1007/s40257-016-0238-9.

Boutros, A., et al. (2024). The treatment of advanced melanoma: Current approaches and new challenges. *Critical Reviews in Oncology/Hematology, 196*, e104276. https://doi.org/10.1016/j.critrevonc.2024.104276.

Broer, L. N., et al. (2024). Monoclonal antibody biosimilars for cancer treatment. *iScience, 27*(6), 110115. https://doi.org/10.1016/j.isci.2024.110115.

Buchbinder, E. I., et al. (2019). Therapy with high-dose Interleukin-2 (HD IL-2) in metastatic melanoma and renal cell carcinoma following PD1 or PDL1 inhibition. *Journal for ImmunoTherapy of Cancer, 7*(1), 49. https://doi.org/10.1186/s40425-019-0522-3.

Buchbinder, E. I., & Desai, A. (2016). CTLA-4 and PD-1 pathways: Similarities, differences, and implications of their inhibition. *American Journal of Clinical Oncology, 39*(1), 98–106. https://doi.org/10.1097/COC.0000000000000239.

Burns, M. C., O'Donnell, A., & Puzanov, I. (2016). Pembrolizumab for the treatment of advanced melanoma. *Expert Opinion on Orphan Drugs, 4*(8), 867–873. https://doi.org/10.1080/21678707.2016.1191348.

Camacho, L. H. (2015). CTLA-4 blockade with ipilimumab: Biology, safety, efficacy, and future considerations. *Cancer medicine, 4*(5), 661–672. https://doi.org/10.1002/cam4.371.

Chatterjee, M., et al. (2016). Systematic evaluation of pembrolizumab dosing in patients with advanced non-small-cell lung cancer. *Annals of Oncology : Official Journal of the European Society for Medical Oncology, 27*(7), 1291–1298. https://doi.org/10.1093/annonc/mdw174.

Chen, P., et al. (2022). Potential pathophysiological mechanisms underlying multiple organ dysfunction in cytokine release syndrome. *Mediators of Inflammation, 2022*, e7137900. https://doi.org/10.1155/2022/7137900.

Cheng, W., et al. (2024). Dual blockade immunotherapy targeting PD-1/PD-L1 and CTLA-4 in lung cancer. *Journal of Hematology & Oncology, 17*(1), 54. https://doi.org/10.1186/s13045-024-01581-2.

Chesney, J., et al. (2022). Efficacy and safety of lifileucel, a one-time autologous tumor-infiltrating lymphocyte (TIL) cell therapy, in patients with advanced melanoma after progression on immune checkpoint inhibitors and targeted therapies: Pooled analysis of consecutive cohorts. *Journal for Immunotherapy of Cancer, 10*(12), e005755. https://doi.org/10.1136/jitc-2022-005755.

Chopra, R., & Lopes, G. (2017). Improving access to cancer treatments: The role of biosimilars. *Journal of Global Oncology, 3*(5), 596–610. https://doi.org/10.1200/JGO.2016.008607.

Dafni, U., et al. (2019). Efficacy of adoptive therapy with tumor-infiltrating lymphocytes and recombinant interleukin-2 in advanced cutaneous melanoma: A systematic review and meta-analysis. *Annals of Oncology, 30*(12), 1902–1913. https://doi.org/10.1093/annonc/mdz398.

Dang, T. O., et al. (2016). Pembrolizumab for the treatment of PD-L1 positive advanced or metastatic non-small cell lung cancer. *Expert review of Anticancer Therapy, 16*(1), 13–20. https://doi.org/10.1586/14737140.2016.1123626.

Davar, D., Tarhini, A. A., & Kirkwood, J. M. (2012). Adjuvant therapy for melanoma. *Cancer Journal, 18*(2), 192–202. https://doi.org/10.1097/PPO.0b013e31824f118b.

De Pinto, G., et al. (2024). Global trends in cutaneous malignant melanoma incidence and mortality. *Melanoma Research, 34*(3), 265–275. https://doi.org/10.1097/CMR.0000000000000959.

Ditani, A. S., et al. (2021). Biosimilars accessible in the market for the treatment of cancer. *Journal of Controlled Release, 336*, 112–129. https://doi.org/10.1016/j.jconrel.2021.06.014.

Dixon, A. J., et al. (2024). Primary cutaneous melanoma-management in 2024. *Journal of Clinical Medicine, 13*(6), e1609. https://doi.org/10.3390/jcm13061607.

Dummer, R., & Mangana, J. (2009). Long-term pegylated interferon-alpha and its potential in the treatment of melanoma. *Biologics : Targets & Therapy, 3*, 169–182. https://doi.org/10.2147/btt.2009.3051.

Dutcher, J. P., et al. (2014). High dose interleukin-2 (Aldesleukin) - expert consensus on best management practices-2014. *Journal for ImmunoTherapy of Cancer, 2*(1), 26. https://doi.org/10.1186/s40425-014-0026-0.

Fang, K. K.-L., Lee, J. B., & Zhang, L. (2022). Adoptive cell therapy for T-Cell malignancies. *Cancers, 15*(1), e94. https://doi.org/10.3390/cancers15010094.

Feldman, S. A., et al. (2015). Adoptive cell therapy–Tumor-infiltrating lymphocytes, T-cell receptors, and chimeric antigen receptors. *Seminars in Oncology, 42*(4), 626–639. https://doi.org/10.1053/j.seminoncol.2015.05.005.

Ferrucci, P. F., et al. (2021). Talimogene laherparepvec (T-VEC): An intralesional cancer immunotherapy for advanced melanoma. *Cancers, 13*(6), 1383. https://doi.org/10.3390/cancers13061383.

Floros, T., & Tarhini, A. A. (2015). Anticancer cytokines: Biology and clinical effects of interferon-α2, interleukin (IL)-2, IL-15, IL-21, and IL-12. *Seminars in Oncology, 42*(4), 539–548. https://doi.org/10.1053/j.seminoncol.2015.05.015.

Formycon, A. G. (2024). *Formycon announces start of clinical phase III trial for Keytruda® biosimilar candidate FYB206*. Formycon AG Press Release https://www.formycon.com/en/blog/press-release/formycon-announces-start-of-clinical-phase-iii-trial-for-keytruda-biosimilar-candidate-fyb206/.

Gao, Y., et al. (2024). mRNA vaccines in tumor targeted therapy: Mechanism, clinical application, and development trends. *Biomarker Research, 12*(1), 93. https://doi.org/10.1186/s40364-024-00644-3.

Géraud, A., et al. (2024). Reactions and adverse events induced by T-cell engagers as anti-cancer immunotherapies, a comprehensive review. *European Journal of Cancer, 205*, e114075. https://doi.org/10.1016/j.ejca.2024.114075.

Gupta, A. K., Bharadwaj, M., & Mehrotra, R. (2016). Skin cancer concerns in people of color: Risk factors and prevention. *Asian Pacific Journal of Cancer Prevention: APJCP, 17*(12), 5257–5264. https://doi.org/10.22034/APJCP.2016.17.12.5257.

Han, Y., Liu, D., & Li, L. (2020). PD-1/PD-L1 pathway: Current researches in cancer. *American Journal of Cancer Research, 10*(3), 727–742.

He, M., et al. (2017). Remarkably similar CTLA-4 binding properties of therapeutic ipilimumab and tremelimumab antibodies. *Oncotarget, 8*(40), 67129–67139. https://doi.org/10.18632/oncotarget.18004.

Hirayama, A. V., et al. (2019). The response to lymphodepletion impacts PFS in patients with aggressive non-Hodgkin lymphoma treated with CD19 CAR T cells. *Blood, 133*(17), 1876–1887. https://doi.org/10.1182/blood-2018-11-887067.

Hong, Y., et al. (2024). Safety of talimogene laherparepvec: A real-world retrospective pharmacovigilance study based on FDA Adverse Event Reporting System (FAERS). *Journal of Pharmaceutical Health Care and Sciences, 10*(1), 79. https://doi.org/10.1186/s40780-024-00388-0.

Huddleston, A. N. (2015). Pembrolizumab for recurrent metastatic melanoma: Review of current use and future implications. *Journal of Hematology Oncology Pharmacy, 5*(4), 116–122.

Hulen, T. M., et al. (2021). ACT Up TIL now: The evolution of tumor-infiltrating lymphocytes in adoptive cell therapy for the treatment of solid tumors. *Immuno, 1*(3), 194–211. https://doi.org/10.3390/immuno1030012.

Husebye, E. S., et al. (2022). Endocrine-related adverse conditions in patients receiving immune checkpoint inhibition: An ESE clinical practice guideline. *European Journal of Endocrinology, 187*(6), G1–G21. https://doi.org/10.1530/EJE-22-0689.

Jiang, M., et al. (2022). Dosing regimens of immune checkpoint inhibitors: Attempts at lower dose, less frequency, shorter course. *Frontiers in Oncology, 12*, e906251. https://doi.org/10.3389/fonc.2022.906251.

Jin, K.-T., et al. (2021). Crosstalk between oncolytic viruses and autophagy in cancer therapy. *Biomedicine & Pharmacotherapy, 134*, e110932. https://doi.org/10.1016/j.biopha.2020.110932.

Johnson, D. B., Peng, C., & Sosman, J. A. (2015). Nivolumab in melanoma: Latest evidence and clinical potential. *Therapeutic Advances in Medical Oncology, 7*(2), 97–106. https://doi.org/10.1177/1758834014567469.

Johnson, D. B., Puzanov, I., & Kelley, M. C. (2015). Talimogene laherparepvec (T-VEC) for the treatment of advanced melanoma. *Immunotherapy, 7*(6), 611–619. https://doi.org/10.2217/imt.15.35.

Joshi, D., et al. (2022). Biosimilars in oncology: Latest trends and regulatory status. *Pharmaceutics, 14*(12), e2721. https://doi.org/10.3390/pharmaceutics14122721.

Jozala, A. F., et al. (2016). Biopharmaceuticals from microorganisms: From production to purification. *Brazilian Journal of Microbiology, 47*(1), 51–63. https://doi.org/10.1016/j.bjm.2016.10.007.

Kaczmarek, M., et al. (2023). Cancer vaccine therapeutics: Limitations and effectiveness: A literature review. *Cells, 12*(17), e2159. https://doi.org/10.3390/cells12172159.

Kaufman, H. L., Shalhout, S. Z., & Iodice, G. (2022). Talimogene laherparepvec: Moving from first-in-class to best-in-class. *Frontiers in Molecular Biosciences, 9*, e834841. https://doi.org/10.3389/fmolb.2022.834841.

Khraishi, M., et al. (2016). Biosimilars: A multidisciplinary perspective. *Clinical Therapeutics, 38*(5), 1238–1249. https://doi.org/10.1016/j.clinthera.2016.02.023.

Knight, A., Karapetyan, L., & Kirkwood, J. M. (2023). Immunotherapy in melanoma: Recent advances and future directions. *Cancers, 15*(4), e1106. https://doi.org/10.3390/cancers15041106.

Ko, B., et al. (2023). Rethinking oncologic treatment strategies with interleukin-2. *Cells, 12*(9), e1316. https://doi.org/10.3390/cells12091316.

König, D., et al. (2024). Adoptive cell therapy with tumor-infiltrating lymphocytes in combination with nivolumab in patients with advanced melanoma. *Immuno-Oncology and Technology, 24*, e100728. https://doi.org/10.1016/j.iotech.2024.100728.

Konstantinidou, S., Papaspiliou, A., & Kokkotou, E. (2020). Current and future roles of biosimilars in oncology practice. *Oncology Letters, 19*(1), 45–51. https://doi.org/10.3892/ol.2019.11105.

Kooshkaki, O., et al. (2020). Combination of ipilimumab and nivolumab in cancers: From clinical practice to ongoing clinical trials. *International Journal of Molecular Sciences, 21*(12), e4427. https://doi.org/10.3390/ijms21124427.

Kumar, M., et al. (2020). Monoclonal antibody-based immunotherapy and its role in the development of cardiac toxicity. *Cancers, 13*(1), e86. https://doi.org/10.3390/cancers13010086.

Lee, H., et al. (2021). Targeting NK cells to enhance melanoma response to immunotherapies. *Cancers, 13*(6), e1363. https://doi.org/10.3390/cancers13061363.

Lemos de Matos, A., Franco, L. S., & McFadden, G. (2020). Oncolytic viruses and the immune system: The dynamic duo. *Molecular Therapy. Methods & Clinical Development, 17*, 349–358. https://doi.org/10.1016/j.omtm.2020.01.001.

Liao, W., Lin, J.-X., & Leonard, W. J. (2013). Interleukin-2 at the crossroads of effector responses, tolerance, and immunotherapy. *Immunity, 38*(1), 13–25. https://doi.org/10.1016/j.immuni.2013.01.004.

Lickefett, B., et al. (2023). Lymphodepletion - an essential but undervalued part of the chimeric antigen receptor T-cell therapy cycle. *Frontiers in Immunology, 14*, e1303935. https://doi.org/10.3389/fimmu.2023.1303935.

Lim, S., et al. (2017). Interplay between immune checkpoint proteins and cellular metabolism. *Cancer Research, 77*(6), 1245–1249. https://doi.org/10.1158/0008-5472.CAN-16-1647.

Lin, S. Y., et al. (2018). Tumor PD-L1 expression and clinical outcomes in advanced-stage non-small cell lung cancer patients treated with nivolumab or pembrolizumab: Real-world data in Taiwan. *Journal of Cancer, 9*(10), 1813–1820. https://doi.org/10.7150/jca.24985.

Lin, X., et al. (2024). Regulatory mechanisms of PD-1/PD-L1 in cancers. *Molecular Cancer, 23*(1), e108. https://doi.org/10.1186/s12943-024-02023-w.

Lipson, E. J., et al. (2025). Relatlimab (RELA) plus nivolumab (NIVO) versus NIVO in first-line advanced melanoma: Primary phase III results from RELATIVITY-047 (CA224-047). *Journal of Clinical Oncology, 39*(15S), e9503. https://doi.org/10.1200/JCO.2021.39.15_suppl.9503.

Lipson, E. J., & Drake, C. G. (2011). Ipilimumab: An anti-CTLA-4 antibody for metastatic melanoma. *Clinical Cancer Research : An Official Journal of the American Association for Cancer Research,, 17*(22), 6958–6962. https://doi.org/10.1158/1078-0432.CCR-11-1595.

Liu, B., et al. (2024). Exploring treatment options in cancer: Tumor treatment strategies. *Signal Transduction and Targeted Therapy, 9*(175), 1–44. https://doi.org/10.1038/s41392-024-01856-7.

Liu, D., et al. (2023). Tumor vaccines: Unleashing the power of the immune system to fight cancer. *Pharmaceuticals (Basel, Switzerland), 16*(10), e1384. https://doi.org/10.3390/ph16101384.

Liu, X., & Yi, Y. (2020). Recent updates on Sintilimab in solid tumor immunotherapy. *Biomarker Research, 8*(1), 69. https://doi.org/10.1186/s40364-020-00250-z.

Liu, Y., & Sheikh, M. S. (2014). Melanoma: Molecular pathogenesis and therapeutic management. *Molecular and cellular pharmacology, 6*(3), 228.

Long, G. V., et al. (2023). Overall survival and response with nivolumab and relatlimab in advanced melanoma. *NEJM Evidence, 2*(4), 1–10. https://doi.org/10.1056/evidoa2200239.

Malakar, S., Gontor, E. N., Dugbaye, M. Y., et al. (2024). Cancer treatment with biosimilar drugs: A review. *Cancer Innovation, 3*(2), e115. https://doi.org/10.1002/cai2.115.

Marin-Acevedo, J. A., et al. (2018). Next generation of immune checkpoint therapy in cancer: New developments and challenges. *Journal of Hematology & Oncology, 11*(1), 39. https://doi.org/10.1186/s13045-018-0582-8.

Martins, F., et al. (2019). Adverse effects of immune-checkpoint inhibitors: Epidemiology, management and surveillance. *Nature Reviews Clinical Oncology, 16*(9), 563–580. https://doi.org/10.1038/s41571-019-0218-0.

Mascarenhas-Melo, F., et al. (2024). An overview of biosimilars-development, quality, regulatory issues, and management in healthcare. *Pharmaceuticals (Basel, Switzerland), 17*(2), e235. https://doi.org/10.3390/ph17020235.

Massi, D., et al. (2014). PD-L1 marks a subset of melanomas with a shorter overall survival and distinct genetic and morphological characteristics. *Annals of Oncology, 25*(12), 2433–2442. https://doi.org/10.1093/annonc/mdu452.

Mehta, A., et al. (2025). Advancements in melanoma treatment: A review of PD-1 Inhibitors, T-VEC, mRNA vaccines, and tumor-infiltrating lymphocyte therapy in an evolving landscape of immunotherapy. *Journal of Clinical Medicine, 14*(4), e1200. https://doi.org/10.3390/jcm14041200.

Miao, L., Zhang, Y., & Huang, L. (2021). mRNA vaccine for cancer immunotherapy. *Molecular Cancer, 20*(1), 41. https://doi.org/10.1186/s12943-021-01335-5.

Mooradian, M. J., & Sullivan, R. J. (2023). Immunotherapy in melanoma: Recent advancements and future directions. *Cancers, 15*(16), e4176. https://doi.org/10.3390/cancers15164176.

Nicholas, C., & Lesinski, G. B. (2011). Immunomodulatory cytokines as therapeutic agents for melanoma. *Immunotherapy, 3*(5), 673–690. https://doi.org/10.2217/imt.11.45.

O'Donoghue, C., Doepker, M. P., & Zager, J. S. (2016). Talimogene laherparepvec: Overview, combination therapy and current practices. *Melanoma Management, 3*(4), 267–272. https://doi.org/10.2217/mmt-2016-0021.

Olivera, I., et al. (2024). Regional and intratumoral adoptive T-cell therapy. *Immuno-Oncology and Technology, 24*, e100715. https://doi.org/10.1016/j.iotech.2024.100715.

Owen, K., et al. (2023). Lymphodepleting chemotherapy practices and effect on safety and efficacy outcomes in patients with solid tumours undergoing T cell receptor-engineered T cell (TCR-T) Therapy: A systematic review and meta-analysis. *Cancer Immunology, Immunotherapy, 72*(4), 805–814. https://doi.org/10.1007/s00262-022-03287-1.

Paluncic, J., et al. (2016). Roads to melanoma: Key pathways and emerging players in melanoma progression and oncogenic signaling. *Biochimica et Biophysica Acta (BBA) - Molecular Cell Research, 1863*(4), 770–784. https://doi.org/10.1016/j.bbamcr.2016.01.025.

Papież, M. A., & Krzyściak, W. (2021). Biological therapies in the treatment of cancer—Update and new directions. *International Journal of Molecular Sciences, 22*(21), 11694. https://doi.org/10.3390/ijms222111694.

Pardoll, D. M. (2012). The blockade of immune checkpoints in cancer immunotherapy. *Nature Reviews. Cancer, 12*(4), 252–264. https://doi.org/10.1038/nrc3239.

Park, S. L., et al. (2012). Risk factors for malignant melanoma in white and non-white/non-African American populations: The multiethnic cohort. *Cancer Prevention Research (Philadelphia, Pa.), 5*(3), 423–434. https://doi.org/10.1158/1940-6207.CAPR-11-0460.

Parums, D. V. (2024). First regulatory approval for adoptive cell therapy with autologous tumor-infiltrating lymphocytes (TILs) - Lifileucel (Amtagvi). In *Medical Science Monitor: International Medical Journal of Experimental and Clinical Research* https://doi.org/10.12659/MSM.944927.

Paul, S., & Lal, G. (2017). The molecular mechanism of natural killer cells function and its importance in cancer immunotherapy. *Frontiers in Immunology, 8*(09), e1124. https://doi.org/10.3389/fimmu.2017.01124.

Perez, M. C., et al. (2019). Observational study of talimogene laherparepvec use for melanoma in clinical practice in the United States (COSMUS-1). *Melanoma Management, 6*(2), eT19. https://doi.org/10.2217/mmt-2019-0012.

Pham, J. V., et al. (2019). A review of the microbial production of bioactive natural products and biologics. *Frontiers in Microbiology, 10*, e1404. https://doi.org/10.3389/fmicb.2019.01404.

Pol, J., Kroemer, G., & Galluzzi, L. (2016). First oncolytic virus approved for melanoma immunotherapy. *Oncoimmunology, 5*(1), e1115641. https://doi.org/10.1080/2162402X.2015.1115641.

Puccetti, M., et al. (2024). Biologics, theranostics, and personalized medicine in drug delivery systems. *Pharmacological Research, 201*, e107086. https://doi.org/10.1016/j.phrs.2024.107086.

Rajan, A., et al. (2016). Nivolumab, anti-programmed death-1 (PD-1) monoclonal antibody immunotherapy: Role in advanced cancers. *Human Vaccines and Immunotherapeutics, 12*(9), 2219–2231. https://doi.org/10.1080/21645515.2016.1175694.

Ramadan, E., Ahmed, A., & Naguib, Y. W. (2024). Advances in mRNA LNP-based cancer vaccines: Mechanisms, formulation aspects, challenges, and future directions. *Journal of Personalized Medicine, 14*(11), e1092. https://doi.org/10.3390/jpm14111092.

Rehman, H., et al. (2016). Into the clinic: Talimogene laherparepvec (T-VEC), a first-in-class intratumoral oncolytic viral therapy. *Journal for Immunotherapy of Cancer, 4*, 53. https://doi.org/10.1186/s40425-016-0158-5.

Riveiro-Barciela, M., et al. (2024). Management of liver and gastrointestinal toxicity induced by immune checkpoint inhibitors: Position statement of the AEEH–AEG–SEPD–SEOM–GETECCU. *Gastroenterología y Hepatología, 47*(4), 401–432. https://doi.org/10.1016/j.gastre.2023.10.003.

Rohaan, M. W., et al. (2018). Adoptive transfer of tumor-infiltrating lymphocytes in melanoma: A viable treatment option. *Journal for ImmunoTherapy of Cancer, 6*(1), 102. https://doi.org/10.1186/s40425-018-0391-1.

Roky, A. H., et al. (2024). Overview of skin cancer types and prevalence rates across continents. *Cancer Pathogenesis and Therapy, 3*(2), 89–100. https://doi.org/10.1016/j.cpt.2024.08.002.

Ross, S. H., & Cantrell, D. A. (2018). Signaling and function of interleukin-2 in T lymphocytes. *Annual Review of Immunology, 36*, 411–433. https://doi.org/10.1146/annurev-immunol-042617-053352.

Rosskopf, S., et al. (2019). CTLA-4 antibody ipilimumab negatively affects CD4(+) T-cell responses in vitro. *Cancer Immunology, Immunotherapy, 68*(8), 1359–1368. https://doi.org/10.1007/s00262-019-02369-x.

Rugo, H. S., et al. (2016). A clinician's guide to biosimilars in oncology. *Cancer Treatment Reviews, 46*, 73–79. https://doi.org/10.1016/j.ctrv.2016.04.003.

Sabbatino, F., et al. (2022). Immune checkpoint inhibitors for the treatment of melanoma. *Expert Opinion on Biological Therapy, 22*(5), 563–576. https://doi.org/10.1080/14712598.2022.2038132.

Safaei, S., et al. (2025). The role of cytokines in shaping the future of cancer immunotherapy. *Cytokine, 189*, e156888. https://doi.org/10.1016/j.cyto.2025.156888.

Saginala, K., et al. (2021). Epidemiology of Melanoma. *Medical Sciences, 9*(63), 1–9. https://doi.org/10.3390/medsci9040063.

Sample, A., & He, Y.-Y. (2018). Mechanisms and prevention of UV-induced melanoma. *Photodermatology, Photoimmunology & Photomedicine, 34*(1), 13–24. https://doi.org/10.1111/phpp.12329.

Satam, H., et al. (2023). Next-generation sequencing technology : Current trends and advancements. *Biology, 12*, e997.

Sato, T. (2014). Tumor necrosis factor-α blockade and development of uveal melanoma: Expected adverse effect or just coincidence? *Mayo Clinic Proceedings, 89*(11), 1467–1470. https://doi.org/10.1016/j.mayocp.2014.09.005.

Savoia, P., Astrua, C., & Fava, P. (2016). Ipilimumab (Anti-Ctla-4 Mab) in the treatment of metastatic melanoma: Effectiveness and toxicity management. *Human Vaccines & Immunotherapeutics, 12*(5), 1092–1101. https://doi.org/10.1080/21645515.2015.1129478.

Senti, M. E., García del Valle, L., & Schiffelers, R. M. (2024). mRNA delivery systems for cancer immunotherapy: Lipid nanoparticles and beyond. *Advanced Drug Delivery Reviews, 206*, e115190. https://doi.org/10.1016/j.addr.2024.115190.

Shalata, W., et al. (2023). Long-lasting therapeutic response following treatment with pembrolizumab in patients with non-small cell lung cancer: A real-world experience. *International Journal of Molecular Sciences, 24*(6), e5938. https://doi.org/10.3390/ijms24065938.

Shanghai Henlius Biotech, Inc. (2023). *Henlius immune checkpoints inhibitor anti-CTLA-4 mAb received IND approval for the treatment of HCC. China: Shanghai.*

Shanghai Henlius Biotech, Inc. (2024). *NMPA Approves IND for Henlius' pembrolizumab biosimilar.* Synapse by Patsnap https://synapse.patsnap.com/article/nmpa-approves-ind-for-henlius-pembrolizumab-biosimilar.

Sharma, P., et al. (2023). Immune checkpoint therapy—Current perspectives and future directions. *Cell, 186*(8), 1652–1669. https://doi.org/10.1016/j.cell.2023.03.006.

Shepard, H. M., et al. (2017). Developments in therapy with monoclonal antibodies and related proteins. *Clinical Medicine, 17*(3), 220–232. https://doi.org/10.7861/clinmedicine.17-3-220.

Sheridan, M., Massich, M., & Ashourian, N. (2024). Biosimilars: From Production to Patient. *Journal of Infusion Nursing : The Official Publication of the Infusion Nurses Society, 47*(1), 19–29. https://doi.org/10.1097/NAN.0000000000000528.

Siegel, R. L., et al. (2023). Cancer statistics, 2023. *CA: A Cancer Journal for Clinicians, 73*(1), 17–48. https://doi.org/10.3322/caac.21763.

Spain, L., et al. (2015). Pembrolizumab in the management of metastatic melanoma. *Melanoma Management, 2*(4), 315–325. https://doi.org/10.2217/mmt.15.33.

Su, J., et al. (2023). Relatlimab: A novel drug targeting immune checkpoint LAG-3 in melanoma therapy. *Frontiers in Pharmacology, 14*, e1349081. https://doi.org/10.3389/fphar.2023.1349081.

Süntar, I., et al. (2021). Bioproduction process of natural products and biopharmaceuticals: Biotechnological aspects. *Biotechnology Advances, 50*, e107768. https://doi.org/10.1016/j.biotechadv.2021.107768.

Switzer, B., et al. (2022). Managing metastatic melanoma in 2022: A clinical review. *JCO Oncology Practice, 18*(5), 335–351. https://doi.org/10.1200/op.21.00686.

Tarhini, A. A., et al. (2020). Phase III study of adjuvant ipilimumab (3 or 10 mg/kg) versus high-dose interferon alfa-2b for resected high-risk melanoma: North American Intergroup E1609. *Journal of Clinical Oncology : Official Journal of the American Society of Clinical Oncology, 38*(6), 567–575. https://doi.org/10.1200/JCO.19.01381.

Tarhini, A., et al. (2018). Neoadjuvant ipilimumab (3 mg/kg or 10 mg/kg) and high dose IFN-α2b in locally/regionally advanced melanoma: Safety, efficacy and impact on T-cell repertoire. *Journal for ImmunoTherapy of Cancer, 6*(1), 112. https://doi.org/10.1186/s40425-018-0428-5.

Tawbi, H. A., et al. (2022). Relatlimab and nivolumab versus nivolumab in untreated advanced melanoma. *New England Journal of Medicine, 386*(1), 24–34. https://doi.org/10.1056/nejmoa2109970.

Timis, T., et al. (2025). Advances in cell and immune therapies for melanoma. *Biomedicines, 13*(1), e98. https://doi.org/10.3390/biomedicines13010098.

Tkaczuk, K. H. R., & Jacobs, I. A. (2014). Biosimilars in oncology: From development to clinical practice. *Seminars in Oncology, 41*, S3–S12. https://doi.org/10.1053/j.seminoncol.2014.03.008.

Topalian, S. L., et al. (2014). Survival, durable tumor remission, and long-term safety in patients with advanced melanoma receiving nivolumab. *Journal of Clinical Oncology, 32*(10), 1020–1030. https://doi.org/10.1200/JCO.2013.53.0105.

Upadhyay, P. R., Ho, T., & Abdel-Malek, Z. A. (2021). Participation of keratinocyte- and fibroblast-derived factors in melanocyte homeostasis, the response to UV, and pigmentary disorders. *Pigment Cell & Melanoma Research, 34*(4), 762–776. https://doi.org/10.1111/pcmr.12985.

Vázquez-Montero, L., de la Gala, M. D. C. Á., & de la Cruz-Merino, L. (2023). Nivolumab plus ipilimumab in metastatic melanoma: A critical appraisal focused on specific subpopulations. *Frontiers in Oncology, 13*, e1187840. https://doi.org/10.3389/fonc.2023.1187840.

Vidal, P. (2020). Interferon α in cancer immunoediting: From elimination to escape. *Scandinavian Journal of Immunology, 91*(5), 1–8. https://doi.org/10.1111/sji.12863.

Vinay, D. S., et al. (2015). Immune evasion in cancer: Mechanistic basis and therapeutic strategies. *Seminars in Cancer Biology, 35*, S185–S198. https://doi.org/10.1016/j.semcancer.2015.03.004.

Walker, L. S. K. (2013). Treg and CTLA-4: Two intertwining pathways to immune tolerance. *Journal of Autoimmunity, 45*(100), 49–57. oi.org/10.1016/j.jaut.2013.06.006.

Walker, L. S. K., & Sansom, D. M. (2015). Confusing signals: Recent progress in CTLA-4 biology. *Trends in Immunology, 36*(2), 63–70. https://doi.org/10.1016/j.it.2014.12.001.

Wang, J., & Chow, S.-C. (2012). On the regulatory approval pathway of biosimilar products. *Pharmaceuticals, 5*(4), 353–368. https://doi.org/10.3390/ph5040353.

Wang, L., et al. (2022). Remodeling the tumor microenvironment by oncolytic viruses: Beyond oncolysis of tumor cells for cancer treatment. *Journal for Immunotherapy of Cancer, 10*(5), e004167. https://doi.org/10.1136/jitc-2021-004167.

Warner, A. B., Corrie, P. G., & Hamid, O. (2023). Tumor-infiltrating lymphocyte therapy in melanoma: Facts to the future. *Clinical Cancer Research : An Official Journal of the American Association for Cancer Research, 29*(10), 1835–1854. https://doi.org/10.1158/1078-0432.CCR-22-1922.

Wei, J., et al. (2024). Current trends in sensitizing immune checkpoint inhibitors for cancer treatment. *Molecular Cancer, 23*(1), e279. https://doi.org/10.1186/s12943-024-02179-5.

Wojtukiewicz, M. Z., et al. (2021). Inhibitors of immune checkpoints-PD-1, PD-L1, CTLA-4-new opportunities for cancer patients and a new challenge for internists and general practitioners. *Cancer Metastasis Reviews, 40*(3), 949–982. https://doi.org/10.1007/s10555-021-09976-0.

Wolchok, J. D., et al. (2013). Nivolumab plus ipilimumab in advanced melanoma. *New England Journal of Medicine, 369*(2), 122–133. https://doi.org/10.1056/nejmoa1302369.

Xiong, F., et al. (2022). Direct and indirect effects of IFN-α2b in malignancy treatment: Not only an archer but also an arrow. *Biomarker Research, 10*(1), 69. https://doi.org/10.1186/s40364-022-00415-y.

Xu, Z., & Fisher, D. E. (2023). mRNA melanoma vaccine revolution spurred by the COVID-19 pandemic. *Frontiers in Immunology, 14*, e1155728. https://doi.org/10.3389/fimmu.2023.1155728.

Yamaguchi, T., & Arato, T. (2011). Quality, safety and efficacy of follow-on biologics in Japan. *Biologicals, 39*(5), 328–332. https://doi.org/10.1016/j.biologicals.2011.06.015.

Yamazaki, N., et al. (2016). Phase I study of pegylated interferon-alpha-2b as an adjuvant therapy in Japanese patients with malignant melanoma. *The Journal of Dermatology, 43*(10), 1146–1153. https://doi.org/10.1111/1346-8138.13338.

Yan, T., et al. (2024a). Achilles' Heel of currently approved immune checkpoint inhibitors: Immune related adverse events. *Frontiers in Immunology, 15*(01), 1–21. https://doi.org/10.3389/fimmu.2024.1292122.

Yan, Z., et al. (2024b). Enhancing cancer therapy: The integration of oncolytic virus therapy with diverse treatments. *Cancer Cell International, 24*(1), 242. https://doi.org/10.1186/s12935-024-03424-z.

Yaremenko, A. V., et al. (2025). Clinical advances of mRNA vaccines for cancer immunotherapy. *Med, 6*(1), e100562. https://doi.org/10.1016/j.medj.2024.11.015.

Zaidi, M. R., Fisher, D. E., & Rizos, H. (2019). Biology of melanocytes and primary melanoma. In C. Balch, et al. (Eds.), *Cutaneous melanoma* (pp. 1–38). Springer International Publishing. https://doi.org/10.1007/978-3-319-46029-1_42-1.

Zeng, C., et al. (2022). Formulation and delivery technologies for mRNA vaccines. *Current Topics in Microbiology and Immunology, 440*, 71–110. https://doi.org/10.1007/82_2020_217.

Zhang, P., et al. (2023). The therapeutic potential of PD-1/PD-L1 pathway on immune-related diseases: Based on the innate and adaptive immune components. *Biomedicine & Pharmacotherapy, 167*, e115569. https://doi.org/10.1016/j.biopha.2023.115569.

Zhao, Y., et al. (2022). Tumor infiltrating lymphocyte (TIL) therapy for solid tumor treatment: Progressions and challenges. *Cancers, 14*(17), e4160. https://doi.org/10.3390/cancers14174160.

CHAPTER 15

Biologics and biosimilars from the bench to the bedside in oncology

Krishna Sanka[a], Sohini Chakraborty[b], Deepthi Cherukuri[c], Kishore Rapolu[d], Ramesh Jadhav[e] and Sravan Kumar Aligeti[f]

[a]*Department of Pharmaceutics, School of Pharmacy and Technology Management, SVKM's Narsee Monjee Institute of Management Studies (NMIMS) (Deemed to be University), Shirpur, Maharashtra, India,* [b]*Department of Biotechnology, School of Applied Sciences, REVA University, Bengaluru, Karnataka, India,* [c]*SRM College of Pharmacy, SRM Institute of Science and Technology, Chengalpattu, Tamil Nadu, India,* [d]*Department of Pharmaceutics, University College of Technology, Osmania University, Hyderabad, Telangana, India,* [e]*Department of Pharmaceutics, Avanthi Institute of Pharmaceutical Sciences, Hyderabad, Telangana, India,* [f]*Department of Pharmaceutics, Vaageswari College of Pharmacy, Karimnagar, Telangana, India*

15.1 Introduction

Biologics have revolutionized the treatment of cancer as the targeted personalized therapies allow better outcomes in patients and reduce systemic toxicity. These living-cell-derived complex drugs include monoclonal antibodies, cytokines, immune checkpoint inhibitors, and therapeutic cancer vaccines, which target cancer-specific pathways rather than normal rapidly dividing cells, unlike traditional chemotherapy. This is a significant improvement of the conventional cytotoxic treatment, as this precise method causes fewer side effects and increases efficacy (Joshi et al., 2022; Peeters et al., 2021; Rugo et al., 2016). Nevertheless, the costs and complexity of manufacturing biologics make them less accessible, particularly in low- and middle-income nations (LMICs), posing a significant economic challenge for global health care systems and patients (Joshi et al., 2022; Liu et al., 2025).

Biosimilars have become affordable alternatives to reference biologics, are similar in safety, efficacy, and quality, and help lower treatment costs and increase patient access (Hung et al., 2017; Markus et al., 2017; Nupur et al., 2022; Wolff-Holz et al., 2019). The regulatory bodies, including the FDA and EMA, impose stringent approval procedures with rigorous analytical, preclinical, and clinical reviews to ensure biosimilars meet strict standards and are clinically similar to originator products (Bloomfield et al., 2022; Joshi et al., 2022; Rugo et al., 2016). Oncology biosimilars are similar to their respective reference biologics in terms of efficacy and safety, and clinical trials and real-world evidence (RWE) demonstrate that oncology biosimilars

Biosimilars and Biologics in Cancer Treatment: An Era in Therapeutic Strategies.
DOI: https://doi.org/10.1016/B978-0-443-43824-0.00017-1

should be allowed to integrate into cancer care (Bloomfield et al., 2022; Lo & Tzeng, 2024; Nikitina et al., 2023).

Biosimilars are new to the market, increasing competition, lowering prices, and reducing out-of-pocket payments for certain patients. However, the monetary gains may differ depending on the healthcare environment. This transition to biosimilars is essential to enhancing sustainable and equitable cancer care worldwide, reallocating resources to more efficient healthcare systems, and investing in innovative treatments (Liu et al., 2025; Nahleh et al., 2022; Robinson & Whaley, 2025). The introduction of biosimilars is one meaningful way forward to strike a balance between costs and the ever-evolving innovation in the biologic treatment of oncology (Joshi et al., 2022; Peeters et al., 2021).

For biologics, the approval process is traditional and requires large-scale preclinical research, randomized clinical trials, and long-term safety trials. Biosimilars, on the other hand, are characterized by a lower regulatory pathway that emphasizes analytical and functional similarity with the reference biologic. The step-by-step screening procedure should demonstrate the most crucial parameters, including structural similarity, pharmacokinetics (PK) and pharmacodynamics (PD), immunogenicity, and clinical efficacy. Biosimilars are not exact copies of generic drugs; instead, they must be as close as possible to the reference product and cannot have any clinically significant differences (Hung et al., 2017; Markus et al., 2017; Wolff-Holz et al., 2019). postmarketing surveillance (PMS) and pharmacovigilance (PV) are key components of the biosimilar regulatory oversight process that enable long-term systematic surveillance of safety and immunogenicity. The establishment of requirements for interchangeability, standard nomenclature, and extrapolation of indications also shapes biosimilar policies. The regulatory landscape is changing, and international harmonization efforts focus on streamlining the approval process and increasing the availability of cost-effective biologic therapeutics, thereby expanding access to state-of-the-art oncologic therapies (Markus et al., 2017; Nupur et al., 2022; Wolff-Holz et al., 2019). Table 15.1 presents key differences in the approvals of biologics and biosimilars.

15.2 Biologics in oncology

15.2.1 Definition and classification of biologics

Biologics represent a heterogeneous group of therapeutic agents derived from living cells and have significantly changed the treatment of oncology by enabling targeted, individualized therapeutic interventions. In contrast to traditional small-molecule-based pharmacotherapies, which are chemically synthesized and have well-defined chemical structures, biologics are complex macromolecules (including proteins, monoclonal antibodies, nucleic acid constructs, and viable, maintained cellular entities) that interact with specific biological pathways and affect disease progression. In the oncologic context, biologics may be systemically grouped by their mechanisms of

Table 15.1 Key differences for the approval of biologics and biosimilars.

Parameter	Biologics	Biosimilars	Citations
Preclinical Studies	Extensive (in vitro, animal)	Focused, may be reduced if similarity is shown	Markus et al. (2017); Nupur et al. (2022)
Clinical Trials	Multiple, large-scale, all indications	Fewer, often in one sensitive population	Markus et al. (2017); Nupur et al. (2022)
Analytical Comparability	Not required vs. reference	Extensive foundation of approval	Markus et al. (2017); Wolff-Holz et al. (2019)
PK/PD Studies	Required for all indications	Required, often pivotal for approval	Nupur et al. (2022); Hung et al. (2017)
Immunogenicity	Full assessment	Comparative, focused on differences	Markus et al. (2017); Nupur et al. (2022)
Efficacy/Safety Trials	Full demonstration	Equivalence, not superiority, may be waived	Markus et al. (2017); Hung et al. (2017)
Long-term Safety	Required	Postmarketing surveillance	Markus et al. (2017)

action and molecular architectures (Goydel & Rader, 2021; Jin et al., 2022; Liu et al., 2024; Papież & Krzyściak, 2021; Riccardi et al., 2023):

15.2.2 Monoclonal antibodies

Engineered proteins that target the surface markers of cancer cells, resulting in direct killing of the cells (cytotoxicity), stimulation of the immune system, or prevention of the growth signals from cancer cells (e.g., trastuzumab for HER2-positive breast cancer).

15.2.2.1 Immune checkpoint inhibitors

Block inhibitory pathways that suppress immune responses, enabling T cells to recognize and attack cancer cells (e.g., pembrolizumab, nivolumab targeting PD-1/PD-L1).

15.2.2.2 Cytokines and growth factors

Proteins involved in the modulation of immune reactions, angiogenesis, and hematopoiesis are critical to aid oncological therapeutic strategies (e.g., interleukin-2, IFNs, granulocyte-colony stimulating factor [G-CSF]).

15.2.2.3 Cancer vaccines

elicit an immune response to tumor-specific antigens to prevent or manage malignancies (e.g., sipuleucel-T for prostate cancer).

Table 15.2 Overview of biologic classes, mechanisms, and examples in oncology.

Biologic type	Mechanism of Action	Example(s)	Citations
Monoclonal Antibodies	Bind specific antigens on cancer cells; direct killing or activate the immune system.	Trastuzumab, Rituximab	Goydel & Rader, (2021); Jin et al. (2022); Papież & Krzyściak, (2021)
Antibody-Drug Conjugates	Deliver cytotoxic drugs directly to tumor cells	Trastuzumab emtansine (T-DM1)	Goydel & Rader, (2021); Jin et al., (2022); Liu et al. (2024)
Bispecific Antibodies	Engage immune cells (e.g., T cells) to attack cancer	Blinatumomab	Goydel & Rader, (2021); Jin et al. (2022)
Cell-Based Therapies	Genetically engineered immune cells target cancer	CAR-T cells	Goydel & Rader, (2021); Liu et al. (2024); Papież & Krzyściak, (2021)
Cytokines	Modulate immune response	IL-2, INFs	Papież & Krzyściak, (2021)
Cancer Vaccines	Stimulate the immune system against tumor antigens	Sipuleucel-T	Papież & Krzyściak, (2021)
Oncolytic Viruses	Infect and lyse cancer cells, stimulate immunity	T-VEC	Papież & Krzyściak, (2021)
Gene Therapies	Modify genetic material to correct or kill cancer cells	Various (in development)	Liu et al. (2024); Papież & Krzyściak, (2021)

15.2.2.4 Oncolytic viruses

Genetically engineered viruses that selectively infect and destroy malignant cells and, at the same time, augment antitumor immune responses (e.g., talimogene laherparepvec [T-VEC] for melanoma).

15.2.2.5 Cell-based therapies

Engineered immune cells, represented by chimeric antigen receptor T cell therapies, such as axicabtagene ciloleucel and tisagenlecleucel, should be used to attack tumor-associated antigens and induce strong antitumor responses selectively.

Ongoing advances in biologics have significantly improved outcomes in oncology, with greater specificity, reduced toxicity, and improved survival. Nonetheless, challenges such as the high production costs, immunogenicity risks, and complex regulatory pathways remain critical hurdles to their large-scale clinical implementation. Table 15.2 demonstrates the biologic classes, mechanisms, and examples in oncology.

15.2.3 Mechanisms of action in cancer treatment

15.2.3.1 Targeted therapy versus traditional chemotherapy

In traditional chemotherapy, the chemotoxic agent is used, and this attacks cells of rapid division that are both cancerous and healthy tissues indiscriminately. This nonselectivity leads to a high level of toxicity, such as bone marrow suppression, gastrointestinal malaise, and alopecia. The chemotherapeutic agents (alkylating agents, such as cyclophosphamide), antimetabolites (5-fluorouracil), and topoisomerase inhibitors (doxorubicin) induce DNA damage, disrupt cell division, and ultimately cause apoptosis. Nevertheless, the drugs usually cause acquired resistance, so other methods have to be applied (Bukowski, Kciuk, & Kontek, 2020).

Instead, targeted therapy aims to disrupt specific molecular pathways that cancer cells rely on to survive. The treatments take advantage of cancer cell-specificities, including receptors that have been overexpressed or mutated signaling proteins, to target cancer cells and increase their efficacy. Examples include TKIs, such as imatinib used in chronic myeloid leukemia, and monoclonal antibodies (mAbs), such as trastuzumab used in breast cancer that has the HER2 receptor. Compared with chemotherapy, targeted therapies are more effective and have less systemic toxicity; however, technical challenges, such as resistance and the development of predictive biomarkers, remain significant (Sarmento-Ribeiro et al., 2019).

15.2.3.2 Monoclonal antibodies and their role in cancer

Monoclonal antibodies (mAbs) are engineered immunoglobulins that can recognize and bind to specific antigens on cancer cells, thereby inhibiting tumor growth in various ways. These are direct tumor cell killing, immune system activation, and cytotoxic agent delivery. Depending on their mechanism of action, mAbs may be grouped into: (Broer et al., 2024; Dabkowska et al., 2024; Dumontet et al., 2023; Pierpont et al., 2018).

15.2.3.2.1 Growth factor receptor blockers

Inhibit oncogenic signaling by targeting receptors like EGFR (cetuximab) and HER2 (trastuzumab).

15.2.3.2.2 Immune-mediated cytotoxicity inducers

Engage immune cells to attack tumors, as seen with rituximab (CD20 targeting).

15.2.3.2.3 Antibody-drug conjugates

Deliver cytotoxic agents specifically to cancer cells (e.g., ado-trastuzumab emtansine). mAbs have revolutionized oncology by offering particular, powerful treatments with fewer off-target effects. Nevertheless, their extremely high production costs and the risk of immune-related side effects require the development of biosimilars to enhance accessibility.

15.2.3.3 Immune checkpoint inhibitors (PD-1/PD-L1, CTLA-4 (cytotoxic t-lymphocyte-associated protein-4)

The immune system is a potent tool against cancer, and cancer treatment with immune ICIs has transformed the field. In most cases, immune checkpoints like PD-1/PD-L1 and CTLA-4 play a role in immune homeostasis, inhibiting unwarranted T-cell activation. To avoid immune detection, however, tumors use these pathways (Shirav& et al., 2022).

15.2.3.3.1 PD-1/PD-L1 Inhibitors

Antitumor immunity and T-cell function are reinstated by the PD-1 receptor blocking drugs, such as pembrolizumab and nivolumab.

15.2.3.3.2 CTLA-4 inhibitors

Agents such as ipilimumab target CTLA-4 to stimulate T-cell activation and expansion to attack the cancer cells.

ICIs have demonstrated exceptional efficacy in melanoma, lung cancer, and other cancers; however, immune-related adverse effects, such as colitis, hepatitis, and endocrinopathies, are still of concern. Ongoing research to optimise combination therapies and biomarker-driven approaches to optimise patient response (Hayashi & Nakagawa, 2020; Shirav& et al., 2022).

15.2.3.4 Cytokine therapy and cancer immunotherapy

Cytokines are signaling proteins that modulate immune responses and have been exploited for cancer immunotherapy (Balkhi et al., 2025; Fagerberg, 1996; García-Ortega et al., 2023; Tanigawa & Redmond, 2025). Key cytokines used in oncology include:

15.2.3.4.1 Interleukin-2

Promotes T-cell proliferation and NK cell activity, historically used in melanoma and renal cell carcinoma. However, in high doses, they cause huge toxicity and thus cannot be used widely.

15.2.3.4.2 Interferons

Interferon-alpha (IFN-α) has been used as a treatment for leukemia and melanoma due to its ability to induce apoptosis, decrease angiogenesis, and increase immunity.

15.2.3.4.3 Granulocyte-macrophage colony-stimulating factor

Promotes antigen presentation and immune cell activation, often used as an adjuvant therapy. Cytokine therapy, despite being efficacious, causes undiscriminate stimulation of immune pathways, which could lead to systemic inflammatory reactions and autoimmune-like adverse events. A combinatorial approach using CPIs or targeted agents could amplify therapeutic response while mitigating toxicity.

15.2.3.5 Cancer vaccines and oncolytic viruses

Cancer vaccines and oncolytic viruses are emerging immunotherapeutic strategies that aim to prime the immune system against malignancies.

15.2.3.5.1 Cancer vaccines

These can be prophylactic (e.g., HPV vaccine for cervical cancer) and therapeutic (e.g., sipuleucel-T for prostate cancer). Therapeutic vaccines also create an immune response against tumor-associated antigens and increase T-cell recognition and attack. Despite promising preclinical data, clinical efficacy is still limited, and combination strategies with immune CPIs are necessary (Fan et al., 2023; Kaczmarek et al., 2023).

15.2.3.5.2 Oncolytic viruses

Genetically modified viruses selectively infect and kill tumor cells and stimulate an antitumor immune response. T-VEC, a modified herpes simplex virus, has been approved for the treatment of melanoma with improved local tumor control (Bommareddy et al., 2018; Tian et al., 2022). Both strategies are promising, less-toxic alternatives to traditional treatment, but problems, including tumor antigen heterogeneity, immune evasion, and delivery limitations, need to be overcome to maximize their clinical impact.

15.2.4 Challenges in the development of biologic therapies

15.2.4.1 Stability and manufacturing complexities

Biologic therapies are highly stable, but their use is affected by manufacturing considerations due to their complex, highly diverse molecular frameworks, and they are vulnerable to environmental conditions, including temperature, pH, and mechanical forces, which might result in denaturation, aggregation, or reduced activity. Unlike small-molecule drugs, biologics, i.e., monoclonal antibodies and recombinant proteins, should be stored under specific conditions to maintain their functional activity and structural integrity. Biofuels require advanced bioprocessing methods, including cell culture, purification, and quality control, to reduce variability in living cell sources and provide uniform batches. Translational changes, especially glycosylation, make analysis difficult and necessitate more sophisticated methods of analysis to confirm biological integrity and activity (Elsayed et al., 2023; Pham & Meng, 2020). Production is also more challenging than regulatory requirements for safety, effectiveness, and immunogenicity, and requires cold-chain logistics to ensure stability in storage and distribution, making it less accessible. The mentioned issues highlight the importance of streamlining the formulation and manufacturing plans for biologics to stabilize them (Elsayed et al., 2023; Pham & Meng, 2020). All these indicate the need for streamlined formulation and production processes to ensure biologic stability and therapeutic efficacy.

15.2.4.2 High cost of development

Biologic drug products require significant financial investments, often amounting to millions of dollars per product. The following factors can explain these high fiscal burdens, long research and development cycles, complex production processes, and high regulatory requirements. Unlike the traditional pharmaceuticals, which can be synthesized through a simple chemical reaction, biologics can only be tested using preclinical and clinical research over a long period to determine safety, efficacy, and immunogenicity (Farid et al., 2020; Paul & Sarkar, 2024).

Biologic clinical trials are more likely to be larger, costly, and time-consuming than small-molecule clinical trials. The high production costs are also driven by the need to apply a complex analytical approach, specialized bioreactors, and strict quality control standards. In addition, branded biologics are protected by intellectual property rights with exclusivity periods, which permit manufacturers to charge high prices to recoup investment costs. It is a financial limitation that restricts the access of the patient to life-saving treatments and places a substantial economic strain on health systems (Deza et al., 2025).

The arrival of biosimilars, cost-efficient analogs of biologic agents, has reduced some of the financial burdens of biopharmaceuticals. Nevertheless, the regulatory approval process of biosimilars still has strict requirements and requires extensive comparability studies to establish equivalence with reference biologics. Consequently, despite being a viable strategy for cost-containment, the development of biosimilars still requires significant fiscal and technical investment. Overcoming the cost-related barriers is critical to increasing the accessibility and affordability of biologic therapies in oncology (Joshi et al., 2022).

15.3 Biosimilars: An alternative to biologics

15.3.1 Defining biosimilars and how they differ from biologics

Biosimilars are biologic medicines intended to be highly similar to existing reference biologics in terms of safety, efficacy, and quality, and that do not have clinically meaningful differences. Unlike generic drugs, which are chemically identical to their brand-name counterparts, biosimilars are made in living systems and inevitably have minor variations due to the complexity of biologic molecules and the proprietary manufacturing process. These variations are carefully evaluated by means of full comparability trials, such as analytical, functional, nonclinical, and clinical trials, to ensure that the treatment is therapeutically equivalent, and to eliminate the possibility of any significant impact on safety or efficacy (particularly immunogenicity, a factor that regulatory agencies place a lot of emphasis upon)—biologic production of biosimilars. The biosimilar process involves reverse-engineering the reference biologic using alternative cell lines and production procedures. It requires complete analytical characterization and process validation to achieve a high level of similarity. Biosimilars receive regulatory approval through a shorter route than the original biologics; they are required to demonstrate similarity but do not undertake any specific, extensive clinical trials. However, confirmatory clinical studies on at least one indication and ongoing PV are obligatory. Although biosimilars are not identical to their reference products, stringent regulations ensure that any differences between the two are not clinically relevant. The introduction of biosimilars has led to increased competition, a decrease in treatment expenses, and increased exposure to sophisticated therapies in patients, especially those with cancer and chronic illnesses, which has contributed to the sustainability of healthcare without reducing patient outcomes (Markus et al., 2017; Mascarenhas-Melo et al., 2024; Ratih et al., 2021).

15.3.2 The scientific basis for biosimilarity

The scientific rationale for biosimilarity is that a biosimilar is highly similar to its reference biologic in terms of structure, function, PK, PD, and clinical performance. Unlike generic drugs, which can be replicated with the same chemical structure and molecules, biosimilars are produced in living systems and thus are bound to exhibit minor differences. Therefore, regulatory agencies require a totality-of-evidence approach, beginning with thorough analytical characterization and supported by targeted clinical studies (Markus et al., 2017; Wolff-Holz et al., 2019).

Analytical comparability is the foundation of biosimilar development. Advanced technologies, including mass spectrometry, chromatography, capillary electrophoresis, and bioassays, are used to determine primary amino acid sequences, higher-order structures, glycosylation patterns, charge variants, and post-translational modifications (PTMs). Functional assays are further used to test the biological activity of monoclonal antibodies, including receptor binding, signal transduction, and effector activity such as antibody-dependent cellular cytotoxicity. These analytical and functional evaluations must show no statistically and clinically significant differences when compared to the reference biologic (Ayoub et al., 2013; Alez-Martin et al., 2026; Cerutti et al., 2019; Visser et al., 2013). Table 15.3 demonstrates the analytical and functional methods in biosimilar comparability.

Subsequently, the nonclinical evaluations (in vitro and in vivo) and clinical studies aim to confirm similarity in PK/PD behavior and immunogenicity, rather than reestablish safety and efficacy from scratch. Clinical trials usually involve sensitive populations and endpoints that are sensitive to any differences that might be present, so that if there is any variation, it will not have any therapeutic effect. This stepwise, comparative methodology helps reduce redundancies in large-scale clinical trials while ensuring that biosimilars achieve the same therapeutic outcomes as reference products. The solid scientific basis of biosimilarity helps provide safe, effective, and more widely accessible biologic therapies in oncology (Schiestl et al., 2020; Stebbing et al., 2020; Wolff-Holz et al., 2019).

15.3.3 Regulatory approval pathways for biosimilars

The regulatory approval of biosimilars by the EMA, the U.S. FDA, and the World Health Organization (WHO) is based on a stringent, evidence-based approach that

Table 15.3 Overview of analytical and functional methods in biosimilar comparability.

Attribute	Technology	Applications	Citations
Primary sequence	MS, LC-MS/MS, peptide mapping	Sequence confirmation, PTMs	Ayoub et al. (2013); Visser et al. (2013)
Higher-order structure	CD, FTIR, NMR, DSC	Folding, stability	Cerutti et al. (2019); Visser et al. (2013)
Glycosylation	MS, HILIC, NP-HPLC, CE-SDS	Glycoform profiling	Ayoub et al. (2013); Cerutti et al. (2019); Visser et al. (2013)
Charge/size variants	CEX, CE, SEC, MCE-MS	Charge heterogeneity, aggregation	Alez-Martin et al. (2026); Ayoub et al. (2013); Visser et al. (2013)
Functional activity	Bioassays, SPR, cell-based assays	ADCC, CDC, receptor binding	Cerutti et al. (2019); Visser et al. (2013)

assesses the totality of the evidence at each step. The EMA made the first step in 2005. It demands a high level of comparative analytical similarity, functional characterization, and customized clinical research to demonstrate that a biosimilar is highly similar to its reference product in quality, safety, and efficacy. The FDA, in the Biologics Price Competition and Innovation Act, has an identical stepwise process: analytical investigations, preclinical evaluations, and comparative clinical tests, particularly PK, PD, and immunogenicity. The WHO has developed global guidelines to harmonize standards, especially for countries without their own biosimilar frameworks, to ensure that biosimilars are subjected to consistent standards for analytical characterization and clinical confirmation before approval (Gherghescu & Delgado-Charro, 2021; Patel et al., 2024).

15.3.4 Interchangeability and substitution of biosimilars

The higher regulatory requirements for interchangeability compared to the U.S. (where the FDA demands additional evidence, including dedicated switching studies) are intended to demonstrate that switching between a biosimilar and its reference biologic does not increase risks or decrease efficacy. Automatic pharmacy-level interchangeable biosimilars can only be substituted for FDA-designated interchangeable biosimilars without prescriber intervention, as is the case with generic drugs. The EMA and WHO, on the other hand, do not grant formal status to interchangeability but leave substitution and switching to national bodies, and focus on physician-led switching based on clinical judgment and patient participation (Afzali et al., 2021; Alvarez et al., 2020; Iskit, 2025).

Dispensing a biosimilar instead of a reference biologic without the express consent of the prescriber is referred to as substitution. Although automatic substitution can be applied to FDA-designated interchangeable biosimilars in some states of the United

States, in most areas, especially Europe, substitution is preferred by physicians. As demonstrated by both real-world and clinical trials, the transition from biosimilars to reference biologics does not pose greater safety risks or lower efficacy. Still, a lack of confidence among patients and providers, education, and strong PV are the factors that enable successful adoption (Alvarez et al., 2020; Herndon et al., 2023).

15.3.5 Cost-effectiveness and access to treatment

With the introduction of biosimilars in oncology, there are tremendous cost savings and subsequent access to life-saving cancer therapies. By providing clinically comparable alternatives to costly biologics, biosimilars have reduced the cost of acquiring drugs, increased the number of patients health care systems can treat, and driven price competition that has lowered the prices of biosimilars and reference biologics. These economic gains are particularly high in high-burden cancers like breast cancer and lymphoma, where the use of biosimilars has generated significant savings for payers and enabled resources to be redirected to innovative treatments and wide-scale cancer care programs (Gilmore et al., 2025; Hübel et al., 2020).

Biosimilars have increased patient access by lowering costs, enabling more people to start and continue therapy. Practical and simulation studies indicate that biosimilars contribute to healthcare sustainability by making treatments more accessible and less expensive. Yet, several issues, including irregular reimbursement, patent challenges, and inconsistent adoption, persist. It is necessary to overcome these obstacles by educating the public and implementing policies that maximize the cost-effectiveness and accessibility of biosimilars in oncology (Gilmore et al., 2025; Hübel et al., 2020)

15.4 Preclinical and clinical development of biosimilars

15.4.1 Preclinical studies: Analytical and functional comparability

High-resolution analytical and functional comparability assays are used in the preclinical testing of a biosimilar, rather than repetitive nonclinical toxicology packages. In a stepwise, comparative exercise, as practiced by developers, analyses orthogonal to one another (e.g., mass spectrometry of primary sequence, peptide mapping, glycan profiling, capillary electrophoresis/ion-exchange of charge variants, and differential scanning calorimetry/circular dichroism of higher-order structure) are performed first. Cell-based functional assays (e.g., ADDC, CDC) of binding, neutralization, receptor activation, signal transduction, or effector functions are then done, based on this analytics-first approach. The totality-of-evidence EMA Guideline on Similar Biological Medicinal Products (2014/2015) sets out the stepwise comparability paradigm, which gives primacy to analytical and in vitro functional information. It explains how the nonclinical and clinical programs should be designed in response to residual uncertainty after analytical work (Nupur et al., 2022).

15.4.2 Pharmacokinetics and pharmacodynamics in biosimilar development

PK and PD trials are designed to ensure that the systemic exposure and pharmacologic activity of the biosimilar are within an acceptable range of equivalence relative to the reference product. Most commonly used first-in-human comparative designs include single-dose, cross-over, or parallel designs, with healthy volunteers when possible and ethical (e.g., many recombinant proteins) or sensitive patient populations when necessary. PD biomarkers (e.g., ANC response to filgrastim, hemoglobin response to ESAs) can serve as additional evidence of a similar biological effect. Regulators expect preset equivalence margins, warranting of the chosen population and endpoints in PK/PD studies, and close consideration of immunogenicity. Consideration of FDA clinical pharmacology provides clear guidelines for conducting PK/PD studies to support the concept of biosimilarity and to define acceptable study populations, endpoints, and statistical equivalence margins. In the guidance, the cases in which the PD endpoints may substitute for or supplement the PK endpoints are explained (Chekka et al., 2025; Strauss et al., 2023).

15.4.3 Clinical trial design for biosimilarity assessment

15.4.3.1 Phase I and Phase III trials for biosimilars

The biosimilars phase I (PK/PD) trials aim to demonstrate that pharmacokinetic equivalence is achieved between the biosimilar and its reference product, usually through area under the curve and maximum concentration, and, where applicable, pharmacodynamic equivalence between the biosimilars. Unless there is a reason to use patient populations, such as safety or the mechanism of action of the drug, these studies usually rely on healthy volunteers to reduce variability. Usually, standard equivalence margins (80%–125% of PK parameters) are used, with close monitoring of immunogenicity. Many studies, including those comparing trastuzumab, denosumab, and nivolumab biosimilars, have confirmed that PK/PD equivalence can be well established in this way and provides a basis for further clinical development (Jeka et al., 2024; Wang et al., 2020).

Biosimilar Phase II confirmatory trials are, in general, smaller and more focused than those associated with originator biologics and are undertaken to clarify any remaining doubts regarding clinical efficacy, safety, and immunogenicity. Such trials involve sensitive endpoints and patient populations with the highest likelihood of showing differences, as observed in oncology biosimilar trials (e.g., CT-P6 and SB3 of trastuzumab), which showed no differences in efficacy, safety, or immunogenicity relative to the reference product (Jeka et al., 2024; Papp et al., 2025). Regulators tend to favor equivalence trial designs, in which the biosimilar is not worse or better than allowed by preset margins, rather than noninferiority designs, since equivalence testing is more effective at ruling out any clinically significant difference. Selection and justification of equivalence margins are critical, must be scientifically supported, and are often based on historical effect sizes and regulatory precedent (Hemmelmann et al., 2025; Kim et al., 2020).

15.4.4 Immunogenicity concerns in biosimilars

The primary safety issue with all therapeutic proteins is immunogenicity, specifically the formation of antidrug antibodies and neutralizing antibodies, which may disrupt efficacy or cause adverse events. Various factors, including protein sequence, glycosylation, aggregation, formulation, container-closure system, impurities, and patient-specific factors such as immune status and comorbidities, influence the risk of immunogenicity. The process is the product principle was illustrated in the Eprex (epoetin alfa) case, where alterations in formulation and syringe components led to a spike in antibody-mediated pure red cell aplasia. Research has found that the rate of PRCA was much higher when Eprex was administered with a polysorbate 80 formulation and an uncoated rubber stopper, and that interventions, including the use of coated rubber stoppers and better handling of the medication, dramatically reduced the risk (Bennett et al., 2004; Boven et al., 2005). The incidents indicate the possibility of severe clinical immunogenicity consequences from even a slight change in the manufacturing/formulation procedure, and the necessity of rigorous comparability exercises and ongoing PV.

Biosimilar development currently requires sensitive, validated immunogenicity testing in clinical trials and a comprehensive PMS program. It has been found by bioanalytical controls of biosimilars of monoclonal antibodies and other biologics that under suitable conditions of adequate development and monitoring, the immunogenicity pattern of biosimilars is likely to be comparable to that of their reference product, with most of the differences in antidrug antibody (ADA) or neutralizing antibody rates falling within a range of 15%–20% (Casadevall et al., 2002; Strand et al., 2020). Consecutive studies, such as those with filgrastim and infliximab biosimilars, have not shown increased immunogenicity when biosimilars are used alongside reference products, and clinical interchangeability is permitted where allowed (Lauret et al., 2020; Strand et al., 2020). These results demonstrate that careful consideration of immunogenicity and postapproval surveillance are necessary to assess the safety and efficacy of both originator and biosimilar therapeutic proteins.

15.4.5 Extrapolation of indications: Challenges and benefits

Extrapolation enables approval of a biosimilar based on one indication of the reference product for other indications without additional clinical trials in those indications. The regulatory authority has stringent criteria that indicate a high level of scientific justification based on the totality of the evidence, including highly similar mechanisms of action, PK/PD, and immunogenicity across indications. This has been successfully applied in oncology, where trastuzumab and bevacizumab biosimilars

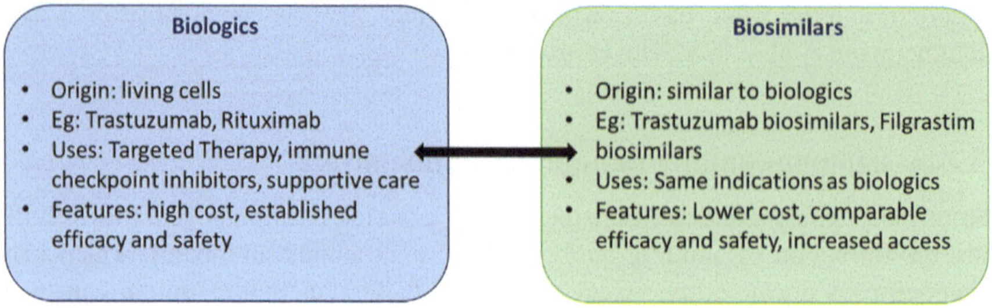

FIGURE 15.1

Comparative overview of biologics and biosimilars in cancer therapy, highlighting their origin, examples, therapeutic uses, key features, and the shared efficacy and safety profile.

were approved, with all reference product indications demonstrating biosimilarity in a sensitive population, and postmarketing studies showing similar efficacy and safety in extrapolated indications (Melosky et al., 2018; Reinisch et al., 2015; Taniguchi et al., 2025).

However, extrapolation does not occur automatically; regulators consider that the mechanism of action will be identical across all indications and that patient populations may not be homogeneous in other aspects that can affect safety or efficacy. For example, the role of the effector as an ADDC may be obligatory in one indication but not the other, or the risk of immunogenicity may differ; additional data may be necessary (Ben-Horin et al., 2016; Ogura et al., 2017; Reinisch et al., 2015). Biosimilars are safe and effective in extrapolated indications, as demonstrated by real-world and long-term follow-up data, despite ongoing PV requirements. Overall, the extrapolation will help reduce unnecessary trials and accelerate access, but in each case, a scientific justification will be required to ensure patient safety and equal access to therapy.

15.5 Biologics and biosimilars in cancer therapy: Current landscape

Biologics remain the center of contemporary oncology practice, and biosimilars accompany the greater availability. Sound underlying evidence, including rigorous regulatory science informed by real-world oncology practice, supports a similar efficacy and safety profile (Konstantinidou et al., 2020). In relation to systematic implementation strategies and open communication, biosimilars can reduce high health system costs, which can be reinvested in patient outcomes and increased access to new therapies Figure 15.1. represents the overall comparison of biologics and biosimilars.

15.5.1 Approved biologics in oncology and their impact

Biologics are large, complex molecules derived from living cells that have transformed cancer treatment over the last two decades. Among them are monoclonal antibodies (mAbs) and ADCs (Zahavi & Weiner, 2020), cytokines and growth factors(Mukai et al., 2018), and cell therapies. The use of mAbs, including trastuzumab (targeted at HER2) (Cortés et al., 2022), rituximab (targeted at CD20) (Gauckler et al., 2021), and bevacizumab (VEGF-A inhibitor) (Garcia et al., 2020), historic example of targeted therapy that set the stage and established it as a standard treatment alongside surgery, radiotherapy, and cytotoxic chemotherapy. They act by selectively interfering with oncogenic signal transduction or tumor-host interactions (e.g., angiogenesis), thereby enhancing response rate, progression-free survival, and, in the majority of indications, overall survival. Biosimilars biosimilabs also include immunomodulatory antibodies (e.g., anti-PD-1/PD-L1, anti-CTLA-4) (Wojtukiewicz et al., 2021) with the restoration of antitumor immunity, sustained remissions have been achieved in melanoma, lung, renal, and other cancers, but have not yet been broadly represented in biosimilar pipelines due to evolving intellectual property and clinical validation requirements.

Supportive-care biologics like granulocyte colony-stimulating factors (G-CSFs such as filgrastim, pegfilgrastim) (Bond et al., 2018) (enabled dose and lower hospitalization) and erythropoiesis-stimulating agents (epoetin alfa) (Aapro et al., 2012) (enabled dose and lower hospitalization) reduce the toxicities that limit the use of these treatments (neutropenia, anemia). The general effect of the biologics has been improved survival in specific molecularly defined cohorts and increased curative intent in early-stage disease (e.g., HER2-positive early-stage breast cancer with perioperative trastuzumab).

The supportive-care biologics (including G-CSFs, such as filgrastim and pegfilgrastim) and erythropoiesis-stimulating proteins (epoetin alfa) reduce treatment-limiting toxicities (neutropenia, anemia) and allow dose intensification and reduced hospitalizations. The net impact of the biologics has been an increase in survival in specific molecularly defined populations, curative intent in early-stage disease (e.g., HER2-positive early-stage breast cancer with perioperative trastuzumab), and increased curative intent.

15.5.2 Key biosimilars approved for cancer treatment

The biosimilars are closely related to reference-approved biologics, and none of the safety, purity, or potency differences are clinically significant. Waves of biosimilar approvals have been observed in oncology, first in supportive care and then in therapeutic mAbs. There are currently many biosimilars of filgrastim and pegfilgrastim (supportive care), and therapeutic mAbs trastuzumab, bevacizumab, and rituximab, in Europe and the U.S. They have had step-by-step analytical, nonclinical, and clinical programs of comparability to determine biosimilarity; in most instances, extrapolations have been made based on regulatory science methods. Regulators and professional societies maintain authoritative lists of a dozen or more oncology-related biosimilars approved and commercialized, and additional new entrants keep appearing as patents expire.

The introduction of those biosimilars has exponentially intensified market competition through price reductions relative to the originators and, more importantly, expanded patients' access, which was limited by extremely high prices (Mountzios & Boukovinas, 2013). Tendering procedures and formulary methods are also used in health systems to achieve further reductions. They are usually supplemented by education campaigns to facilitate their adoption by health care professionals and patients (Maniadakis et al., 2018). Guidelines on scientific criteria and their proper use have been published by professional organizations such as ESMO (Gyawali et al., 2021) and emphasize that approved biosimilars are supported by high-quality evidence.

15.5.3 Case studies of successful biosimilar adoption

Various empirical illustrations highlight the potential of biosimilars in widening access and unlocking resources for innovations:

15.5.3.1 The United Kingdom (NHS England)

Systematic efforts to transition to more valuable biologic therapies, supported by commissioning frameworks, outcome monitoring, and stakeholder engagement, led to significant cost savings (Russell & Galloway, 2024). Real-world experience demonstrates that the hundreds of millions of pounds saved annually in individual fiscal years result from switching to biosimilars rather than originator drugs across multiple compounds, with oncology medicines being the most prominent; more recent estimates suggest this could save a potential billion pounds over 5 years (Seelig, 2020). Such savings were, in the majority of cases, returned to the first line of healthcare delivery, novel treatments, and service enhancement.

15.5.3.2 HER2-positive breast cancer (Europe/Asia)

Medical centres reported successful transitions of the reference product to biosimilars in neoadjuvant, adjuvant, and metastatic settings, with no additional safety issues noted. However, stable pathologic complete response rates were observed. The similar efficacy and cardiac safety of biosimilar trastuzumab (Pronzato & Rondini, 2006), systematic reviews, and real-world evaluations confirmed the reference product. Thus, broader and earlier use of HER2-targeted therapy was approved in resource-limited settings (Mukherjee & Bandyopadhyay, 2024).

15.5.3.3 India and other low- and middle-income nations

The availability of trastuzumab biosimilars has been associated with more patients taking up treatment, enabling HER2-positive early breast cancer patients to receive the standard-of-care treatment they would otherwise not receive due to cost. The Retrospective series report increased the use of similar outcomes, thereby highlighting the equity in biosimilar use.

These examples have certain common success factors: clear national or institutional policy on switching, open PV, effective prescribers and patient education, and harmonization of incentives on clinical value and other factors.

15.5.4 Patient outcomes and safety considerations

The regulatory rationale for biosimilarity is predicated on an overarching evidence-based approach: a candidate's high analytical and functional similarity is sufficient to support limited but accurate clinical investigations to rule out significant differences. Post-approval PV and RWE provide further verification of safety and efficacy (Kumar & Baldi, 2013; Lucas et al., 2022). In oncology, comparative trials and observational studies of trastuzumab, bevacizumab, and rituximab biosimilars routinely demonstrate noninferior clinical outcomes (e.g., response rates and event-free survival surrogates) and safety profiles that are comparable, including class-specific adverse events like cardiotoxicity associated with HER2-targeted therapies and infusion reactions involving anti-CD20 antibodies (Oflazoglu & Audoly, 2010). Although no biologic is immunogenicity-free, stringent manufacturing controls and comparability testing keep ADA levels in line with those of reference products.

Indication extrapolation is permitted by regulators when the mechanisms of action and clinical pharmacology are appropriate and sufficiently similar across tumor types and therapy lines; this is more a matter of education than an evidence gap (Feagan et al., 2014; Tesser et al., 2017).

Nonmedical switches due to cost/availability are standard and, when managed through communication and monitoring, have not been associated with adverse clinical outcomes in oncology patients. Jurisdiction specific "interchangeability" designations (e.g., the U.S.) are regulatory definitions that are more closely related to substitution guidelines at the pharmacy level than to relative safety differences (Roediger et al., 2017). Distinguishable names and batch tracking facilitate signal detection for PV.

Overall, the cumulative evidence supports the benefit–risk similarity of approved oncology biosimilars with reference biologics when used as labeled and in accordance with clinical practice guidelines. Professional society statements affirm those findings and call for responsible use.

15.5.5 Physician and patient perceptions of biosimilars

Perceptions drive uptake. In the early days of the biosimilar era, caution was driven by concerns about immunogenicity, extrapolation, and the adequacy of clinical data. High-quality regulatory decisions, the maturation of RWE, and apparent system-level cost savings over the years have contributed to increased confidence (Aiyegbusi et al., 2022; Preminger & Mittal, 2017). Patient education brochures highlight that inherent, tiny batch-to-batch variability is common across all biologics (including originators) and that a biosimilar is only approved after a rigorous comparative evaluation; these clarifications help dispel myths and build confidence (Biddle et al., 2024).

When institutional guidelines provide clear guidance on initiating therapy with a biosimilar and on switching stable patients, physician acceptance is most significant. This clearly shares local practice safety and outcome data, including that from pharmacy-led reviews and tumor board evaluations. Formularies have easy access channels (order sets, infusion protocols) and payers align reimbursement without administrative conflict. The physical investment of savings in oncology services (e.g., making more new treatments available, hiring support staff, improving diagnostics) creates a positive feedback loop. Health system case studies demonstrate the importance of coordinating clinical leadership, pharmacy, and commissioning bodies to expedite culture change (Burgess et al., 2021; Hydery & Reddy, 2024; Mysler et al., 2021).

15.6 Manufacturing and quality control of biologics and biosimilars

15.6.1 Bioprocessing and manufacturing of biologics

Biologics are advanced therapeutic proteins produced using living cells and require precise, controlled production. The gene-to-final-product process involves several crucial steps that influence the quality, safety, and efficacy of the biologic.

15.6.1.1 Cell line development

This is triggered by the process of the choice and genetic engineering of host cells (typically of mammalian [e.g., CHO cells], yeast, or bacterial systems) to express the therapeutic protein of interest. Next-generation sequencing and omics technologies, also known as cell line engineering technologies, enable precise manipulation of integration, expression levels, and product quality (Garvey, 2022; Grassi et al., 2025). Host selection influences protein folding, glycosylation, and biological activity.

15.6.1.2 Upstream processing: Cell culture and fermentation

Engineered cells are cultured in precisely controlled bioreactor environments (temperature, pH, oxygen, nutrients). Bio-processing today deals with continuous (perfusion) and batch processes. Continuous bioprocessing supports single-use technologies and automation, which are more productive, less expensive, and offer better product consistency than traditional batch methods (Bautista-Vanegas et al., 2025; Zydney, 2016). Nevertheless, achieving long-term sterility and genetic stability in continuous systems remains a challenge.

15.6.1.3 Downstream processing: Purification

Following cell culture, filtration, chromatography, and polishing are among the purification steps used to separate the protein product from cells and impurities. New technologies, such as microfluidic and continuous-flow purification systems, enable rapid, combined, and highly efficient purification, supporting the production of high-purity biologics and enabling real-time process monitoring (Sharma et al., 2024).

15.6.1.4 Process control and quality assurance

Even slight modifications to process parameters, raw materials, or equipment can significantly affect critical quality attributes (CQAs) such as glycosylation, folding, and aggregation. Real-time monitoring and correction of essential process parameters are enabled by advanced process analytical technologies (PAT), automation, and AI-driven control systems, ensuring batch-to-batch consistency and compliance with Good Manufacturing Practice. Quality by Design is a growing approach to systematically designing robust processes (Bautista-Vanegas et al., 2025; Niazi, 2025; Tripathi & Shrivastava, 2019).

The field is moving toward full integration, end-to-end continuous manufacturing, and the use of single-use systems, digital twins, and innovative manufacturing platforms. Such improvements will be more agile, cost-efficient, and environmentally friendly, but will require continued progress in process integration, regulatory adjustment, and employee training (Bautista-Vanegas et al., 2025; Niazi, 2025; Zydney, 2016).

15.6.2 Biosimilar manufacturing: Challenges in achieving similarity

Biosimilars are intended to be very similar, though not identical, to their reference biologic products. This is because biologics are large, structurally complex proteins synthesized in biological cells. Biosimilarity will require substantial scientific and technical challenges to be overcome so that any disparities do not affect clinical safety or efficacy.

15.6.2.1 Inherent variability and the "Process is the Product"

Biosimilars, unlike small-molecule generics, cannot be exact copies, as biological systems are inherently variable at every stage of production. Changes in cell lines, culture conditions, or manufacturing processes can alter CQAs of glycosylation, folding, and aggregation, which, in turn, can modify the drug's functionality, safety, or immunogenicity. This implies that the manufacturing process itself is a significant determinant of the final product, and that tight control of the process is necessary to reduce variability and ensure it is similar to the reference biologic (Duivelshof et al., 2019; Nupur et al., 2022; Ratih et al., 2021; Vulto & Jaquez, 2017).

15.6.2.2 Analytical and functional comparability

Biosimilarity has been shown using a step-by-step, totality-of-evidence strategy, starting with a comprehensive analytical characterization. The primary sequence, higher-order structure, glycosylation, and biological activity of the biosimilar and reference product are compared using sophisticated orthogonal methods, including high-resolution mass spectrometry, nuclear magnetic resonance (NMR), chromatography, and bioassays. These analytical techniques are more delicate than clinical trials in the identification of molecular differences and are vital in creating a substantial similarity fingerprint (Duivelshof et al., 2019; Nupur et al., 2022; Pujahari et al., 2025; Vulto & Jaquez, 2017).

15.6.2.3 Regulatory and manufacturing challenges

The manufacturers are expected to characterize the variability of the reference product across several batches to determine acceptable ranges for CQAs. Reverse engineering is then used to develop a biosimilar that closely matches these properties. Regulatory authorities insist that the perceived differences be clinically insignificant, as demonstrated by a combination of analytical, functional, and, in some cases, clinical tests. Constant monitoring should be conducted to prevent deterioration in product quality over time (Kim et al., 2017; Nupur et al., 2022; Vulto & Jaquez, 2017).

Comparative analyses of biosimilars with rituximab and trastuzumab have revealed that, despite many features being closely similar, some differences in glycosylation or charge formation are possible. They can be allowed provided they do not affect potency, safety, or efficacy. Forced degradation and stability studies also ensure that biosimilars are similar to one another under stress conditions (Kim et al., 2017).

15.6.3 Analytical characterization of biologics and biosimilars: Structural and functional comparability

Analytical characterization is the backbone of biosimilar development, as it allows biosimilars and reference biologics to be compared in detail with respect to structural and functional properties. Multilevel analysis of primary to quaternary structure, purity, charge variants, and biological activity is complex and relies on a wide range of orthogonal methods, such as high-resolution mass spectrometry, NMR, chromatography, capillary electrophoresis, and bioassays. The methods are more sensitive than clinical trials for detecting molecular differences, which is why totality-of-evidence should be used to support regulatory approval (Castel et al., 2023; Dakshinamurthy & Gothandam, 2022; Nupur et al., 2022).

15.6.3.1 Structural and functional comparability

Primary Structure: Peptide mapping and mass spectrometry confirm amino acid sequence identity.

Higher-Order Structure: Circular dichroism, NMR, and advanced mass spectrometry (e.g., HDX-MS, native MS) assess secondary, tertiary, and quaternary structures.

Purity and Variants: Chromatography and capillary electrophoresis detect size and charge variants, aggregates, and impurities.

Biological Activity: Bioassays, including cell-based and receptor-binding assays, confirm functional similarity.

Recent advances include automated, image-based statistical analysis of NMR data, multidimensional chromatographic approaches, and high-throughput mass spectrometry, which enhance accuracy and throughput in similarity assessments.

15.6.3.2 Glycosylation and posttranslational modifications

Glycosylation is a vital quality characteristic that affects immunogenicity, efficacy, and PK. Its heterogeneity, which is predetermined by cell line and process conditions, is a serious challenge to analysis (Duivelshof et al., 2019; Gan et al., 2024; Hu et al., 2025). State-of-the-art techniques for glycan profiling include:

15.6.3.2.1 Hydrophilic interaction chromatography (HILIC) and Hydrophilic interaction chromatography-Mass Spectroscopy (HILIC-MS)

For detailed glycan and glycoform analysis at protein, glycopeptide, and released glycan levels.

15.6.3.2.2 High-performance anion exchange chromatography (HPAEC-PAD)

For monosaccharide and sialic acid content.

15.6.3.2.3 FTIR spectroscopy

For rapid, global glycosylation fingerprinting and batch-to-batch comparison.

Site-specific glycan analysis and quantification of multiple PTMs (e.g., deamidation, oxidation) are essential to ensure biosimilar safety and efficacy. Differences in glycosylation must be thoroughly characterized and shown not to impact clinical performance.

15.6.4 Stability and storage considerations

The stability test is also necessary to ensure that biologics and biosimilars retain their structural integrity and biological activity throughout their shelf life and during clinical use. Recent studies show that many biosimilars can be stored under diverse conditions, including temperature variations and light exposure, provided they remain sterile after dilution. Table 15.4 indicates how biosimilars are stable at different storage conditions.

15.6.4.1 Stability under real-world and stress conditions

Temperature Excursions: It has been demonstrated that biosimilars such as bevacizumab, rituximab, trastuzumab, ranibizumab, adalimumab, and eculizumab can remain stable (physicochemically and biologically) for days or weeks at high temperatures (up to 25°C–30°C), even after their shelf life has expired (Kaiser et al., 2023; Park et al., 2020; Tak et al., 2023; Yun et al., 2019).

Table 15.4 Stability of biosimilars under various storage conditions.

Biosimilar	Stable conditions (Temp/Duration)	Key findings	Citations
Bevacizumab	2°C–8°C (3 mo), 25°C–30°C (72 h), diluted	No loss of integrity or activity	Park et al. (2020); Vieillard & Paul, (2023)
Rituximab	25°C–30°C (14°C–21°C), diluted	Maintains quality, even after shelf life	Borišek et al. (2022, 2024)
Trastuzumab	4°C (90 d), 25°C (3 d), diluted	Stable after reconstitution and dilution	Yun et al. (2019)
Ranibizumab	5°C (30 d), 25°C–30°C (48 h), diluted	Stable in syringes, supports advanced prep	Kaiser et al. (2023)
Adalimumab	25°C (4 wks), freeze–thaw cycles	No significant changes in quality attributes	Borišek et al. (2022)
Eculizumab	5°C (3 mo), 30°C (72 h), diluted	Stable in saline, not in 5% dextrose	Tak et al. (2023)

Dilution and in-use Stability: Biosimilars diluted in saline or dextrose and infused as bags or syringes remain stable for extended periods (up to 3 months at 4°C or a few days at room temperature), which can be helpful for preparation and flexible hospital practice (Borišek et al., 2024; Kaiser et al., 2023; Tak et al., 2023; Yun et al., 2019).

Light Sensitivity: Most biosimilars should be protected from light, and some may be safely exposed to short-term light conditions without compromising quality (Kim et al., 2023).

Freeze-Thaw Cycles: Some biosimilars, e.g., adalimumab, are stable after multiple freeze-thaw cycles, with no substantial changes in purity, strength, or safety.

15.6.4.2 Analytical methods and quality assurance

Stability, degradation, and preservation of biological activity and safety are monitored through advanced analytical techniques (e.g., size-exclusion chromatography, ion-exchange chromatography, dynamic light scattering, bioassays) (Kaiser et al., 2023; Park et al., 2020; Tak et al., 2023; Yun et al., 2019).

Sterility: Although physicochemical and biological stability can be ensured, sterility must be maintained, particularly for opened or diluted products (Park et al., 2020; Vieillard & Paul, 2023).

15.6.5 Quality control and regulatory compliance in biosimilars

Quality control in the manufacture of biosimilars is closely integrated throughout the process. The quality attributes are continuously monitored, and the international regulatory requirements (EMA, FDA, WHO) are strictly followed (Bas, 2025;

Janghorban et al., 2023; Patel et al., 2024; Ratih et al., 2021; Tran et al., 2024). Regulatory approval is based on a solid analytical comparability exercise, buttressed by specific clinical studies when needed, and on continued PV support for long-term safety and efficacy.

15.6.5.1 Regulatory framework and the "Totality of Evidence" approach

Gradual Development: Regulatory authorities need a gradual process, with broad analytical and functional capabilities to begin, followed by specific clinical research in the event of residual uncertainty (Bas, 2025; Niazi, 2025; Ratih et al., 2021).

Totality of Evidence: Approval depends on the totality of evidence, a complete body of structural, functional, PK, and clinical evidence that there are no clinically significant differences between the approval and the reference product (Gaylis et al., 2024).

Global Harmonization: EMA, FDA, and WHO are aligning their guidelines, and the flexibility and less demanding clinical trials are increasing as robust analytical data become available (Bas, 2025; Gaylis et al., 2024; Ratih et al., 2021).

15.6.5.2 Quality control and lifecycle management

15.6.5.2.1 Real-time critical quality attributes monitoring

Real-time monitoring of CQAs is carried out with the help of advanced analytical tools, the PAT, to make sure that there is a batch-to-batch consistency and that deviations are quickly identified (Bas, 2025; Janghorban et al., 2023; Tran et al., 2024).

15.6.5.2.2 Pharmacovigilance

Continuous safety monitoring and combination with manufacturing quality systems are both necessary to identify and prevent risks during the lifecycle of the product (Tran et al., 2024).

15.6.5.2.3 Regulatory flexibility

Agencies allow scientifically justified formulation changes to improve stability or minimize immunogenicity, provided comparability is maintained (Bas, 2025; Niazi, 2025).

15.7 Pharmacovigilance and post-marketing surveillance of biosimilars

Biosimilars have expanded access to biologic therapies, which are highly in demand for conditions such as oncology, autoimmune disease, and other chronic conditions. Although biosimilars have been preapproved to a high degree to assess their similarity to reference biologics, the nature of real-world clinical practice requires ongoing follow-up. PV and PMS are essential safety barriers for long-term safety, the maintenance of confidence among healthcare professionals and patients, and the continuation of biosimilars uptake (Lucas et al., 2022).

15.7.1 Importance of long-term safety monitoring

While small-molecule generics are large molecules, complex proteins are generated in a living system. Even slight modifications in manufacturing conditions (culture medium, cell line, purification steps) can imperceptibly alter their structure, post-translational modifications, or impurity profile. While regulatory agencies like the U.S. FDA (Kandemir, 2023) and the EMA (Liu et al., 2022). Require thorough comparability testing before licensure; some rare or late safety events may emerge only after large-scale patient use. There are various rationales for long-term monitoring.

15.7.1.1 Immunogenicity risks

Biosimilars, like reference biologics, also have the potential to cause ADA formation (Lv et al., 2025). Even though preapproval testing assesses immunogenicity, unexpected events may occur more frequently in broader, more variable populations. For example, neutralizing antibodies can dilute the therapeutic effect or precipitate unforeseen side reactions such as infusion-related hypersensitivity (Bhattacharya et al., 2022).

15.7.1.2 Population diversity

Biosimilar trials for treatment are typically limited to specific indications and patient populations. Biosimilars, once approved, can be used across various cancer indications, therapeutic areas, and comorbid populations, thereby potentially leading to new safety signals (Lawrence & Fraser, 2020; Schmitz et al., 2013).

15.7.1.3 Extrapolated indications

Authorities very commonly approve all reference biologic indications, regardless of whether there have been any clinical trials for a given disease (Weise et al., 2014). Long-term safety monitoring confirms the safety in real-world use for these extrapolated applications (Otremba et al., 2020).

15.7.1.4 Batch-to-batch variability

All biologics, including reference products, retain some manufacturing variability. Postmarketing monitoring ensures that long-term manufacturing changes will not affect patient safety (Mülhopt et al., 2018).

PV thus offers an additional safeguard after regulatory approval, ensuring that biosimilars maintain a comparable benefit–risk profile throughout their entire life cycle.

15.7.2 Real-world evidence in biosimilar adoption

RWE (Dang, 2023; Schad & Thronicke, 2022)—Evidence collected outside the clinical trial environment through randomized controlled trials has become a necessity in biosimilar PV. Electronic health records (EHRs) (Evans, 2016). Insurance claims databases, patient registries, and observational cohort studies are sources.

15.7.2.1 Validation of effectiveness

Biosimilars are approved based on comparative clinical and analytical data rather than extensive phase III efficacy trials for every indication. There is RWE that switching from originator biologics to biosimilars does not affect treatment outcomes. Indicatively, several studies conducted in European registries show that there were no significant differences in survival rates or disease management when patients diagnosed with breast cancer or lymphoma were switched to trastuzumab or rituximab biosimilars (Giusti et al., 2023).

15.7.2.2 Safety assurance

Postapproval data aid in identifying rare adverse events (AEs) by being specifically observed during postapproval trials. Monoclonal antibody biosimilars are associated with infusion-related reactions and unexpected immunogenicity patterns (Kakemam et al., 2021).

15.7.2.3 Health economics and access

RWE also exhibits system-level impact. The UK and Norway have recorded significant cost reductions following the adoption of biosimilars, which have made treatment more accessible to more patients and also enabled reinvestment in oncology treatment. These facts provide comfort between physicians and policymakers (Khosla et al., 2018).

15.7.2.4 Switching and interchangeability investigations

Clinical switching investigations offer confidence that switching stable patients from a reference biologic to a biosimilar would not enhance immunogenicity risk or diminish treatment efficacy (Nagase et al., 2023). This kind of evidence is particularly significant in oncology, for example, because treatment protocols are essential for survival.

15.7.2.5 Global relevance

RWE in LMICs demonstrates that biosimilars are expanding access to treatment in cases where originators are expensive and inaccessible (Schad & Thronicke, 2022). This feature of equity depicts the role of the postmarketing research in becoming relevant to the world health policy.

Hence, RWE helps fill the evidence-to-regulation gap, ensuring that biosimilars are accepted as evidence-based across varied patient populations.

15.7.3 Reporting and managing adverse events

One characteristic of PV is the systematic detection, recording, and assessment of AEs (Murthy et al., 2020). For biosimilars, this activity requires careful traceability within a therapy class to distinguish among products.

15.7.3.1 Unsought reporting mechanisms

All suspected adverse drug reactions (ADRs) should be reported to national PV databases (e.g., the Feedback system on the FDA and EudraVigilance of the EMA) (Cesana & Biganzoli, 2018). Product identifiers (e.g., brand, manufacturer, and batch number) must be included in submissions to enable product tracing.

15.7.3.2 Manufacturer liability

The manufacturers of biosimilars should develop PV systems to support continuous collection and analysis of safety data for submission to the authorities. These should contain regular safety update reports (PSURs) and risk assessment reports.

15.7.3.3 Signal detection

The application of mature analytical techniques, such as disproportionality analyses, enables the identification of safety signals that are noncharacteristic adverse event patterns for a specific biosimilar. They are also examined using epidemiological methods or registry data.

15.7.3.4 Management of adverse events in practice

Infusion reactions are managed by premedication, slowing the infusion rate, or switching therapy.

15.7.3.5 Concerns about immunogenicity

Monitoring of ADA levels during research or testing of loss of efficacy in practice.

15.7.3.6 Stereoselective interferences

Unanticipated toxicities by virtue of (Boni et al., 2023).

Reporting adverse events is particularly crucial in patients switched multiple times between various biosimilars or between biosimilars and originators. Traceability guarantees that causality assessments are correct (Braun & Kudrin, 2016).

By fostering a healthy culture that emphasizes timely reporting and frank communication, healthcare organizations build patient trust while also managing risks.

15.7.4 Risk management strategies in biosimilar use

Risk management for biosimilars involves active methods to identify, prevent, and control potential safety issues, enabling adoption while ensuring patient well-being (Zuñiga & Calvo, 2010). Essential strategies comprise:

15.7.4.1 Regulatory Risk Management Plans

Regulators mandate RMPs that outline known and potential safety concerns, PV activities, and mitigation strategies. For oncology biosimilars, this often includes immunogenicity monitoring, management of infusion reactions, and pregnancy/lactation registries (Butler et al., 2021).

15.7.4.2 Product traceability measures

Unique brand names, suffix-based INNs (in the US) (Jones et al., 2016). Batch number documentation is required for accurate attribution of AEs. Electronic prescribing and barcoding systems improve tracking and auditing across healthcare systems.

15.7.4.3 Protocols and guidelines for switching

Hospitals and oncological centers design standardized switching protocols—such as between physicians and patients, informed consent, and observation—with the intention of smooth transitioning (Jones et al., 2016). Clear protocols help prevent nocebo responses, in which patients perceive adverse outcomes due to psychosocial concerns.

15.7.4.4 Education and communication

Education for physicians, nurses, pharmacists, and patients is vital. Addressing misconceptions—such as "biosimilars are less safe than originators"—helps promote confidence (Pawłowska et al., 2019). Patient-friendly information sheets and open discussions about regulatory rigor support adherence.

15.7.4.5 Collaborative registries

National and international registries collect real-world outcome data for biosimilars, facilitating signal detection and comparative effectiveness research (Fan & Qiao, 2011). Oncology-specific registries (e.g., HER2 breast cancer cohorts on trastuzumab biosimilars) have been instrumental in demonstrating long-term equivalence (Ribelles et al., 2021).

15.7.4.6 Economic and policy implications

Reinvestment of savings from biosimilar use in oncology services can help improve stakeholder acceptance. Risk management, then, is more than safety; it is systematic trust building (Rakha et al., 2011).

PV and PMS are the biosimilar safety assurance. Continued surveillance addresses the inherent challenges of biologics, whereas RWE determines effectiveness and ensures credibility across different demographic groups. The early recognition

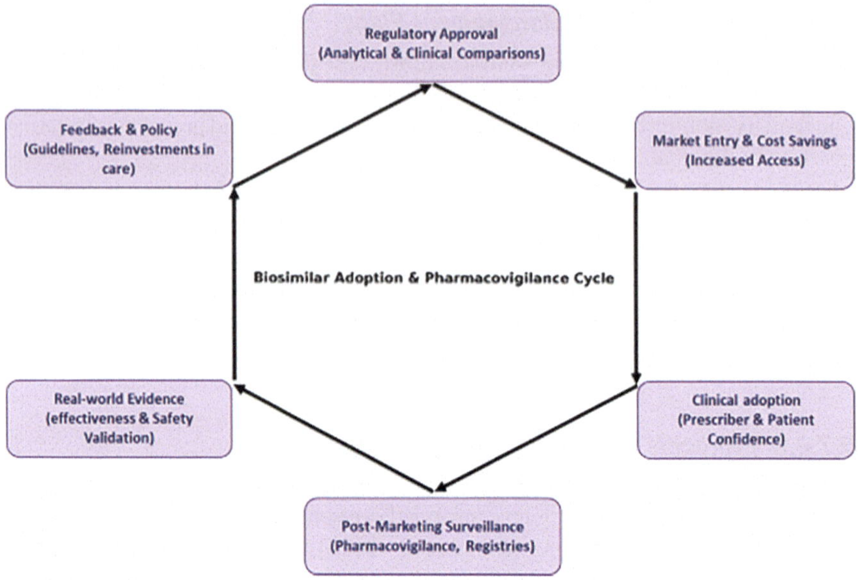

FIGURE 15.2

The biosimilar adoption and PV cycle, illustrating the continuous process from regulatory approval and market entry to clinical adoption, postmarketing surveillance, real-world evidence generation, and feedback-driven policy implementation.

and proper management of rare or unexpected toxicities are followed by systematic systems of reporting spontaneous adverse events. Executed well-organized risk management plans integrating regulatory control, clinical education, track-and-trace for products, and patient involvement are accountable for securing confidence in biosimilars and unlocking their potential for the provision of affordable and equal cancer therapy (Figure 15.2).

PV is more than a regulatory obligation; it is a shared responsibility among regulators, manufacturers, healthcare practitioners, and patients. By adopting effective PV and postmarketing safety monitoring systems, the healthcare system ensures that biosimilars reach their full potential, providing high-quality cancer therapies that are safe and readily available worldwide.

15.8 Future perspectives in biologics and biosimilars for oncology

Rapid changes in enabling technologies, biosimilars, and biologics are moulding the future of oncology. Each subtopic is discussed in detail below, including an overview of recent research, key challenges, and opportunities.

15.8.1 Next-generation biologics and emerging therapies

15.8.1.1 Bispecific antibodies and antibody-drug conjugates

Bispecific antibodies (bsAbs) are transforming the oncology treatment paradigm by enabling interaction with tumor-associated antigens or immune effector cells, amplifying treatments, and enabling novel therapeutic actions. By 2023, 11 bsAbs are approved in cancer with numerous others under development, including trispecifics and bsAb-drug conjugates. They can interact with T cells and natural killer cells and, by targeting signaling pathways, offer new therapeutic opportunities compared to traditional monoclonal antibodies (Klein et al., 2024). ADCs also expand the arsenal by delivering cytotoxic agents to tumor cells, thereby reducing off-target effects to a minor degree.

15.8.1.2 Cell and gene therapy

Cell-based therapies (CAR-T cells) and gene therapies have been increasingly used in hematologic and solid tumors. Newer innovations are incorporated, such as gene editing using the clustered regularly interspaced short palindromic repeats (CRISPR or CAS9) to precisely correct mutations, armored CAR-T cells to overcome tumor microenvironment resistance, and allogeneic "off-the-shelf" CAR-T products for broader access. Regulatory frameworks such as RMAT (US) and ATMP (EU) are evolving to fast-track these therapies, though issues in tumor heterogeneity, immune evasion, and ethics persist (Youssef, Fletcher, & Palmer, 2025).

15.8.1.3 RNA-based therapeutics

mRNA- and RNAi-based treatments are becoming significant pillars of oncologic therapy. mRNA vaccines and therapeutics can express tumor antigens, cytokines, or gene-editing proteins with multiple candidates in clinical trials. RNAi-based drugs (siRNA, miRNA) provide highly specific gene silencing, and recent FDA approvals signal the start of a new era in targeted, personalized cancer therapy. Nanoparticle-based delivery systems are essential for improving tumor targeting and minimizing side effects (Kara et al., 2022; Liu et al., 2023).

15.8.2 The role of artificial intelligence in biologic drug development

AI is transforming oncology drug discovery and biosimilar development by:

- Accelerating molecular modeling, compound screening, and toxicity prediction
- Optimizing manufacturing processes and quality control
- Enhancing clinical trial design, patient stratification, and real-world data analysis
- Supporting regulatory submissions and postmarket surveillance

AI-driven approaches are expected to reduce development time and costs, improve precision, and facilitate the introduction of innovative therapies (Albani et al., 2025; Bas & Duarte, 2024).

15.8.3 Global market trends and the future of biosimilars

The biosimilar industry is expanding at an exceptionally high rate worldwide due to patent expirations and low prices for cancer treatments. Payers are increasingly adopting biosimilars because they are safe, effective, and affordable, and because manufacturers are entering into agreements and the close relationships help adoption. Market growth is also affected by regulatory harmonization, supply reliability, and payer confidence (Yadav et al., 2025).

15.8.4 Policy and regulatory changes impacting biosimilar adoption

Regulatory schemes converge, but significant regional differences remain in approval, substitution, and reimbursement. To illustrate, some countries in the European Union allow pharmacist substitution, whereas in the United States and Japan, the same is not permitted. An established way to increase biosimilar adoption, including radolitinibs, in Japan is through policy incentives, such as reimbursement add-ons. However, the degree of change will vary by product and geography. Further efforts to align regulations and tailor policy interventions are necessary to achieve maximum adoption of biosimilars (Bennett et al., 2020; Renwick et al., 2016).

15.8.5 Improving access to biologics and biosimilars in low- and middle-income countries

Biosimilars play a vital role in enhancing access to cancer treatment in limited-resource environments by offering high-cost biologics at lower costs. Such hindrances include a lack of awareness, negative attitudes, and supply chain challenges. It is necessary to increase adoption and equity access to life-saving therapies through education, regulatory harmonization, and specific procurement strategies (Bachu et al., 2022; Joshi et al., 2022).

15.9 Conclusion

Biologics have taken over the oncology industry and have given us specific, softer, and more efficient ways of fighting cancer compared to the outdated chemotherapies. More innovative drugs that have saved patients from side effects, increased their survival, and overall quality of life have been developed in the last 20 years in the case of breast cancer, lymphoma, colorectal cancer, and melanoma. The patents for these novel biologics are expiring, and we are now experiencing a flood of biosimilars that are identical copies of the original and also safe, purified, and potent. These biosimilars are not sham off-label products; instead, they are subject to thorough laboratory testing, practical trials, and clinical trials, supported by robust regulatory frameworks at the FDA, EMA, WHO, and other regulatory agencies. The evidence is accumulating - the experience of life has confirmed that biosimilars are just as good as the reference drugs, and they offer clinicians and patients one additional reliable option.

Yet a few roadblocks remain. Biologics are so complex that they can also trigger immune reactions. That is why a biosimilar must be closely pharmacovigilanced and monitored throughout its life. More than that, differences in country laws regarding interchangeability, substitution, and extrapolation may cause misunderstandings. Then there is the fact that some doctors and patients still have reservations about the so-called copycat medicines, which can slow adoption. That is why it is necessary to learn throughout life and communicate actively. Finally, biosimilars have long-established safety records, yet extrapolated applications remain in development, and more real-world studies will help clarify best practices.

The upside is enormous. Biosimilars are cost-effective and highly priced, thereby reducing the economic burden on the healthcare system and expanding access for patients who were unable to afford biologics, especially in low- and middle-income countries where access has been limited. More advanced analysis software, improved manufacturing science, and simpler regulatory procedures are hastening the development of biosimilars and making it more predictable. In practice, it will mean shifting to value-based oncology: achieving the maximum utility with minimal budgets, without compromising quality.

The use of biosimilars will be fundamental for the future treatment of cancer worldwide. Products will enter the market faster, prices will be even lower, and clinicians will feel freer to prescribe them. Their impact will be increased by investment in training, a solid evidence base, and alignment with international regulations. Lastly, biosimilars might be seen as a more democratic, equitable, and sustainable future for global cancer treatment.

References

Aapro, M., Jelkmann, W., Constantinescu, S. N., & Leyland-Jones, B. (2012). Effects of erythropoietin receptors and erythropoiesis-stimulating agents on disease progression in cancer. *British Journal of Cancer, 106*(7), 1249–1258. https://doi.org/10.1038/bjc.2012.42.

Afzali, A., Furtner, D., Melsheimer, R., & Molloy, P. J. (2021). The automatic substitution of biosimilars: Definitions of interchangeability are not interchangeable. *Advances in Therapy, 38*(5), 2077–2093. https://doi.org/10.1007/s12325-021-01688-9.

Aiyegbusi, O. L., Roydhouse, J., Rivera, S. C., Kamudoni, P., Schache, P., Wilson, R., Stephens, R., & Calvert, M. (2022). Key considerations to reduce or address respondent burden in patient-reported outcome (PRO) data collection. *Nature Communications, 13*(1). https://doi.org/10.1038/s41467-022-33826-4.

Albani, F. G., Alghamdi, S. S., Almutairi, M. M., & Alqahtani, T. (2025). Artificial intelligence-driven innovations in oncology drug discovery: Transforming traditional pipelines and enhancing drug design. *Drug Design, Development and Therapy, 19*, 5685–5707. https://doi.org/10.2147/DDDT.S509769.

Alez-Martin, L., Houzé, P., Joomun, R., Mignet, N., François, Y.-N., & Gahoual, R. (2026). Biosimilarity and advanced structural characterization of monoclonal antibodies charge variants using capillary zone electrophoresis and mass spectrometry. *Talanta, 296*, 128498. https://doi.org/10.1016/j.talanta.2025.128498.

Alvarez, D. F., Wolbink, G., Cronenberger, C., Orazem, J., & Kay, J. (2020). Interchangeability of biosimilars: What level of clinical evidence is needed to support the interchangeability designation in the United States? *BioDrugs, 34*(6), 723–732. https://doi.org/10.1007/s40259-020-00446-7.

Ayoub, D., Jabs, W., Resemann, A., Evers, W., Evans, C., Main, L., Baessmann, C., Wagner-Rousset, E., Suckau, D., & Beck, A. (2013). Correct primary structure assessment and extensive glyco-profiling of cetuximab by a combination of intact, middle-up, middle-down and bottom-up ESI and MALDI mass spectrometry techniques. *mAbs, 5*(5), 699–710. https://doi.org/10.4161/mabs.25423France.

Bachu, R. D., Abou-Dahech, M., Balaji, S., Boddu, S. H. S., Amos, S., Singh, V., Babu, R. J., & Tiwari, A. K. (2022). Oncology biosimilars: New developments and future directions. *Cancer Reports, 5*(11). https://doi.org/10.1002/cnr2.1720.

Balkhi, S., Bilato, G., De Lerma Barbaro, A., Orecchia, P., Poggi, A., & Mortara, L. (2025). Efficacy of anti-cancer immune responses elicited using tumor-targeted IL-2 cytokine and its derivatives in combined preclinical therapies. *Vaccines, 13*(1). https://doi.org/10.3390/vaccines13010069.

Bas, T. G. (2025). Innovative formulation strategies for biosimilars: Trends focused on buffer-free systems, safety, regulatory alignment, and intellectual property challenges. *Pharmaceuticals, 18*(6). https://doi.org/10.3390/ph18060908.

Bas, T. G., & Duarte, V. (2024). Biosimilars in the era of artificial intelligence—International regulations and the use in oncological treatments. *Pharmaceuticals, 17*(7). https://doi.org/10.3390/ph17070925.

Bautista-Vanegas, F. E., Diaz-Guerrero, J. L., Cabezas-Soliz, I. N., Apaza-Huanca, B., Fernández, E. E. V., Auza-Santivañez, J. C., Gamboa, T. K. O., Rosales, R. R. G., Gómez, L. M. T., Carías, P., & Soliz-Mendoza, R. M. A. (2025). Bioprocess engineering: Advances in cell culture systems, reactor design, scale-up strategies, and intensification processes for the production of biological and bioactive compounds. *eVitroKhem, 4*, 149. https://doi.org/10.56294/evk2025149.

Ben-Horin, S., Vande Casteele, N., Schreiber, S., & Lakatos, P. L. (2016). Biosimilars in inflammatory bowel disease: Facts and fears of extrapolation. *Clinical Gastroenterology and Hepatology, 14*(12), 1685–1696. https://doi.org/10.1016/j.cgh.2016.05.023.

Bennett, C. L., Schoen, M. W., Hoque, S., Witherspoon, B. J., Aboulafia, D. M., Hwang, C. S., Ray, P., Yarnold, P. R., Chen, B. K., Schooley, B., Taylor, M. A., Wyatt, M. D., Hrushesky, W. J., & Tony Yang, Y. (2020). Improving oncology biosimilar launches in the EU, the USA, and Japan: An updated policy review from the Southern network on adverse reactions. *The Lancet Oncology, 21*(12), e575. https://doi.org/10.1016/s1470-2045(20)30485-x.

Bennett, C. L., Luminari, S., Nissenson, A. R., Tallman, M. S., Klinge, S. A., McWilliams, N., McKoy, J. M., Kim, B., Lyons, E. A., Trifilio, S. M., Raisch, D. W., Evens, A. M., Kuzel, T. M., Schumock, G. T., Belknap, S. M., Locatelli, F., Rossert, J. ô, & Casadevall, N. (2004). Pure Red-Cell aplasia and epoetin therapy. *New England Journal of Medicine., 351*(14), 1403–1408. https://doi.org/10.1056/nejmoa040528.

Bhattacharya, M., Chatterjee, S., Mallik, B., Sharma, A. R., & Chakraborty, C. (2022). Therapeutic role of neutralizing antibody for the treatment against SARS-CoV-2 and its emerging variants: A clinical and pre-clinical perspective. *Vaccines, 10*(10). https://doi.org/10.3390/vaccines10101612.

Biddle, M., Stylianou, P., Rekas, M., Wright, A., Sousa, J., Ruddy, D., Stefana, M. I., Kmiecik, K., Bandrowski, A., Kahn, R. A., Laflamme, C., Krockow, E. M., & Virk, H. S. (2024). Improving the integrity and reproducibility of research that uses antibodies: A technical, data sharing, behavioral and policy challenge. *mAbs, 16*(1). https://doi.org/10.1080/19420862.2024.2323706.

Bloomfield, D., D'Andrea, E., Nagar, S., & Kesselheim, A. (2022). Characteristics of clinical trials evaluating biosimilars in the treatment of cancer: A systematic review and meta-analysis. *JAMA Oncology, 8*(4), 537–545. https://doi.org/10.1001/jamaoncol.2021.7230.

Bommareddy, P. K., Shettigar, M., & Kaufman, H. L. (2018). Integrating oncolytic viruses in combination cancer immunotherapy. *Nature Reviews Immunology, 18*(8), 498–513. https://doi.org/10.1038/s41577-018-0014-6.

Bond, T. C., Szabo, E., Gabriel, S., Klastersky, J., Tomey, O., Udo, M., Schwartzberg, L., & Tang, B. (2018). Meta-analysis and indirect treatment comparison of lipegfilgrastim with pegfilgrastim and filgrastim for the reduction of chemotherapy-induced neutropenia-related events. *Journal of Oncology Pharmacy Practice, 24*(6), 412–423. https://doi.org/10.1177/1078155217714859.

Boni, Y. T., Vaitla, J., & Davies, H. M. L. (2023). Catalyst controlled site- and stereoselective Rhodium(II) Carbene C(sp3)-H functionalization of allyl boronates. *Organic Letters, 25*(1), 5–10. https://doi.org/10.1021/acs.orglett.2c03335.

Borišek, R., Mischo, A. é, & Foierl, T. (2024). In-use physicochemical stability of sandoz rituximab biosimilar in 0.9% sodium chloride solution after prolonged storage at room temperature conditions. *Drugs in R&D, 24*(4), 553–561. https://doi.org/10.1007/s40268-024-00496-0.

Borišek, R., Mischo, A. é, & Šmid, I. (2022). Study of the stability of sandoz rituximab biosimilar Rixathon®/Riximyo® when subjected for up to 21 Days to ambient storage. *Drugs in R&D, 22*(3), 225–234. https://doi.org/10.1007/s40268-022-00393-4.

Boven, K., Knight, J., Bader, F., Rossert, J., Eckardt, K. U., & Casadevall, N. (2005). Epoetin-associated pure red cell aplasia in patients with chronic kidney disease: Solving the mystery. *Nephrology Dialysis Transplantation, 20*(3), iii33. https://doi.org/10.1093/ndt/gfh1072.

Braun, J. ü, & Kudrin, A. (2016). Switching to biosimilar infliximab (CT-P13): Evidence of clinical safety, effectiveness, and impact on public health. *Biologicals, 44*(4), 257–266. https://doi.org/10.1016/j.biologicals.2016.03.006.

Broer, L. N., Knapen, D. G., de Groot, D. J. A., Mol, P. G. M., Kosterink, J. G. W., de Vries, E. G. E., & Lub-de Hooge, M. N. (2024). Monoclonal antibody biosimilars for cancer treatment. *iScience., 27*(6). https://doi.org/10.1016/j.isci.2024.110115.

Bukowski, Karol, Kciuk, Mateusz, & Kontek, Renata (2020). Mechanisms of Multidrug Resistance in Cancer Chemotherapy. *International Journal of Molecular Sciences, 21*(9), 3233. https://doi.org/10.3390/ijms21093233.

Burgess, L. H., Kramer, J., Castelein, C., Parra, J. M., Timmons, V., Pickens, S., Fraker, S., & Skinner, C. C. (2021). Pharmacy-led medication reconciliation program reduces adverse drug events and improves satisfaction in a community hospital. *HCA Healthcare Journal of Medicine, 2*(6). https://doi.org/10.36518/2689-0216.1295.

Butler, D., Vucic, K., Straus, S., Cupelli, A., Micallef, B., Serracino-Inglott, A., & Borg, J.-J. (2021). Regulatory experience of handling risk management plans (RMPs) for medicinal products in the EU. *Expert Opinion on Drug Safety, 20*(7), 815–826. https://doi.org/10.1080/14740338.2021.1909569.

Casadevall, N., Nataf, J., Viron, B. é, Kolta, A., Kiladjian, J.-J., Martin-Dupont, P., Michaud, P., Papo, T., Ugo, V. é, Teyssandier, I. rè, Varet, B., & Mayeux, P. (2002). Pure Red-Cell aplasia and antierythropoietin antibodies in patients treated with recombinant erythropoietin. *New England Journal of Medicine, 346*(7), 469–475. https://doi.org/10.1056/nejmoa011931.

Castel, J. érô, Delaux, S., Hernandez-Alba, O., & Cianférani, S. (2023). Recent advances in structural mass spectrometry methods in the context of biosimilarity assessment: From sequence heterogeneities to higher order structures. *Journal of Pharmaceutical and Biomedical Analysis, 236*, 115696. https://doi.org/10.1016/j.jpba.2023.115696.

Cerutti, M. íaL., Pesce, A. ía, Bès, C. é, & Seigelchifer, M. (2019). Physicochemical and biological characterization of RTXM83, a new rituximab biosimilar. *BioDrugs, 33*(3), 307–319. https://doi.org/10.1007/s40259-019-00349-2.

Cesana, B. M., & Biganzoli, E. M. (2018). Phase IV studies: Some insights, clarifications, and issues. *Current Clinical Pharmacology, 13*(1), 14–20. https://doi.org/10.2174/1574884713666180412152949.

Chekka, L. M. S., Samarth, D. P., Guo, Y., Mohamed, E. G., Decker, E., Matta, M. K., Sun, Q., Wheeler, W., Sanabria, C., Wommack, J., Gogain, J., Schrieber, S. J., Florian, J., Wang, Y.-M., Strauss, D. G., & Hyland, P. L. (2025). Characterization of proteomic pharmacodynamic biomarkers of IFNβ -1a biologics to inform potential utility in biosimilar development. *Clinical Pharmacology & Therapeutics, 118*(4), 935–945. https://doi.org/10.1002/cpt.3754.

Cortés, J., Kim, S. B., Chung, W. P., Im, S. A., Park, Y. H., Hegg, R., Kim, M. H., Tseng, L. M., Petry, V., Chung, C. F., Iwata, H., Hamilton, E., Curigliano, G., Xu, B., Huang, C. S., Kim, J. H., Chiu, J. W. Y., Pedrini, J. L., Lee, C., … Hurvitz, S. A. (2022). Trastuzumab deruxtecan versus trastuzumab emtansine for breast cancer. *New England Journal of Medicine., 386*(12), 1142–1154. https://doi.org/10.1056/NEJMoa2115022.

Dabkowska, A., Domka, K., & Firczuk, M. (2024). Advancements in cancer immunotherapies targeting CD20: From pioneering monoclonal antibodies to chimeric antigen receptor-modified T cells. *Frontiers in Immunology, 15*. https://doi.org/10.3389/fimmu.2024.1363102.

Dakshinamurthy, P. K., & Gothandam, K. M. (2022). Comprehensive analytical characterization of the proposed biosimilar trastuzumab. *Indian Journal of Pharmaceutical Education and Research, 56*(2), 438–447. https://doi.org/10.5530/ijper.56.2.64.

Dang, A. (2023). Real-world evidence: A primer. *Pharmaceutical Medicine, 37*(1), 25–36. https://doi.org/10.1007/s40290-022-00456-6.

Deza, D. C., Domínguez, C. L., Oliver, A. P., Cuadro, C. P., Caballero, C. Y., Michelena, L. A., Esteban, A. R., Gracia, A. C., Lidón, R. V., Roche, P. C., Usán, E. C., Sirbu, V. S., Ruiz, C. P., & López, S. G. (2025). The patients included in clinical trials of biological drugs for inflammatory bowel disease do not represent the real-world population. *American Journal of Gastroenterology*. https://doi.org/10.14309/ajg.0000000000003670.

Duivelshof, B. L., Jiskoot, W., Beck, A., Veuthey, J.-L., Guillarme, D., & D'Atri, V. (2019). Glycosylation of biosimilars: Recent advances in analytical characterization and clinical implications. *Analytica Chimica Acta, 1089*, 1–18. https://doi.org/10.1016/j.aca.2019.08.044.

Dumontet, C., Reichert, J. M., Senter, P. D., Lambert, J. M., & Beck, A. (2023). Antibody–drug conjugates come of age in oncology. *Nature Reviews Drug Discovery, 22*(8), 641–661. https://doi.org/10.1038/s41573-023-00709-2.

Elsayed, A., Jaber, N., Al-Remawi, M., & Abu-Salah, K. (2023). From cell factories to patients: Stability challenges in biopharmaceuticals manufacturing and administration with mitigation strategies. *International Journal of Pharmaceutics, 645*, 123360. https://doi.org/10.1016/j.ijpharm.2023.123360.

Evans, R. S. (2016). Electronic health records: Then, now, and in the future. *Yearbook of medical informatics*, S48. https://doi.org/10.15265/IYS-2016-s006.

Fagerberg, J. (1996). Granulocyte-macrophage colony-stimulating factor as an adjuvant in tumor immunotherapy. *Medical Oncology, 13*(3), 155–160.

Fan, T., Zhang, M., Yang, J., Zhu, Z., Cao, W., & Dong, C. (2023). Therapeutic cancer vaccines: Advancements, challenges, and prospects. *Signal Transduction and Targeted Therapy, 8*(1). https://doi.org/10.1038/s41392-023-01674-3.

Fan, W., & Qiao, P. (2011). Vibration-based damage identification methods: A review and comparative study. *Structural Health Monitoring, 10*(1), 83–111. https://doi.org/10.1177/1475921710365419.

Farid, S. S., Baron, M., Stamatis, C., Nie, W., & Coffman, J. (2020). Benchmarking biopharmaceutical process development and manufacturing cost contributions to R&D. *mAbs, 12*(1). https://doi.org/10.1080/19420862.2020.1754999.

Feagan, B. G., Choquette, D., Ghosh, S., Gladman, D. D., Ho, V., Meibohm, B., Zou, G., Xu, Z., Shankar, G., Sealey, D. C., & Russell, A. S. (2014). The challenge of indication extrapolation for infliximab biosimilars. *Biologicals, 42*(4), 177–183. https://doi.org/10.1016/j.biologicals.2014.05.005.

Gan, Y., Lippold, S., Stobaugh, J., Schöneich, C., & Yang, F. (2024). Expanding the structural resolution of glycosylation microheterogeneity in therapeutic proteins by salt-free hydrophilic interaction liquid chromatography tandem mass spectrometry. *mAbs, 16*(1). https://doi.org/10.1080/19420862.2024.2395503.

Garcia, J., Hurwitz, H. I., Sandler, A. B., Miles, D., Coleman, R. L., Deurloo, R., & Chinot, O. L. (2020). Bevacizumab (Avastin®) in cancer treatment: A review of 15 years of clinical experience and future outlook. *Cancer Treatment Reviews, 86*. https://doi.org/10.1016/j.ctrv.2020.102017.

García-Ortega, M. B., Aparicio, E., Griñán-Lisón, C., Jiménez, G., López-Ruiz, E., Palacios, J. L., Ruiz-Alcalá, G., Alba, C., Martínez, A., Boulaiz, H., Perán, M., Hackenberg, M., Bragança, J., Calado, S. M., Marchal, J. A., & García, M. Á. (2023). Interferon-alpha decreases cancer stem cell properties and modulates exosomes in malignant melanoma. *Cancers, 15*(14). https://doi.org/10.3390/cancers15143666.

Garvey, M. (2022). Non-mammalian eukaryotic expression systems yeast and fungi in the production of biologics. *Journal of Fungi, 8*(11), 1179. https://doi.org/10.3390/jof8111179.

Gauckler, P., Shin, J. I., Alberici, F., Audard, V., Bruchfeld, A., Busch, M., Cheung, C. K., Crnogorac, M., Delbarba, E., Eller, K., Faguer, S., Galesic, K., Griffin, S., van den Hoogen, M. W. F., Hrušková, Z., Jeyabalan, A., Karras, A., King, C., Kohli, H. S., … Kronbichler, A. (2021). Rituximab in membranous nephropathy. *Kidney International Reports, 6*(4), 881–893. https://doi.org/10.1016/j.ekir.2020.12.035.

Gaylis, N., Both, C., Lemke, L., Richter, O., & Yamauchi, P. (2024). 'Totality of evidence' approach in the development of GP2017, an approved adalimumab biosimilar. *Advances in Therapy, 41*(5), 1795–1814. https://doi.org/10.1007/s12325-024-02809-w.

Gherghescu, I., & Delgado-Charro, M. B. ña (2021). The biosimilar landscape: An overview of regulatory approvals by the EMA and FDA. *Pharmaceutics, 13*(1), 48. https://doi.org/10.3390/pharmaceutics13010048.

Gilmore, J. W., Indurlal, P., Rainey, A. M., Sabo-Wagner, S., Rahman, A., & Swart, S. (2025). Cost-effective cancer care: The role of oncology biosimilars in generating cost savings.

Journal of Clinical Oncology, 43(16_suppl), 11033. https://doi.org/10.1200/jco.2025.43.16_suppl.11033.

Giusti, F., Martos, C., Trama, A., Bettio, M., Sanvisens, A., Audisio, R., Arndt, V., Francisci, S., Dochez, C., Ribes, J., Fernández, L. P., Gavin, A., Gatta, G., Marcos-Gragera, R., Lievens, Y., Allemani, C., De Angelis, R., Visser, O., & Eycken, L. V. (2023). Cancer treatment data available in European cancer registries: Where are we and where are we going? *Frontiers in Oncology, 13*. https://doi.org/10.3389/fonc.2023.1109978.

Goydel, R. S., & Rader, C. (2021). Antibody-based cancer therapy. *Oncogene, 40*(21), 3655–3664. https://doi.org/10.1038/s41388-021-01811-8.

Grassi, L., Harris, C., Zhu, J., Hatton, D., & Dunn, S. (2025). Next-generation sequencing: A powerful multi-purpose tool in cell line development for biologics production. *Computational and Structural Biotechnology Journal, 27*, 1511–1517. https://doi.org/10.1016/j.csbj.2025.04.006.

Gyawali, B., de Vries, E. G. E., Dafni, U., Amaral, T., Barriuso, J., Bogaerts, J., Calles, A., Curigliano, G., Gomez-Roca, C., Kiesewetter, B., Oosting, S., Passaro, A., Pentheroudakis, G., Piccart, M., Roitberg, F., Tabernero, J., Tarazona, N., Trapani, D., Wester, R., … Cherny, N. I. (2021). Biases in study design, implementation, and data analysis that distort the appraisal of clinical benefit and ESMO-magnitude of clinical benefit scale (ESMO-MCBS) scoring. *ESMO Open, 6*(3), 100117. https://doi.org/10.1016/j.esmoop.2021.100117.

Hayashi, H., & Nakagawa, K. (2020). Combination therapy with PD-1 or PD-L1 inhibitors for cancer. *International Journal of Clinical Oncology, 25*(5), 818–830. https://doi.org/10.1007/s10147-019-01548-1.

Hemmelmann, C., Galta, R. E., Wang, J., Schmitt, S., & Arani, R. (2025). Implementation of the anchor-based indirect comparison method for equivalence margin derivation in biosimilar development. *Pharmaceuticals, 18*(3), 285. https://doi.org/10.3390/ph18030285.

Herndon, T. M., Ausin, C., Brahme, N. N., Schrieber, S. J., Luo, M., Andrada, F. C., Kim, C., Sun, W., Zhou, L., Grosser, S., Yim, S., & Ricci, M. S. (2023). Safety outcomes when switching between biosimilars and reference biologics: A systematic review and meta-analysis. *PLoS ONE, 18*(10). https://doi.org/10.1371/journal.pone.0292231.

Hu, X., Xiang, Y., Zhang, X., Sun, Y., Li, Y. i, Wang, L., Lv, P., Long, Z., Liang, C., & Li, J. (2025). Analysis strategy for identifying the O-Linked glycan profile and O-glycosylation sites on recombinant human follicle stimulating hormone-c-terminal peptide (rhFSH-CTP). *Molecules, 30*(10), 2141. https://doi.org/10.3390/molecules30102141.

Hübel, K., Kron, F., & Patrick Lux, M. (2020). Biosimilars in oncology: Effects on economy and therapeutic innovations. *European Journal of Cancer, 139*, 10–19. https://doi.org/10.1016/j.ejca.2020.07.037.

Hung, A., Vu, Q., & Mostovoy, L. (2017). A systematic review of U.S. biosimilar approvals: What evidence does the FDA require and how are manufacturers responding? *Journal of Managed Care & Specialty Pharmacy, 23*(12), 1234–1244. https://doi.org/10.18553/jmcp.2017.23.12.1234.

Hydery, T., & Reddy, V. (2024). A primer on formulary structures and strategies. *Journal of Managed Care and Specialty Pharmacy, 30*(2), 206–210. https://doi.org/10.18553/jmcp.2024.30.2.206.

Iskit, A. B. (2025). Biosimilars and interchangeability: Regulatory, scientific, and global perspectives. *European Journal of Pharmaceutical Sciences, 213*, 107224. https://doi.org/10.1016/j.ejps.2025.107224.

Janghorban, M., Kazemi, S., Tormon, R., Ngaju, P., & Pandey, R. (2023). Methods and analysis of biological contaminants in the biomanufacturing industry. *Chemosensors, 11*(5), 298. https://doi.org/10.3390/chemosensors11050298.

Jeka, S. ł, Dokoupilová, E., Kivitz, A., Żuchowski, P. ł, Vogg, B., Krivtsova, N., Sekhar, S., Banerjee, S., Schwebig, A., Poetzl, J., Body, J.-J., & Eastell, R. (2024). Equivalence trial of proposed denosumab biosimilar GP2411 and reference denosumab in postmenopausal osteoporosis: The ROSALIA study. *Journal of Bone and Mineral Research, 39*(3), 202–210. https://doi.org/10.1093/jbmr/zjae016.

Jin, S., Sun, Y., Liang, X., Gu, X., Ning, J., Xu, Y., Chen, S., & Pan, L. (2022). Emerging new therapeutic antibody derivatives for cancer treatment. *Signal Transduction and Targeted Therapy, 7*(1). https://doi.org/10.1038/s41392-021-00868-x.

Jones, T. D., Carter, P. J., Plückthun, A., Vásquez, M., Holgate, R. G. E., Hötzel, I., Popplewell, A. G., Parren, P. W. H. I., Enzelberger, M., Rademaker, H. J., Clark, M. R., Lowe, D. C., Dahiyat, B. I., Smith, V., Lambert, J. M., Wu, H., Reilly, M., Haurum, J. S., Dübel, S., … Baker, M. P. (2016). The INNs and outs of antibody nonproprietary names. *mAbs, 8*(1), 1–9. https://doi.org/10.1080/19420862.2015.1114320.

Joshi, D., Khursheed, R., Gupta, S., Wadhwa, D., Singh, T. G., Sharma, S., Porwal, S., Gauniyal, S., Vishwas, S., Goyal, S., Gupta, G., Eri, R. D., Williams, K. A., Dua, K., & Singh, S. K. (2022). Biosimilars in oncology: Latest trends and regulatory status. *Pharmaceutics., 14*(12). https://doi.org/10.3390/pharmaceutics14122721.

Kaczmarek, M., Poznańska, J., Fechner, F., Michalska, N., Paszkowska, S., Napierała, A., & Mackiewicz, A. (2023). Cancer vaccine therapeutics: Limitations and effectiveness—A literature review. *Cells, 12*(17), 2159. https://doi.org/10.3390/cells12172159.

Kaiser, P. K., Yun, J., Kim, S., Kim, J., & Park, S. J. (2023). Physicochemical and biological stability assessment of SB11 (Ranibizumab Biosimilar) under ambient and in-use storage for intravitreal administration. *Ophthalmology and Therapy, 12*(2), 985–998. https://doi.org/10.1007/s40123-022-00645-1.

Kakemam, E., Gharaee, H., Rajabi, M. R., Nadernejad, M., Khakdel, Z., Raeissi, P., & Kalhor, R. (2021). Nurses' perception of patient safety culture and its relationship with adverse events: A national questionnaire survey in Iran. *BMC Nursing, 20*(1). https://doi.org/10.1186/s12912-021-00571-w.

Kandemir, E. A. (2023). Review of novel and supplemental approvals of the targeted cancer drugs by the Food and Drug Administration in 2021. *Journal of Oncology Pharmacy Practice, 29*(1), 191–207. https://doi.org/10.1177/10781552221112015.

Kara, G., Calin, G. A., & Ozpolat, B. (2022). RNAi-based therapeutics and tumor targeted delivery in cancer. *Advanced Drug Delivery Reviews, 182*, 114113. https://doi.org/10.1016/j.addr.2022.114113.

Khosla, S., White, R., Medina, J. ús, Ouwens, M., Emmas, C., Koder, T., Male, G., & Leonard, S. (2018). Real world evidence (RWE)—A disruptive innovation or the quiet evolution of medical evidence generation? *F1000Research, 7*, 111. https://doi.org/10.12688/f1000research.13585.1.

Kim, J. J., Pritts, J. D., Ngo, M., Estoll, C. R., & Rao, V. A. (2023). Trends in light and temperature sensitivity recommendations among licensed biotechnology drug products. *Pharmaceutical Research, 40*(6), 1491–1505. https://doi.org/10.1007/s11095-023-03494-0.

Kim, S., Kim, S., & Lee, H. (2020). A critical review of the United States regulatory pathways for determining the equivalence of efficacy between CT-P13 and original infliximab (Remicade®). *Drug Design, Development and Therapy, 14*, 2831–2840. https://doi.org/10.2147/DDDT.S254776.

Kim, S., Song, J., Park, S., Ham, S., Paek, K., Kang, M., Chae, Y., Seo, H., Kim, H.-C., & Flores, M. (2017). Drifts in ADCC-related quality attributes of Herceptin®: Impact on development of a trastuzumab biosimilar. *mAbs, 9*(4), 704–714. https://doi.org/10.1080/19420862.2017.1305530.

Klein, C., Brinkmann, U., Reichert, J. M., & Kontermann, R. E. (2024). The present and future of bispecific antibodies for cancer therapy. *Nature Reviews Drug Discovery, 23*(4), 301–319. https://doi.org/10.1038/s41573-024-00896-6.

Konstantinidou, S., Papaspiliou, A., & Kokkotou, E. (2020). Current and future roles of biosimilars in oncology practice (Review). *Oncology Letters, 19*(1), 45–51. https://doi.org/10.3892/ol.2019.11105.

Kumar, S., & Baldi, A. (2013). Pharmacovigilance in India: Perspectives and prospects. *Journal of Drug Delivery and Therapeutics, 3*(4). https://doi.org/10.22270/jddt.v3i4.619.

Lauret, A., Moltó, A., Abitbol, V., Gutermann, L., Conort, O., Chast, F. ç, Goulvestre, C., Jeunne, C. L. e, Chaussade, S., Roux, C., Batteux, F. rédé, Dougados, M., Allanore, Y., & Avouac, J. érô (2020). Effects of successive switches to different biosimilars infliximab on immunogenicity in chronic inflammatory diseases in daily clinical practice. *Seminars in Arthritis and Rheumatism, 50*(6), 1449–1456. https://doi.org/10.1016/j.semarthrit.2020.02.007.

Lawrence, E. R., & Fraser, D. J. (2020). Latitudinal biodiversity gradients at three levels: Linking species richness, population richness and genetic diversity. *Global Ecology and Biogeography, 29*(5), 770–788. https://doi.org/10.1111/geb.13075.

Liu, B., Zhou, H., Tan, L., Siu, K. T. H., & Guan, X. Y. (2024). Exploring treatment options in cancer: Tumor treatment strategies. *Signal Transduction and Targeted Therapy, 9*(1). https://doi.org/10.1038/s41392-024-01856-7.

Liu, C., Shi, Q., Huang, X., Koo, S., Kong, N. a, & Tao, W. (2023). mRNA-based cancer therapeutics. *Nature Reviews Cancer, 23*(8), 526–543. https://doi.org/10.1038/s41568-023-00586-2.

Liu, P., Chen, G., & Zhang, J. (2022). A review of liposomes as a drug delivery system: Current status of approved products, regulatory environments, and future perspectives. *Molecules, 27*(4), 1372. https://doi.org/10.3390/molecules27041372.

Liu, X., Xu, X., Lu, Z., & Shih, Y. a-C. T. (2025). Assessing biosimilar entry in the market for biologic cancer drugs. *Journal of Clinical Oncology, 43*(16_suppl), 11162. https://doi.org/10.1200/jco.2025.43.16_suppl.11162.

Lo, Y. i-H., & Tzeng, Y.-D. (2024). Abstract PO1-17-03: Real-world evidence of trastuzumab biosimilars in HER2-positive breast cancer: Evaluating utilization, efficacy, and safety in Taiwan. *Cancer Research, 84*(9_Supplement). https://doi.org/10.1158/1538-7445.sabcs23-po1-17-03.

Lucas, S., Ailani, J., Smith, T. R., Abdrabboh, A., Xue, F., & Navetta, M. S. (2022). Pharmacovigilance: Reporting requirements throughout a product's lifecycle. *Therapeutic Advances in Drug Safety, 13*. https://doi.org/10.1177/20420986221125006.

Lv, Y., Cui, X., Li, T., Liu, C., Wang, A., Wang, T., Zhou, X., Li, R., Zhang, F., Hu, Y., Zhang, T., & Liu, Z. (2025). Mechanism of action and future perspectives of ADCs in combination with immune checkpoint inhibitors for solid tumors. *Clinical and Experimental Medicine, 25*(1). https://doi.org/10.1007/s10238-025-01655-6.

Maniadakis, N., Holtorf, A. P., Otávio Corrêa, J., Gialama, F., & Wijaya, K. (2018). Shaping pharmaceutical tenders for effectiveness and sustainability in countries with expanding healthcare coverage. *Applied Health Economics and Health Policy, 16*(5), 591–607. https://doi.org/10.1007/s40258-018-0405-7.

Markus, R., Liu, J., Ramchandani, M., Landa, D., Born, T., & Kaur, P. (2017). Developing the totality of evidence for biosimilars: Regulatory considerations and building confidence for the healthcare community. *BioDrugs, 31*(3), 175–187. https://doi.org/10.1007/s40259-017-0218-5.

Mascarenhas-Melo, F., Diaz, M., Gonçalves, M. B. S., Vieira, P., Bell, V., Viana, S., Nunes, S., Paiva-Santos, A. C., & Veiga, F. (2024). An overview of biosimilars—Development, quality, regulatory issues, and management in healthcare. *Pharmaceuticals, 17*(2). https://doi.org/10.3390/ph17020235.

Melosky, B., Reardon, D. A., Nixon, A. B., Subramanian, J., Bair, A. H., Jacobs, I., & Biosimilars, B. (2018). Scientific justification for extrapolation of indications. *Future Oncology, 14*(24), 2507–2520. https://doi.org/10.2217/fon-2018-0051.

Mountzios, G., & Boukovinas, I. (2013). Biosimilars: Facing the challenge. *Forum of Clinical Oncology, 4*(4), 26–28.

Mukai, K., Tsai, M., Saito, H., & Galli, S. J. (2018). Mast cells as sources of cytokines, chemokines, and growth factors. *Immunological Reviews, 282*(1), 121–150. https://doi.org/10.1111/imr.12634.

Mukherjee, A., & Bandyopadhyay, D. (2024). Targeted therapy in breast cancer: Advantages and advancements of antibody–drug conjugates, a type of chemo-biologic hybrid drugs. *Cancers, 16*(20). https://doi.org/10.3390/cancers16203517.

Mülhopt, S., Diabaté, S., Dilger, M., Adelhelm, C., Anderlohr, C., Bergfeldt, T., Gómez de la Torre, J., Jiang, Y., Valsami-Jones, E., Langevin, D., Lynch, I., Mahon, E., Nelissen, I., Piella, J., Puntes, V., Ray, S., Schneider, R., Wilkins, T., Weiss, C., & Paur, H.-R. (2018). Characterization of nanoparticle batch-to-batch variability. *Nanomaterials., 8*(5), 311. https://doi.org/10.3390/nano8050311.

Murthy, R. K., Loi, S., Okines, A., Paplomata, E., Hamilton, E., Hurvitz, S. A., Lin, N. U., Borges, V., Abramson, V., Anders, C., Bedard, P. L., Oliveira, M., Jakobsen, E., Bachelot, T., Shachar, S. S., Muller, V., Braga, S., Duhoux, F. P., Greil, R., … Winer, E. P. (2020). Tucatinib, trastuzumab, and capecitabine for HER2-positive metastatic breast cancer. *New England Journal of Medicine, 382*(7), 597–609. https://doi.org/10.1056/NEJMoa1914609.

Mysler, F., Azevedo, V. F., Danese, S., Alvarez, D., Iikuni, N., Ingram, B., Mueller, M., & Peyrin-Biroulet, L. (2021). Biosimilar-to-biosimilar switching: What is the rationale and current experience? *Drugs, 81*(16), 1859–1879. https://doi.org/10.1007/s40265-021-01610-1.

Nagase, H., Suzukawa, M., Oishi, K., & Matsunaga, K. (2023). Biologics for severe asthma: The real-world evidence, effectiveness of switching, and prediction factors for the efficacy. *Allergology International, 72*(1), 11–23. https://doi.org/10.1016/j.alit.2022.11.008.

Nahleh, Z., Lyman, G. H., Schilsky, R. L., Peterson, D. E., Tagawa, S. T., Chavez-MacGregor, M., Rumble, R. B., & Gupta, S. (2022). Use of biosimilar medications in oncology. *JCO Oncology Practice, 18*(3), 177–186. https://doi.org/10.1200/OP.21.00771.

Niazi, S. K. (2025). Continuous manufacturing of recombinant drugs: comprehensive analysis of cost reduction strategies, regulatory pathways, and global implementation. *Pharmaceuticals, 18*(8). https://doi.org/10.3390/ph18081157.

Nikitina, V., Santi Laurini, G., Montanaro, N., & Motola, D. (2023). Comparative safety profiles of oncology biosimilars vs. originators in Europe: An analysis of the eudravigilance database. *Cancers, 15*(14), 3680. https://doi.org/10.3390/cancers15143680.

Nupur, N., Joshi, S., Gulliarme, D., & Rathore, A. S. (2022). Analytical similarity assessment of biosimilars: Global regulatory landscape, recent studies and major advancements in orthogonal platforms. *Frontiers in Bioengineering and Biotechnology, 10*. https://doi.org/10.3389/fbioe.2022.832059.

Oflazoglu, E., & Audoly, L. P. (2010). Evolution of anti-CD20 monoclonal antibody therapeutics in oncology. *mAbs, 2*(1), 14–19. https://doi.org/10.4161/mabs.2.1.10789.

Ogura, M., Coiffier, B., Kwon, H. C., & Yoon, S. W. (2017). Scientific rationale for extrapolation of biosimilar data across cancer indications: Case study of CT-P10. *Future Oncology, 13*(15s), 45–53. https://doi.org/10.2217/fon-2017-0156.

Otremba, B., Borchardt, J., Kuske, A., Hollnagel-Schmitz, M., & Losch, F. O. (2020). Real-world use and acceptance of rituximab biosimilars in non-Hodgkin lymphoma in an oncologist network in Germany. *Future Oncology, 16*(15), 1001–1012. https://doi.org/10.2217/fon-2020-0180.

Papież, M. A., & Krzyściak, W. (2021). Biological therapies in the treatment of cancer—Update and new directions. *International Journal of Molecular Sciences, 22*(21), 11694. https://doi.org/10.3390/ijms222111694.

Papp, K., Balser, S., Nopora, K., Rewerski, P., Freudensprung, B., Trieb, M., & Randomised, A. (2025). Double-blind trial to compare the efficacy, safety, and immunogenicity of the biosimilar ustekinumab FYB202 with reference ustekinumab in patients with moderate-to-severe plaque psoriasis. *Advances in Therapy, 42*(5), 2135–2149. https://doi.org/10.1007/s12325-025-03138-2.

Park, D., Kim, J., Yun, J., & Park, S. J. (2020). Evaluation of the physico-chemical and biological stability of SB8 (Aybintio), a proposed biosimilar to bevacizumab, under ambient and in-use conditions. *Advances in Therapy, 37*(10), 4308–4324. https://doi.org/10.1007/s12325-020-01465-0.

Patel, N., Prajapati, H., Joshi, N., Movaliya, V., Kanaki, N., Deshpande, S., & Zaveri, M. (2024). Regulatory requirements for approval and registration procedure of biosimilar in US and European Union (EU). *International Journal of Drug Regulatory Affairs, 12*(1), 70–82. https://doi.org/10.22270/ijdra.v12i1.655.

Paul, S., & Sarkar, A. (2024). Pharmaceutical and clinical challenges of biological medicines: Ongoing hurdles from drug development to therapeutic applications. *Asian Journal of Pharmaceutical and Clinical Research, 17*(8), 9–21. https://doi.org/10.22159/ajpcr.2024v17i8.51398.

Pawłowska, I., Pawłowski, L., Krzyżaniak, N., & Kocić, I. (2019). Perspectives of hospital pharmacists towards biosimilar medicines: A survey of polish pharmacy practice in general hospitals. *BioDrugs, 33*(2), 183–191. https://doi.org/10.1007/s40259-019-00341-w.

Peeters, M., Planchard, D., Pegram, M., Gonçalves, J. oão, Bocquet, F. ç, & Jang, H. (2021). Biosimilars in an era of rising oncology treatment options. *Future Oncology, 17*(29), 3881–3892. https://doi.org/10.2217/fon-2021-0546.

Pham, N. B., & Meng, W. S. (2020). Protein aggregation and immunogenicity of biotherapeutics. *International Journal of Pharmaceutics, 585*, 119523. https://doi.org/10.1016/j.ijpharm.2020.119523.

Pierpont, T. M., Limper, C. B., & Richards, K. L. (2018). Past, present, and future of rituximab—The world's first oncology monoclonal antibody therapy. *Frontiers in Oncology, 8*. https://doi.org/10.3389/fonc.2018.00163.

Preminger, M. W., & Mittal, S. (2017). Leadless pacing meets the real world: The maturation of clinical evidence behind a miniaturized pacemaker. *Heart Rhythm, 14*(9), 1380–1381. https://doi.org/10.1016/j.hrthm.2017.06.008.

Pronzato, P., & Rondini, M. (2006). First line chemotherapy of metastatic breast cancer. *Annals of Oncology, 17*(5), v165. https://doi.org/10.1093/annonc/mdj974.

Pujahari, S. R., Engla, S., Soni, R., Patra, S., Hanawal, M. K., & Kumar, A. (2025). Structural similarity of biological drugs using statistical signal processing and nuclear magnetic resonance spectral pattern analysis. *Molecular Pharmaceutics, 22*(5), 2684–2693. https://doi.org/10.1021/acs.molpharmaceut.5c00108.

Rakha, E. A., Gandhi, N., Climent, F., Van Deurzen, C. H. M., Haider, S. A., Dunk, L., Lee, A. H. S., Macmillan, D., & Ellis, I. O. (2011). Encapsulated papillary carcinoma of the breast: An invasive tumor with excellent prognosis. *American Journal of Surgical Pathology, 35*(8), 1093–1103. https://doi.org/10.1097/PAS.0b013e31821b3f65.

Ratih, R., Asmari, M., Abdel-Megied, A. M., Elbarbry, F., & El Deeb, S. (2021). Biosimilars: Review of regulatory, manufacturing, analytical aspects and beyond. *Microchemical Journal, 165*. https://doi.org/10.1016/j.microc.2021.106143.

Reinisch, W., Louis, E., & Danese, S. (2015). The scientific and regulatory rationale for indication extrapolation: A case study based on the infliximab biosimilar CT-P13. *Expert Review of Gastroenterology & Hepatology, 9*(sup1), 17–26. https://doi.org/10.1586/17474124.2015.1091306.

Renwick, M. J., Smolina, K., Gladstone, E. J., Weymann, D., & Morgan, S. G. (2016). Postmarket policy considerations for biosimilar oncology drugs. *The Lancet Oncology, 17*(1), e31. https://doi.org/10.1016/S1470-2045(15)00381-2.

Ribelles, N., Alvarez-Lopez, I., Arcusa, A., Chacon, J. I., de la Haba, J., García-Corbacho, J., Garcia-Mata, J., Jara, C., Jerez, J. M., Lázaro-Quintela, M., Leon-Mateos, L., Ramirez-Merino, N., Tibau, A., & Garcia-Palomo, A. (2021). Electronic health records and patient registries in medical oncology departments in Spain. *Clinical and Translational Oncology, 23*(10), 2099–2108. https://doi.org/10.1007/s12094-021-02614-9.

Riccardi, F., Bo, M. D., Macor, P., & Toffoli, G. (2023). A comprehensive overview on antibody-drug conjugates: From the conceptualization to cancer therapy. *Frontiers in Pharmacology, 14*. https://doi.org/10.3389/fphar.2023.1274088.

Robinson, J. C., & Whaley, C. M. (2025). Site-neutral payment and biosimilars competition are complementary purchaser strategies for cancer biologics. *Health Affairs, 44*(6), 677–683. https://doi.org/10.1377/hlthaff.2024.01549.

Roediger, A., Freischem, B., & Reiland, J. B. (2017). What pricing and reimbursement policies to use for off-patent biologicals in Europe? Results from the second EBE biological medicines policy survey. *GaBI Journal, 6*(2), 61–78. https://doi.org/10.5639/gabij.2017.0602.014.

Rugo, H. S., Linton, K. M., Cervi, P., Rosenberg, J. A., & Jacobs, I. (2016). A clinician's guide to biosimilars in oncology. *Cancer Treatment Reviews, 46*, 73–79. https://doi.org/10.1016/j.ctrv.2016.04.003.

Russell, M. D., & Galloway, J. B. (2024). Driving down the cost of biologics: Lessons from a nationalised health-care system. *The Lancet, 404*(10464), 1723–1724. https://doi.org/10.1016/S0140-6736(24)02151-2.

Sarmento-Ribeiro, A. B., Scorilas, A., Gonçalves, A. C., Efferth, T., & Trougakos, I. P. (2019). The emergence of drug resistance to targeted cancer therapies: Clinical evidence. *Drug Resistance Updates, 47*, 100646. https://doi.org/10.1016/j.drup.2019.100646.

Schad, F., & Thronicke, A. (2022). Real-world evidence—Current developments and perspectives. *International Journal of Environmental Research and Public Health, 19*(16), 10159. https://doi.org/10.3390/ijerph191610159.

Schiestl, M., Ranganna, G., Watson, K., Jung, B., Roth, K., Capsius, B. jö, Trieb, M., Bias, P., & Maréchal-Jamil, J. (2020). The path towards a tailored clinical biosimilar development. *BioDrugs, 34*(3), 297–306. https://doi.org/10.1007/s40259-020-00422-1.

Schmitz, R. J., Schultz, M. D., Urich, M. A., Nery, J. R., Pelizzola, M., Libiger, O., Alix, A., McCosh, R. B., Chen, H., Schork, N. J., & Ecker, J. R. (2013). Patterns of population epigenomic diversity. *Nature, 495*(7440), 193–198. https://doi.org/10.1038/nature11968.

Seelig, A. (2020). P-Glycoprotein: One mechanism, many tasks and the consequences for pharmacotherapy of cancers. *Frontiers in Oncology, 10*. https://doi.org/10.3389/fonc.2020.576559.

Sharma, V., Mottafegh, A., Joo, J. U., Kang, J. H., Wang, L., & Kim, D. P. (2024). Toward microfluidic continuous-flow and intelligent downstream processing of biopharmaceuticals. *Lab on a Chip, 24*(11), 2861–2882. https://doi.org/10.1039/d3lc01097j.

Shiravand, Y., Khodadadi, F., Kashani, S. M. A., Hosseini-Fard, S. R., Hosseini, S., Sadeghirad, H., Ladwa, R., O'byrne, K., & Kulasinghe, A. (2022). Immune checkpoint inhibitors in cancer therapy. *Current Oncology, 29*(5), 3044–3060. https://doi.org/10.3390/curroncol29050247.

Stebbing, J., Mainwaring, P. N., Curigliano, G., Pegram, M., Latymer, M., Bair, A. H., & Rugo, H. S. (2020). Understanding the role of comparative clinical studies in the development of oncology biosimilars. *Journal of Clinical Oncology, 38*(10), 1070–1080. https://doi.org/10.1200/jco.19.02953.

Strand, V., Gonçalves, J., Hickling, T. P., Jones, H. E., Marshall, L., & Isaacs, J. D. (2020). Immunogenicity of biosimilars for rheumatic diseases, plaque psoriasis, and inflammatory bowel disease: A review from clinical trials and regulatory documents. *BioDrugs, 34*(1), 27–37. https://doi.org/10.1007/s40259-019-00394-x.

Strauss, D. G., Wang, Y.-M., Florian, J., & Zineh, I. (2023). Pharmacodynamic biomarkers evidentiary considerations for biosimilar development and approval. *Clinical Pharmacology & Therapeutics, 113*(1), 55–61. https://doi.org/10.1002/cpt.2761.

Tak, M., Jeong, H., Yun, J., Kim, J., Kim, S., Lee, Y., & Park, S. J. (2023). In-use stability of SB12 (eculizumab, soliris biosimilar) diluted in saline and dextrose infusion solution after an extended storage period. *Drugs in R&D, 23*(4), 363–375. https://doi.org/10.1007/s40268-023-00433-7.

Tanigawa, K., & Redmond, W. L. (2025). Current landscape and future prospects of interleukin-2 receptor (IL-2R) agonists in cancer immunotherapy. *OncoImmunology, 14*(1). https://doi.org/10.1080/2162402X.2025.2452654.

Taniguchi, H., Nishikawa, K., Haneji, T., Izawa, N., Imamura, H., & Yamaguchi, H. (2025). Prospective cohort study of trastuzumab biosimilar CT-P6 in HER2-positive gastric cancer: Japanese real-world outcomes. *Oncology and Therapy, 13*(2), 485–503. https://doi.org/10.1007/s40487-025-00341-7.

Tesser, J. R. P., Furst, D. E., & Jacobs, I. (2017). Biosimilars and the extrapolation of indications for inflammatory conditions. *Biologics: Targets and Therapy, 11*, 5–11. https://doi.org/10.2147/BTT.S124476.

Tian, Y., Xie, D., & Yang, L. i (2022). Engineering strategies to enhance oncolytic viruses in cancer immunotherapy. *Signal Transduction and Targeted Therapy, 7*(1). https://doi.org/10.1038/s41392-022-00951-x.

Tran, R., Fraser, G., Fisher, A. C., Lee, S. L., Boam, A., Tsinontides, S., Maguire, J., Yu, L. X., Rosencrance, S., Kozlowski, S., & Henry, D. (2024). A network of regulatory innovations to improve FDA quality assessments of human drug applications. *International Journal of Pharmaceutics: X, 7*. https://doi.org/10.1016/j.ijpx.2024.100239.

Tripathi, N. K., & Shrivastava, A. (2019). Recent developments in bioprocessing of recombinant proteins: Expression hosts and process development. *Frontiers in Bioengineering and Biotechnology, 7*. https://doi.org/10.3389/fbioe.2019.00420.

Vieillard, V., & Paul, M. (2023). Physicochemical stability study of a biosimilar of bevacizumab in vials and after dilution in 0.9% NaCl in polyolefin intravenous bags. *Pharmaceutical Technology in Hospital Pharmacy, 8*(1). https://doi.org/10.1515/pthp-2022-0007.

Visser, J., Feuerstein, I., Stangler, T., Schmiederer, T., Fritsch, C., & Schiestl, M. (2013). Physicochemical and functional comparability between the proposed biosimilar rituximab GP2013 and originator rituximab. *BioDrugs, 27*(5), 495–507. https://doi.org/10.1007/s40259-013-0036-3.

Vulto, A. G., & Jaquez, O. A. (2017). The process defines the product: What really matters in biosimilar design and production? *Rheumatology, 56*(suppl_4), iv14. https://doi.org/10.1093/rheumatology/kex278.

Wang, J., Niu, S., Dong, W., Wei, L. i, Ou, L., Zhang, T., Zhang, L., Nie, X., Wang, Q., Shen, T., Wang, Q. i, Xia, L., Liu, G., Jin, J., Zheng, Q., Song, H., & Fang, Y. i (2020). A randomized phase I clinical trial comparing the pharmacokinetic, safety, and immunogenicity of potential biosimilar recombinant human HER2 monoclonal antibody for injection and trastuzumab in healthy Chinese adults. *Expert Opinion on Investigational Drugs, 29*(7), 755–762. https://doi.org/10.1080/13543784.2020.1770226.

Weise, M., Kurki, P., Wolff-Holz, E., Bielsky, M. C., & Schneider, C. K. (2014). Biosimilars: The science of extrapolation. *Blood, 124*(22), 3191–3196. https://doi.org/10.1182/blood-2014-06-583617.

Wojtukiewicz, M. Z., Rek, M. M., Karpowicz, K., Górska, M., Polityńska, B., Wojtukiewicz, A. M., Moniuszko, M., Radziwon, P., Tucker, S. C., & Honn, K. V. (2021). Inhibitors of immune checkpoints—PD-1, PD-L1, CTLA-4—New opportunities for cancer patients and a new challenge for internists and general practitioners. *Cancer and Metastasis Reviews, 40*(3), 949–982. https://doi.org/10.1007/s10555-021-09976-0.

Wolff-Holz, E., Tiitso, K., Vleminckx, C., & Weise, M. (2019). Evolution of the EU biosimilar framework: Past and future. *BioDrugs, 33*(6), 621–634. https://doi.org/10.1007/s40259-019-00377-y.

Yadav, P., Keluskar, A., Mathuria, M., & Corridon, C. (2025). Payers' perspectives on biosimilars in the US: Past, present, and future. *Journal of Clinical Oncology, 43*(16_suppl). https://doi.org/10.1200/jco.2025.43.16_suppl.e13510.

Youssef, Emile, Fletcher, Brandon, & Palmer, Dannelle (2025). Enhancing precision in cancer treatment: the role of gene therapy and immune modulation in oncology. *Frontiers in Medicine, 11*, 1527600. doi:10.3389/fmed.2024.1527600.

Yun, J., Kim, J., Chung, J., Hwang, S. J., & Park, S. J. (2019). Extended stability of reconstituted and diluted SB3 (trastuzumab biosimilar) assessed by physicochemical and biological properties. *Advances in Therapy, 36*(7), 1700–1714. https://doi.org/10.1007/s12325-019-00973-y.

Zahavi, D., & Weiner, L. (2020). Monoclonal antibodies in cancer therapy. *Antibodies, 9*(3), 34. https://doi.org/10.3390/antib9030034.

Zuñiga, L., & Calvo, B. ña (2010). Biosimilars: Pharmacovigilance and risk management. *Pharmacoepidemiology and Drug Safety, 19*(7), 661–669. https://doi.org/10.1002/pds.1948.

Zydney, A. L. (2016). Continuous downstream processing for high value biological products: A Review. *Biotechnology and Bioengineering, 113*(3), 465–475. https://doi.org/10.1002/bit.25695.

CHAPTER

16 Recent evidence of biologics and biosimilars and its perspectives on clinical advances and emerging approaches in cancer

Murugesan Vanangamudi[a], Sagar Pamu[b], Deevan Paul Amarnath[c], Vijayaraj Surendran[d], Raghavendra Kumar Gunda[e] and Prathap Madeswara Guptha[f]

[a]*Institute of Pharmaceutical Research, GLA University, Mathura, Uttar Pradesh, India,* [b]*Department of Pharmacy Practice, Institute of Pharmacy, Nirma University, Ahmedabad, Gujarat, India,* [c]*Department of Pharmaceutics, Sri Ramachandra Faculty of Pharmacy, Sri Ramachandra Institute of Higher Education and Research (Deemed to be University), Porur, Chennai, Tamil Nadu, India,* [d]*Department of Pharmaceutical Analysis and Chemistry, Dr. Kalam College of Pharmacy, Thanjavur, Tamil Nadu, India,* [e]*Department of Pharmaceutics, Narasaraopeta Institute of Pharmaceutical Sciences (Autonomous), Narasaraopet, Andhra Pradesh, India,* [f]*Department of Pharmaceutical Sciences, School of Biotechnology and Pharmaceutical Sciences, Vignan's Foundation for Science, Technology and Research (Deemed to be University), Guntur, Andhra Pradesh, India*

16.1 Introduction

Cancer is one of the biggest health issues worldwide, and so far, novel advancements in the treatment of the disease are required. Conventional treatments include chemotherapy and radiation therapy, which can be efficacious but exhibit significant toxicity and lack specificity. Biologics are products derived from live cells that have transformed cancer treatment. The agents present a site-specific, personalized therapy with better efficacy and lesser toxicity (Debela et al., 2021).

16.1.1 The role of biologics in cancer treatment

Biologics are large-molecule medicinal products synthesised by living organisms, including bacteria, yeast, or mammalian cells. They specifically target cancer cell components, which minimizes the side effects on healthy tissues. These approaches have transformed cancer treatment and are primarily based on immune modulation, pathway inhibition, and direct tumor destruction to benefit the patient (Phyo et al., 2021). Biologics are significant pharmaceuticals designed based on a patient's genetic

Biosimilars and Biologics in Cancer Treatment: An Era in Therapeutic Strategies.
DOI: https://doi.org/10.1016/B978-0-443-43824-0.00005-5

and molecular characteristics, thereby optimising efficacy and minimising adverse effects. The most common biologics in oncology include monoclonal antibodies (mAbs), cytokines, growth factors, and Chimeric antigen receptor (CAR) T-cell therapy. These therapies have greatly improved survival in breast, lung, colorectal, and hematologic malignancies. Inhibitors of immunological checkpoints improve the body's natural capacity to detect and destroy tumors, and antibody-drug conjugates (ADCs) combine mAbs with cytotoxic chemicals to kill cancer cells specifically (Papież & Krzyściak, 2021). Despite these advantages, biologics have weaknesses, such as high production and manufacturing complexity. To enhance accessibility, cost-effective alternatives called biosimilars have been made available with comparative safety and efficacy (Kabir et al., 2019).

16.1.2 Significance of biosimilars in cancer care

Biological production is highly costly and complex, which limits access to cancer treatment. Biosimilars are cost-effective alternatives that provide equal safety, efficacy, and quality to reference biologics. They increase competition and reduce cost, making it possible to expand patient access to necessary cancer therapies for equitable treatment availability (Chopra & Lopes, 2017). Biosimilars undergo thorough evaluation via analytical, preclinical, and clinical investigations to ascertain their comparability to reference biologics. The U.S. Food and Drug Administration (FDA) and European Medicines Agency (EMA) ensure strict approval processes by regulatory agencies. Biosimilars differ from generic drugs because they must be supported by extensive comparability studies, including pharmacokinetic (PK) and pharmacodynamic (PD) assessments, to establish equivalent clinical outcomes (Khraishi et al., 2016). In oncology, biosimilars have been developed for trastuzumab (HER2 inhibitors), cetuximab (EGFR inhibitors), and bevacizumab (VEGF inhibitors), hugely benefiting breast, lung, colorectal, and hematologic cancers while significantly saving healthcare (Miller & Schwartzberg, 2019). In addition to saving costs, biosimilars spur innovation and competition, ensuring quality control and drug delivery. They provide more flexibility to oncologists, which means more people will have access to effective treatments. However, physician awareness, regulatory hurdles, and immunogenicity concerns stand in the way. As the clinical evidence grows for their safety and efficacy, biosimilars are slowly gaining acceptance in oncology. The next chapter discusses the advancement of biological therapies, which include immunotherapy, checkpoint inhibitors, and CAR T-cell therapy (Buchler et al., 2019; Joshi et al., 2022).

16.2 Recent clinical advances with biologics in cancer

Biologic treatments have transformed cancer treatment from chemotherapy to targeted and immune-based methods. Unlike other traditional treatments, biologics strengthen the natural body defense and provide more significant efficacy with fewer side effects. This elaborates on clinical developments in the recent past involving biologic therapy, beginning with immunotherapy—the most groundbreaking development in oncology this century (Papież & Krzyściak, 2021).

16.2.1 Immunotherapy

Immunotherapy stimulates the immune system to recognise and eradicate cancer cells. This treatment facilitates continuous immune cell assaults, leading to extended remission in malignancies such as melanoma, lung cancer, leukaemia, and lymphoma, unlike chemotherapy and radiation (Kciuk et al., 2023). Significant advancements encompass immune checkpoint inhibitors, namely the inhibition of the PD-1/PD-L1 pathway by pembrolizumab and nivolumab, as well as the blockade of CTLA-4 by ipilimumab, which facilitate T-cell activation and improve survival rates in melanoma, lung, and bladder malignancies. CAR T-cell therapy, particularly efficacious in acute lymphoblastic leukaemia (ALL) and diffuse large B-cell lymphoma (DLBCL), reprograms T cells to recognise and eradicate cancer cells; ongoing research is concentrated on its utilisation in solid tumours (Buchbinder & Desai, 2016). With remarkable efficacy in melanoma and hematologic cancers, cytokine and interferon-based therapies (IL-2) and oncolytic virus therapy (T-VEC) are still plagued by toxicity. In addition, progress in CRISPR-based gene editing is being made for better outcomes with immunotherapy (Chehelgerdi et al., 2024). Combination strategies continue, including ICIs with chemotherapy, targeted therapy, and neoantigen vaccines for intensified immune recognition. Immunotherapy has revolutionized the whole approach to cancer treatment, providing renewed hope and salvage for better survival in cancer patients (Garg et al., 2024).

16.2.2 Immune checkpoint inhibitors: PD-1/PD-L1 and CTLA-4 inhibitors

The discovery of immune checkpoint inhibitors has revolutionized cancer therapy. These chemicals indeed restore T-cell functionality and augment the body's capacity to combat cancer. Immune checkpoints (ICI) use cancer cells' capacity to avoid detection but are inhibited by ICIs, hence providing a continuous immune response. The drugs aim at the Programmed cell death protein 1 (PD-1) receptor on T cells or the PD-L1 ligand on neoplastic cells to obstruct immune suppression. The most commonly used drugs include Pembrolizumab (Keytruda) and Nivolumab (Opdivo) for melanoma, NSCLC, RCC, and Hodgkin lymphoma. Atezolizumab, Durvalumab, and Avelumab are all PD-L1 inhibitors. These have been quite effective in high-mutation cancers such as melanoma and lung cancer (Chen et al., 2024; Shiravand et al., 2022). CTLA-4 inhibitors include ipilimumab, sold under the brand name Yervoy—these function by interfering with another checkpoint that inhibits T-cell activation and are typically combined with PD-1 inhibitors. The use of nivolumab and ipilimumab has achieved a response rate of 50% in patients with melanoma and nonsmall cell lung cancer (NSCLC) (Buchbinder & Desai, 2016). Despite their

Table 16.1 Overview of biologics and biosimilars in cancer treatment.

Category	Description	Examples	Advantages
Biologics	Biologics, large-molecule drugs originating from living organisms, designed to target specific oncogenic pathways.	mAbs (Trastuzumab, Bevacizumab), Immunotherapy agents (Pembrolizumab, Nivolumab), Cytokines (Interleukin-2, Interferon-alpha).	High specificity, reduced toxicity, targeted action, and improved patient survival.
Biosimilars	Biologic products closely resemble an approved reference biologic, exhibiting no clinically significant variations in safety and efficacy.	Trastuzumab biosimilars (Herzuma, Ogivri), Bevacizumab biosimilars (Mvasi, Zirabev), Rituximab biosimilars (Truxima, Ruxience).	Cost-effective alternative, increased patient access, comparable efficacy to biologics, and potential for reducing healthcare costs.
Combination therapies	The use of biologics or biosimilars with chemotherapy, radiation, or immunotherapy to enhance treatment effectiveness.	Nivolumab + Ipilimumab (Checkpoint inhibitor combination), Bevacizumab + Chemotherapy (Colorectal cancer).	Synergistic effects, improved tumor response, reduced resistance development, and prolonged survival benefits.
Challenges in biologics and biosimilars	Factors that limit the widespread use and effectiveness of these therapies in oncology.	High production cost, immunogenicity concerns, regulatory approval complexities, and accessibility limitations.	Ongoing research for next-generation biologics, emerging biosimilar policies, and development of personalized treatment strategies.

success, ICIs still have resistance in some patients. Research continues on predictive biomarkers (PD-L1, TMB), next-generation ICIs (LAG-3, TIGIT), and combinations with cancer vaccines and adoptive cell therapies to improve outcomes further (Wei et al., 2024a). Table 16.1 compares biologics and biosimilars, indicating their functions, advantages, and disadvantages in the treatment of cancer. Table 16.1 compares biologics and biosimilars, indicating their functions, advantages, and disadvantages in the treatment of cancer.

16.2.3 CAR T-cell therapy

CAR T-cell therapy is a groundbreaking treatment that genetically modifies a patient's T cells to identify and target cancer cells. It is a long-lasting immune response, hence very effective in hematologic malignancies compared to chemotherapy or mAbs (Pavelescu et al., 2024). This approach includes collecting T cells, genetically altering them to produce chimeric antigen receptors that target cancer antigens, and subsequently cultivating them in the laboratory. Upon reinfusion into the patients, these modified T cells may identify and eliminate B-cell cancers by targeting CD19 (Davila et al., 2012). FDA-sanctioned CAR T-cell treatments encompass Tisagenlecleucel (Kymriah) for ALL and DLBCL, as well as Axicabtagene ciloleucel (Yescarta) for DLBCL. Alternative therapies, such as Idecabtagene vicleucel (Abecma) and Ciltacabtagene autoleucel (Carvykti), specifically target BCMA in multiple myeloma. Challenges include cytokine release syndrome (CRS), neurotoxicity, high cost, and limited efficacy in solid tumors (Wang & Wang, 2023). Ongoing research in next-generation CAR T cells, dual-targeting CARs, and CAR NK/macrophage therapies will make therapy available for the wider population (Fig. 16.1). Advancements in CRISPR and synthetic biology will take their toll, and the efficacy and accessibility of this therapy will improve soon enough to make CAR T-cell therapy a conventional cancer treatment (Marofi et al., 2021).

16.2.4 Targeted therapies

Targeted therapies are precision drugs that can modulate major molecular pathways that enable cancer cells to survive. Unlike chemotherapy, such therapies target and destroy both normal and cancer cells, thus significantly enhancing efficacy while minimizing toxicities. With a greater understanding of genomics and molecular profiling, oncogenic pathway inhibition can now be applied broadly across cancers ranging from breast to lung, colorectal, and many hematologic malignancies (Min & Lee, 2022). Small-molecule inhibitors are orally administered drugs that block tumor-promoting proteins. Tyrosine kinase inhibitors (TKIs) like Gefitinib and Osimertinib treat NSCLC, while ALK inhibitors (Crizotinib, Lorlatinib) target ALK-positive lung cancer. BRAF and MEK inhibitors (e.g., Vemurafenib, Dabrafenib, and Trametinib) treat BRAF-mutated melanoma, and angiogenesis inhibitors (Bevacizumab) prevent tumor blood vessel formation (Liu et al., 2022). mAbs trigger immune destruction or block tumor growth. HER2-targeted therapies (Trastuzumab, Pertuzumab) help in HER2-positive cancers, and CD20-targeted mAbs (Rituximab, Obinutuzumab) treat B-cell lymphomas (Zahavi & Weiner, 2020). Combination strategies, such as EGFR inhibitors with chemotherapy and BRAF + MEK inhibitors, are steadily improving patient outcomes. Research remains focused on developing next-generation inhibitors and ADCs, thus making precision oncology even more accessible and effective (Gouda & Subbiah, 2023).

16.2.5 Antibody-drug conjugates

The latest class of targeted cancer therapies has been developed based on the novel concept of an antibody-drug conjugate combining mAbs with cytotoxic chemotherapy: drugs are transported precisely to cancer cells while minimizing systemic toxicity. ADCs target tumor-specific antigens for more excellent selectivity and

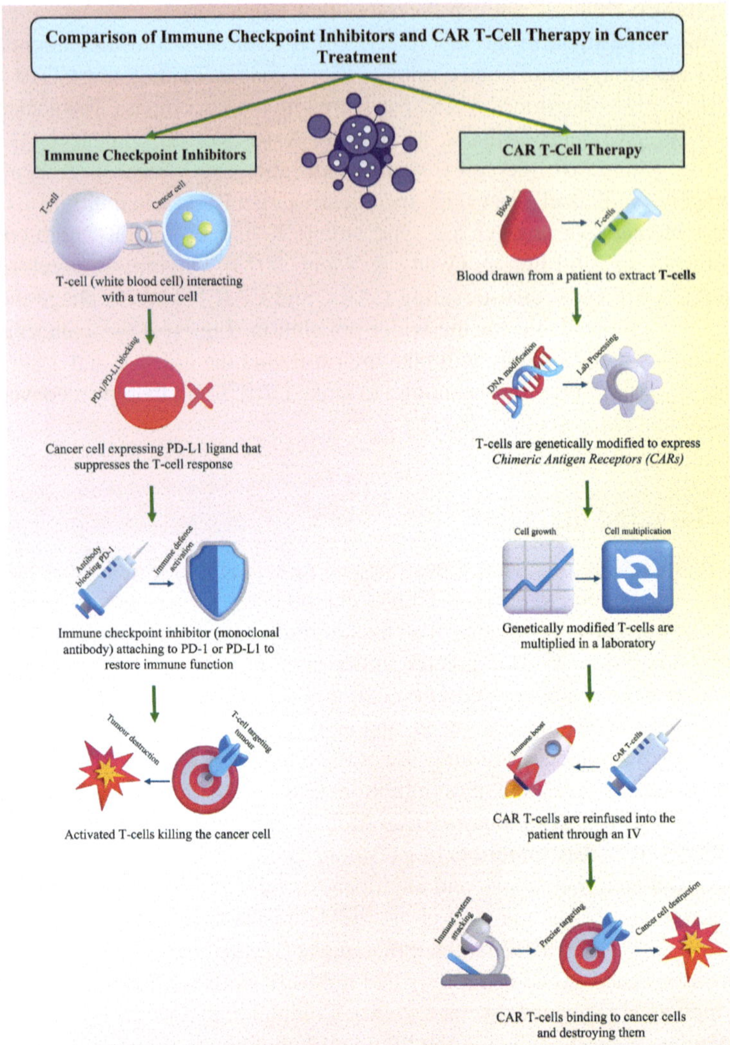

FIGURE 16.1 Comparison of immune checkpoint inhibitors and CAR-T cells therapy in cancer cells.

This figure illustrates the mechanisms and clinical processes of two prominent immunotherapies. On the left, immune checkpoint inhibitors block PD-1/PD-L1 interactions between T cells and cancer cells, restoring immune function and enabling T cells to kill tumors. On the right, CAR T-cell therapy involves extracting a patient's T cells, genetically modifying them to express Chimeric Antigen Receptors (CARs), expanding them in the lab, and reinfusing them to target and destroy cancer cells. Both therapies aim to enhance immune-mediated tumor destruction but differ in approach—checkpoint inhibitors release natural immune brakes, while CAR T-cell therapy engineers targeted immune responses.

effectiveness than traditional therapies. ADCs include three components: mAbs, a cytotoxic payload, and a linker linking them together. Once bound to a tumor cell, ADCs undergo endocytosis by releasing the cytotoxic drug within the cell to induce apoptosis (Aggarwal et al., 2023). Several ADCs have shown clinical efficacy and have been approved by the FDA. T-DM1 is used to treat breast cancer that is HER2-positive. Brentuximab vedotin is used to treat Hodgkin lymphoma and anaplastic large cell lymphoma, and it targets CD30. Enfortumab vedotin is used to treat urothelial carcinoma, and it targets Nectin-4. Sacituzumab govitecan is used to treat (TNBC), and it targets Trop-2 (Gogia et al., 2023). Despite ADCs' promise, challenges have emerged, such as drug resistance, off-target toxicity, and antigen downregulation. These issues are now being addressed in developing next-generation ADCs featuring optimized drug-to-antibody ratios, improved linkers, and innovative payloads, such as immune-modulating agents and radiotherapeutics. As such, ADCs are likely to be an essential component of the arsenal of treatments in personalized therapies for cancer patients with hard-to-treat conditions (Gu et al., 2024).

16.2.6 Cytokine and interferon-based therapies

Cytokine and interferon-based therapies use natural signaling proteins to enhance the immune response against tumor cells. Cytokines are agents modulating immune activity, and interferons, being antiviral and immunomodulatory, have therapeutic potential. In addition to hematologic malignancies and melanoma, much attention is being paid to the wider application of these forms of therapy (Bracarda et al., 2010). It favors the proliferation of T cells and NK cells. It was one of the first immunotherapies approved for metastatic RCC and melanoma. However, toxicities were significant, resulting in safer variants of IL-2. GM-CSF stimulates dendritic cell and macrophage activation and is employed in cancer vaccines and combination therapies for melanoma and leukemia (Mooradian & Sullivan, 2023). Interferon-alpha (IFN-α) aids immune surveillance and tumor suppression. It is approved for hairy cell leukemia, chronic myeloid leukemia (CML), melanoma, and Kaposi's sarcoma, though side effects limit its use (Xiong et al., 2022). Recent advances include engineered IL-2 variants (Bempegaldesleukin) and IL-15-based therapies to enhance efficacy with reduced toxicity. Combination strategies with cytokines and checkpoint inhibitors, cancer vaccines, or adoptive cell therapies are improving outcomes. Challenges exist, but cytokine and interferon-based therapies remain integral to immunotherapy, with continued research into personalized approaches for improved cancer treatment (Kim et al., 2021).

16.2.7 Oncolytic viruses and gene therapy

Oncolytic viruses therapy and gene therapy are new forms of advanced therapy that selectively kill the tumor cells by stimulating an immune response. OVs cause tumor cell lysis through infection and replication in cancer cells. The FDA has approved T-VEC, genetically modified HSV-1, for treatment in advanced melanoma. Onyx-015 has shown great potential for selectively infecting and destroying p53-deficient tumor cells, while Reovirus or Reolysin has potential applications for treating glioblastoma and lung cancers. OVs will be incredibly beneficial if combined with checkpoint inhibitors or CAR T-cell therapy (Chiocca & Rabkin, 2014). Gene therapy alters genetic material to correct mutations, enhance immune responses, or even kill tumor cells. In replacement gene therapy, tumor suppressor genes are reconstituted. In suicide gene therapy, such as HSV-tk/Ganciclovir, one can selectively destroy cancer cells. RNAi and CRISPR-Cas9 offer alternatives to silence oncogenes and boost immunity (Wang et al., 2004). Challenges include efficient delivery, immune balance, and tumor heterogeneity. Advances in nanotechnology, viral vectors, and combination therapies are improving efficacy. OVs and gene therapy are expected to become key components of personalized cancer treatment (Chen et al., 2023).

16.3 Biosimilars: Overview and development, definition, and regulations

Biologics are medications that have their basis in or contain elements of live organisms. Medicines that are already in existence, known as "reference originator biologics," have been approved for sale. Biosimilars are biological medications that are both similar to and less expensive than these. Biologics and biosimilars are not like conventional pharmaceuticals or their generic equivalents in that they are created in living cells. Biosimilar candidates and originator biologics cannot be identical, even though they share the same amino acid sequence, because of the intrinsic diversity of complex biological molecules. A biosimilar's active ingredient is physically and functionally comparable to, but not an exact replica of, the reference biologic. Biosimilars and biologics include a wide variety of products, including hormones, tiny proteins, immunisations, fusions, and mAbs. joining forces (Tabernero et al., 2016). The most common inflammatory cytokine that mAbs and their derivatives, biosimilars and biologics, target is tumour necrosis factor alpha. A biosimilar, which is a biological medicine that is highly comparable to another biological medicine that has previously been licensed in the EU, is clinically equivalent to the reference medication in terms of safety, quality, and efficacy. successful completion. A biologic product can only be deemed a biosimilar in the US if it closely resembles an already-approved biologic and there are no obvious differences in safety or efficacy between the two (O'Callaghan et al., 2019). Modifications to inert components are sufficient for a product to be classified as a biosimilar. The following entry: biologic, follow-on biologic, and follow-on protein are some alternative names for biosimilars. It should be mentioned that reference items authorised by the US and the EU are not always seen as identical (Gherghescu & Delgado-Charro, 2021). For patients dealing with long-term and often crippling conditions, including cancer, diabetes, and autoimmune diseases, biological medicines (both original and biosimilar) offer therapeutic options. Biologics have a 12-year US exclusivity period and an 11-year EU exclusivity period. The US period includes 8 years of data exclusivity, 2 years of commercial exclusivity, and a 1-year extension for a new indication, whereas the EU period is 10 years for new biologics. Once biosimilarity has been proven, a biosimilar candidate can be manufactured and sold at a lower cost than the original biologic. This is because of the simplified clinical development process that relies primarily on the original biologic's established safety and efficacy data. As a result, companies may gain an advantage by specializing in biosimilar development, and people with chronic conditions may profit from having access to biosimilar treatments at prices lower than their originator biologics. After receiving their first approval in 2006, biosimilars have been on the market in the European Union for 13 years. In the United States, they have been available for 4 years following the initial FDA authorisation in 2015 (Cordeiro et al., 2024). Biosimilars are subject to stringent regulations during the approval process because of the high risk of biological function loss and serious, as-yet-undiscovered, adverse effects caused by variability (both qualitative and quantitative). The US and the EU use different approaches to this problem. Historically, the development of biosimilars has been closely followed by regulatory mandates in the US and EU. In 2003, the European Union put in place the rules that govern biosimilars. In 2005, the CHMP produced a comprehensive guideline on biosimilars; in 2015, a new edition was put into effect. With the growing body of knowledge about biosimilars, the main guideline and its supplemental guidelines have been updated in recent years to cover quality, nonclinical, and clinical topics. Over the past few years, the US FDA has been quite busy providing parties with advice and developing criteria for biosimilar development. "Similar biotherapeutic products" were the subject of a 2010 WHO recommendation. Movements to establish global standards are still in their early stages. More than that, there are a number of health agencies that use slightly different terminology at the moment. Internal best practices should be established by pharmaceutical enterprises, taking into account recommendations from their regulatory affairs departments for terminology that is appropriate in the EU, US, and worldwide (Gherghescu & Delgado-Charro, 2021). An original manufacturer of biological goods has challenges when trying to alter the production process and must prove equivalency, for example, for different versions of the same drug, because of the inherent variability of biologics. A comprehensive analytical (biochemical and biophysical) comparative testing program may be used to prove similarity or equivalence, which was inspired by changes in regulatory requirements that sought to improve and streamline biologics production processes. This idea is known as the biochemical bridge. The biochemical bridge was crucial in beginning to define differences and their links to physiological and clinical impacts, and it also made it easier to investigate potential biosimilars. Studies meant to verify similarity must use the reference biologic that corresponds to the intended target location since reference products approved in the EU and the US are not necessarily considered to be identical. This usually means that the biosimilar candidate is evaluated for equivalence to both reference products in the same clinical trial, as is common in lean development programs (Pan et al., 2021).

16.3.1 Comparison of biosimilars and originator biologics

Biologic therapies have transformed the treatment of numerous chronic diseases, particularly autoimmune and inflammatory disorders, malignancies, and uncommon genetic problems. With the expiration of patents on numerous original biologics, also known as originator or reference biologics, biosimilars have surfaced as economical replacements. Nonetheless, despite their clinical resemblance, originators and biosimilars exhibit substantial differences in structure, manufacturing processes, regulatory frameworks, immunogenicity, and clinical use (Blandizzi et al., 2017).

16.3.2 Structural and manufacturing differences

Biologics are substantial, structurally intricate substances generated by recombinant DNA technology and other biotechnological methods utilizing living cells. They typically fall into three classifications: (1) biologics that reproduce natural proteins for replacement therapy, (2) mAbs, and (3) fusion proteins that imitate receptors. These molecules are generally 200 to 1000 times larger than small-molecule medications, and their pharmacological activity relies not only on amino acid sequences but also on intricate secondary and tertiary structures, posttranslational changes such as glycosylation, and protein folding (Mendes de Abreu et al., 2014; Stevenson, 2015). Originator biologics are manufactured using patented cell-based systems that encompass specialized host cells, nutritional conditions, purification techniques, and environmental controls. The production process is so essential to the result that the phrase "the process is the product" is frequently referenced in biologic manufacturing. Minor alterations in the production process—such as the purification technique or the cellular clone—can lead to physiologically and clinically important variations. Biosimilars, in contrast, are produced without access to the original manufacturing cell lines or secret methodologies. Consequently, they are reverse-engineered to closely resemble the reference biologic regarding structure, activity, and clinical behavior. Nonetheless, owing to inherent biological variability and manufacturing inconsistencies, biosimilars can never be precise replicas. Regulatory bodies permit slight variations in inactive ingredients, as long as they lack clinical significance (Blandizzi et al., 2017).

16.3.3 Regulatory pathways and comparability exercises

EMA and the FDA have developed rigorous and flexible procedures for the approval of biosimilars, reflecting their understanding of the complexities of biologics. The foundation of biosimilar development is the "comparability exercise," a systematic assessment that contrasts the biosimilar with the originator across analytical, pre-clinical, and clinical parameters. The assessments commence with physicochemical characterisation and functional experiments, subsequently progressing to in vitro and in vivo examinations. Clinical comparability trials often encompass PK and PD evaluations, along with a minimum of one randomized controlled trial that establishes equal efficacy and safety. A crucial regulatory distinction is that while originators undergo extensive clinical development, biosimilars must demonstrate resemblance to their reference rather than independently validating efficacy and safety. The inaugural biosimilar in Europe received permission in 2006, and current regulatory rules permit the extrapolation of indications—granting approval for other diseases based on data from a single indication—if scientifically substantiated. In the United States, extrapolation is allowed but necessitates comprehensive justification, especially when the mechanism of action varies among indications (Florian et al., 2023; Lee et al., 2012).

16.3.4 Immunogenicity and safety considerations

Immunogenicity—the capacity of a biologic to provoke an immune response—is a significant concern for both originator biologics and biosimilars. immunological responses can lead to the production of antidrug antibodies (ADAs), which may diminish efficacy (by neutralizing drug action or increasing clearance) or induce undesirable effects (e.g., allergic reactions, immunological complex formation). Factors influencing immunogenicity include manufacturing impurities, protein conformation, glycosylation patterns, and host-related variables. Despite the modest alterations in original biologics over time due to manufacturing variability, their long-term immunogenicity and safety have been thoroughly described via years of clinical application (Baert et al., 2003; Vincent et al., 2013). Biosimilars, conversely, must demonstrate comparable immunogenic characteristics via direct clinical trials. The PLANETRA and PLANETAS studies assessed the infliximab biosimilar CT-P13 in rheumatoid arthritis and ankylosing spondylitis, respectively. The trials demonstrated comparable efficacy and incidence of ADA relative to the originator, hence endorsing the use of biosimilars for these indications. Experts, however, warn that small patient cohorts and restricted follow-up periods may overlook infrequent or delayed immune responses. Therefore, postmarketing surveillance and risk management strategies are essential for biosimilars (Park et al., 2013; Yoo et al., 2017).

16.3.5 Switching, interchangeability, and substitution

Switching denotes the transition of a patient's treatment from an innovator drug to its biosimilar counterpart. Interchangeability denotes that a biosimilar may be substituted without the prescriber's consent, and substitution permits pharmacists to execute this change autonomously (McKinnon et al., 2018). The EMA permits switching under medical supervision, but does not govern interchangeability, which is the responsibility of member states. Conversely, the U.S. FDA mandates supplementary studies to classify a biosimilar as interchangeable, encompassing evidence on multiple swaps demonstrating no effect on safety or efficacy. Clinical switching, due to factors such as inadequate response or harmful effects, is broadly acknowledged; nevertheless, nonclinical switching, driven by cost-saving policies, remains contentious. Numerous professional organizations, such as the American College of Rheumatology and the European League Against Rheumatism (EULAR), oppose nonmedical switching without the approval of both patients and physicians. Concerns focus on the potential immunogenic effects of recurrent switching and complexities in pharmacovigilance arising from monitoring difficulties (Blandizzi et al., 2017; Skingle, 2015).

16.3.6 Nomenclature and traceability

Clear nomenclature and traceability are essential for ensuring clarity in prescribing and adverse event reporting. The utilization of unique brand names, batch numbers, and identifiers for biosimilars is advocated. The EMA requires brand-name reporting in pharmacovigilance; nevertheless, surveys indicate that batch numbers are frequently unrecorded, impeding traceability. The FDA has suggested employing a distinctive four-letter suffix with the nonproprietary name (e.g., infliximab-dyyb) to differentiate biosimilars from their originators. This approach enhances oversight and mitigates ambiguity in therapeutic environments (Chao et al., 2015).

16.3.7 Patient perspectives and education

Patients frequently articulate apprehensions regarding the efficacy, safety, and underlying motives for the use of biosimilars. There is concern that biosimilars are being prescribed mainly for economic reasons rather than for their clinical advantages. Entities such as EULAR's Standing Committee of People with Arthritis/Rheumatism in Europe and the UK National Rheumatoid Arthritis Society underscore the necessity for educated decision-making, clear communication, and the consideration of specific patient circumstances. Individuals with chronic diseases, pediatric cohorts, or those who have attained disease stability may be especially susceptible to the dangers associated with switching. Biosimilar incorporation into care should be directed by collaborative decision-making between physicians and patients (Oskouei & Kusmierczyk, 2021).

16.4 The process of biosimilar development

In order to ensure the safety and effectiveness of biologic medicines, doctors and scientists spend years perfecting their creation, conducting research, and testing them in clinical trials. A biologic drug can take several months to make, so every step of the process must be carefully monitored and regulated. Making a biosimilar is quite similar to making the original biologic. Before they may be made available to patients, biosimilars must meet a rigorous scientific standard of similarity and not have any discernible deviations from the original product, according to the FDA. The scientific approach is used to confirm that there are no noticeable variations between the original product and the biosimilar. This guarantees that the biosimilar will work the same way in a patient's body as the reference biologic. Throughout the development process, similarity assessments are frequently performed. Characterisation and optimisation of the method, validation of biosimilarity, and regulatory approval are the three main stages of the biosimilar development process (Mascarenhas-Melo et al., 2024).

16.4.1 Developing the product: Process descriptive analysis and optimization

In the first stage of development, the reference biologic is thoroughly studied to understand its structure and function. This process is called characterisation. Phase two entails developing the production technique that yields the medicinally close relative once this data has been collected. A biosimilar molecule that matches the quality of the medicine discovered during characterization is methodically designed using state-of-the-art biological development technology and extremely sensitive analytical tools. As the process iteratively refines each step of the manufacturing operation, it becomes cyclical. This keeps happening until the manufacturing process consistently produces a molecular structure that closely matches the reference medication. The second step is to confirm the biosimilarity through studies and regulatory cooperation. The next step is to analyse and test the biosimilar to see how similar it is to the reference biologic medication. In order to confirm biosimilarity and interchangeability, the FDA reviews all available information and decides whether further clinical and nonclinical studies are required. In heavily regulated markets, such as those in the EU, US, Japan, Canada, and Australia, biosimilar clearance is usually subject to clinical trials. Results from biosimilar clinical trials will determine their breadth and depth. A tailored clinical trial design might be a better way to prove biosimilarity and interchangeability when strong and convincing analytical evidence already exists, and further data is needed. The third step is approval. Biosimilar firms work with the FDA in the early stages of development to establish a product development strategy and ways to provide enough scientific rationales for assessment. The FDA checks if all the necessary scientific criteria have been met before authorising a product for human use by reviewing the entire body of evidence and carefully analysing all data. It is possible to produce and distribute biosimilars in the US once they have received approval. In terms of safety and effectiveness, patients taking biosimilars can be certain that they are on par with those taking their brand-name biologics. According to research conducted all across the world, patients get the same positive effects from both the original biologic and its biosimilar versions. Aiming to educate lawmakers, patients, and providers on biosimilars and speed up their development, the Biosimilars Council has also worked to pass legislation that is inclusive of biosimilars (Tsiftsoglou et al., 2013).

16.5 Current status of biosimilars in oncology

Biosimilars are assuming a progressively significant role in contemporary cancer treatment owing to their cost-effectiveness and therapeutic efficacy. As expensive biologics persist in leading cancer therapy, biosimilars present a cost-efficient option while ensuring equivalent efficacy and safety standards. The increasing incidence of cancer globally has heightened the need for accessible treatments, which biosimilars are addressing. In oncology, biosimilars, including filgrastim, trastuzumab, rituximab, bevacizumab, epoetin alfa, denosumab, and pegfilgrastim, have been developed and sanctioned by several international regulatory bodies. Filgrastim biosimilars, such as Zarxio (approved in 2015), were among the initial products launched to assist in the management of neutropenia in cancer patients after chemotherapy. Bevacizumab-awwb, authorized in 2017, was the inaugural biosimilar for the treatment of several malignancies, including brain, colorectal, and cervical cancers. Trastuzumab-dkst and pegfilgrastim-jmdb have been utilized for breast and nonmyeloid malignancies, respectively, signifying notable clinical progress. Regulatory approval is strong, with the EMA, US FDA, and World Health Organization (WHO) offering extensive frameworks. Since 2018, the EMA has granted numerous oncology biosimilars, and by May 2021, it had approved 39 patents for essential cancer biosimilars. Regulatory authorities underscore the necessity of comprehensive analytical, clinical, and postmarketing surveillance investigations to ascertain the safety, effectiveness, and immunogenicity comparability of biosimilars with their originators (Joshi et al., 2022; Malakar et al., 2024). Biosimilars exert a significant economic influence. Their debut has resulted in a substantial decrease in cancer treatment expenses. The cost-to-value ratio for biosimilars ranges from 59.4% to 86.0% in comparison to biologics. The projected monthly treatment expenses for biosimilars vary between $322 and $7424, in contrast to elevated costs associated with reference biologics. Europe is projected to save in excess of €10 billion from 2016 to 2022 as a result of biosimilar utilization. In the United States, the introduction of further biosimilars for medications such as Avastin and Herceptin could potentially decrease healthcare expenditures by $250 billion over a 5-year period. Notwithstanding these advancements, obstacles remain. Global disparities exist in patient and healthcare provider awareness, substitute protocols, and biosimilar changeover regulations. In certain countries, biosimilar substitution is allowed without the prescriber's consent, whilst others mandate specific medical approval. Concerns regarding safety persist in relation to immunogenicity; nevertheless, comprehensive postmarketing surveillance and real-world evidence are progressively enhancing trust. The acceptance of healthcare systems is increasing. Biosimilars are now commonly incorporated into the treatment of malignancies, including breast, stomach, cervical, and colorectal. Their function in mitigating side effects such as chemotherapy-induced neutropenia enhances their significance. The efficacy of biosimilars in oncology is significantly influenced by regulatory approval, price strategies, physician confidence, and patient awareness (Tandulje et al., 2024).

16.6 Biosimilars in cancer treatment

Over time, cancer therapy has changed dramatically, and biologic medications are now essential for bettering patient outcomes. Biosimilars, which are physiologically similar versions of reference biologic medications that have already received approval, have been developed as a result of the high cost of biologics restricting their accessibility. Biosimilars provide an affordable substitute with similar safety and effectiveness features. Because of their intricate biological production methods, biosimilars are inherently variable in contrast to generic medications, which are precise chemical replicas of their reference counterparts (Kabir et al., 2019).

16.6.1 Economic impact and relevance of cancer biosimilars

By offering a less expensive substitute for pricey biologics, the advent of biosimilars lowers total healthcare spending. As a result, healthcare systems are able to invest in more patient care services and manage resources more efficiently. Healthcare spending has significantly decreased after the introduction of biosimilars (Humphreys, 2022). By 2026, for instance, it is anticipated that the use of biosimilars will save the American healthcare system between $25 billion and $150 billion. Both reference biologics and biosimilars see price reductions as a result of the increased competition in the pharmaceutical industry brought about by biosimilars. Both biosimilars and their reference biologics see price reductions as a result of the increased competition in the pharmaceutical sector brought about by biosimilars. It has been noted that this competitive climate lowers the price of biologic treatments, making them more widely available. The availability of less costly biosimilars reduces the financial strain on patients and, by reducing out-of-pocket expenses, enhances adherence to treatment programs. This is particularly crucial in the field of cancer, where the cost of therapy may be prohibitive. As a consequence, the landscape of cancer therapy is more economical and sustainable. Patients' out-of-pocket expenses are decreased as a result of lower treatment costs, which improves affordability and therapy compliance (Kvien et al., 2022). The biosimilar industry fosters economic growth by encouraging biopharmaceutical research and development, creating jobs, and driving technological advancements in drug manufacturing. This leads to job creation and technological advancements in drug manufacturing, further enhancing the healthcare sector's efficiency. The launch of biosimilars into oncology has had a significant impact on the cost of cancer treatment, even while it has improved patient access to essential treatments and resulted in significant cost savings.

16.6.2 Influence of biosimilars on oncology treatment

Biosimilars have a significant influence on oncology treatment, impacting various aspects of patient care, economics, and healthcare systems:

16.6.2.1 Increased access to treatment

As patent protections for reference biologics expire, biosimilars emerge as lower-cost alternatives. This can enhance patient access to essential cancer therapies, potentially leading to improved treatment outcomes for a broader range of patients, especially those in underserved populations. A biosimilar is a biological medication that closely resembles an established biopharmaceutical benchmark product in terms of quality, efficacy, and safety (Bas & Duarte, 2024). Biosimilars have the potential to significantly improve patient access to biologic medicines in this respect (Peeters et al., 2021). This price reduction reduces the financial burden and increases the number of prescriptions for drugs. After the filgrastim biosimilar was introduced, filgrastim uptake in the UK increased by 104%. As epoetin biosimilars became accessible, treatment volume increased by 263%, and prices decreased by 50% in three European countries (Bulgaria, the Czech Republic, and Romania) between 2007 and 2014. The primary patents for a number of cancer biologics have expired, and numerous biosimilars have been approved. Based on zx EMA sources, about 39 biosimilars were approved as of May 2021, including rituximab, bevacizumab, trastuzumab, pegfilgrastim, and filgrastim (Bachu et al., 2022). As of May 2021, the EMA has authorized six trastuzumab biosimilars since the first one was introduced in Europe in 2018. Phase III studies of several trastuzumab biosimilars in terms of clinical development showed some variations in clinical study design. While MYL-1401O (Ogivri; Mylan), PF-05280014 (Trazimera; Pfizer), and HLX02 (Zercepac; Shanghai Henlius Biotech) underwent Phase III trials in HER2-positive early breast carcinoma with pathologic complete response as their primary end point, ABP 980 (Kanjinti; Amgen), CT-P6 (Herzuma; Celltrion), and SB3 (Ontruzant; Samsung Bioepis) had ORR as the main clinical outcome of phase III studies carried out in the context of first-line metastatic breast cancer (Thill, 2019).

16.6.2.2 Cost-effectiveness

When compared to their reference biologics, biosimilars are frequently less expensive, which helps ease the financial strain on healthcare systems. More patients can obtain therapy because of this cost decrease, which may also assist in reallocating funds to other oncology-related issues that require attention. Biosimilars with lower prices might potentially save the healthcare system money. According to one estimate, the introduction of 11 biosimilars, such as Rituxan (rituximab), Herceptin (trastuzumab), and Avastin (bevacizumab), might save the United States $250 billion between 2014 and 2024 (Mestre-Ferrandiz et al., 2024). A study by the IQVIA Institute estimates that biosimilars will save over €10 billion in costs in the EU5 (France, Germany, Italy, Spain, and the UK) alone between 2016 and 2020 (Clarke et al., 2024). The reduced cost of biosimilars may result in decreased healthcare costs and improved access to cancer biologic treatment. The cost ratio of monoclonal antibody biosimilars to reference products now varies between 59.4% and 86.0% (Huang et al., 2020). In addition, monoclonal antibody biosimilars are predicted to save around $322–$7424 for a single patient over the course of 1 month, while they are expected to save approximately $17,517–$38,923 for each patient over the duration of therapy.

16.6.2.3 Regulatory approvals and guidelines

The increasing use of biosimilars has prompted regulatory agencies to develop stringent guidelines to ensure their safety, efficacy, and quality. This has led to a better regulatory environment for oncologists and healthcare professionals, allowing them to make informed choices about biosimilar use in clinical settings. The body of evidence serves as the foundation for the regulatory approval process for biosimilars (Mascarenhas-Melo et al., 2024). A clinical immunogenicity assessment and clinical studies showing comparable effectiveness with the standard biologic product in a randomly assigned Phase III head-to-head comparison trial follow this methodical approach, which begins with the foundation of analytical studies and progresses to preclinical animal studies and clinical PKs (Rahalkar et al., 2018). In order to show similarity between the biosimilar and the reference product, regulators assess the entirety of the evidence at each stage to determine whether additional research is required at any point to remove any remaining ambiguity between the biosimilar and the reference biologic (Markus et al., 2017). Guidelines committees like the National Comprehensive Cancer Network have recognized that "an FDA-approved biosimilar is an appropriate substitute for trastuzumab," demonstrating the belief with which drug indications can be extrapolated based upon the complete body of proof backing biosimilar development (Socinski et al., 2015). This recognition is based on the accuracy with which the whole of verification leads to regulatory approval for the biosimilar drug. Clinical investigations are the last step in the sequential process to show that the biosimilar and the reference biologic product do not differ clinically meaningfully (Isakov et al., 2016). Generally speaking, the goal of a clinical efficacy study is to verify clinical similarity between the suggested biosimilar and its reference product based on predetermined margins, as well as comparable safety and immunogenicity; de novo efficacy and safety are not the goals of these studies.

16.6.2.4 Clinical efficacy and safety concerns

Although biosimilars are intended to closely resemble their reference drugs, safety and immunogenicity issues persist. There are apprehensions among healthcare professionals about switching from reference biologics to biosimilars due to uncertainties regarding the extrapolation of indications and the potential for adverse events. Ongoing studies and real-world evidence are vital to address these concerns and build confidence in biosimilars among clinicians (Herndon et al., 2023). A major regulatory criteria of biosimilarity is that there are no clinically relevant variations in the competing biologic product (i.e., biosimilar) and the reference product with regard to effectiveness, safety, or quality (Kirchhoff et al., 2017). A "biosimilar" cannot be authorized as such if it does not satisfy this requirement. Patient safety is unquestionably a top priority for the US and EU biosimilar development process, which continues with postmarketing safety monitoring.

16.6.2.5 Educational efforts

More patient and healthcare professional education and understanding on the advantages and disadvantages of biosimilars is desperately needed. The first biosimilar to be authorized worldwide was somatropin, a human growth hormone, which was approved by the EMA in 2006. The importance of biosimilar education didn't begin to take hold until the EMA authorized the very first monoclonal antibody biosimilars in 2013 (to infliximab), a drug that presumably addresses a smaller patient group. Prescribers' comfort with the new medications would be a major factor in adoption, according to early studies to measure clinicians' impression of biosimilars. Just weeks after the approval of the infliximab biosimilar, for instance, a very early survey of randomly chosen members of the European Crohn's and Colitis Organization (ECCO) was conducted in the fall of 2013. The results showed that 61% of responding clinicians had little to no confidence in using biosimilars in routine clinical practice, 5% were completely confident, and the remaining respondents fell somewhere in the middle. This opinion was not exclusive to European practitioners, as evidenced by the findings of a poll conducted in February 2014 among Canadian Rheumatology Association members (Danese et al., 2016). A total of 72% of medicos would be reluctant or extremely unlikely to choose a biosimilar as their first option if a patient who has never received treatment were a perfect candidate for a biologic and cost was not a concern (Bombardier et al., 2012). There were similarities among providers who were hesitant to be early adopters of biosimilars, according to these early market observations. Adoption is nevertheless hampered by certain healthcare practitioners' ignorance of or lack of faith in biosimilars; this problem is made worse by the existence of false information and denigration. Knowledge and education will need to be at the forefront in order to tackle any possible reluctance to biosimilars from a clinical or scientific viewpoint. Improved understanding can foster greater acceptance and incorporation of biosimilars in routine oncological care, ultimately enhancing patient outcomes (Leonard et al., 2019).

16.6.2.6 Clinical evidence supporting biosimilar efficacy

The efficiency and safety of biosimilars in treating a variety of cancers have been solidly established by an increasing amount of clinical data. Biosimilars of important biologics, such as trastuzumab, bevacizumab, rituximab, filgrastim, pegfilgrastim, and epoetin alfa, have been shown in numerous Phase III randomized, double-blind trials to have clinical equivalence to their reference products in terms of efficacy, PKs, and safety profiles. For example, trastuzumab biosimilars such as CT-P6 and SB3 have shown comparable overall response rates and pathologic complete response in patients with HER2-positive breast cancer, while bevacizumab biosimilars such as ABP 215 and PF-06439535 have replicated the efficacy of Avastin in NSCLC. Similarly, rituximab biosimilars, like GP2013 and CT-P10, have demonstrated noninferiority in treating follicular lymphoma, reflecting the clinical response of the reference product Rituxan; supportive agents, like filgrastim (EP2006) and pegfilgrastim biosimilars

(LA-EP2006), have also shown equivalent effectiveness in managing chemotherapy-induced neutropenia; epoetin alfa biosimilars have matched the therapeutic outcomes of their originator in the management of anemia; and infliximab biosimilars, like CT-P13, have shown promising results in real-world studies for managing immune-related adverse events (IrAEs) associated with cancer immunotherapy. All of these studies (Table 16.2) highlight the dependability of biosimilars as clinically sound and cost-effective alternatives in contemporary oncology practice.

16.7 Emerging approaches in cancer treatment

16.7.1 Cell and gene therapies

Novel therapeutic strategies, such as cell-based and gene therapies, have been reported to be beneficial by targeting cancer cells specifically, with minimal collateral damage to healthy tissues. It is an area where cell-based therapies, for example, Tumor-Infiltrating Lymphocytes (TILs) and CAR-T-cell therapy, are growing in research aimed at curing breast and lung cancer (Garg et al., 2024; Liu et al., 2024). Moreover, gene therapies like suicide gene transfer allow for the precise targeting of the tumor. New possibilities in the field arise, as one might expect, for instance, new virus vectors are used, and new methods of gene delivery are developed. As a result, the side effects of gene therapy decrease, and the treatment of the disease is greatly successful. However, only a comprehensive investigation that will take into account challenges such as tumor heterogeneity and immune evasion, thereby ensuring the safety of patients from such experimental treatments, may actually pave the way for the therapeutic use of new drugs (Navarro et al., 2016). Breakthroughs in genome editing tools like CRISPR-Cas9 technology provide a helping hand in mostly used gene therapy by allowing precise modifications in cancer-related genes. Such a technology is very useful in positive gene alterations, or the turning off of harmful genes, and can also activate the immune cells of the body against the cancer cells. Moreover, gene therapy, when combined with immunotherapy, is one way of finding a cure for cancer by transforming T cells to recognize and kill cancer cells in an efficient manner (Garbayo et al., 2020). Another of the several problems out there they are facing is that, although these treatments are promising, the issue of ethics, potential long-term genetic alterations, and insertional mutagenesis are paramount challenges that should necessitate thorough clinical evaluation. However, despite this wide room for scientific advancement, the risks are many.

16.7.2 Novel delivery systems for biologics

Cancer treatment advancements increasingly favour ultra-modern particulate delivery systems for gene therapy using nanomedicine and nanotechnology to heighten efficiency and precision. Such technologies are focused on challenges such as bioavailability and therapy breakdown. Electrospray nano- and micro-fibers play a controlled

Table 16.2 Clinical evidence from pivotal trials supporting the efficacy and safety of biosimilars in the treatment of cancer.

Biosimilar	Reference Biologic	Indication	Study Design	Key Findings
Trastuzumab biosimilar (CT-P6)	Trastuzumab (Herceptin)	HER2-positive breast cancer	Phase III, randomized, double-blind	Comparable efficacy and safety; equivalent overall response rate
Trastuzumab biosimilar (SB3)	Trastuzumab (Herceptin)	HER2-positive early breast cancer	Phase III, randomized, Double-Blind	Equivalent pathologic complete response and comparable safety
Bevacizumab biosimilar (ABP 215)	Bevacizumab (Avastin)	NSCLC	Phase III, randomized, double-blind	Equivalent ORR and safety profile
Bevacizumab biosimilar (PF-06439535)	Bevacizumab (Avastin)	NSCLC	Phase III, randomized, double-blind	Demonstrated clinical equivalence in efficacy and safety
Rituximab biosimilar (GP2013)	Rituximab (Rituxan)	Follicular lymphoma	Phase III, randomized, double-blind	Demonstrated comparable overall response and safety
Rituximab biosimilar (CT-P10)	Rituximab (Rituxan)	Advanced follicular lymphoma	Phase III, randomized, double-blind	Equivalent overall response rate and similar safety
Filgrastim biosimilar (EP2006)	Filgrastim (Neupogen)	Chemotherapy-Induced neutropenia	Phase III, randomized, double-blind	Equivalent duration of severe neutropenia
Pegfilgrastim biosimilar (LA-EP2006)	Pegfilgrastim (Neulasta)	Chemotherapy-Induced neutropenia	Phase III, randomized, double-blind	Noninferior efficacy and similar safety profile
Epoetin alfa biosimilar (HX575)	Epoetin Alfa (Eprex)	Chemotherapy-Induced anemia	Phase III, randomized, double-blind	Equivalent hemoglobin response and safety
Infliximab biosimilar (CT-P13)	Infliximab (Remicade)	IrAEs in cancer immunotherapy	Observational, postmarketing surveillance	Demonstrated efficacy and safety in real-world settings

role in drug and biologic delivery, especially in ocular treatments (Mishra et al., 2023). Microneedles, in a noninvasive manner, provide drug delivery, which in turn enhances the cell delivery in cancer therapies. Delivered through a short peptide that is known for its large protein and small molecule participation, these systems offer a fast and precisely targeted cancer treatment. In general, the threshold of these most innovative systems, including nanoparticles, liposomes, microneedles, and peptides, is shifting cancer therapy's paradigms toward improved targeted delivery, lower adverse effects, and better outcomes (Mbituyimana et al., 2023). In addition to this, smart drug delivery systems that utilize stimuli-responsive nanoparticles are geared toward therapy support in cancer as well. Such systems react to the conditions of their surroundings, namely pH, temperature, or enzymatic activity, which ensures targeted drug delivery (Conibear et al., 2020). On the other hand, pH-sensitive liposomes are a case in point where they release their drug load only in the acidic microenvironment of tumors, which prevents systemic toxicity. Moreover, the innovative multifunctional nanocarriers are made with the purpose of being able to deliver multiple treatment substances, such as chemotherapeutics, gene therapies, and immunomodulators, alongside each other, to enhance synergistic treatment effects. But these are the newer edges for the drug releases that face the issues like large-scale manufacturing along with approval from regulatory bodies, and also longer-term biocompatibility, which are, by a large proportion, areas of current research (Gupta & Kim, 2021; Jia et al., 2023).

16.7.3 Immunotherapy advances

Cancer treatment has been completely altered by immunotherapy to make the immune system fight cancer cells. Recent advances, especially nanoparticle-based delivery systems, are boosting the targeting and the efficiency of immune-activating factors (Runcie & Dallos, 2021). One appealing preclinical study has been the use of the nanotechnology-based photodynamic therapy and immunotherapy combination. Besides, the combination therapy that the researchers develop and apply, such as chemotherapy and surgery, is still a better choice. Nevertheless, much of the work that should be done, like a clinical trial for the drug in humans, is still left, and the main obstacle is the resistance or ineffectiveness of the drug in pancreatic ductal adenocarcinoma cancer (Dziubańska-Kusibab et al., 2024). Research in progress is laser-focused on this particular mission, comprising the directions of antigen-specific treatments, the conditions when immunosuppressive tumor microenvironment is hit, and nano-vaccines and multimodal nanoparticles.br Another treatment method to offer better results is the use of CAR-T therapies, which are recently developed cell-based approaches. Furthermore, novel practices such as adoptive cell transfer (ACT) and immune checkpoint inhibitors are significantly augmenting the array of immunotherapy techniques (La'ah & Chiou, 2024). ACT is the process of generating numpty immune cells, such as T cells, for instance, by genetically engineering the patient's T cells in the lab and then reintroducing them to the patient's body to enable them to combat cancer. This approach is particularly successful in blood cancer, as displayed by some of the ongoing trials that are showing its potential not only on a stand-alone level but in solid tumors as well. What it refers to are BiTE products and a cytokine panel that has been modified to generate new solutions for enhancing the immune system activation against tumors. However, management of IrAEs, for example, the CRS, and the autoimmune response (AR), are still major obstacles that need more investigation in order to optimize the safety and the efficacy of the treatment (Bilotta et al., 2022; Jachowski et al., 2023).

16.7.4 Targeted and combination therapies

New kinds of cancer treatment, such as the combined and targeted therapies, have brought about a revolution in patient care and resistance to the drugs that are used for that purpose. Rapid drug screening methods have recently unequivocally demonstrated the lifestyle immune system's unearthed responses to the infections, therefore, discharging the resistant bacteria that had been using these genes as protective mechanisms against the antimicrobials (El-Tanani et al., 2024). Receptor Tyrosine Kinase Inhibitors not only have a higher efficacy, leading to fewer side effects, but have also been able to be off late brought to the spotlight in oncology treatments because these agents, namely m-Tors, are good therapy for multiple sclerosis. With the most recent advancements in cancer research, such drug resistance can now be checked by a combination of target molecules and immunotherapy as an already established treatment model, hence making cancer homogeneity and side effects be at a decreased level (Yin et al., 2019). The immune system becomes capable of killing cancer cells by acting as a team player when the targeted therapies are administered together with immunotherapy. On the other hand, the progress of studies has been directed at ameliorating therapeutic utility and, at the same time, promoting the well-being of patients in real-life situations (Shields et al., 2021). The years to come will bring advancements in genome sequencing, enabling doctors to prescribe drugs that have been tailored to the individual genome of the patient, which is personalized medicine. The newly discovered next-generation sequencing and liquid biopsy methods provide up-to-the-minute ultrasonic tumor tracking and continuous alterations of the patient's tumor management approaches (Ma et al., 2023). Additionally, the introduction of ADCs marks a major milestone in the treatment of cancerous cells, with the cytotoxic agents being delivered to cancer cells while at the same time, normal tissues are being saved from suffering. Despite the potential benefits that these methods may have, existing therapies that overcome the resistance to target drugs and the need for patient selection based on personalized needs need to be tested (Séguier et al., 2024).

16.7.5 Tumor microenvironment modulation using biologics

Tumour microenvironment (TME) has been well-established as a key target for cancer treatment, with biologics emerging as central players in changing this complex network. Changing the TME is a promising new way to make anticancer drugs more effective and allow them to overcome drug resistance, as has been pointed out in the most recent studies. The different biological strategies for the correct modulation of the TME are representative of cell growth factors, platelet-rich plasma, cell stem therapy, and finally, the local environment of tumours, which is influenced by scaffolds (Chen et al., 2023). Interpretive biological medications such as anti-TNF-α and anti-IL-1 agents, which have proven to be effective in arthritis treatment, are still an untied territory in oncology; hence, research on their potential in oncology is still needed. State-of-the-art interventions, such as reprogramming the stromal, modulating the immune microenvironment, and targeting the extracellular matrix, are the moves while targeting the TME (Patrick et al., 2023). On the other hand, the latest evidence seems to confirm the idea that the gut microbiota, metabolic pathways, biomechanics, as well as the neuroimmunology axis, are the common factors arising in cancer progression, thus the justification of the whole picture in the fight against cancer. The immune cells' metabolic reprogramming in the tumor microenvironment is a cutting-edge point in cancer immunotherapy that allows the biologics to be optimized. Biologic therapy in TME administration can serve as an attractive niche in cancer treatment (Fig. 16.2), focusing on the size and variety of the TME. By using not only advanced but also innovative treatment approaches and utilizing the dynamics of the TME, these techniques could be able to provide the characteristics of more substantive and personalized cancer treatment methods. However, inevitable future studies should be aimed at unravelling the complex assemblage inside the TME as well as at determining reliable biomarkers for the prediction of therapeutic responses (Kiran et al., 2024; Peterson et al., 2022; Sooi et al., 2023).

16.8 Future perspective

Table 16.3 summarizes key immunotherapy strategies in cancer treatment. Immune checkpoint inhibitors and CAR T-cell therapy have shown remarkable success, particularly in melanoma, lung cancer, and blood cancers. Cancer vaccines and OVs aim to stimulate tumor-specific immune responses, with varying degrees of success. Delivery innovations like intratumoral systems, nanobiomaterials, and carbon nanomaterials enhance therapeutic precision and immune activation, especially in resistant solid tumors. Bispecific T-cell engagers offer targeted killing of cancer cells and show promise in early hematologic trials. Overall, these approaches represent a diverse and rapidly advancing field aiming to improve survival and reduce toxicity in cancer patients.

16.8.1 Personalized medicine and biologics in cancer treatment

Personalized cancer treatments refer to the use of data that precisely suits individual patients and allows for the creation of therapies that suit them, using their molecular profile and the presence of biomarkers. The use of artificial intelligence (AI) and

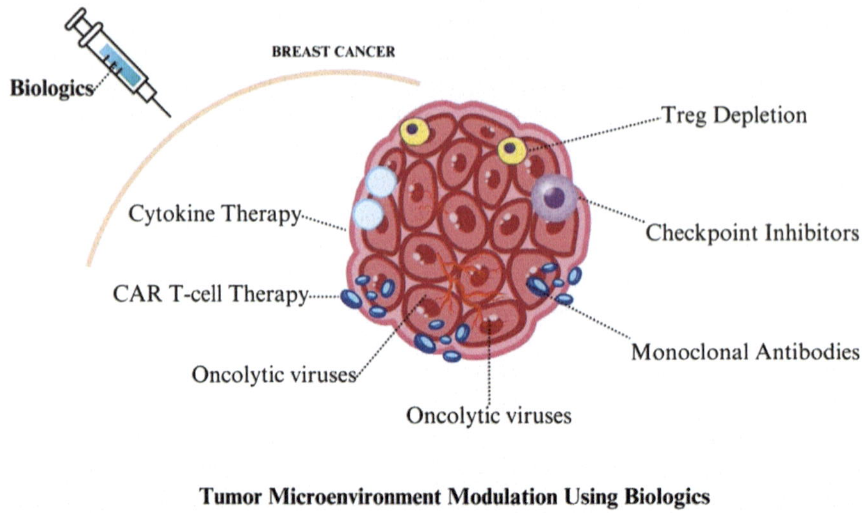

FIGURE 16.2 Tumor microenvironment modulation using biologics.

This illustration demonstrates how various biologic therapies modulate the tumor microenvironment in breast cancer. Biologics—including cytokine therapy, CAR T-cell therapy, checkpoint inhibitors, mAbs, and oncolytic viruses—target the tumor and its surrounding environment to enhance immune-mediated tumor destruction. These agents work by depleting regulatory T cells (Tregs), improving T-cell activation, and directly lysing tumor cells. Oncolytic viruses also disrupt tumor integrity, while cytokines and antibodies boost immune signaling. Collectively, these strategies aim to overcome tumor-induced immunosuppression and improve therapeutic outcomes.

machine learning (ML) in personal care has brought a big change by analysing multiomics data and predicting treatment responses, and early detection development. Liquid biopsies that monitor treatment processes and also find real-time resistance are the ones that enable a true personalized therapy. Apart from that, the application of nanotechnology, which is an important part of this process, is something that makes it possible to implement highly precise treatment (Srivastava et al., 2024; Zhou et al., 2024). However, despite the fact that the technology has so far advanced greatly, there remain some problems to be sorted out, for instance, tumor heterogeneity, the integration of personalized approaches into routine care, and the handling of ethical issues. In addition to that, future technologies such as 3D bio-printing and advanced imaging techniques would really contribute to cancer treatment precision (Riedl et al., 2024). Moreover, progression in the field of single-cell sequencing and spatial transcriptomics has brought a previously unattainable level of detail and understanding in the study of tumor evolution and treatment resistance. These technologies make it possible to monitor the instantaneous process of cellular adaptation to treatments that, in turn, oncologists can adjust based on the occurrence of new cellular dynamics. Organoid models that are also derived from patient tumors do play

Table 16.3 Overview of emerging immunotherapy strategies in cancer treatment.

Immunotherapy Type	Mechanism of Action	Cancer Type	Clinical Outcomes/Advances
Immune checkpoint inhibitors	Blocks immune checkpoints (e.g., PD-1/PD-L1) to enhance T-cell activation	Nonsmall cell lung cancer, melanoma	Single-agent to quadruple therapy options available; significantly improved survival rates
CAR T-cell therapy	Genetically engineered T cells to recognize and kill tumor cells	Acute lymphoblastic leukemia, lymphoma	High remission rates in hematologic malignancies
Cancer vaccines	Induces immune response against tumor-specific antigens	Prostate cancer, melanoma	Limited success in 'cold' tumors like prostate cancer; potential enhanced efficacy in combinations
Intratumoral delivery systems	Direct injection of immunotherapies into tumors	Solid tumors	Localized delivery reduces systemic side effects and increases therapeutic concentrations
Nanobiomaterial-based delivery systems	Enhances delivery of immunotherapeutics and modulates tumor microenvironment	Pancreatic cancer, solid tumors	Improved drug co-delivery and modulation of antitumor immune responses
Carbon nanomaterials	Modulates antitumor immunity and enhances immune activation	Solid tumors	Potential for improving immunotherapy outcomes in resistant cancers
Oncolytic viruses	Genetically engineered viruses selectively infect and destroy cancer cells, activating immunity	Advanced melanoma, glioblastoma	Dual action: direct tumor lysis and immune activation; ongoing clinical trials
Bispecific T-cell engagers (BiTEs)	Redirects T cells to engage and kill tumor cells	Early trials in hematologic malignancies	Demonstrated promising efficacy in targeting cancer cells in preliminary studies

their role in increasing drug screening and therapy customization (Joshi et al., 2022). In that way, they further decrease the number of trial and error in the treatment of cancer. Nonetheless, the lack of data privacy, regulatory approvals, and the high costs associated with genomic profiling are some difficulties that personalized medicine must face before becoming a commonly used clinical treatment approach (Aldea et al., 2023).

16.8.2 Combination therapies with biologics and other modalities

The treatment of cancer becomes more and more successful with the combination of biologics and alternative therapies, as these treatments show, with an increase in their efficacy and the breaking of resistance mechanisms. There are currently ongoing clinical trials that are specifically focusing on the use of biologic combinations to increase the therapeutic outcome and thus make treatment more effective (Scattolin et al., 2024). Over the last few years, there has been a significant breakthrough in molecular diagnostics and cancer biology studies, which resulted in the shift to personalized, targeted therapies. Nevertheless, defining the molecular pattern of the cells, tumor heterogeneity, and the use of these methods in regular care are the major obstacles to be overcome (Wei et al., 2024b). These promising new approaches are the result of the combination of immunotherapy with nanoparticle-based hyperthermia, which will lead to improving the outcomes and reducing the side effects. Furthermore, therapies that exploit ROS cascades along with the combination of photodynamic and hemodynamic treatments offer a significant promise for long-lasting solutions (Fu et al., 2023). Recent research has also brought forward the strong possibility of using biologic and epigenetic therapies together in order to modify the tumor microenvironment and enhance treatment response. Drugs of this type of epigenetic nature, like the so-called DNA methylation inhibitors and histone deacetylase inhibitors, can reverse cancer-associated gene silencing and thus make the tumors more vulnerable to immune checkpoint inhibitors and targeted therapies. Moreover, the inclusion of AI in the creation of drugs is speeding up the process of finding new therapeutic regimens, optimizing the dosing schedules, and minimizing negative effects. Nevertheless, implementation of these innovations in routine clinical practice has to be verified through large-scale clinical trials and different steps of regulatory approval (Douez et al., 2023).

16.8.3 Challenges and opportunities in biosimilar development in cancer treatment

Cancer management with biosimilar medicines is a two-way street, as it can greatly aid in the process of enhancing patient care and reducing the healthcare budget. These are the artificially designed alternatives that are similar to the reference biologics, which can be a good option, without us having to spend more. Because the antibodies or mAbs that are used to treat drug-resistant diseases have different characteristics, the production of biosimilars is a problem. The FDA approval of 17 biosimilars for cancer and their endorsement by ASCO prove their promptness in clinical practice, making them a supportive part of underdeveloped areas. The introduction of real-world data and postmarketing surveillance has been of great help in predicting the long-term safety and efficacy of biosimilars. Pharmacovigilance programs are set up to keep tabs on all the cases, especially those one out of a thousand that might come up when the medicine is on the market. It is also necessary to raise awareness and education among the healthcare professionals and patients with a joint focus group composed mainly of professional people from those regions were moving from reference biologics is still unacceptable. Success in this field will only come through the combined efforts of regulatory agencies, industry stakeholders, and healthcare professionals who advocate for more widespread application (Nahleh et al., 2022; Rodriguez et al., 2023).

16.9 Conclusion

Biologics and biosimilars are fundamental components of contemporary cancer therapy, providing focused strategies that markedly improve treatment effectiveness and patient outcomes. Recent advancements in biologics, including mAbs, immune checkpoint inhibitors, and CAR-T cell treatments, have revolutionized cancer treatment, especially for hematologic malignancies and solid tumours. These novel treatments utilize the body's immune system to specifically target cancer cells, leading to considerable improvements in survival rates and better management of unwanted effects. The advancement and incorporation of biosimilars in oncology provide both prospects and obstacles. Biosimilars have the potential to decrease healthcare expenses and enhance patient access to vital medications; nevertheless, the intricate process of manufacturing these alternatives to reference biologics is a considerable challenge. Notwithstanding these limitations, the FDA's authorization of 17 biosimilars for cancer and their support by ASCO highlight their increasing significance in clinical practice, particularly in resource-constrained environments. Ongoing endeavours in postmarketing surveillance, Pharmacovigilance, and education for healthcare professionals and patients are essential for guaranteeing the safe and effective integration of biosimilars, necessitating collaborative efforts from regulatory authorities, industry stakeholders, and healthcare providers to expand their utilization.

AI disclosure

During the preparation of this work, the author(s) used grammarly tool in order to language editing and grammatical correction during the preparation of the manuscript. After using this tool/service, the author(s) reviewed and edited the content as needed and take(s) full responsibility for the content of the publication.

References

Aggarwal, D., Yang, J., Salam, M. dA., Sengupta, S., Al-Amin, M. dY., Mustafa, S., Khan, M. A., Huang, X., & Pawar, J. S. (2023). Antibody-drug conjugates: The paradigm shifts in the targeted cancer therapy. *Frontiers in Immunology, 14*, 1203073. https://doi.org/10.3389/fimmu.2023.1203073.

Aldea, F., Apcher, J., Mosele, S., & Soria, N. ,A. é (2023). Precision medicine in the era of multi-omics: Can the data tsunami guide rational treatment decision? *ESMO Open, 8*(5), 101627. https://doi.org/10.1016/j.esmoop.2023.101642.

Bachu, R. D., Abou-Dahech, M., Balaji, S., Boddu, S. H. S., Amos, S., Singh, V., Babu, R. J., & Tiwari, A. K. (2022). Oncology biosimilars: New developments and future directions. *Cancer Reports, 5*(11), e1617. https://doi.org/10.1002/cnr2.1720.

Baert, F., Noman, M., Vermeire, S., Assche, G. V., Haens, G. D. ', Carbonez, A. n, & Rutgeerts, P. (2003). Influence of immunogenicity on the long-term efficacy of infliximab in Crohn's disease. *The New England Journal of Medicine, 348*(7), 601–608.

Bas, T. G., & Duarte, V. (2024). Biosimilars in the era of artificial intelligence—International regulations and the use in oncological treatments. *Pharmaceuticals, 17*(7), 875. https://doi.org/10.3390/ph17070925.

Bilotta, M. T., Antignani, A., & Fitzgerald, D. J. (2022). Managing the TME to improve the efficacy of cancer therapy. *Frontiers in Immunology, 13*, 902342. https://doi.org/10.3389/fimmu.2022.954992.

Blandizzi, C., Meroni, P. L., & Lapadula, G. (2017). Comparing originator biologics and biosimilars: A review of the relevant issues. *Clinical Therapeutics, 39*(5), 1026–1039. https://doi.org/10.1016/j.clinthera.2017.03.014.

Bombardier, C., Hazlewood, G. S., Akhavan, P., Schieir, O., Dooley, A., Haraoui, B., Khraishi, M., Leclercq, S. A., Légaré, J., Mosher, D. P., Pencharz, J., Pope, J. E., Thomson, J., Thorne, C., Zummer, M., Gardam, M. A., Askling, J., & Bykerk, V. (2012). Canadian rheumatology association recommendations for the pharmacological management of rheumatoid arthritis with traditional and biologic disease-modifying antirheumatic drugs: Part II safety. *Journal of Rheumatology, 39*(8), 1583–1602. https://doi.org/10.3899/jrheum.120165.

Bracarda, S., Eggermont, A. M. M., & Samuelsson, J. (2010). Redefining the role of interferon in the treatment of malignant diseases. *European Journal of Cancer, 46*(2), 284–297. https://doi.org/10.1016/j.ejca.2009.10.013.

Buchbinder, E. I., & Desai, A. (2016). CTLA-4 and PD-1 pathways: Similarities, differences, and implications of their inhibition. *American Journal of Clinical Oncology, 39*(1), 98–106. https://doi.org/10.1097/COC.0000000000000239.

Buchler, T., Chloupkova, R., Poprach, A., Fiala, O., Kiss, I., Kopeckova, K., Dusek, L., Veskrnova, V., Slavicek, L., Kohoutek, M., Finek, J., Svoboda, M., Petruzelka, L., & Melichar, B. (2019). Sequential therapy with bevacizumab and EGFR inhibitors for metastatic colorectal carcinoma: A national registry-based analysis. *Cancer Management and Research, 11*, 359–368. https://doi.org/10.2147/CMAR.S183093.

Chao, J., Skup, M., Alexander, E., Tundia, N., Macaulay, D., Wu, E., & Mulani, P. (2015). Nomenclature and traceability debate for biosimilars: small-molecule surrogates lend support for distinguishable nonproprietary names. *Advances in Therapy, 32*(3), 270–283. https://doi.org/10.1007/s12325-015-0193-5.

Chehelgerdi, M., Chehelgerdi, M., Khorramian-Ghahfarokhi, M., Shafieizadeh, M., Mahmoudi, E., Eskandari, F., Rashidi, M., Arshi, A., & Mokhtari-Farsani, A. (2024). Comprehensive review of CRISPR-based gene editing: Mechanisms, challenges, and applications in cancer therapy. *Molecular Cancer, 23*(1), 160. https://doi.org/10.1186/s12943-023-01925-5.

Chen, L., Zuo, M., Zhou, Q., & Wang, Y. (2023). Oncolytic virotherapy in cancer treatment: Challenges and optimization prospects. *Frontiers in Immunology, 14*, 1183821. https://doi.org/10.3389/fimmu.2023.1308890.

Chen, W., Li, C., & Jiang, X. (2023). Advanced biomaterials with intrinsic immunomodulation effects for cancer immunotherapy. *Small Methods, 7*(5), 2201508. https://doi.org/10.1002/smtd.202201404.

Chen, X., Chen, L.-J., Peng, X.-F., Deng, L., Wang, Y., Li, J.-J., Guo, D.-L. i, & Niu, X.-H. (2024). Anti-PD-1/PD-L1 therapy for colorectal cancer: Clinical implications and future considerations. *Translational Oncology, 40*, 101852.

Chiocca, E. A., & Rabkin, S. D. (2014). Oncolytic viruses and their application to cancer immunotherapy. *Cancer Immunology Research, 2*(4), 295–300. https://doi.org/10.1158/2326-6066.CIR-14-0015.

Chopra, R., & Lopes, G. (2017). Improving access to cancer treatments: The role of biosimilars. *Journal of Global Oncology, 3*(5), 596–610. https://doi.org/10.1200/JGO.2016.008607.

Clarke, K., Ainslie-Garcia, M., Ferko, N., & Shastri, K. (2024). Modelling the opportunity for cost-savings or patient access with biosimilar adalimumab and tocilizumab: A European perspective. *Journal of Medical Economics, 27*(1), 952–962. https://doi.org/10.1080/13696998.2024.2379212.

Conibear, A. C., Schmid, A., Kamalov, M., Becker, C. F. W., & Bello, C. (2020). Recent advances in peptide-based approaches for cancer treatment. *Current Medicinal Chemistry, 27*(8), 1174–1205. https://doi.org/10.2174/0929867325666171123204851.

Cordeiro, M. A., Vitorino, C., Sinogas, C., & Sousa, J. oãoJ. (2024). A regulatory perspective on biosimilar medicines. *Pharmaceutics, 16*(3), 343. https://doi.org/10.3390/pharmaceutics16030321.

Danese, S., Fiorino, G., & Michetti, P. (2016). Changes in biosimilar knowledge among European Crohn's Colitis Organization [ECCO] members: An updated survey. *Journal of Crohn's and Colitis, 10*(11), 1362–1365. https://doi.org/10.1093/ecco-jcc/jjw090.

Davila, M. L., Brentjens, R., Wang, X., Rivière, I., & Sadelain, M. (2012). How do CARs work?: Early insights from recent clinical studies targeting CD19. *Oncoimmunology, 1*(9), 1577–1583.

Debela, D. T., Gy Muzazu, S., Digamo Heraro, K., Ndalama, M. T., Woldemedhin Mesele, B., Chilot Haile, D., Kitui, S. K., & Manyazewal, T. (2021). New approaches and procedures for cancer treatment: Current perspectives. *SAGE Open Medicine, 9*, 20503121211034366. https://doi.org/10.1177/20503121211034366.

Douez, E., D'Atri, V., Guillarme, D., Antier, D., Guerriaud, M., Beck, A., Watier, H é, & Foucault-Fruchard, L. (2023). Why is there no biosimilar of Erbitux®? *Journal of Pharmaceutical and Biomedical Analysis, 234*, 115478. https://doi.org/10.1016/j.jpba.2023.115544.

Dziubańska-Kusibab, P. J., Nevedomskaya, E., & Haendler, B. (2024). Preclinical anticipation of on- and off-target resistance mechanisms to anti-cancer drugs: A systematic review. *International Journal of Molecular Sciences, 25*(2), 1023. https://doi.org/10.3390/ijms25020705.

El-Tanani, M., Rabbani, S. A., Babiker, R., Rangraze, I., Kapre, S., Palakurthi, S. S., Alnuqaydan, A. M., Aljabali, A. A., Rizzo, M., El-Tanani, Y., & Tambuwala, M. M. (2024). Unraveling the tumor microenvironment: Insights into cancer metastasis and therapeutic strategies. *Cancer Letters, 591*, 216847. https://doi.org/10.1016/j.canlet.2024.216894.

Florian, J., Sun, Q., Schrieber, S. J., White, R., Shubow, S., Johnson-Williams, B. E., Sheikhy, M., Harrison, N. R., Parker, V. J., Wang, Y.-M., & Strauss, D. G. (2023). Pharmacodynamic biomarkers for biosimilar development and approval: A workshop summary. *Clinical Pharmacology and Therapeutics, 113*(5), 1030–1035. https://doi.org/10.1002/cpt.2795.

Fu, Q., Yu, L., Zhang, M., Li, S., & Liu, L. (2023). Engineering nanosystems for ROS-bridged cancer cascade therapy. *Chemical Engineering Journal, 473*, 145556. https://doi.org/10.1016/j.cej.2023.145415.

Garbayo, E., Pascual-Gil, S., Rodríguez-Nogales, C., Saludas, L., Mendoza, A. E.-H., & Blanco-Prieto, M. J. (2020). Nanomedicine and drug delivery systems in cancer and regenerative medicine. *Wiley Interdisciplinary Reviews: Nanomedicine and Nanobiotechnology, 12*(5), e1637. https://doi.org/10.1002/wnan.1637.

Garg, P., Singhal, S., Kulkarni, P., Horne, D., Malhotra, J., Salgia, R., & Singhal, S. S. (2024). Advances in non-small cell lung cancer: Current insights and future directions. *Journal of Clinical Medicine, 13*(14), 4060. https://doi.org/10.3390/jcm13144189.

Garg, P., Pareek, S., Kulkarni, P., Horne, D., Salgia, R., & Singhal, S. S. (2024). Next-generation immunotherapy: Advancing clinical applications in cancer treatment. *Journal of Clinical Medicine, 13*(21), 6328. https://doi.org/10.3390/jcm13216537.

Gherghescu, I., & Delgado-Charro, M. B. ña (2021). The biosimilar landscape: An overview of regulatory approvals by the EMA and FDA. *Pharmaceutics, 13*(1), 1–16. https://doi.org/10.3390/pharmaceutics13010048.

Gogia, P., Ashraf, H., Bhasin, S., & Xu, Y. (2023). Antibody–drug conjugates: A review of approved drugs and their clinical level of evidence. *Cancers, 15*(15), 3897. https://doi.org/10.3390/cancers15153886.

Gouda, M. A., & Subbiah, V. (2023). Precision oncology for BRAF-mutant cancers with BRAF and MEK inhibitors: From melanoma to tissue-agnostic therapy. *ESMO Open, 8*(2), 100819.

Gu, Y., Wang, Z., & Wang, Y. (2024). Bispecific antibody drug conjugates: Making 1+1>2. *Acta Pharmaceutica Sinica B, 14*(5), 1965–1986. https://doi.org/10.1016/j.apsb.2024.01.009.

Gupta, B., & Kim, J. O. h (2021). Recent progress in cancer immunotherapy approaches based on nanoparticle delivery devices. *Journal of Pharmaceutical Investigation, 51*(4), 399–412. https://doi.org/10.1007/s40005-021-00527-x.

Herndon, T. M., Ausin, C., Brahme, N. N., Schrieber, S. J., Luo, M., Andrada, F. C., Kim, C., Sun, W., Zhou, L., Grosser, S., Yim, S., & Ricci, M. S. (2023). Safety outcomes when switching between biosimilars and reference biologics: A systematic review and meta-analysis. *PLoS ONE, 18*(10 October), e0292817. https://doi.org/10.1371/journal.pone.0292231.

Huang, H.-Y., Liu, C.-C., Yu, Y., Wang, L. e, Wu, D. a-W., Guo, L.-W., Wang, S.-H., Fang, H., Bai, Y., Fang, Y., Fan, Q. i, Sun, C., Wu, Y., Shi, J. u-F., Ma, F., Tang, Y. u, Dai, M., & Li, N. (2020). Pharmacoeconomic evaluation of cancer biosimilars worldwide: A systematic review. *Frontiers in Pharmacology, 11*, 601569. https://doi.org/10.3389/fphar.2020.572569.

Humphreys, S. Z. (2022). Real-world evidence of a successful biosimilar adoption program. *Future Oncology, 18*(16), 1997–2006. https://doi.org/10.2217/fon-2021-1584.

Isakov, L., Jin, B. o, & Jacobs, I. A. (2016). Statistical primer on biosimilar clinical development. *American Journal of Therapeutics, 23*(6), e1903–e1910. https://doi.org/10.1097/MJT.0000000000000391.

Jachowski, A., Marcinkowski, M. ł, Szydłowski, J., Grabarczyk, O., Nogaj, Z., Marcin, Ł, Pławski, A., Jagodziński, P. łP., & Słowikowski, B. K. (2023). Modern therapies of nonsmall cell lung cancer. *Journal of Applied Genetics, 64*(4), 695–711. https://doi.org/10.1007/s13353-023-00786-4.

Jia, J., Wu, X., Long, G., Yu, J., He, W., Zhang, H., Wang, D., Ye, Z., & Tian, J. (2023). Revolutionizing cancer treatment: Nanotechnology-enabled photodynamic therapy and immunotherapy with advanced photosensitizers. *Frontiers in Immunology, 14*, 1185998. https://doi.org/10.3389/fimmu.2023.1219785.

Joshi, D., Khursheed, R., Gupta, S., Wadhwa, D., Singh, T. G., Sharma, S., Porwal, S., Gauniyal, S., Vishwas, S., Goyal, S., Gupta, G., Eri, R. D., Williams, K. A., Dua, K., & Singh, S. K. (2022). Biosimilars in oncology: Latest trends and regulatory status. *Pharmaceutics., 14*(12), 2727. https://doi.org/10.3390/pharmaceutics14122721.

Kabir, E. R., Moreino, S. S., & Siam, M. K. S. (2019). The breakthrough of biosimilars: A twist in the narrative of biological therapy. *Biomolecules, 9*(9), 410. https://doi.org/10.3390/biom9090410.

Kciuk, M., Bashir Yahya, E., Mohamed, M. M. I., Rashid, S., Omer Iqbal, M., Kontek, R., Abdulsamad, M. A., & Allaq, A. A. (2023). Recent advances in molecular mechanisms of cancer immunotherapy. *Cancers, 15*(10), 2684. https://doi.org/10.3390/cancers15102721.

Khraishi, M., Stead, D., Lukas, M., Scotte, F., & Schmid, H. (2016). Biosimilars: A multidisciplinary perspective. *Clinical Therapeutics, 38*(5), 1238–1249.

Kim, C. G., Sang, Y. B., Lee, J. iH., & Chon, H. J. (2021). Combining cancer vaccines with immunotherapy: Establishing a new immunological approach. *International Journal of Molecular Sciences, 22*(15), 8031. https://doi.org/10.3390/ijms22158035.

Kiran, N. S., Yashaswini, C., Maheshwari, R., Bhattacharya, S., & Prajapati, B. G. (2024). Advances in precision medicine approaches for colorectal cancer: From molecular profiling to targeted therapies. *ACS Pharmacology and Translational Science, 7*(4), 967–990. https://doi.org/10.1021/acsptsci.4c00008.

Kirchhoff, C. F., Wang, X.-Z. M., Conlon, H. D., Anderson, S., Ryan, A. M., & Bose, A. (2017). Biosimilars: Key regulatory considerations and similarity assessment tools. *Biotechnology and Bioengineering, 114*(12), 2696–2705. https://doi.org/10.1002/bit.26438.

Kvien, T. K., Patel, K., & Strand, V. (2022). The cost savings of biosimilars can help increase patient access and lift the financial burden of health care systems. *Seminars in Arthritis and Rheumatism, 52*, 151939.

La'ah, A. S., & Chiou, S. H. (2024). Cutting-edge therapies for lung cancer. *Cells., 13*(5), 404. https://doi.org/10.3390/cells13050436.

Lee, J. F., Litten, J. B., & Grampp, G. (2012). Comparability and biosimilarity: Considerations for the healthcare provider. *Current Medical Research and Opinion, 28*(6), 1053–1058. https://doi.org/10.1185/03007995.2012.686902.

Leonard, E., Wascovich, M., Oskouei, S., Gurz, P., & Carpenter, D. (2019). Factors affecting health care provider knowledge and acceptance of biosimilar medicines: A systematic review. *Journal of Managed Care and Specialty Pharmacy, 25*(1), 102–112. https://doi.org/10.18553/jmcp.2019.25.1.102.

Liu, B., Zhou, H., Tan, L., Siu, K. T. oH., & Guan, X.-Y. (2024). Exploring treatment options in cancer: Tumor treatment strategies. *Signal Transduction and Targeted Therapy, 9*(1), 74. https://doi.org/10.1038/s41392-024-01856-7.

Liu, G.-H., Chen, T., Zhang, X., Ma, X.-L., & Shi, H.-S. (2022). Small molecule inhibitors targeting the cancers. *MedComm, 3*(4), e173. https://doi.org/10.1002/mco2.181.

Ma, R., Li, Z., Chiocca, E. A., Caligiuri, M. A., & Yu, J. (2023). The emerging field of oncolytic virus-based cancer immunotherapy. *Trends in Cancer, 9*(2), 122–139. https://doi.org/10.1016/j.trecan.2022.10.003.

Malakar, S., Gontor, E. N., Dugbaye, M. Y., Shah, K., Sinha, S., Sutaoney, P., & Singh Chauhan, N. (2024). Cancer treatment with biosimilar drugs: A review. *Cancer Innovation, 3*(2), e129. https://doi.org/10.1002/cai2.115.

Markus, R., Liu, J., Ramchandani, M., Landa, D., Born, T., & Kaur, P. (2017). Developing the totality of evidence for biosimilars: Regulatory considerations and building confidence for the healthcare community. *BioDrugs, 31*(3), 175–187. https://doi.org/10.1007/s40259-017-0218-5.

Marofi, F., Motavalli, R., Safonov, V. A., Thangavelu, L., Valerievich Yumashev, A., Alexander, M., Shomali, N., Chartrand, M. S., Pathak, Y., Jarahian, M., Izadi, S., Hassanzadeh, A., Shirafkan, N., Tahmasebi, S., & Khiavi, F. H. (2021). CAR T cells in solid tumors: Challenges and opportunities. *Stem Cell Research & Therapy, 12*(1), 81. https://doi.org/10.1186/s13287-020-02128-1.

Mascarenhas-Melo, F., Diaz, M., Gonçalves, M. B. S., Vieira, P., Bell, V., Viana, S., Nunes, S., Paiva-Santos, A. C. lá, & Veiga, F. (2024). An overview of biosimilars—Development, quality, regulatory issues, and management in healthcare. *Pharmaceuticals, 17*(2), 217. https://doi.org/10.3390/ph17020235.

Mbituyimana, B., Adhikari, M., Qi, F., Shi, Z., Fu, L., & Yang, G. (2023). Microneedle-based cell delivery and cell sampling for biomedical applications. *Journal of Controlled Release, 362*, 692–714. https://doi.org/10.1016/j.jconrel.2023.09.013.

McKinnon, R. A., Cook, M., Liauw, W., Marabani, M., Marschner, I. C., Packer, N. H., & Prins, J. B. (2018). Biosimilarity and interchangeability: Principles and evidence: A systematic review. *BioDrugs: Clinical Immunotherapeutics, Biopharmaceuticals and Gene Therapy., 32*(1), 27–52. https://doi.org/10.1007/s40259-017-0256-z.

Mendes de Abreu, M., Strand, V., Levy, R. A., & Vianna Araujo, D. (2014). Putting the value into biosimilar decision making. The judgment value criteria. *Autoimmunity Reviews, 13*(6), 678–684. https://doi.org/10.1016/j.autrev.2014.01.051.

Mestre-Ferrandiz, J., Czech, M., Smolen, J. S., Cornes, P., Aapro, M. S., Danese, S., Deitch, S., Tyldsley, H., Foster, W., Shah, P., Latymer, M., & Vulto, A. G. (2024). Capturing the holistic value of biosimilars in Europe–part 1: A historical perspective. *Expert Review of Pharmacoeconomics and Outcomes Research, 24*(2), 237–250. https://doi.org/10.1080/14737167.2023.2297926.

Miller, E. M., & Schwartzberg, L. S. (2019). Biosimilars for breast cancer: A review of HER2-targeted antibodies in the United States. *Therapeutic Advances in Medical Oncology, 11*, 1758835919851744. https://doi.org/10.1177/1758835919887044.

Min, H.-Y., & Lee, H. o-Y. (2022). Molecular targeted therapy for anticancer treatment. *Experimental & Molecular Medicine, 54*(10), 1670–1694. https://doi.org/10.1038/s12276-022-00864-3.

Mishra, D., Gade, S., Pathak, V., Vora, L. K., Mcloughlin, K., Medina, R., Donnelly, R. F., & Thakur, R. R. S. (2023). Ocular application of electrospun materials for drug delivery and cellular therapies. *Drug Discovery Today, 28*(9), 103659. https://doi.org/10.1016/j.drudis.2023.103676.

Mooradian, M. J., & Sullivan, R. J. (2023). Immunotherapy in melanoma: Recent advancements and future directions. *Cancers, 15*(16), 4096. https://doi.org/10.3390/cancers15164176.

Nahleh, Z., Lyman, G. H., Schilsky, R. L., Peterson, D. E., Tagawa, S. T., Chavez-MacGregor, M., Rumble, R. B., & Gupta, S. (2022). Use of biosimilar medications in oncology. *JCO Oncology Practice, 18*(3), 177–186. https://doi.org/10.1200/OP.21.00771.

Navarro, S., Carrillo, E., Griñán-Lisón, C., Martín, A., Perán, M., Marchal, J. A., & Boulaiz, H. (2016). Cancer suicide gene therapy: A patent review. *Expert Opinion on Therapeutic Patents, 26*(9), 1095–1104. https://doi.org/10.1080/13543776.2016.1211640.

O'Callaghan, J., Barry, S. P., Bermingham, M., Morris, J. M., & Griffin, B. T. (2019). Regulation of biosimilar medicines and current perspectives on interchangeability and policy. *European Journal of Clinical Pharmacology, 75*(1), 1–11. https://doi.org/10.1007/s00228-018-2542-1.

Oskouei, S. T., Kusmierczyk, A. R., & Uptake, B. (2021). The importance of healthcare provider education. *Pharmaceutical Medicine, 35*(4), 215–224. https://doi.org/10.1007/s40290-021-00396-7.

Pan, Z., Huang, W., Pan, J., & Koch, G. (2021). Design and analysis of a biosimilar bridging study with a prediction interval-based consistency test. *Therapeutic Innovation and Regulatory Science, 55*(4), 765–772. https://doi.org/10.1007/s43441-021-00280-1.

Papież, M. A., & Krzyściak, W. (2021). Biological therapies in the treatment of cancer—Update and new directions. *International Journal of Molecular Sciences, 22*(21), 1170. https://doi.org/10.3390/ijms222111694.

Park, W., Hrycaj, P., Jeka, S., Kovalenko, V., Lysenko, G., Miranda, P., Mikazane, H., Gutierrez-Ureña, S., Lim, M. J., Lee, Y., Lee, S. J., Kim, H. oU., Yoo, D. H., & Braun, J. ü (2013). A randomised, double-blind, multicentre, parallel-group, prospective study comparing the pharmacokinetics, safety, and efficacy of CT-P13 and innovator infliximab in patients with ankylosing spondylitis: The PLANETAS study. *Annals of the Rheumatic Diseases, 72*(10), 1605–1612. https://doi.org/10.1136/annrheumdis-2012-203091.

Patrick, B., Akhtar, T., Kousar, R., Huang, C.-C., Li, X.-G., & Nanomaterials, C. (2023). Emerging roles in immuno-oncology. *International Journal of Molecular Sciences, 24*(7). https://doi.org/10.3390/ijms24076600.

Pavelescu, L. A., Enache, R. M., Roşu, O. A., Profir, M., Creţoiu, S. M., & Gaspar, B. S. (2024). Predictive biomarkers and resistance mechanisms of checkpoint inhibitors in malignant solid tumors. *International Journal of Molecular Sciences, 25*(17), 9364. https://doi.org/10.3390/ijms25179659.

Peeters, M., Planchard, D., Pegram, M., Gonçalves, J. oão, Bocquet, F. ç, & Jang, H. (2021). Biosimilars in an era of rising oncology treatment options. *Future Oncology, 17*(29), 3881–3892. https://doi.org/10.2217/fon-2021-0546.

Peterson, C., Denlinger, N., & Yang, Y. (2022). Recent advances and challenges in cancer immunotherapy. *Cancers, 14*(16), 4027. https://doi.org/10.3390/cancers14163972.

Phyo, P., Zhao, X. i, Templeton, A. C., Xu, W., Cheung, J. K., & Su, Y. (2021). Understanding molecular mechanisms of biologics drug delivery and stability from NMR spectroscopy. *Advanced Drug Delivery Reviews, 174*, 1–29.

Rahalkar, H., Cetintas, H. C., & Salek, S. (2018). Quality, non-clinical and clinical considerations for biosimilar monoclonal antibody development: EU, WHO, USA, Canada, and BRICS-TM regulatory guidelines. *Frontiers in Pharmacology, 9*, 1079. https://doi.org/10.3389/fphar.2018.01079.

Riedl, J. M., Moik, F., Esterl, T., Kostmann, S. M., Gerger, A., & Jost, P. J. (2024). Molecular diagnostics tailoring personalized cancer therapy—An oncologist's view. *Virchows Archiv, 484*(2), 169–179. https://doi.org/10.1007/s00428-023-03702-7.

Rodriguez, G., Mancuso, J., Lyman, G. H., Cardoso, F., Nahleh, Z., Vose, J. M., Gralow, J. R., Francisco, M., & Sherwood, S. (2023). ASCO policy statement on biosimilar and interchangeable products in oncology. *JCO Oncology Practice, 19*(7), 411–419. https://doi.org/10.1200/OP.22.00783.

Runcie, K. D., & Dallos, M. C. (2021). Prostate cancer immunotherapy—Finally in from the cold? *Current Oncology Reports, 23*(8), 94. https://doi.org/10.1007/s11912-021-01084-0.

Scattolin, T., Tonon, G., Botter, E., Canale, V. C., Hasanzadeh, M., Cuscela, D. M., Buschini, A., Zarepour, A., Khosravi, A., Cordani, M., Rizzolio, F., & Zarrabi, A. (2024). Synergistic applications of cyclodextrin-based systems and metal-organic frameworks in transdermal drug delivery for skin cancer therapy. *Journal of Materials Chemistry B, 12*(16), 3807–3839. https://doi.org/10.1039/d4tb00312h.

Séguier, D., Adams, E. S., Kotamarti, S., D'Anniballe, V., Michael, Z. D., Deivasigamani, S., Olivier, J., Villers, A., Hoimes, C., & Polascik, T. J. (2024). Intratumoural immunotherapy plus focal thermal ablation for localized prostate cancer. *Nature Reviews Urology., 21*(5), 290–302. https://doi.org/10.1038/s41585-023-00834-y.

Shields, M. D., Marin-Acevedo, J. A., & Pellini, B. (2021). Immunotherapy for advanced non-small cell lung cancer: A decade of progress. *American Society of Clinical Oncology., 41*, 105–127. https://doi.org/10.1200/EDBK_321483.

Shiravand, Y., Khodadadi, F., Kashani, S. M. A., Hosseini-Fard, S. R., Hosseini, S., Sadeghi-rad, H., Ladwa, R., O'Byrne, K., & Kulasinghe, A. (2022). Immune checkpoint inhibitors in cancer therapy. *Current Oncology (Toronto, Ont.)., 29*(5), 3044–3060. https://doi.org/10.3390/curroncol29050247.

Skingle, D. (2015). Biosimilars: What do patients need to consider? *RMD Open, 1*(1). https://doi.org/10.1136/rmdopen-2015-000141.

Socinski, M. A., Curigliano, G., Jacobs, I., Gumbiner, B., MacDonald, J., & Thomas, D. (2015). Clinical considerations for the development of biosimilars in oncology. *mAbs, 7*(2), 286–293. https://doi.org/10.1080/19420862.2015.1008346.

Sooi, K., Walsh, R., Kumarakulasinghe, N., Wong, A., & Ngoi, N. (2023). A review of strategies to overcome immune resistance in the treatment of advanced prostate cancer. *Cancer Drug Resistance, 6*(3), 656–673. https://doi.org/10.20517/cdr.2023.48.

Srivastava, S., Jayaswal, N., Kumar, S., Sharma, P. K., Behl, T., Khalid, A., Mohan, S., Najmi, A., Zoghebi, K., & Alhazmi, H. A. (2024). Unveiling the potential of proteomic and genetic signatures for precision therapeutics in lung cancer management. *Cellular Signalling, 113*, 110997. https://doi.org/10.1016/j.cellsig.2023.110932.

Stevenson, J. G. (2015). Clinical data and regulatory issues of biosimilar products. *The American Journal of Managed Care, 21*(16), s331–s340 Suppl.

Tabernero, J., Vyas, M., Giuliani, R., Arnold, D., Cardoso, F., Casali, P. G., Cervantes, A., Eggermont, A. M. M., Eniu, A., Jassem, J., Pentheroudakis, G., Peters, S., Rauh, S., Zielinski, C. C., Stahel, R. A., Voest, E., Douillard, J.-Y., McGregor, K., & Ciardiello, F. (2016). Biosimilars: A position paper of the European Society for Medical Oncology, with particular reference to oncology prescribers. *ESMO Open, 1*(6), e000142.

Tandulje, A., Varpe, P., Chaugule, P., Singh Raghuvanshi, R., & Srivastava, S. (2024). Deciphering the US regulatory framework: Comparison between oncology biosimilars and reference biologics. *Seminars in Oncology, 51*(5-6), 135–141. https://doi.org/10.1053/j.seminoncol.2024.08.002.

Thill, M. (2019). Biosimilar trastuzumab in clinical trials: Differences or not? *Breast Care, 14*(1), 17–22. https://doi.org/10.1159/000496503.

Tsiftsoglou, A. S., Ruiz, S., & Schneider, C. K. (2013). Development and regulation of biosimilars: Current status and future challenges. *BioDrugs, 27*(3), 203–211. https://doi.org/10.1007/s40259-013-0020-y.

Vincent, F. B., Morand, E. F., Murphy, K., Mackay, F., Mariette, X., & Marcelli, C. (2013). Antidrug antibodies (ADAb) to tumour necrosis factor (TNF)-specific neutralising agents in chronic inflammatory diseases: A real issue, a clinical perspective. *Annals of the Rheumatic Diseases, 72*(2), 165–178. https://doi.org/10.1136/annrheumdis-2012-202545.

Wang, J., Lu, X.-X., Chen, D.-Z., Li, S.-F., & Zhang, L. i-S. (2004). Herpes simplex virus thymidine kinase and ganciclovir suicide gene therapy for human pancreatic cancer. *World Journal of Gastroenterology, 10*(3), 400–403. https://doi.org/10.3748/wjg.v10.i3.400.

Wang, J.-Y. i, & Wang, L. (2023). CAR-T cell therapy: Where are we now, and where are we heading? *Blood Science (Baltimore, Md.), 5*(4), 237–248. https://doi.org/10.1097/BS9.0000000000000173.

Wei, J., Li, W., Zhang, P., Guo, F., & Liu, M. (2024a). Current trends in sensitizing immune checkpoint inhibitors for cancer treatment. *Molecular Cancer, 23*(1). https://doi.org/10.1186/s12943-024-02179-5.

Wei, Q., Li, P., Yang, T., Zhu, J., Sun, L. u, Zhang, Z., Wang, L. u, Tian, X., Chen, J., Hu, C., Xue, J., Ma, L., Shimura, T., Fang, J., Ying, J., Guo, P., & Cheng, X. (2024b). The promise and challenges of combination therapies with antibody-drug conjugates in solid tumors. *Journal of Hematology and Oncology, 17*(1). https://doi.org/10.1186/s13045-023-01509-2.

Xiong, F., Wang, Q. i, Wu, G.-H., Liu, W.-Z., & Wang, B. (2022). Yong-Jun Chen, Direct and indirect effects of IFN-α2b in malignancy treatment: Not only an archer but also an arrow. *Biomarker Research, 10*(1), 55. https://doi.org/10.1186/s40364-022-00415-y.

Yin, Z., Bai, L., Li, W., Zeng, T., Tian, H., & Cui, J. (2019). Targeting T cell metabolism in the tumor microenvironment: An anti-cancer therapeutic strategy. *Journal of Experimental & Clinical Cancer Research: CR, 38*(1), 403. https://doi.org/10.1186/s13046-019-1409-3.

Yoo, D. H., Prodanovic, N., Jaworski, J., Miranda, P., Ramiterre, E., Lanzon, A., Baranauskaite, A., Wiland, P., Abud-Mendoza, C., Oparanov, B., Smiyan, S., Kim, H. oU., Lee, S. J., & Kim, S. uY. (2017). Won Park, Efficacy and safety of CT-P13 (biosimilar infliximab) in patients with rheumatoid arthritis: Comparison between switching from reference infliximab to CT-P13 and continuing CT-P13 in the PLANETRA extension study. *Annals of the rheumatic diseases, 76*(2), 355–363. https://doi.org/10.1136/annrheumdis-2015-208786.

Zahavi, D., & Weiner, L. (2020). Monoclonal antibodies in cancer therapy. *Antibodies, 9*(3), 1–20. https://doi.org/10.3390/antib9030034.

Zhou, Y., Peng, S., Wang, H., Cai, X., & Wang, Q. (2024). Review of personalized medicine and pharmacogenomics of anti cancer compounds and natural products. *Genes, 15*(4), 458. https://doi.org/10.3390/genes15040468.

CHAPTER

17 Marching of biologics and biosimilars in artificial intelligence and machine learning as tools to detect various cancers

Neelufar Shama Shaik[a], Srinivasan Ramamurthy[b], Harika Balya[b], Balaji Pandiyan[c], Saravanan Govindaraj[d] and Shanmugasundaram Palani[e]

[a] *Department of Pharmacognosy, Scient Institute of Pharmacy, Ibrahimpatnam, Telangana, India,* [b] *Department of Pharmacy, College of Dentistry & Health Sciences, Fujairah University, Fujairah, United Arab Emirates,* [c] *Department of Pharmacology, School of Pharmaceutical Sciences, Vels Institute of Science, Technology and Advanced Studies, Chennai, Tamil Nadu, India,* [d] *Department of Pharmaceutical Chemistry, Vel Pharmacy College, VISTAS Off Campus Centre, Vels Institute of Science, Technology and Advanced Studies, Tiruvallur, Tamil Nadu, India,* [e] *Department of Pharmaceutical Chemistry and Analysis, School of Pharmaceutical Sciences, Vels Institute of Science, Technology and Advanced Studies, Chennai, Tamil Nadu, India*

17.1 Introduction

The development of cancer therapy is a milestone achieved in modern biomedical research. For many years, therapeutic strategies relied on small-molecule chemotherapeutic agents that acted by interfering with cell replication and triggering cell death. Though they were successful, these agents often showed poor selectivity, caused severe adverse reactions, restricted maximum dosing, and led to the development of drug resistance (Khan & Spicer, 2019). The development of biologics or large complex molecules, such as monoclonal antibodies, recombinant proteins, cytokines, and nucleic acid-based molecules, has completely modified the field of cancer therapy. Biologics showed improved selectivity by specifically targeting tumor receptors and molecular pathways, thereby improving efficacy and reducing adverse effects (Seebacher et al., 2019).

Biologics have emerged as a leading player during the past two decades in the cancer therapeutic drug market. The data from U.S. Food and Drug Administration (FDA) from 2009 to 2023 revealed that out of 599 new therapeutic agents approved, 159 (or 27%) were biologics and 440 (or 73%) were small molecule drugs. The median time for development of both biologics and small molecules was almost equal, at 12.6 years and 12.7 years, respectively, but showed higher success rates in all clinical

Biosimilars and Biologics in Cancer Treatment: An Era in Therapeutic Strategies.
DOI: https://doi.org/10.1016/B978-0-443-43824-0.00012-2

stages (Mullard, 2024). The median estimated cost of biologic development was about US$3.0 billion, compared with US$2.1 billion for small molecules. Biologics have been shown to obtain better intellectual property rights, with an average of 14 patents per biologic and market exclusivity of over 20 years. In the cancer therapy field, biologics now account for more than half of the total number of marketed drugs, especially in the treatment of solid tumors, thus emphasizing their importance in cancer therapy (Wong et al., 2018).

Biosimilars are biologic agents that share similar characteristics with approved biologic references, but they are not exactly the same. The development of biosimilars has broadened the scope of access to innovative therapies by providing similar efficacy, safety, and immunogenicity profiles and also with a significantly reduced cost, thus overcoming the affordability crisis in cancer treatment globally (Weise et al., 2012). The regulatory bodies, such as the FDA, EMA, and CDSCO, have developed strict guidelines that define the criteria for comparability, which will ensure the safety and efficacy of these agents (Jois et al., 2020). Biosimilars from major biologic agents such as trastuzumab, bevacizumab, and rituximab reduce the economic burden on the healthcare system and also improve access to life-saving therapies for various populations (Cohen et al., 2018).

Cancer is a major global health problem. The International Agency for Research on Cancer reported in 2022 that approximately 20 million new cancer cases were diagnosed and 9.7 million deaths occurred worldwide. Lung cancer comprised 12.4% of new cases, followed by 11.6% for breast cancer and 9.6% for colorectal cancer. Rare cancers together comprised about 26.7% of all new cases, and nearly 30% of cancer deaths were attributed to rare cancers. Moreover, the occurrence of cancers before the age of 50 years increased by nearly 79% from 1990 to 2019, while the mortality rate in this group increased by 27.7%. These findings indicate an immediate need for improved early detection and precision therapeutic approaches to resolve the growing complexity of oncologic disease patterns (Sung et al., 2021; Zhao et al., 2023).

Artificial intelligence (AI) and machine learning (ML) are advanced technologies in cancer research and pharmaceutical development, along with advances in biomedical research. The incorporation of AI and ML into biopharmaceutical research enables the analysis of large sets of data from genomics, proteomics, transcriptomics, radiomics, and histopathology (Topol, 2019). ML algorithms are widely used for the identification of predictive biomarkers, cancer subtyping, drug formulation optimization, and therapeutic response prediction. AI has also contributed to the development of biologics by enabling protein structure prediction, antibody affinity optimization, immunogenicity evaluation, and biomanufacturing process control. AI technologies have improved analytical comparability, pharmacokinetic modeling, and stability prediction as a part of biosimilar development with the goal of reducing development time and costs without compromising on quality (Paul et al., 2020).

The objective of this chapter is to explore evidence-based integration of biologics and biosimilars with AI and ML applications in oncology. It discusses the role of these technologies in cancer detection, biomarker identification, and treatment enhancement. It also emphasizes various case studies from scientific literature, highlighting therapeutic significance over progress. The chapter addresses the regulatory, ethical, and technical difficulties related to the implementation of AI-driven models in the field of biology and cancer diagnosis. This study aims to bring together current scientific data to provide a scientifically validated perspective on how AI and ML are supporting, rather than transforming, the advancement of biologics and biosimilars in the fight against cancer.

17.2 Biologics and biosimilars: A new frontier in cancer therapeutics

17.2.1 Overview of biologics in oncology

The advent of biologics has basically changed the landscape of oncology by introducing highly specific, mechanism-driven therapies that are capable of selectively targeting tumor cells and minimizing off-target toxicity. Biologics, comprising of monoclonal antibodies, fusion proteins, cytokines, and cellular therapies, operate through molecular precision, modulating immune responses or directly engaging tumor-associated antigens, unlike small-molecule chemotherapeutic agents that act by broadly interfering with cell division (Table 17.1) (Weiner et al., 2010).

Monoclonal antibodies represent the most widely used class of biologics in cancer therapy. Since the approval of rituximab in 1997 for non-Hodgkin lymphoma, more than 35 monoclonal antibodies have been approved by the U.S. Food and Drug Administration (FDA) for oncologic use. They include trastuzumab against HER2-positive breast cancer, bevacizumab inhibiting VEGF in colorectal and lung cancers, and pembrolizumab and nivolumab, which impede programmed death-1 (PD-1) immune checkpoints to enhance T-cell-mediated tumor clearance. The mechanism of action of monoclonal antibodies may involve direct binding to cell surface antigens, ADCC, complement activation, or even the blockade of growth factor signaling pathways crucial for tumor survival. More importantly, ADCs, such as ado-trastuzumab emtansine and brentuximab vedotin, integrate cytotoxic payloads with antibody specificity and thus deliver chemotherapeutic agents directly to malignant cells while sparing normal tissues (Scott et al., 2012).

Other biologics that have gained important therapeutic relevance in oncology are fusion proteins. These engineered molecules combine the functional domains contributed by different proteins to execute dual biological activities. Examples include aflibercept, a recombinant fusion protein consisting of the extracellular domains of VEGF receptors 1 and 2 fused to the Fc region of human IgG1. Working as a soluble decoy receptor, aflibercept sequesters VEGF ligands and inhibits angiogenesis. Denileukin diftitox is a fusion protein of interleukin-2 with fragments of diphtheria toxin that exerts selective cytotoxicity against cells expressing the IL-2 receptor (Gencsoy Eker et al., 2024).

Cytokine-based therapies take advantage of endogenous immunomodulatory proteins, including interferons, interleukins, and colony-stimulating factors, to modulate immune function and hematopoiesis in cancer patients. Interleukin-2 (aldesleukin) and interferon-α were among the first biologic immunotherapies, which were

Table 17.1 FDA-approved biologics in oncology.

Drug (Brand)	Generic Name	Principal Target/Mechanism	First U.S. Approval (year)	Principal Oncology Indication (at First Approval)	References
Herceptin	Trastuzumab	HER2 receptor antagonist; inhibits HER2 signaling, mediates ADCC	1998	HER2-positive metastatic breast cancer	Verma et al. (2012)
Rituxan	Rituximab	Anti-CD20 monoclonal antibody; B-cell depletion via ADCC and complement	1997	Relapsed/refractory $CD20^+$ B-cell non-Hodgkin's lymphoma	Maloney et al. (1997)
Avastin	Bevacizumab	Anti-VEGF antibody blocks angiogenesis	2004	Metastatic colorectal cancer	Ferrara et al. (2004)
Erbitux	Cetuximab	Anti-EGFR mAb; inhibits EGFR signaling and mediates ADCC	2004	EGFR-expressing metastatic colorectal cancer	Cunningham et al. (2004)
Keytruda	Pembrolizumab	Anti-PD-1 immune-checkpoint inhibitor; restores T-cell activity	2014	Unresectable/metastatic melanoma after progression	Robert et al. (2015)
Opdivo	Nivolumab	Anti-PD-1 mAb; reinvigorates cytotoxic T-cell response	2014	Unresectable/metastatic melanoma	Weber et al. (2015)
Kadcyla	Ado-trastuzumab emtansine (T-DM1)	HER2-targeted antibody–drug conjugate (trastuzumab + DM1 payload)	2013	HER2-positive metastatic breast cancer	Jois et al. (2020)
Kymriah	Tisagenlecleucel (CAR-T)	Autologous CD19-directed CAR-T cells; redirect T cells to $CD19^+$ B cells	2017	B-cell precursor acute lymphoblastic leukemia (pediatric/young adult r/r)	Cohen et al. (2018)
Yescarta	Axicabtagene ciloleucel (CAR-T)	Autologous CD19-directed CAR-T therapy; engineered T cells against CD19	2017	Relapsed/refractory large B-cell lymphoma	Sung et al. (2021)
Darzalex	Daratumumab	Anti-CD38 monoclonal antibody; induces direct oncolysis and immune modulation	2015	Heavily pretreated multiple myeloma	Zhao et al. (2023)

approved for metastatic melanoma and renal cell carcinoma. Pegylated interferon formulations and granulocyte colony-stimulating factors have also more recently been incorporated into routine practice to prevent chemotherapy-induced neutropenia and to support immune recovery. The cytokine therapies have been associated with enhancing tumor antigen presentation and promoting cytotoxic lymphocyte activation. However, cytokine therapies are often limited clinically by systemic inflammatory toxicity and thus require careful dosing and monitoring of patients (Waldmann, 2018).

Cellular therapies, including chimeric antigen receptor T-cell and T-cell receptor-modified therapies, are indeed the most advanced frontier of biologic oncology. The first FDA-approved CAR T-cell product, tisagenlecleucel, targets the CD19 antigen on B-cell malignancies and has shown durable remissions in refractory acute lymphoblastic leukemia. Various other products, such as axicabtagene ciloleucel and idecabtagene vicleucel, have widened the reach of CAR T-cell therapy into lymphomas and multiple myeloma, respectively (June et al., 2018). Cellular immunotherapies harness patient-derived T cells, reprogramming them to recognize and destroy tumor cells with unprecedented specificity. However, associated challenges, such as cytokine release syndrome, neurotoxicity, and antigen escape, require optimization in clinical practice.

The mechanism of the biologic action in oncology depends on their ability to modulate key cellular and molecular pathways that control tumor initiation, progression, and immune evasion. Biologics can:

1. Inhibit aberrant receptor tyrosine kinase signaling, for instance, the inhibition of HER2 by trastuzumab (Verma et al., 2012).
2. Inhibit angiogenesis necessary for tumor growth, such as bevacizumab that targets VEGF-A (Ferrara et al., 2004).
3. Immune checkpoints: reestablishing antitumor immunity, for example, inhibition of PD-1 by pembrolizumab (Topalian et al., 2015).
4. Deliver cytotoxic or radioactive payloads directly to tumor cells, for example, ADCs and radiolabeled antibodies (Beck et al., 2017).
5. Improve immune system recruitment to tumor sites via Fc-mediated mechanisms (Nimmerjahn & Ravetch, 2008).

This targeted precision enhances therapeutic efficacy and also epitomizes the principles of precision oncology by which the therapy is tailored to tumor-specific molecular signatures rather than to tissue origin alone. Thus, the integration of biologics into oncology has redefined treatment algorithms across a wide range of malignancies, from those of hematologic origin to solid tumors. Indeed, as of 2024, biologic-based therapies account for more than half of global oncology drug revenues, underlining their clinical and economic impact in modern cancer management.

17.2.2 Evolution and regulatory perspective of biosimilars

The regulatory pathway for biosimilars evolved as a response to the scientific complexity of biologic medicines and the need to create an abbreviated but rigorous route to market for highly similar products. In the European Union, the regulatory architecture for biosimilars was established earlier than in many other jurisdictions, with formal guidance and a pathway for marketing authorization implemented by the European Medicines Agency (EMA) in the mid-2000s. The EMA's experience led to one of the largest bodies of regulatory precedent worldwide and established the core principle that a totality-of-evidence approach anchored in high-resolution analytical comparability as the foundation, supported by pharmacokinetic/pharmacodynamic (PK/PD) and, when necessary, comparative clinical data can demonstrate biosimilarity without requiring a full repeat of the originator's clinical development program. Regulatory evolution in Europe has continued through periodic guidance updates and reflection papers that clarify scenarios where a tailored clinical dataset can suffice when analytical and functional similarities are robustly demonstrated (Gherghescu & Delgado-Charro, 2020)

The United States codified an abbreviated licensure pathway with the Biologics Price Competition and Innovation Act; the U.S. Food and Drug Administration (FDA) has since approved multiple biosimilars across therapeutic areas and publishes an up-to-date list of approved biosimilar and interchangeable products. The first biosimilar approved specifically for oncology indications in the United States was a bevacizumab biosimilar (Mvasi; approval September 2017), marking the beginning of biosimilar competition in cancer therapeutics in the U.S. market. Subsequent FDA approvals have included biosimilars to rituximab (first rituximab biosimilar approved in the U.S. in November 2018) and multiple trastuzumab biosimilars approved in 2017–2019, reflecting an expansion of oncology biosimilar availability in the American market (Miller & Schwartzberg, 2019).

Regulatory practice and timelines differ across jurisdictions. The EMA has a mature, centralized marketing-authorization process and substantial accumulated experience in evaluating biosimilar dossiers; historically, the EMA received and approved a large number of biosimilar applications during the first decade of biosimilar regulation. The FDA's regulatory program has been active since the late 2010s and continues to refine its guidance on interchangeability, labeling, and postmarketing considerations; FDA guidance emphasizes the central role of analytical similarity while providing detailed recommendations for nonclinical, PK/PD, and clinical assessments where necessary (Barlas, 2017). National regulators in emerging markets, notably India (CDSCO/DCGI), have developed their own regulatory frameworks for "similar biologics," approving a substantial number of locally developed and imported products intended to improve access; recent CDSCO compilations document multiple r-DNA origin biologics and similar biologics approved for manufacture and marketing in India through the 2020–2024 period.

Key oncology biosimilar examples illustrate how these regulatory frameworks have been applied. Bevacizumab biosimilars were among the earliest oncology biosimilars approved in major markets (first U.S. oncology biosimilar approval in 2017), and additional bevacizumab biosimilars received approvals in 2019 and later years. Rituximab biosimilars entered major markets beginning in 2018, expanding therapeutic options for CD20-positive hematologic malignancies and autoimmune indications. Trastuzumab biosimilars (first approvals in 2017–2019 across regions) provide replicated HER2-targeted therapy at potentially lower cost, with multiple distinct products approved after demonstration of analytical and clinical comparability to the Herceptin reference product (Sharma et al., 2020).

The scientific underpinnings of biosimilar regulation rest on a stepwise comparability paradigm. Developers are expected to perform comprehensive structural and functional characterization (including primary, secondary, and higher-order structure, posttranslational modifications, and functional bioassays) to establish high-resolution analytical similarity. Comparative nonclinical studies and PK/PD assessments follow, and the need for comparative clinical efficacy trials is determined by residual uncertainty after analytical and PK/PD evaluations. Where appropriate, sensitive clinical endpoints or immunogenicity-focused studies are required; regulatory agencies have increasingly recognized circumstances in which robust analytical comparability plus comparative PK data may obviate extensive efficacy trials, provided scientific justification is supplied. Postmarketing pharmacovigilance and real-world evidence collection remain integral to biosimilar stewardship due to the complexity of biologics and the need to monitor immunogenicity and rare adverse events in broader populations (Markus et al., 2017).

17.2.3 Economic and accessibility impact of biosimilars

The advancement of biosimilars has effectively modified the international oncology field by improving the affordability and availability of costly biologic therapies. Biologic anticancer therapies, such as monoclonal antibodies and fusion proteins, represent some of the most expensive pharmaceuticals in oncology, often constituting a significant portion of total treatment costs (Table 17.2). The IQVIA Institute (2023) reports that in 2022, global expenditures on biologic medications exceeded US$ 390 billion. This accounts for approximately 45% of total pharmaceutical expenditure in major countries. However, biologic medicines remain prohibitively costly for numerous cancer patients in low- and middle-income countries (LMICs), where out-of-pocket healthcare expenses continue to be substantial (Kvien et al., 2021).

Biosimilars have emerged as an option to these economic obstacles. According to comparative analyses conducted in the US and the EU, competition among biosimilars has resulted in price reductions of 20%–40% when compared to reference biologics. But there are significant differences between treatment classes and healthcare systems in this regard. Mostly due to competition in immunology and cancer, the European Medicines Agency (EMA) and the European Commission reported in 2022 that the introduction of biosimilars to Europe had resulted in savings of over €30 billion over the previous ten years (Mascarenhas-Melo et al., 2024). Similarly, data from the FDA and IQVIA (2023) in the United States show that the introduction of biosimilars reduced healthcare expenditures by over US$21 billion, with oncology biosimilars such as trastuzumab, bevacizumab, and rituximab accounting for a significant portion of these savings.

In addition to cost savings, biosimilars have significantly enhanced patient access to life-saving treatments. Postlaunch use studies in Europe indicated that following the approvals of biosimilar trastuzumab and bevacizumab, treatment uptake among eligible breast and colorectal cancer patients rose by 25%–35% in specific national

Table 17.2 Economic and access impact of key oncology biosimilars.

Reference Biologic	Biosimilar Examples (First Approval Year)	Average Price Reduction vs. Innovator (%)	Observed Increase in Patient Access (%)	Region / Market	Primary Data Sources	References
Trastuzumab (Herceptin)	Ogivri (2017, EMA), Kanjinti (2019, FDA), Trazimera (2019, EMA)	25%–35%	30%–40% increase in HER2+ breast cancer patients receiving therapy	EU, U.S., India	EMA 2022; IQVIA 2023; Inotai et al., 2022	Verma et al. (2012)
Bevacizumab (Avastin)	Mvasi (2017, FDA), Zirabev (2019, EMA/FDA), Alymsys (2021, EMA)	20%–30%	20–25% higher treatment inclusion for colorectal and lung cancer	EU, U.S.	FDA 2023; EMA 2022; IQVIA 2023	Ferrara et al. (2004)
Rituximab (Rituxan/MabThera)	Truxima (2018, FDA), Rixathon (2017, EMA), Reditux (2007, DCGI)	30%–50%	25%–35% increase in treated B-cell lymphoma patients	EU, India	EMA 2022; CDSCO 2024; Kakkar & Dahiya, 2021	Maloney et al. (1997)
Filgrastim (Neupogen) *(supportive oncology care)*	Zarzio (2009, EMA), Nivestym (2018, FDA)	35%–65%	40%–60% broader prophylactic G-CSF coverage	EU, U.S.	EMA 2022; IQVIA 2023; Moorkens et al., 2021	Moorkens et al. (2017)
Epoetin alfa (Erythropoietin) *(supportive oncology care)*	Binocrit (2007, EMA), Retacrit (2018, FDA)	25%–50%	30%–40% higher utilization in chemotherapy-induced anemia	EU, U.S., India	EMA 2022; WHO 2021; IQVIA 2023	Jones et al. (2004)

health systems (Kvien et al., 2021). In the developing markets like India, the Central Drugs Standard Control Organization (CDSCO) has sanctioned over 70 comparable biologics since 2010. The accessibility to biosimilars has made it possible to introduce biologic treatments in government-funded cancer campaigns and this initiation has enhanced therapeutic access in urban and semiurban oncology facilities.

From the perspective of healthcare system, the expanded use of biosimilars has provided cost flexibility that allows for reinvestment in innovative therapies and precision medicine diagnostics. The national reimbursement authorities in the European Union (for example, NICE, HAS, G-BA) have recognized that the cost savings realized through the use of biosimilars represent an acceptable mechanism to facilitate the development of innovative next-generation cancer therapies and biomarker-driven treatment approaches (Mullard, 2024; Zhao et al., 2023). In cancer care models that focus on value-based outcomes, biosimilars are consistent with the ideals of cost-effectiveness and equality in access, making it possible for doctors to prescribe biologic drugs earlier in the disease process without unbalancing the healthcare budget.

In personalized cancer medicine, the economic value of biosimilars indirectly promotes personalized medicine by unburdening the budget for more advanced diagnostic modalities, such as next-generation sequencing (NGS) and molecular characterization. A 2021 World Health Organization assessment of cancer treatment in LMICs showed that adding biosimilars to essential drugs could increase access to biologic therapy by up to about two-thirds more patients, especially those with breast or hematologic malignancies (Topol, 2019). Biosimilars have enabled the expansion of oncology biologic therapy administration, allowing secondary and tertiary hospitals, which were previously resource-limited, to provide high-cost biologic therapies more easily.

The economic value of biosimilars is market-dependent. Factors such as regulatory efficiency, intellectual property regulation, prescriber confidence, and interchangeability regulation affect market penetration and subsequent economic value. Biosimilar market penetration has surpassed 80% in areas such as the Nordic countries and the United Kingdom, where there is a well-developed regulatory infrastructure and competitive procurement systems. In contrast, in areas with inadequate regulatory harmonization and procurement systems, biosimilar market penetration remains below 30% (Mullard, 2024; Paul et al., 2020; Seebacher et al., 2019). To improve market penetration, regulators, payers, and clinicians must work together to develop standardized interchangeability policies, implement educational programs, and provide incentives for biosimilar prescribing.

17.3 Artificial intelligence and machine learning in cancer detection and diagnosis

17.3.1 Artificial intelligence/machine learning concepts in biomedical research

AI and ML have enabled the use of data-driven elements by which biomedical research is being advanced beyond the conventional hypothesis-driven models. AI refers to computational systems that mimic human intelligence to perform tasks such as reasoning, learning, and problem-solving. ML is a subset of AI that involves algorithms that learn patterns from a large dataset to make predictions or classifications without the extensive requirement of programming. Deep Learning is a branch of ML that uses multilayered artificial neural networks to extract hierarchical features from complicated datasets like genomic data or medical images.

This shift from hypothesis-driven methodologies, in which specific biological mechanisms were proposed and tested by the researcher, to data-driven discovery has been revolutionary for cancer research. Data-driven AI models can analyze extensive datasets from genomics, radiomics, and pathology to uncover latent correlations that may be missed by traditional analysis. Deep learning models, particularly convolutional neural networks (CNNs), have exhibited enhanced accuracy in image-based cancer diagnostics relative to human pathologists in certain contexts, including the histopathology of breast and lung cancers (Jiang et al., 2024).

AI and ML are used in oncology for early detection and diagnosis, and also for drug research, personalized therapy, and predicting the extent to which a patient will respond to treatment in the future. Modern cancer biology generates huge databases with digital pathology, radiological imaging, and multiomics data (genomics, transcriptomics, and proteomics). Further, ML tools like random forests, support vector machines (SVMs), and gradient boosting models are being employed to classify tumor subtypes, predict how treatments will be effective, and use molecular signatures for accurate oncology (Kourou et al., 2014).

Computer models and biological analysis are working together progressively, indicating the paradigm shift toward integrative oncology. In this field, AI doesn't replace human knowledge, but rather works with humans to make decisions more accurately. Because AI systems are flexible and can learn new things continuously, they can be used to build strong predictive models to diagnose cancer. This is particularly relevant when biologics and replicated data are integrated into these models for translational research (Esteva et al., 2019).

17.3.2 Applications of artificial intelligence/machine learning in oncology

AI and ML are now established components of contemporary cancer research and clinical diagnostics, with validated applications in medical imaging (radiomics), digital pathology (pathomics), multiomics biomarker discovery, and automated histopathological tumor classification. The literature contains multiple reproducible examples where ML methods provide clinically relevant information—either by improving diagnostic sensitivity/specificity, extracting prognostic signatures from imaging, or predicting molecular features from routine diagnostic material (Aerts et al., 2014).

17.3.2.1 Early detection using imaging (radiomics and pathomics)

Radiomics formalizes the extraction of large numbers of quantitative imaging features (intensity, shape, texture, wavelet decompositions) from standard clinical imaging and uses ML to relate those features to tumor phenotype, prognosis, or treatment response. A landmark radiomics study analyzed 440 quantitative CT features from 878 patients with lung or head-and-neck cancer and showed that radiomic signatures correlate with underlying gene-expression patterns and are prognostic for survival independent of established clinical covariates. Subsequent systematic reviews and application studies have reproduced the utility of radiomic features across CT, MRI, and PET for risk stratification and response prediction in multiple tumor types (McKinney et al., 2020; Parmar et al., 2015). Radiomics pipelines typically include image harmonization, region-of-interest segmentation, reproducible feature extraction (including test–retest stability), and supervised or unsupervised modeling with appropriate cross-validation and external validation sets.

Pathomics applies deep learning to digitized whole-slide images (WSIs). Weakly supervised and multiple-instance learning approaches allow training at the slide level using routine diagnostic labels without exhaustive pixel-level annotation. Campanella et al. (2019) developed a clinical-grade, weakly supervised deep-learning (DL) system trained on tens of thousands of WSIs, which achieved clinically useful performance across prostate, breast, and skin datasets and demonstrated feasibility for large-scale deployment in diagnostic workflows (Campanella et al., 2019). In large multicenter challenges (e.g., CAMELYON), convolutional neural networks (CNNs) trained on annotated WSIs have reached near-human performance for detecting lymph node metastases, with high area under the receiver-operating curve (AUC) and clinically relevant sensitivities at specified false-positive rates (Liu et al., 2018).

17.3.2.2 Predictive modeling for genetic and proteomic biomarkers (multiomics integration)

ML methods have been used to mine genomic, transcriptomic, and proteomic datasets to identify diagnostic and predictive biomarkers. Coudray et al. (2018) trained CNNs on hematoxylin-eosin histology to predict common somatic mutations in lung adenocarcinoma and reported AUCs ranging approximately from 0.73 to 0.86 for several mutations (STK11, EGFR, KRAS, TP53, among others), demonstrating that morphological patterns in routine slides may carry signal about molecular status (Coudray et al., 2018). Proteogenomic studies have integrated mass-spectrometry proteomics with genomic annotation to reveal protein signatures linked to actionable pathways; these high-dimensional datasets have been successfully mined with ML to nominate candidate biomarkers and therapeutic targets in breast and other cancers (Mertins et al., 2016). Variational autoencoders and other DL latent-space methods trained on The Cancer Genome Atlas (TCGA) and related datasets have produced biologically meaningful representations used for subtype discovery and response prediction, and semisupervised approaches have been applied to predict chemotherapy response from transcriptomic profiles with cross-validated performance (Way & Greene, 2018).

17.3.2.3 *Artificial intelligence-based histopathological analysis for tumor classification*

Automated histopathology systems using deep CNNs have demonstrated high diagnostic accuracy in multiple contexts. In the Camelyon16/17 benchmark and related work, models such as LYNA (Lymph Node Assistant) achieved slide-level AUCs near 0.99 with tumor-level sensitivities above 90% at clinically relevant false-positive rates, and have, in some instances, identified micrometastases missed in initial human review (Budginaite et al., 2024). Multiple groups have reproduced high performance for tumor detection, mitosis identification, grading, and subtyping tasks; however, generalisability across institutions and staining/scanner variation remains a central technical and implementation challenge that must be addressed through careful external validation and domain-adaptation strategies (Table 17.3) (Coudray et al., 2018).

17.3.3 Integration of multiomics and real-world data

One of the most promising evolutions in oncology is the fusion of genomics, transcriptomics, proteomics, metabolomics, and real-world clinical data, integrated via AI and ML to enhance diagnostic accuracy, therapeutic stratification, and dynamic monitoring. Multiomics approaches enable a layered view of tumor biology—from DNA alterations (genomics) through RNA expression (transcriptomics), protein abundance and modifications (proteomics), to small-molecule metabolites (metabolomics)—thereby capturing tumor heterogeneity, pathway deregulation, and microenvironmental interactions more comprehensively than single-omics alone (Biswas & Chakrabarti, 2020; Ozaki et al., 2024).

Recent reviews report that in the period 2020–2023, over 80 studies utilized combinations of two or more omics types together with AI/ML models for cancer prognosis and diagnosis, with eight studies specifically integrating genomics, transcriptomics, epigenomics, and proteomics concurrently (Ozaki et al., 2024). ML models (like deep neural networks, autoencoders, and graph-neural networks) have done better than single-omics baselines at classifying tumor subtypes, predicting drug responses, and finding biomarkers in this combined approach (Ali, 2023). For example, DL-based multiomics frameworks have been shown to improve the discrimination of cancer subtypes and predict survival outcomes by learning latent representations that reflect meaningful biological heterogeneity (Zhang et al., 2025).

In parallel, the concept of the "digital twin" in oncology is gaining traction—a virtual, dynamic model of the individual patient that assimilates multimodal and longitudinal data (imaging, omics, pathology, clinical, wearable or sensor data) to simulate tumor evolution and response to therapy. These digital twin frameworks rely on continuous data assimilation, model calibration, and predictive simulation, enabling in-silico experimentation of treatment scenarios and personalized therapy optimization (Giansanti & Morelli, 2025; Shen et al., 2025). A narrative review

Table 17.3 Various studies related to AI-based histopathological analysis for tumor classification.

Modality/Task	Dataset/Scale	Main Result (Selected Metric)	References
CT radiomics; prognostic signature	1019 patients (lung and head-neck)	440 image features; radiomic signature prognostic independent of clinical covariates	Aerts et al. (2014)
Mammography screening (DL)	Large UK and US cohorts (multi-site)	AI system reduced false positives by 5.7% (US) and 1.2% (UK); reduced false negatives by 9.4% (US) and 2.7% (UK) in evaluation sets	McKinney et al. (2020)
Weakly supervised WSI analysis	Tens of thousands of whole-slide images across tumor types	Clinical-grade performance; near-pathologist accuracy without pixel annotations	Campanella et al. (2019)
Lymph node metastasis detection	CAMELYON benchmark sets	Slide-level AUC ≈0.99; tumor-level sensitivity ≈91% at 1 false positive/patient	Liu et al. (2018)
Histology → mutation prediction (CNN)	TCGA LUAD slides	Predicted several mutations with AUCs 0.73–0.86 (held-out test)	Coudray et al. (2018)
Proteogenomics (MS)	105 genomically annotated breast cancers	Integrated proteomic signatures linked somatic mutations to pathway activation; candidate biomarkers identified	Mertins et al. (2016)
Transcriptome latent representations (VAE)	TCGA pan-cancer datasets	Learned a biologically meaningful latent space useful for subtype/response studies	Way and Greene (2018)

collating 17 prior systematic reviews highlighted that digital twin methods in oncology offer advantages such as improved decision-making and tailored treatment regimens, but also note significant barriers in data integration, modeling complexity, and validation (Giansanti & Morelli, 2025). In terms of data-flow architecture, real-world data (RWD) from electronic health records, imaging archives, lab results, and multiomics repositories are being standardized (e.g., OMOP Common Data Model, HL7 FHIR) and streamed into analytical pipelines for twin modeling and ML training.

The integration of real-time omics data and clinical endpoints allows models to learn trajectories of disease evolution, resistance emergence, and therapeutic adaptation. Studies of multiomics-based digital twins show that only about 15% of those polled have used both genomes and proteomics data. Metabolomics and single-cell omics data are still not widely used, which is likely because of ongoing technical and data-availability issues (Shen et al., 2025).

From a translational perspective, multiomics + RWD integration enhances biologic- and biosimilar-driven oncology in several ways. First, it supports biomarker discovery relevant to biologic targets (e.g., monoclonal antibodies, fusion proteins) by mapping omic signatures to target antigens or signaling pathways. Second, patient-specific profiles derived from integrated datasets can inform biosimilar comparability and patient stratification for biologic therapies by linking molecular phenotype to expected therapeutic response. Third, real-world longitudinal omics and clinical data enable monitoring of biologic or biosimilar therapy effectiveness, immune response, and emergence of resistance—thus extending the value-chain of biologic development into postmarketing and clinical algorithm domains.

Nevertheless, challenges persist: high dimensionality of multiomics datasets, missing data, heterogeneity of platforms, reproducibility of ML models, requirement for robust external validation, interpretability of ML-derived features, and regulatory acceptance of digital twin simulations. Ethical, legal, and data-governance issues (consent, data sharing, transparency) are especially acute when real-world multimodal data are fused in predictive oncology systems (Giansanti & Morelli, 2025).

17.4 Interface of biologics/biosimilars with artificial intelligence and machine-learning: A new therapeutic intelligence

17.4.1 Artificial intelligence-guided design and development of biologics

Deep learning and ML methods are now an established part of contemporary biologic design workflows. High-accuracy protein-structure prediction by AlphaFold2 demonstrated near-experimental performance in CASP14 and has since enabled proteome-scale structural models that inform epitope mapping and antibody–antigen interface analysis. ML approaches that preceded and complemented AlphaFold2 (for example, work on distance-prediction networks) established core methods for predicting residue-level contacts and distance constraints that underpin modern models (Jumper et al., 2021; Senior et al., 2020). In antibody engineering, peer-reviewed reviews and method papers document successful integration of ML with high-throughput sequence and structure data to accelerate lead identification and affinity maturation, to predict developability attributes (aggregation propensity, solubility, viscosity predictors), and to triage candidates before

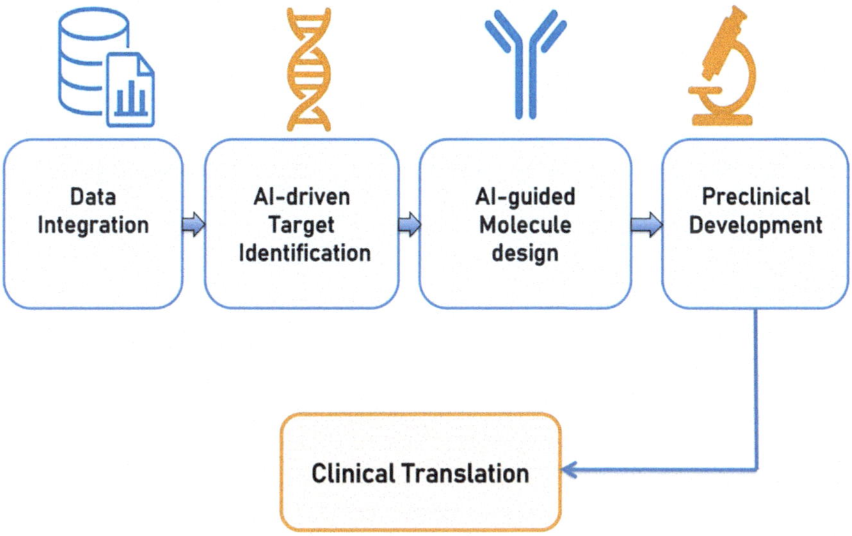

FIGURE 17.1

AI-driven biologic development workflow.

extensive wet-lab work (Mason et al., 2021). Predictive immunogenicity and developability models—combining sequence, structural features, and physicochemical descriptors—are used to flag T-cell epitope risk and biophysical liabilities during candidate selection, thereby reducing downstream clinical risk and supporting manufacturability decisions (Sormanni et al., 2015). Collectively, these validated computational capabilities (structure prediction, sequence–function modeling, and developability/immunogenicity analytics) are applied within biologic and biosimilar pipelines to shorten design cycles, focus experimental effort, and provide stronger preclinical evidence in support of candidate selection and comparability assessments (Fig. 17.1) (Gromiha & Huang, 2011).

17.4.2 Biosimilar characterization and analytical comparability using artificial intelligence

ML methods are increasingly applied to the high-dimensional analytical datasets that underpin biosimilar comparability exercises. High-resolution physicochemical and functional characterization of biologics generates mass spectra, chromatograms, glycan profiles, peptide maps, NMR fingerprints, and orthogonal bioassay readouts; ML approaches (unsupervised clustering, dimensionality-reduction, supervised classification, and anomaly detection) have been used to detect subtle lot-to-lot differences, to integrate orthogonal assay outputs into composite fingerprints, and to prioritize critical quality attributes for further evaluation. In proteomics and peptidomics, studies have shown that supervised ML models trained on large MS datasets

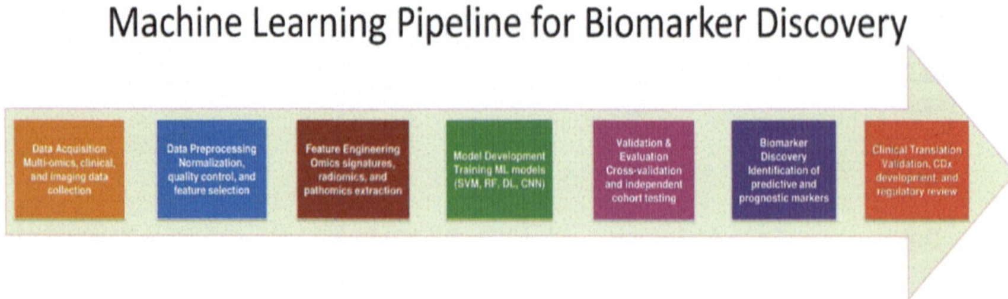

FIGURE 17.2

Machine learning pipeline for biomarker discovery.

can predict peptide candidates and classify spectral features with better sensitivity than manual inspection alone, supporting lead identification and characterisation workflows relevant to biologic comparability (Fig. 17.2) (Madsen et al., 2022).

AI/ML also supports spectral-data interpretation and fingerprinting workflows by improving signal deconvolution, peak annotation, and feature alignment across platforms and laboratories; these capabilities reduce analytical uncertainty in comparability assessments and help quantify multivariate similarity metrics. Industry and academic reports have demonstrated ML-enhanced pipelines that combine chromatographic and MS features into reproducible comparative visualizations and similarity scores used during regulatory dossiers. Regulatory discussion papers from authorities note both opportunities and considerations for incorporating AI/ML into drug development activities, emphasising that ML outputs intended to inform regulatory decisions should be transparent, validated on representative datasets, and complemented by orthogonal experimental evidence (Sibieude et al., 2022).

Predictive analytics using ML are being explored to augment biosimilarity assessments by linking analytical fingerprints to downstream PK/PD endpoints. Hybrid approaches that merge mechanistic PBPK/PKPD modeling with ML (AI-PBPK or data-driven surrogates) have been reported in the literature as promising for predicting population PK variability and for simulating scenarios that might reduce residual clinical uncertainty. Such models can assist in study design and in interpreting whether observed analytical differences are likely to be clinically meaningful, although the literature and regulatory guidance underscore that these methods are currently complementary tools rather than replacements for established comparability studies and clinical/PK data where required (Huang et al., 2024).

17.5 Computational discovery of onco-biomarkers

The explosion of high-dimensional data from genomics, transcriptomics, proteomics, and metabolomics has transformed biomarker discovery in oncology. Traditional candidate-biomarker approaches are increasingly supplemented by ML and AI pipelines that can integrate multiomics data, imaging features, and clinical metadata to identify diagnostic, prognostic, and predictive biomarkers. A systematic review on nonsmall cell lung cancer (NSCLC) identified 215 studies using AI/ML to discover potential biomarkers from molecular, histologic, radiologic, and multimodal data sources. In parallel, a recent paper reviewing ML models for the identification of prognostic and predictive cancer biomarkers found that ML models have been used broadly to identify candidate biomarkers across multiple tumor types and suggested that greater integration of omics, feature-selection methods, and external validation will be needed (Al-Tashi et al., 2023; Çalışkan & Tazaki, 2023).

ML algorithms also facilitate the discovery of druggable targets within cancer-pathway networks by analysing correlation structures, mutation-expression associations, and downstream effect signatures. For instance, AI-driven methods applied to transcriptomic/proteomic datasets can identify dysregulated pathways and nominate hubs (e.g., kinases, epigenetic regulators) that are amenable to therapeutic intervention. A recent review described how multiomics integration by AI enables such discovery, moving beyond single-gene biomarker paradigms to network-based target identification.

Nevertheless, challenges remain. The interpretability of ML models is a frequent limitation: many high-performance models act as "black-box" predictors rather than mechanistically transparent tools. The quality of the underlying data—batch effects, small sample sizes, incomplete annotation—and the absence of independent external validation cohorts limit clinical translation. Despite these challenges, AI-assisted biomarker and target discovery is rapidly developing and is becoming an important aspect in precision oncology pipelines (Alum, 2025).

17.5.1 Linking biologics/biosimilars to cancer signatures

The integration of therapeutic biologics and biosimilars with AI-driven cancer signature profiling offers novel opportunities for precision oncology. Monoclonal antibodies (mAbs) and biosimilars targeted at specific antigens (e.g., HER2, EGFR, PD-1/PD-L1) can be more effectively deployed when AI/ML models stratify patients by molecular subtype, biomarker expression, or predictive response patterns. For example, in advanced NSCLC treated with immune-checkpoint inhibitors, an ML algorithm built from a nationwide electronic health-record database (N = 38,048) achieved AUCs of 0.701 for 12-month progression-free survival and 0.718 for 12-month overall survival, outperforming PD-L1 status alone (HR low-risk vs. high-risk $\approx$ 0.47 for progression; $\approx$ 0.31 for mortality) (Ahluwalia & Parikh, 2025). Such stratification models enable clinicians to identify subsets of patients most likely to respond to biologics (Table 17.4).

ML methods have also been applied to multiomics data to define molecular subtypes that map onto biologic targets. A systematic review of ML in oncology noted that unsupervised clustering and supervised classification techniques are increasingly used to delineate tumor phenotypes that carry prognostic or predictive value for

Table 17.4 Examples of AI/ML studies linking biologic targets or therapies to AI-predicted cancer signatures/stratification.

Cancer Type	Biologic/Target Relevance	AI/ML Method	Main Finding (As Reported)	References
Nonsmall cell lung cancer (NSCLC)	Actionable mutations (EGFR, KRAS, STK11, TP53, etc.) — relevant to targeted therapies	CNN (Inception-v3) on H&E whole-slide images	Several somatic mutations (e.g., EGFR, KRAS, STK11, TP53) were predictable from routine H&E with AUCs $\approx$0.73–0.86 on held-out sets, indicating morphology carries a molecular signal relevant to targeted therapy selection.	Coudray et al. (2018)
Breast cancer	Hormone-receptor status (ER) — informs endocrine vs biologic decisions	Deep convolutional neural network on tissue microarrays / WSIs	Model predicted estrogen-receptor status from H&E with accuracy reported as non-inferior to standard immunohistochemistry in the study cohorts (supports molecular surrogate from morphology).	Shamai et al. (2019)
Breast cancer	HER2 status — direct target for trastuzumab/Her2-directed biologics	Community challenge: multiple DL models on curated H&E dataset (509 slides)	Multiple teams developed models to predict HER2 from H&E; the challenge established benchmark datasets and methods for HER2 prediction from routine slides.	Conde-Sousa et al. (2022)
Breast cancer	HER2 status prediction — implications for trastuzumab eligibility	HAHNet: multiscale CNN with attention mechanisms on H&E WSIs	Reported improved classification performance for HER2 status from H&E vs prior computational methods, demonstrating the feasibility of morphology-based HER2 inference.	Wang et al. (2023)

(continued on next page)

Multiple (lung, gastric, others)	PD-L1 expression — key biomarker for anti-PD-1/PD-L1 biologics	Deep learning on H&E images/weakly supervised approaches	Demonstrated that PD-L1 expression can be predicted from H&E with clinically useful performance in selected cohorts, supporting image-based surrogate biomarkers for immunotherapy selection.	Shamai et al. (2022)
NSCLC	Immunotherapy response (ICI) associations — informs use of anti-PD-1/PD-L1 agents	Subregional radiomics model (SRRM) on CT	SRRM outperformed conventional radiomics, TMB, and PD-L1 in predicting immunotherapy response in their cohorts and identified mutation associations (e.g., EGFR) in subgroups.	Peng et al. (2024)
Various advanced/rare cancers	Pembrolizumab (anti-PD-1) response prediction	Radiomics + ML models on CT/PET imaging	Developed a radiomics model that predicted pembrolizumab response in a cohort of advanced rare cancers; reported predictive performance supporting imaging-based stratification for immunotherapy.	Colen et al. (2021)
Gastric/breast cancer (several groups)	HER2 / trastuzumab response modeling	DL on H&E and multimodal inputs (US/MR/WSI)	Several studies (including external validation sets and challenge results) have shown DL models can predict HER2 status and, in select cohorts, correlate with response to trastuzumab or HER2-directed therapy. These reports provide validated examples of mapping AI-predicted signatures to biologic therapy eligibility.	Wu et al. (2025)

biologic therapy selection (Kourou et al., 2021). For instance, radiomic and transcriptomic ML studies in glioma subtype classification demonstrated median AUCs >0.80 for predicting IDH mutation or MGMT promoter methylation from imaging features alone (Lost et al., 2023). These approaches support a workflow where AI models first classify the tumor into a subtype, then select the biologic or biosimilar treatment best aligned with the signature.

From the biosimilar perspective, AI-guided signature matching increases confidence in prescribing biosimilars for populations whose molecular profiles mirror those originally studied for the innovator biologic. As biosimilars target the same antigen or pathway, AI-derived stratification ensures that the molecular context is appropriate, thereby optimizing therapeutic equivalence in real-world usage. By combining patient subtype classifications, biomarker expression levels, and AI-predicted response likelihood, it becomes feasible to personalize biologic/biosimilar selection, monitor for early resistance, and re-assign the therapeutics dynamically.

17.5.2 Case studies

Case study 1: AI-based HER2 detection and implications for trastuzumab/biosimilars: multiple peer-reviewed reports have shown that DL models trained on routine hematoxylin & eosin (H&E) histology can predict HER2 status and, in some cohorts, correlate with trastuzumab response. Farahmand et al. (2022) trained a convolutional neural network on manually annotated tumor regions and reported slide-level AUCs of 0.90 in cross-validation and 0.81 on an independent TCGA test set for HER2 status; when the same framework was applied to pretreatment samples from patients who subsequently received trastuzumab, the model showed an AUC of 0.80 for predicting treatment response (Farahmand et al., 2022). Subsequent large, multi-institutional and method papers have confirmed feasibility and refined architectures: HAHNet (Wang et al., 2023) reported improved accuracy and multiscale attention mechanisms for HER2 classification from H&E (Wang et al., 2023), and a multicentre weakly-supervised study demonstrated the ability to identify HER2-low subgroups directly from H&E (Valieris et al., 2024). These studies consistently emphasize that H&E-based AI can recover morphology-linked molecular signals relevant to HER2 and may serve as a low-cost adjunct for screening or quality assurance; however, authors also note dependence on reliable ground truth (IHC/ISH) and call for prospective validation before clinical deployment in decisions about trastuzumab or biosimilars.

Case study 2: DL-guided immunotherapy personalization: foundational genomic studies established biological correlates of checkpoint-inhibitor response (high tumor mutational burden and specific transcriptomic signatures). Rizvi et al. (2015) showed that higher somatic mutational burden in NSCLC is associated with improved response to anti-PD-1 therapy. Rizvi et al. (2015) and Hugo et al. (2016) identified the transcriptomic patterns that are associated with anti-PD-1 sensitivity in melanoma (Hugo et al., 2016). Building on these biological findings, a large and growing body of AI work integrates imaging, multiomics, and clinical data to predict immunotherapy outcomes. A multiinstitutional radiomics study in advanced NSCLC found that imaging-based prediction for PD-L1 and progression-free survival (PFS) was modest but reproducible (AUCs ≈0.59-0.64 in validation cohorts). The study also identified feature-selection and ML algorithms that generalized across sites (Yolchuyeva et al., 2023). Liao et al. (2024) used CT radiomics and DL models to predict PFS for lung cancer immunotherapy. Their results showed time-dependent AUCs of ~0.76–0.80 and a C-index of ≈0.83 in their cohort. Other studies have connected dynamic CT radiomic changes and peritumoral immune-related features to survival and response to pembrolizumab and other ICIs. Taken together, these lines of evidence show (1) biological plausibility from genomics/transcriptomics and (2) practical predictive performance of AI models using radiologic and multimodal inputs; investigators consistently stress the need for external validation, transparent reporting, and integration with established biomarkers (PD-L1, TMB) in prospective trials before replacing standard clinical decision pathways (Liao et al., 2024).

17.6 Translational oncology: Integration of artificial intelligence with clinical biologic data

17.6.1 Predictive oncology platforms

Predictive oncology platforms combine curated real-world clinical datasets (electronic health records, registries, insurance claims), clinico-genomic repositories, digital pathology, and imaging archives to create interoperable data resources that support biologic-therapy optimization (Fig. 17.3). Large curated oncology RWD resources such as Flatiron's clinico-genomic datasets and AACR Project GENIE provide linked treatment, outcome and genomic data that have been used to identify molecularly defined cohorts, evaluate real-world effectiveness, and generate hypotheses about biomarker-driven use of targeted biologics (André et al., 2017). These platforms enable retrospective analyses that can inform population-level treatment patterns, identify underserved subgroups, and supply external control arms or real-world comparators for trials.

Regulatory and industry frameworks increasingly recognize the utility of RWD/RWE for drug development and postmarketing assessment. Reviews and regulatory papers describe how RWE can inform trial design, refine eligibility criteria, and provide supporting evidence for regulatory and payer decisions; however, they also emphasize that fitness-for-purpose (data completeness, ascertainment bias, standardization) must be demonstrated for each intended use. Data quality and provenance assessment are therefore an essential component of predictive-platform deployment in biologic development.

AI/ML methods are embedded within predictive platforms to derive digital biomarkers, to perform patient matching, and to generate near-real-time risk or response predictions that can be operationalized in clinical or trial settings. For adaptive and platform trial designs (for example, I-SPY2 in breast cancer),

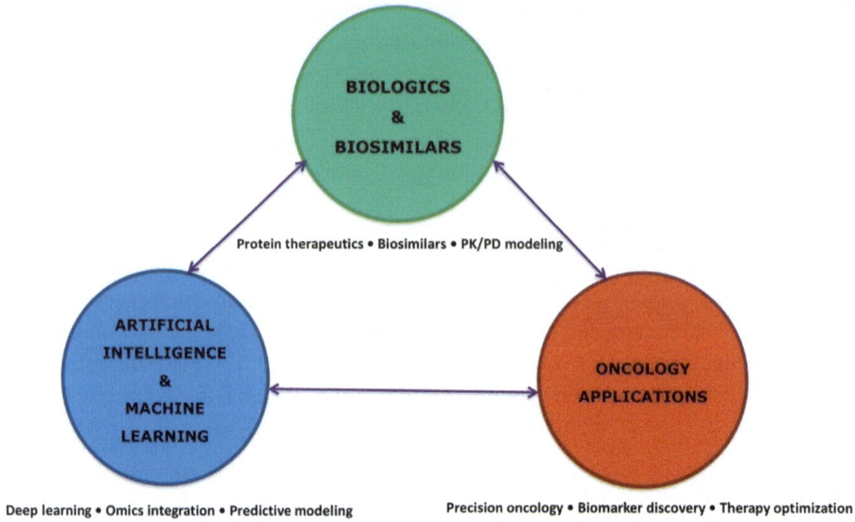

FIGURE 17.3

Integrated biologic-AI-oncology framework.

ML-enabled biomarker classifiers and continuously updated Bayesian models have been used to assign patients to experimental arms with a higher prior probability of benefit, accelerating go/no-go decisions and enriching trials for likely responders. AI is also used for operational efficiencies in trial conduct—automated eligibility screening from EHRs, dynamic cohort balancing, and centralized monitoring of safety signals. These applications are documented in the adaptive-trial literature and in methodological reviews of ML in clinical trials.

For response monitoring, predictive platforms integrate serial data streams—radiology (radiomics), digital pathology, circulating tumor DNA (ctDNA), and routine labs—applying ML models to detect early signs of response, minimal residual disease, or resistance. Industry and academic groups have reported RWD studies using clinico-genomic datasets to evaluate ctDNA testing adoption and to relate molecular dynamics to outcomes; similarly, radiomics models trained on large imaging cohorts have been used to predict early treatment response to immunotherapies and targeted biologics. These approaches support more frequent, individualized monitoring than is feasible with conventional trial schedules, but they require rigorous external validation and prespecified performance thresholds before influencing treatment changes (Prelaj et al., 2024).

17.6.2 Digital biomarkers and companion diagnostics

The rise of digital biomarkers, quantitative, algorithm-derived indicators of biological processes, has accelerated the evolution of companion diagnostics (CDx) in oncology.

AI and ML have become critical tools for extracting biomarkers from imaging, molecular, and liquid biopsy datasets, allowing for more precise biologic and biosimilar therapy decisions.

17.6.2.1 Artificial intelligence-enabled liquid biopsy interpretation

Liquid biopsy technologies have enhanced noninvasive cancer monitoring and diagnosis by analyzing circulating tumor cells (CTCs), exosomes, and circulating tumor DNA (ctDNA). Together with ML and AI algorithms, these technologies may offer increased sensitivity for early cancer detection and treatment response monitoring. ML algorithms, for instance, have been trained on approximately 140,000 methylation profiles from cfDNA samples of 50 cancer types to detect tumor-derived signals in Grail's GalleriTM multicancer early detection (MCED) test. According to the findings, the stage I–III sensitivity was 67.3%, and the overall specificity was 99.5% (Klein et al., 2021). This is further demonstrated by the ShieldTM and GuardantINFORMTM platforms from Guardant Health, which have employed AI-driven noise suppression algorithms for longitudinal analysis and ctDNA variant detection in over 250,000 patient samples. The signal separation for this was significantly enhanced (Guardant Health, 2020).

According to the CancerSEEK trial, an AI-based liquid biopsy platform that combined circulating protein biomarkers and cfDNA mutation analysis in 1005 patients with eight distinct cancer types achieved a median sensitivity of 70% and specificity of 99% for early-stage cancer detection (Cohen et al., 2018). AI-enhanced liquid biopsies can be used as companion diagnostics for the selection of biologic therapies, according to this integration approach's basic model. A study by Nguyen et al. (2023) separated tumor-derived fragments based on end motifs and sequence patterns using deep neural networks for cfDNA fragmentomics, increasing the accuracy of cancer type prediction in multicancer samples by more than 85% (Nguyen et al., 2023). These advances help to improve digital companion diagnostics for monoclonal antibodies and immune checkpoint inhibitors, in which the clearance of ctDNA is associated with treatment response and survival rates.

17.6.2.2 Machine learning-driven integration of imaging and molecular diagnostics

Radiogenomics refers to the combination of radiomics and genetics. It can be regarded as a major advance in the field of companion diagnostics with AI. ML algorithms are able to predict relevant genetic changes that direct biologic therapy without the use of invasive methods.

CNN-based models were able to correctly predict EGFR and ALK mutations using CT scan data obtained from The Cancer Imaging Archive (TCIA) for NSCLC patients. This molecular profiling can help guide the administration of EGFR biologics such as Necitumumab and Cetuximab, as well as inhibitors of ALK, and hence allow for customized treatment and achieve significant reductions in biopsies (Rakaee et al., 2024).

The same trends in convergence have also been witnessed in HER2-positive breast cancer. AI-based imaging biomarkers showed significant correlation with HER2 immunohistochemistry and fluorescence in situ hybridization (FISH) results. It has achieved an area under the curve (AUC) value of more than 0.90, which helped in predicting the status of HER2 (Fan et al., 2025). This result suggests the potential use of AI-assisted digital companion diagnostics in the management of trastuzumab and its biosimilars.

In addition, pathomics, the use of AI in the evaluation of histopathology slides, has also been found to be highly accurate in estimating the levels of PD-L1 expression, a critical biomarker for determining the eligibility of patients for anti-PD-L1 therapy. Cheng et al. (2022) used deep learning to evaluate more than 1288 whole-slide images to predict PD-L1 positivity, achieving an accuracy of 96% compared with immunohistochemistry (Cheng et al., 2022). These imaging-molecular diagnostic tools are increasingly being used within the regulatory framework for the approval of biologics and biosimilars.

17.6.2.3 Digital companion diagnostics in regulatory and clinical adoption

Since 2019, more than 50 AI/ML-based medical devices for imaging analysis have been approved by the FDA, some of which are being used as digital companion diagnostics in oncology. Of note, the Thermo Fisher Oncomine Dx Target Test and FoundationOne CDx have combined ML-driven variant interpretation pipelines to inform biologic therapy decisions in NSCLC and colorectal cancer (Hummel et al., 2024).

Moreover, AI-enabled digital pathology systems like Paige.AI and PathAI are being deployed in CLIA-certified laboratories for diagnostic evaluation, with FDA breakthrough device designation for prostate and breast cancer biomarker detection (Bera et al., 2019). These tools bridge imaging, molecular, and clinical data to create adaptive companion diagnostic platforms that evolve with incoming patient data and emerging biologics.

17.6.3 Personalized biologic therapy through predictive algorithms

Precision oncology has advanced significantly with the use of AI-driven predictive algorithms in biologic therapy selection. These algorithms are integrated with multi-omic, clinical, imaging, and pharmacological data, and hence it enables to personalize biologic dose, predicting toxicity, and allowing real-time therapeutic modifications, which are crucial for maximizing efficacy and reducing adverse events.

17.6.3.1 Artificial intelligence-guided therapy selection and precision dosing

The personalization of biologic therapy is now increasingly guided by ML models capable of analyzing extensive clinical and molecular datasets to predict treatment efficacy. AI-driven platforms for therapy selection, such as Tempus xT, Caris CODEai, and IBM Watson for Oncology, combine genomic and transcriptomic data with electronic health records (EHRs) to produce ranked treatment recommendations for oncologists.

DL pharmacokinetic/pharmacodynamic (PK/PD) models have been developed to optimize monoclonal antibody (mAb) dosing. For instance,(Yang et al., 2021) tested recurrent neural networks that were trained on over 1000 patient datasets receiving nivolumab or pembrolizumab. The study predicted optimal dosage schedules with a mean absolute error (MAE) of less than 10% in comparison to therapeutic drug monitoring (Yang et al., 2021). Research was conducted to study these types of adaptive dosing approaches for trastuzumab, bevacizumab, and rituximab to maintain therapeutic levels and lower immune-related adverse effects.

AI-assisted decision-making tools are useful in evaluating the effect of biologics and chemotherapeutic or targeted therapies together. The DeepSynergy model employed chemical and genomic information as input information, a normalization strategy to account for input data heterogeneity, and conical layers to model drug synergies. The dataset comprises 23,062 samples, where each sample consists of two compounds and a cell line. Thereby, the dataset covers 583 distinct combinations, each tested against 39 human cancer cell lines derived from seven different tissue types. DeepSynergy was able to provide the best predictions in a cross-validation setting with external test sets, outperforming other methods by a wide margin. Prioritizing drug combinations on the basis of the predictions of DeepSynergy at an AUC of 0.90 could already decrease the time and costs spent on experimental validation (Preuer et al., 2017). These computer programs are increasingly integral to the processes employed in the discovery of novel combination medicines in cancer research.

17.6.3.2 Toxicity prediction and immune-related adverse events

AI and ML algorithms are increasingly used to predict immune-related adverse events (irAEs) associated with immune checkpoint inhibitors (ICIs) and other biologics. An ML model was developed using pretreatment clinical and laboratory data to predict irAEs in patients receiving immune checkpoint inhibitors using a large Cedars-Sinai cohort, which demonstrated the best performance (AUC 0.64) and identified key predictors including age, sex, ethnicity, race, and alcohol use, in which patients in the highest risk quartile were more than twice as likely to develop irAEs as those in the lowest quartile, supporting the feasibility of MLA-based pretreatment risk stratification to enable earlier intervention (Waissengrin et al., 2025).

In a prospective study, 34 advanced cancer patients receiving ICIs had their electronic patient-reported outcome (ePRO) data subjected to ML in order to predict irAEs. With a 97% accuracy rate and an AUC of 0–99, the extreme gradient boosting (XGBoost) model was able to forecast the occurrence of irAEs. Additionally, it had a 96% accuracy rate and an AUC of 0.93 in predicting the onset of irAEs, indicating that ML algorithms can efficiently use clinical data and symptomatology to identify patients who are at risk for irAEs. According to Iivanainen et al. (2021), these findings suggest that predictive models utilizing AI and ML may soon be able to detect toxicity in patients receiving immune checkpoint inhibitor treatment (Iivanainen et al., 2021).

17.6.3.3 Ethical and regulatory considerations

Predictive AI models can be effectively employed, but there are some ethical, therapeutic, and regulatory issues to consider. Among these are interpretability, bias, and transparency. The results of AI algorithms may be less beneficial for underrepresented ethnic or genetic groups if they are created using datasets that represent a particular census. This could result in inconsistencies in the recommendations for therapeutic interventions (Rajkomar et al., 2019).

Regulatory bodies like the European Medicines Agency (EMA) and the U.S. Food and Drug Administration (FDA) have developed guidelines that focus particularly on Good Machine Learning Practice (GMLP) to address these challenges and decrease the associated risks. The guidelines have mandated that ethical AI models adhere to HIPAA and GDPR rules in case of patient privacy. This ensures the proper utilization and anonymization of genetic and electronic health records. Ethical aspects indicate that artificial intelligence should function only as a supplementary tool, but not as an independent entity intended to make decision processes and hence keep the final decisions and approvals with oncologists. ASCO and ESMO endorse the implementation of "human-in-the-loop" AI systems, indicating that healthcare professionals retain responsibility for assessing the outputs of the AI model within a clinical setting (ESMO, 2023).

17.7 Challenges, limitations, and ethical considerations

Using AI and ML to make biologics and biosimilars could greatly improve cancer research, but there are still big problems with data quality, interpretability, and ethics that make it hard to do so. A significant limitation stems from the restricted availability and heterogeneity of biologic-specific datasets. While genomic and transcriptomic databases like The Cancer Genome Atlas (TCGA) and the GENIE consortium have yielded significant cancer datasets, analogous comprehensive resources for biological interactions, encompassing antibody/antigen complexes, glycosylation patterns, and immunogenicity profiles, are currently lacking. AI models suffer from repeatability and generalizability (Schneider, 2018). Rules for how to style, annotate, and gather data vary by research institutes and pharmaceutical companies, leading to batch effects and noise that render models less reliable. The FDA Sentinel Program and Project DataSphere are taking steps to address this through a push for systematic data collection and use of standardized frameworks for real-world biologic data. A major concern is the transparency of deep learning models, which are often black boxes and, therefore, difficult to verify when their predictions impact clinical dosing or toxicity assessments (Popejoy & Fullerton, 2016). Explainable AI (XAI) models are supported by the U.S. Food and Drug Administration (FDA) and the European Medicines Agency (EMA), as they facilitate algorithmic decision-making.

The use of artificial intelligence is becoming increasingly challenging due to ethical and regulatory concerns. Patient privacy, data ownership, and algorithmic bias are concerns when using massive genomics and clinical data to alter biological outcomes. It is challenging to extend models to other ethnic groups because the majority of the current data in cancer genomics comes from patients of European ancestry (Popejoy & Fullerton, 2016). Global regulations such as the Health Insurance Portability and Accountability Act (HIPAA) and the General Data Protection Regulation (GDPR) ensure that algorithms operate equitably and safeguard data privacy. Human control, accountability, and involvement are emphasized in the rules and guidelines that the WHO has established for the use of AI in healthcare. Therefore, guidelines for ongoing assessment and validation must be created. Biologics testing can now be done using standard protocols thanks to AI.

To increase the reliability of results, the FDA-CDER AI Framework and the IMI-BIGPICTURE program both promote the use of validated data and open benchmarking techniques. The IEEE Standards Association has created guidelines for the application of AI in biopharmaceutical systems at the same time (IEEE P2801-2022). With the FDA's assistance, the BioCompute Object (BCO) standard also facilitates the documentation of computational operations, which increases regulator confidence in AI-driven evaluations of biologics and facilitates their traceability (Patel et al., 2021). The responsible application of AI and ML in biologics and biosimilars requires a simultaneous assessment of technical, ethical, and scientific concerns. The primary objective in precision oncology is to promote innovation while protecting patient health.

17.8 Future perspectives and opportunities

The development of biologics and biosimilars with the use of AI is a major step forward in precision oncology. The AI-integrated approaches are key tools for biologics discovery, expanding data, and improving computational functions. These tools will help in integrating structural biology, omics, and real-world outcomes in clinical settings through which new biological entities can be identified rapidly. AI tools, like AlphaFold2 and Meta's ESMFold, have performed well in protein structure prediction. Their accuracy is almost similar to that of experimental methods. With these advancements, it is easier to design antibodies and enzymes for cancer treatment. Reinforcement learning's application in biological optimization, utilizing de novo peptide and antibody design, has proven useful in the initial stages of discovery by reducing the number of wet lab procedures and their associated funding (Lin et al., 2023).

The combination of quantum computing and federated learning in biosimilar analysis is a breakthrough in the area. The use of quantum ML (QML) to model complex molecular interactions is growing, particularly for the effective simulation of conformational dynamics and protein-ligand binding. Federated learning tools preserve patient privacy while enhancing biosimilar comparability studies and generalizability by enabling AI models to learn from decentralized, multi-institutional datasets. These tools are extremely valuable in the field of oncology biologics because they facilitate data exchange between clinical institutions and pharmaceutical companies, which is necessary for regulatory approval and predictive modeling (Sheller et al., 2020).

Predictive oncology platforms that incorporate biologics, digital twins, and AI-based decision support tools are emerging as a promising trend. Digital twins are digital replicas of each patient that integrate genetic, proteomic, and pharmacodynamic data to permit virtual simulations of biological responses and toxicity profiles before clinical use; in silico modeling tools for simulating patient-specific responses to monoclonal antibodies have been established by the Avicenna Alliance and the Virtual Physiological Human (VPH) Institute to enable adaptive therapy planning and postmarket surveillance, and multimodal tools may bridge the gap between computational design, clinical implementation, and pharmacovigilance, thus revolutionizing therapeutic stratification and dose response.

In the fields of precision and preventive oncology, AI-based biologics are expected to promote self-learning biologic systems that are able to predict and retard the progression of diseases. AI systems that are ethically sound and interoperable are essential for the development of personalized cancer treatments in the future because they make it easier to integrate clinical, genetic, and environmental data. For equal access and public acceptance, explainable algorithms, standardized regulatory frameworks, and rigorous adherence to ethical standards are essential Viceconti et al. (2016). It is anticipated that the development of digital-biological synergies will make it easier to incorporate AI, biologics, and biosimilars as fundamental tools for precision oncology.

17.9 Conclusion

By improving cancer detection, characterization, and treatment, the integration of AI, ML, and biologic therapies has improved the field of cancer treatment. A variety of data sources, practical experience, and knowledge are being used by AI-based systems to help develop new biomarkers, targets, and biosimilars. Physicians and scientists can improve personalized cancer therapy by using these computational systems to evaluate patient responses, modify drug dosage, and lessen side effects.

The use of predictive models, adaptive algorithms, and digital twins in biological therapy processes aids in the creation of individualized treatment plans and makes it easier to track the effectiveness of treatments. Biologic treatments become more accessible and affordable when biosimilars and AI analytics are used together, particularly in low- and middle-income nations. By overcoming challenges like data scarcity, ethical concerns, and algorithmic biases, the application of AI to cancer treatment is rapidly advancing the field toward patient-centered, predictive, and preventive therapy models. The integration of biological agents and artificial intelligence is a crucial component of oncologic practice in the future. To ensure that cancer treatment becomes dynamic, this synergy envisions a future in which intelligent, adaptive, and ethically informed computational systems will direct the development, evaluation, and delivery of biologics.

AI disclosure

During the preparation of this work, the author(s) used Grammarly to check grammar and language editing. After using this tool/service, the author(s) reviewed and edited the content as needed and take(s) full responsibility for the content of the publication.

References

Aerts, H. J. W. L., Velazquez, E. R., Leijenaar, R. T. H., Parmar, C., Grossmann, P., Carvalho, S., Bussink, J., Monshouwer, R. é, Haibe-Kains, B., Rietveld, D., Hoebers, F., Rietbergen, M. M., Leemans, C. R. é, Dekker, A., Quackenbush, J., Gillies, R. J., & Lambin, P. (2014). Correction: Corrigendum: Decoding tumour phenotype by noninvasive imaging using a quantitative radiomics approach. *Nature Communications, 5*(1). https://doi.org/10.1038/ncomms5644.

Ahluwalia, V. S., & Parikh, R. B. (2025). Explainable machine learning to predict treatment response in advanced non–small cell lung cancer. *JCO Clinical Cancer Informatics, 9*. https://doi.org/10.1200/CCI-24-00157.

Ali, H. (2023). Artificial intelligence in multi-omics data integration: Advancing precision medicine, biomarker discovery and genomic-driven disease interventions. *International Journal of Science and Research Archive, 8*(1), 1012–1030. https://doi.org/10.30574/ijsra.2023.8.1.0189.

Al-Tashi, Q., Saad, M. B., Muneer, A., Qureshi, R., Mirjalili, S., Sheshadri, A., Le, X., Vokes, N. I., Zhang, J., & Wu, J. (2023). Machine learning models for the identification of prognostic and predictive cancer biomarkers: A systematic review. *International Journal of Molecular Sciences, 24*(9), 7781. https://doi.org/10.3390/ijms24097781.

Alum, E. U. (2025). AI-driven biomarker discovery: Enhancing precision in cancer diagnosis and prognosis. *Discover Oncology, 16*(1). https://doi.org/10.1007/s12672-025-02064-7.

André, F., Arnedos, M., Baras, A. S., Baselga, J. é, Bedard, P. L., Berger, M. F., Bierkens, M., Calvo, F., Cerami, E., Chakravarty, D., Dang, K. K., Davidson, N. E., Fitz, C. D. V., Dogan, S., DuBois, R. N., Ducar, M. D., Futreal, P. A., Gao, J., Garcia, F., … Zhang, H. (2017). AACR Project GENIE: Powering precision medicine through an International Consortium. *Cancer Discovery, 7*(8), 818–831. https://doi.org/10.1158/2159-8290.cd-17-0151.

Barlas, S. (2017). FDA guidance on biosimilar interchangeability elicits diverse views: Current and potential marketers complain about too-high hurdles. *P & T : A Peer-Reviewed Journal for Formulary Management, 42*(8), 509–512.

Beck, A., Goetsch, L., Dumontet, C., & Corvaïa, N. (2017). Strategies and challenges for the next generation of antibody–drug conjugates. *Nature Reviews Drug Discovery, 16*(5), 315–337. https://doi.org/10.1038/nrd.2016.268.

Bera, K., Schalper, K. A., Rimm, D. L., Velcheti, V., & Madabhushi, A. (2019). Artificial intelligence in digital pathology — new tools for diagnosis and precision oncology. *Nature Reviews Clinical Oncology, 16*(11), 703–715. https://doi.org/10.1038/s41571-019-0252-y.

Biswas, N., & Chakrabarti, S. (2020). Artificial intelligence (AI)-based systems biology approaches in multi-omics data analysis of cancer. *Frontiers in Oncology, 10*. https://doi.org/10.3389/fonc.2020.588221.

Budginaite, E., Magee, D. R., Kloft, M., Woodruff, H. C., & Grabsch, H. I. (2024). Computational methods for metastasis detection in lymph nodes and characterization of the metastasis-free lymph node microarchitecture: A systematic-narrative hybrid review. *Journal of Pathology Informatics, 15*. https://doi.org/10.1016/j.jpi.2024.100367.

Çalışkan, M., & Tazaki, K. (2023). AI/ML advances in non-small cell lung cancer biomarker discovery. *Frontiers in Oncology, 13*. https://doi.org/10.3389/fonc.2023.1260374.

Campanella, G., Hanna, M. G., Geneslaw, L., Miraflor, A., Krauss Silva, V. W., Busam, K. J., Brogi, E., Reuter, V. E., Klimstra, D. S., & Fuchs, T. J. (2019). Clinical-grade computational pathology using weakly supervised deep learning on whole slide images. *Nature Medicine, 25*(8), 1301–1309. https://doi.org/10.1038/s41591-019-0508-1.

Cheng, G., Zhang, F., Xing, Y., Hu, X., Zhang, H. e, Chen, S., Li, M., Peng, C., Ding, G., Zhang, D., Chen, P., Xia, Q., & Wu, M. (2022). Artificial intelligence-assisted score analysis for predicting the expression of the immunotherapy biomarker PD-L1 in lung cancer. *Frontiers in Immunology, 13*. https://doi.org/10.3389/fimmu.2022.893198.

Cohen, H. P., Blauvelt, A., Rifkin, R. M., Danese, S., Gokhale, S. B., & Woollett, G. (2018). Switching reference medicines to biosimilars: A systematic literature review of clinical outcomes. *Drugs, 78*(4), 463–478. https://doi.org/10.1007/s40265-018-0881-y.

Colen, R. R., Rolfo, C., Ak, M., Ayoub, M., Ahmed, S., Elshafeey, N., Mamindla, P., Zinn, P. O., Ng, C., Vikram, R., Bakas, S., Peterson, C. B., Ahnert, J. R., Subbiah, V., Karp, D. D., Stephen, B., Hajjar, J., & Naing, A. (2021). Radiomics analysis for predicting pembrolizumab response in patients with advanced rare cancers. *Journal for ImmunoTherapy of Cancer, 9*(4), e001752. https://doi.org/10.1136/jitc-2020-001752.

Conde-Sousa, E., Vale, J., Feng, M., Xu, K., Wang, Y., Della Mea, V., La Barbera, D., Montahaei, E., Baghshah, M., Turzynski, A., Gildenblat, J., Klaiman, E., Hong, Y., Aresta, G., Araújo, T., Aguiar, P., Eloy, C., & Polónia, A. (2022). HEROHE Challenge: Predicting HER2 status in breast cancer from hematoxylin–eosin whole-slide imaging. *Journal of Imaging, 8*(8), 213. https://doi.org/10.3390/jimaging8080213.

Coudray, N., Ocampo, P. S., Sakellaropoulos, T., Narula, N., Snuderl, M., Fenyö, D., … Tsirigos, A. (2018). Classification and mutation prediction from non–small cell lung cancer histopathology images using deep learning. *Nature Medicine, 24*(10), 1559–1567. https://doi.org/10.1038/s41591-018-0177-5.

Cunningham, D., Humblet, Y., Siena, S., Khayat, D., Bleiberg, H., Santoro, A., Bets, D., Mueser, M., Harstrick, A., Verslype, C., Chau, I., & Van Cutsem, E. (2004). Cetuximab monotherapy and cetuximab plus irinotecan in irinotecan-refractory metastatic colorectal cancer. *New England Journal of Medicine, 351*(4), 337–345. https://doi.org/10.1056/NEJMoa033025.

Esteva, A., Robicquet, A., Ramsundar, B., Kuleshov, V., DePristo, M., Chou, K., Cui, C., Corrado, G., Thrun, S., & Dean, J. (2019). A guide to deep learning in healthcare. *Nature Medicine, 25*(1), 24–29. https://doi.org/10.1038/s41591-018-0316-z.

Fan, Y., Sun, K., Xiao, Y., Zhong, P., Meng, Y., Yang, Y., Du, Z., & Fang, J. (2025). Deep learning predicts HER2 status in invasive breast cancer from multimodal ultrasound and MRI. *Biomolecules and Biomedicine, 25*(10), 2243–2251. https://doi.org/10.17305/bb.2025.12475.

Farahmand, S., Fernandez, A. I., Ahmed, F. S., Rimm, D. L., Chuang, J. H., Reisenbichler, E., & Zarringhalam, K. (2022). Deep learning trained on hematoxylin and eosin tumor region of

Interest predicts HER2 status and trastuzumab treatment response in HER2+ breast cancer. *Modern Pathology, 35*(1), 44–51. https://doi.org/10.1038/s41379-021-00911-w.

Ferrara, N., Hillan, K. J., Gerber, H. P., & Novotny, W. (2004). Discovery and development of bevacizumab, an anti-VEGF antibody for treating cancer. *Nature Reviews Drug Discovery, 3*(5), 391–400. https://doi.org/10.1038/nrd1381.

Gencsoy Eker, S., Inetas Yengin, G., Tatar, C., & Oktem, G. (2024). A Comprehensive Review of the Mechanisms and Clinical Development of Monoclonal Antibodies in Cancer Therapy. In K Turksen (Ed.), *Cell Biology and Translational Medicine*: 24. Cham: Springer.

Gherghescu, I., & Delgado-Charro, M. B. (2020). The Biosimilar Landscape: An Overview of Regulatory Approvals by the EMA and FDA. *Pharmaceutics, 13*(1), 48. doi:10.3390/pharmaceutics13010048.

Giansanti, D., & Morelli, S. (2025). Exploring the potential of digital twins in cancer treatment: A narrative review of reviews. *Journal of Clinical Medicine, 14*(10), 3574. https://doi.org/10.3390/jcm14103574.

Guardant Health. GuardantINFORM™ Real-World Clinico-Genomic Platform Overview. https://investors.guardanthealth.com/press-releases/press-releases/2020/Guardant-Health-Launches-Real-World-Clinical-Genomic-Platform-to-Accelerate-Precision-Oncology-Drug-Development/default.aspx.

Huang, Z., Denti, P., Mistry, H., & Kloprogge, F. (2024). Machine learning and artificial intelligence in PK-PD modeling: Fad, friend, or foe? *Clinical Pharmacology & Therapeutics, 115*(4), 652–654. https://doi.org/10.1002/cpt.3165.

Hugo, W., Zaretsky, J. M., Sun, L., Song, C., Moreno, B. H., Hu-Lieskovan, S., Berent-Maoz, B., Pang, J., Chmielowski, B., Cherry, G., Seja, E., Lomeli, S., Kong, X., Kelley, M. C., Sosman, J. A., Johnson, D. B., Ribas, A., & Lo, R. S. (2016). Genomic and transcriptomic features of response to anti-PD-1 therapy in metastatic melanoma. *Cell, 165*(1), 35–44. https://doi.org/10.1016/j.cell.2016.02.065.

Hummel, J. J., Liu, D., Tallon, E., Snyder, J., Warren, W., Shyu, C. R., Mitchem, J., & Cortese, R. (2024). Identification of genomic signatures for colorectal cancer survival using exploratory data mining. *International Journal of Molecular Sciences, 25*(6). https://doi.org/10.3390/ijms25063220.

Iivanainen, S., Ekstrom, J., Virtanen, H., Kataja, V. V., & Koivunen, J. P. (2021). Electronic patient-reported outcomes and machine learning in predicting immune-related adverse events of immune checkpoint inhibitor therapies. *BMC Medical Informatics and Decision Making, 21*(1), 205. https://doi.org/10.1186/s12911-021-01564-0.

IQVIA. *The Global Use of Medicines 2023*. https://www.iqvia.com/-/media/iqvia/pdfs/library/presentations/presentation_global_meds_2023_webinar.pdf.

Jiang, B., Bao, L., He, S., Chen, X., Jin, Z., & Ye, Y. (2024). Deep learning applications in breast cancer histopathological imaging: Diagnosis, treatment, and prognosis. *Breast Cancer Research, 26*(1). https://doi.org/10.1186/s13058-024-01895-6.

Jois, R., Mukherjee, S., Rajeswari, S., Rath, P., Goyal, V., & Gupta, D. (2020). Similar biologics in India: A story of access or potential for compromise? *Indian Journal of Medical Research, 152*(5), 456–467. https://doi.org/10.4103/ijmr.ijmr_43_18.

Jones, M., Schenkel, B., Just, J., & Fallowfield, L. (2004). Epoetin alfa improves quality of life in patients with cancer: Results of a metaanalysis. *Cancer, 101*(8), 1720–1732. https://doi.org/10.1002/cncr.20569.

Jumper, J., Evans, R., Pritzel, A., Green, T., Figurnov, M., Ronneberger, O., Tunyasuvunakool, K., Bates, R., Žídek, A., Potapenko, A., Bridgland, A., Meyer, C., Kohl, S. A. A.,

Ballard, A. J., Cowie, A., Romera-Paredes, B., Nikolov, S., Jain, R., Adler, J., … Hassabis, D. (2021). Highly accurate protein structure prediction with AlphaFold. *Nature, 596*(7873), 583–589. https://doi.org/10.1038/s41586-021-03819-2.

June, C. H., O'Connor, R. S., Kawalekar, O. U., Ghassemi, S., & Milone, M. C. (2018). CAR T cell immunotherapy for human cancer. *Science, 359*(6382), 1361–1365. https://doi.org/10.1126/science.aar6711.

Khan, M., & Spicer, J. (2019). The evolving landscape of cancer therapeutics. *Handbook of Experimental Pharmacology, 260*, 43–79. https://doi.org/10.1007/164_2019_312.

Klein, E. A., Richards, D., Cohn, A., Tummala, M., Lapham, R., Cosgrove, D., Chung, G., Clement, J., Gao, J., Hunkapiller, N., Jamshidi, A., Kurtzman, K. N., Seiden, M. V., Swanton, C., & Liu, M. C. (2021). Clinical validation of a targeted methylation-based multi-cancer early detection test using an independent validation set. *Annals of Oncology, 32*(9), 1167–1177. https://doi.org/10.1016/j.annonc.2021.05.806.

Kourou, K., Exarchos, T. P., Exarchos, K. P., Karamouzis, M. V., & Fotiadis, D. I. (2014). Machine learning applications in cancer prognosis and prediction. *Comput Struct Biotechnol J, 13*, 8–17. doi:10.1016/j.csbj.2014.11.005.

Kourou, K., Exarchos, K. P., Papaloukas, C., Sakaloglou, P., Exarchos, T., & Fotiadis, D. I. (2021). Applied machine learning in cancer research: A systematic review for patient diagnosis, classification and prognosis. *Computational and Structural Biotechnology Journal, 19*, 5546–5555. https://doi.org/10.1016/j.csbj.2021.10.006.

Kvien, T. K., Patel, K., & Strand, V. (2021). The cost savings of biosimilars can help increase patient access and lift the financial burden of health care systems. *Semin Arthritis Rheum, 52*, 151939. doi:10.1016/j.semarthrit.2021.11.009.

Liao, C. Y., Chen, Y. M., Wu, Y. T., Chao, H. S., Chiu, H. Y., Wang, T. W., Chen, J. R., Shiao, T. H., & Lu, C. F. (2024). Personalized prediction of immunotherapy response in lung cancer patients using advanced radiomics and deep learning. *Cancer Imaging, 24*(1). https://doi.org/10.1186/s40644-024-00779-4.

Lin, Z., Akin, H., Rao, R., Hie, B., Zhu, Z., Lu, W., Smetanin, N., Verkuil, R., Kabeli, O., Shmueli, Y., dos Santos Costa, A., Fazel-Zarandi, M., Sercu, T., Candido, S., & Rives, A. (2023). Evolutionary-scale prediction of atomic-level protein structure with a language model. *Science, 379*(6637), 1123–1130. https://doi.org/10.1126/science.ade2574.

Liu, Y., Kohlberger, T., Norouzi, M., Dahl, G. E., Smith, J. L., Mohtashamian, A., … Stumpe, M. C. (2018). Artificial intelligence–based breast cancer nodal metastasis detection: Insights into the black box for pathologists. *Archives of Pathology & Laboratory Medicine, 143*(7), 859–868. https://doi.org/10.5858/arpa.2018-0147-oa.

Lost, J., Verma, T., Jekel, L., von Reppert, M., Tillmanns, N., Merkaj, S., Cassinelli Petersen, G., Bahar, R., Gordem, A., Haider, M. A., Subramanian, H., Brim, W., Ikuta, I., Omuro, A., Marco Conte, G., Marquez-Nostra, B. V., Avesta, A., Bousabarah, K., Nabavizadeh, A., … Aboian, M. (2023). Systematic literature review of machine learning algorithms using pretherapy radiologic imaging for glioma molecular subtype prediction. *American Journal of Neuroradiology, 44*(10), 1126–1134. https://doi.org/10.3174/ajnr.A8000.

Madsen, C. T., Refsgaard, J. C., Teufel, F. G., Kjærulff, S. K., Wang, Z., Meng, G., Jessen, C., Heljo, P., Jiang, Q., Zhao, X., Wu, B., Zhou, X., Tang, Y., Jeppesen, J. F., Kelstrup, C. D., Buckley, S. T., Tullin, S., Nygaard-Jensen, J., Chen, X., … de Lichtenberg, U. (2022). Combining mass spectrometry and machine learning to discover bioactive peptides. *Nature Communications, 13*(1). https://doi.org/10.1038/s41467-022-34031-z.

Maloney, D. G., Grillo-López, A. J., White, C. A., Bodkin, D., Schilder, R. J., Neidhart, J. A., Janakiraman, N., Foon, K. A., Liles, T. M., Dallaire, B. K., Wey, K., Royston, I., Davis, T., & Levy, R. (1997). IDEC-C2B8 (rituximab) anti-CD20 monoclonal antibody therapy in patients with relapsed low-grade non-Hodgkin's lymphoma. *Blood, 90*(6), 2188–2195. https://doi.org/10.1182/blood.v90.6.2188.

Markus, R., Liu, J., Ramchandani, M., Landa, D., Born, T., & Kaur, P. (2017). Developing the totality of evidence for biosimilars: Regulatory considerations and building confidence for the healthcare community. *BioDrugs, 31*(3), 175–187. https://doi.org/10.1007/s40259-017-0218-5.

Mascarenhas-Melo, F., Diaz, M., Gonçalves, M. B. S., Vieira, P., Bell, V., Viana, S., Nunes, S., Paiva-Santos, A. C., & Veiga, F. (2024). An overview of biosimilars—Development, quality, regulatory issues, and management in healthcare. *Pharmaceuticals, 17*(2). https://doi.org/10.3390/ph17020235.

Mason, D. M., Friedensohn, S., Weber, C. R., Jordi, C., Wagner, B., Meng, S. M., Ehling, R. A., Bonati, L., Dahinden, J., Gainza, P., Correia, B. E., & Reddy, S. T. (2021). Optimization of therapeutic antibodies by predicting antigen specificity from antibody sequence via deep learning. *Nature Biomedical Engineering, 5*(6), 600–612. https://doi.org/10.1038/s41551-021-00699-9.

McKinney, S. M., Sieniek, M., Godbole, V., Godwin, J., Antropova, N., Ashrafian, H., Back, T., Chesus, M., Corrado, G. C., Darzi, A., Etemadi, M., Garcia-Vicente, F., Gilbert, F. J., Halling-Brown, M., Hassabis, D., Jansen, S., Karthikesalingam, A., Kelly, C. J., King, D., … Shetty, S. (2020). International evaluation of an AI system for breast cancer screening. *Nature, 577*(7788), 89–94. https://doi.org/10.1038/s41586-019-1799-6.

Mertins, P., Mani, D. R., Ruggles, K. V., Gillette, M. A., Clauser, K. R., Wang, P., Wang, X., Qiao, J. W., Cao, S., Petralia, F., Kawaler, E., Mundt, F., Krug, K., Tu, Z., Lei, J. T., Gatza, M. L., Wilkerson, M., Perou, C. M., Yellapantula, V., … Carr, S. A. (2016). Proteogenomics connects somatic mutations to signalling in breast cancer. *Nature, 534*(7605), 55–62. https://doi.org/10.1038/nature18003.

Gromiha, M. M., & Huang, L.-T. (2011). Machine learning algorithms for predicting protein folding rates and stability of mutant proteins: Comparison with statistical methods. *Current Protein & Peptide Science, 12*(6), 490–502. https://doi.org/10.2174/138920311796957630.

Miller, E. M., & Schwartzberg, L. S. (2019). Biosimilars for breast cancer: A review of HER2-targeted antibodies in the United States. *Therapeutic Advances in Medical Oncology, 11*. https://doi.org/10.1177/1758835919887044.

Moorkens, E., Vulto, A. G., Huys, I., Dylst, P., Godman, B., Keuerleber, S., Claus, B., Dimitrova, M., Petrova, G., Sović-Brkičić, L., Slabý, J., Šebesta, R., Laius, O., Karr, A., Beck, M., Martikainen, J. E., Selke, G. W., Spillane, S., McCullagh, L., … Bochenek, T. (2017). Policies for biosimilar uptake in Europe: An overview. *PLOS ONE, 12*(12), e0190147. https://doi.org/10.1371/journal.pone.0190147.

Mullard, A. (2024). 2023 FDA approvals. *Nature Reviews Drug Discovery, 23*(2), 88–95. https://doi.org/10.1038/d41573-024-00001-x.

Nguyen, V. T. C., Nguyen, T. H., Doan, N. N. T., Pham, T. M. Q., Nguyen, G. T. H., Nguyen, T. D., Tran, T. T. T., Vo, D. L., Phan, T. H., Jasmine, T. X., Nguyen, V. C., Nguyen, H. T., Nguyen, T. V., Nguyen, T. H. H., Huynh, L., Tran, T. H., Dang, Q. T., Doan, T. N., Tran, A. M., … Tran, L. (2023). Multimodal analysis of methylomics and fragmentomics in plasma cell-free DNA for multi-cancer early detection and localization. *eLife, 12*. https://doi.org/10.7554/elife.89083.

Nimmerjahn, F., & Ravetch, J. V. (2008). Fcγ receptors as regulators of immune responses. *Nature Reviews Immunology, 8*(1), 34–47. https://doi.org/10.1038/nri2206.

Ozaki, Y., Broughton, P., Abdollahi, H., Valafar, H., & Blenda, A. V. (2024). Integrating omics data and AI for cancer diagnosis and prognosis. *Cancers, 16*(13). https://doi.org/10.3390/cancers16132448.

Parmar, C., Leijenaar, R. T. H., Grossmann, P., Velazquez, E. R., Bussink, J., Rietveld, D., Rietbergen, M. M., Haibe-Kains, B., Lambin, P., & Aerts, H. J. W. L. (2015). Radiomic feature clusters and prognostic signatures specific for lung and head & neck cancer. *Scientific Reports, 5*. https://doi.org/10.1038/srep11044.

Patel, J. A., Dean, D. A., King, C. H., Xiao, N., Koc, S., Minina, E., Golikov, A., Brooks, P., Kahsay, R., Navelkar, R., Ray, M., Roberson, D., Armstrong, C., Mazumder, R., & Keeney, J. (2021). Bioinformatics tools developed to support BioCompute Objects. *Database, 2021*. https://doi.org/10.1093/database/baab008.

Paul, D., Sanap, G., Shenoy, S., Kalyane, D., Kalia, K., & Tekade, R. K. (2020). Artificial intelligence in drug discovery and development. *Drug Discovery Today, 26*(1), 80–93.

Peng, J., Zou, D., Zhang, X., Ma, H., Han, L., & Yao, B. (2024). A novel sub-regional radiomics model to predict immunotherapy response in non-small cell lung carcinoma. *Journal of Translational Medicine, 22*(1). https://doi.org/10.1186/s12967-024-04904-6.

Popejoy, A. B., & Fullerton, S. M. (2016). Genomics is failing on diversity. *Nature, 538*(7624), 161–164. https://doi.org/10.1038/538161a.

Prelaj, A., Miskovic, V., Zanitti, M., Trovo, F., Genova, C., Viscardi, G., et al. (2024). Artificial intelligence for predictive biomarker discovery in immuno-oncology: a systematic review. *Ann Oncol, 35*(1), 29–65. doi:10.1016/j.annonc.2023.10.125.

Preuer, K., Lewis, R. P. I., Hochreiter, S., Bender, A., Bulusu, K. C., & Klambauer, G. (2017). DeepSynergy: predicting anti-cancer drug synergy with Deep Learning. *Bioinformatics, 34*(9), 1538–1546. doi:10.1093/bioinformatics/btx806.

Rajkomar, A., Dean, J., & Kohane, I. (2019). Machine Learning in Medicine. *N Engl J Med, 380*(14), 1347–1358. doi:10.1056/nejmra1814259.

Rakaee, M., Tafavvoghi, M., Ricciuti, B., Alessi, J. V., Cortellini, A., Citarella, F., et al. (2024). Deep Learning Model for Predicting Immunotherapy Response in Advanced Non−Small Cell Lung Cancer. *JAMA Oncology, 11*(2), 109–118. doi:10.1001/jamaoncol.2024.5356.

Rizvi, N. A., Hellmann, M. D., Snyder, A., Kvistborg, P., Makarov, V., Havel, J. J., Lee, W., Yuan, J., Wong, P., Ho, T. S., Miller, M. L., Rekhtman, N., Moreira, A. L., Ibrahim, F., Bruggeman, C., Gasmi, B., Zappasodi, R., Maeda, Y., Sander, C., … Chan, T. A. (2015). Mutational landscape determines sensitivity to PD-1 blockade in non-small cell lung cancer. *Science, 348*(6230), 124–128. https://doi.org/10.1126/science.aaa1348.

Robert, C., Schachter, J., Long, G. V., Arance, A., Grob, J. J., Mortier, L., Daud, A., Carlino, M. S., McNeil, C., Lotem, M., Larkin, J., Lorigan, P., Neyns, B., Blank, C. U., Hamid, O., Mateus, C., Shapira-Frommer, R., Kosh, M., Zhou, H., … Ribas, A. (2015). Pembrolizumab versus ipilimumab in advanced melanoma. *New England Journal of Medicine, 372*(26), 2521–2532. https://doi.org/10.1056/NEJMoa1503093.

Schneider, G. (2018). Automating drug discovery. *Nature Reviews Drug Discovery, 17*(2), 97–113. https://doi.org/10.1038/nrd.2017.232.

Scott, A. M., Wolchok, J. D., & Old, L. J. (2012). Antibody therapy of cancer. *Nature Reviews Cancer, 12*(4), 278–287. doi:10.1038/nrc3236.

Seebacher, N. A., Stacy, A. E., Porter, G. M., & Merlot, A. M. (2019). Clinical development of targeted and immune based anti-cancer therapies. *Journal of*

Experimental & Clinical Cancer Research, 38(1). https://doi.org/10.1186/s13046-019-1094-2.

Senior, A. W., Evans, R., Jumper, J., Kirkpatrick, J., Sifre, L., Green, T., Qin, C., Žídek, A., Nelson, A. W. R., Bridgland, A., Penedones, H., Petersen, S., Simonyan, K., Crossan, S., Kohli, P., Jones, D. T., Silver, D., Kavukcuoglu, K., & Hassabis, D. (2020). Improved protein structure prediction using potentials from deep learning. *Nature, 577*(7792), 706–710. https://doi.org/10.1038/s41586-019-1923-7.

Shamai, G., Binenbaum, Y., Slossberg, R., Duek, I., Gil, Z., & Kimmel, R. (2019). Artificial intelligence algorithms to assess hormonal status from tissue microarrays in patients with breast cancer. *JAMA Network Open, 2*(7), e197700. https://doi.org/10.1001/jamanetworkopen.2019.7700.

Shamai, G., Livne, A., Polónia, A. ó, Sabo, E., Cretu, A., Bar-Sela, G., & Kimmel, R. (2022). Deep learning-based image analysis predicts PD-L1 status from H&E-stained histopathology images in breast cancer. *Nature Communications, 13*(1). https://doi.org/10.1038/s41467-022-34275-9.

Sharma, A., Khante, S., Mahadik, K. R., & Gaikwad, V. L. (2020). Regulatory perspective of international agencies for development of biosimilar products (monoclonal antibodies): An overview. *Therapeutic Innovation and Regulatory Science, 54*(5), 965–977. https://doi.org/10.1007/s43441-020-00112-8.

Sheller, M. J., Edwards, B., Reina, G. A., Martin, J., Pati, S., Kotrotsou, A., Milchenko, M., Xu, W., Marcus, D., Colen, R. R., & Bakas, S. (2020). Federated learning in medicine: Facilitating multi-institutional collaborations without sharing patient data. *Scientific Reports, 10*(1). https://doi.org/10.1038/s41598-020-69250-1.

Shen, S., Qi, W., Liu, X., Zeng, J., Li, S., Zhu, X., Dong, C., Wang, B., Shi, Y., Yao, J., Wang, B., Jing, L., Cao, S., & Liang, G. (2025). From virtual to reality: Innovative practices of digital twins in tumor therapy. *Journal of Translational Medicine, 23*(1). https://doi.org/10.1186/s12967-025-06371-z.

Sibieude, E., Khandelwal, A., Girard, P., Hesthaven, J. S., & Terranova, N. (2022). Population Pharmacokinetic Model Selection Assisted by Machine Learning. *J. Pharmacokinet. Pharmacodyn, 49*, 257–270. doi:10.1007/s10928-021-09793-6.

Sormanni, P., Aprile, F. A., & Vendruscolo, M. (2015). Rational design of antibodies targeting specific epitopes within intrinsically disordered proteins. *Proc Natl Acad Sci USA, 112*(32), 9902–9907. doi:10.1073/pnas.1422401112.

Sung, H., Ferlay, J., Siegel, R. L., Laversanne, M., Soerjomataram, I., Jemal, A., & Bray, F. (2021). Global Cancer Statistics 2020: GLOBOCAN estimates of incidence and mortality worldwide for 36 cancers in 185 countries. *CA Cancer Journal for Clinicians, 71*(3), 209–249. https://doi.org/10.3322/caac.21660.

Topalian, S., Drake, C., & Pardoll, D. (2015). Immune Checkpoint Blockade: A Common Denominator Approach to Cancer Therapy. *Cancer Cell, 27*(4), 450–461.

Topol, E. J. (2019). High-performance medicine: The convergence of human and artificial intelligence. *Nature Medicine, 25*(1), 44–56. https://doi.org/10.1038/s41591-018-0300-7.

Valieris, R., Martins, L., Defelicibus, A., Bueno, A. P., de Toledo Osorio, C. A. B., Carraro, D., Dias-Neto, E., Rosales, R. A., de Figueiredo, J. M. B., & Silva, I. T. d. (2024). Weakly-supervised deep learning models enable HER2-low prediction from H &E stained slides. *Breast Cancer Research, 26*(1). https://doi.org/10.1186/s13058-024-01863-0.

Verma, S., Miles, D., Gianni, L., Krop, I. E., Welslau, M., Baselga, J., Pegram, M., Oh, D. Y., Diéras, V., Guardino, E., Fang, L., Lu, M. W., Olsen, S., & Blackwell, K. (2012). Trastuzumab emtansine for HER2-positive advanced breast cancer. *New England Journal of Medicine, 367*(19), 1783–1791. https://doi.org/10.1056/NEJMoa1209124.

Viceconti, M., Henney, A., & Morley-Fletcher, E. (2016). In silico clinical trials: how computer simulation will transform the biomedical industry. *Int J Clin Trials, 3*(2), 37–46. doi:10.18203/2349-3259.ijct20161408.

Waissengrin, B., Fatapour, Y., Acitores, J., Friedrich, N., Tatonetti, N., Figueiredo, J., & Reckamp, K. L. (2025). Machine learning to predict immune-related adverse events in cancer patients treated with ICIs. *Journal of Clinical Oncology, 43*(16_suppl). https://doi.org/10.1200/jco.2025.43.16_suppl.e13620.

Waldmann, T. A. (2018). Cytokines in cancer immunotherapy. *Cold Spring Harbor Perspectives in Biology, 10*(12), a028472. https://doi.org/10.1101/cshperspect.a028472.

Wang, J., Zhu, X., Chen, K., Hao, L., & Liu, Y. (2023). HAHNet: A convolutional neural network for HER2 status classification of breast cancer. *BMC Bioinformatics, 24*(1). https://doi.org/10.1186/s12859-023-05474-y.

Way, G. P., & Greene, C. S. (2018). Extracting a biologically relevant latent space from cancer transcriptomes with variational autoencoders. *Pacific Symposium on Biocomputing., 23*, 80–91. doi:10.1142/9789813235533_0008.

Weber, J. S., D'Angelo, S. P., Minor, D., Hodi, F. S., Gutzmer, R., Neyns, B., Hoeller, C., Khushalani, N. I., Miller, W. H., Lao, C. D., Linette, G. P., Thomas, L., Lorigan, P., Grossmann, K. F., Hassel, J. C., Maio, M., Sznol, M., Ascierto, P. A., Mohr, P., … Larkin, J. (2015). Nivolumab versus chemotherapy in patients with advanced melanoma who progressed after anti-CTLA-4 treatment (CheckMate 037): A randomised, controlled, open-label, phase 3 trial. *The Lancet Oncology, 16*(4), 375–384. https://doi.org/10.1016/S1470-2045(15)70076-8.

Weiner, L. M., Surana, R., & Wang, S. (2010). Monoclonal antibodies: versatile platforms for cancer immunotherapy. *Nature Reviews Immunology, 10*(5), 317–327. doi:10.1038/nri2744.

Weise, M., Bielsky, M. C., De Smet, K., Ehmann, F., Ekman, N., Giezen, T. J., Gravanis, I., Heim, H. K., Heinonen, E., Ho, K., Moreau, A., Narayanan, G., Kruse, N. A., Reichmann, G., Thorpe, R., Van Aerts, L., Vleminckx, C., Wadhwa, M., & Schneider, C. K. (2012). Biosimilars: What clinicians should know. *Blood, 120*(26), 5111–5117. https://doi.org/10.1182/blood-2012-04-425744.

Wu, Z., Wang, T., Lan, J., Wang, J., Chen, G., Tong, T., & Zhang, H. (2025). Deep learning-based prediction of HER2 status and trastuzumab treatment efficacy of gastric adenocarcinoma based on morphological features. *Journal of Translational Medicine, 23*(1). https://doi.org/10.1186/s12967-024-06034-5.

Yang, Y., Yang, J., Shen, L., Chen, J., Xia, L., Ni, B., Ge, L., Wang, Y., & Lu, S. (2021). A multi-omics-based serial deep learning approach to predict clinical outcomes of single-agent anti-PD-1/PD-L1 immunotherapy in advanced stage non-small-cell lung cancer. *American Journal of Translational Research, 13*(2), 743–756.

Yolchuyeva, S., Giacomazzi, E., Tonneau, M., Lamaze, F., Orain, M., Coulombe, F. ç, Malo, J., Belkaid, W., Routy, B., Joubert, P., & Manem, V. S. K. (2023). Radiomics approaches to predict PD-L1 and PFS in advanced non-small cell lung patients treated with immunotherapy: A multi-institutional study. *Scientific Reports, 13*(1). https://doi.org/10.1038/s41598-023-38076-y.

Zhang, J., Che, Y., Liu, R., Wang, Z., & Liu, W. (2025). Deep learning–driven multi-omics analysis: Enhancing cancer diagnostics and therapeutics. *Briefings in Bioinformatics, 26*(4). https://doi.org/10.1093/bib/bbaf440.

Zhao, J., Xu, L., Sun, J., Song, M., Wang, L., Yuan, S., Zhu, Y., Wan, Z., Larsson, S., Tsilidis, K., Dunlop, M., Campbell, H., Rudan, I., Song, P., Theodoratou, E., Ding, K., & Li, X. (2023). Global trends in incidence, death, burden and risk factors of early-onset cancer from 1990 to 2019. *BMJ Oncology, 2*(1), e000049. https://doi.org/10.1136/bmjonc-2023-000049.

Index

Page numbers followed by "*f*" and "*t*" indicate, figures and tables respectively.

Printed and bound by CPI Group (UK) Ltd, Croydon, CR0 4YY
21/09/2026
02219915-0012